Linear Systems

lege, relege, labora et invenies

Linear Systems

Henri Bourlès

iSTE

WILEY

First published 2010 in Great Britain and the United States by ISTE Ltd and John Wiley & Sons, Inc. Adapted and updated from *Systèmes linéaires* published 2006 in France by Hermes Science/Lavoisier

ISTE Ltd
27-37 St George's Road
London SW19 4EU
UK

www.iste.co.uk

John Wiley & Sons, Inc.
111 River Street
Hoboken, NJ 07030
USA

www.wiley.com

Library of Congress Cataloging-in-Publication Data

Bourlès, Henri.
[Systèmes linéaires. English]
Linear systems / Henri Bourlès.
p. cm.
Includes bibliographical references and index.
ISBN 978-1-84821-162-9
1. Linear systems. I. Title.
QA402.B62713 2010
003'.74--dc22

2010016973

British Library Cataloguing-in-Publication Data
A CIP record for this book is available from the British Library
ISBN 978-1-84821-162-9

Printed and bound in Great Britain by CPI Antony Rowe, Chippenham and Eastbourne.

MIX
Paper from responsible sources
FSC® C013604

Contents

Preface

The notion of system

The notion of *system* is the basic concept of *control theory*. What is a system? It is quite difficult to address this question in all its aspects. We can say somewhat loosely that it is an entity consisting of interacting parts; an entity that is itself, most often, also interacting with other systems. The solar system, a computer system, etc. are examples of a system.

The control system analyst is interested in systems which are, at least in part, designed and constructed by man, in order to be utilized – a car engine or the entire automobile; an alternator in a power station or the "power system" in its entirety (consisting of production centers, lines of energy transport and consumption centers); an airplane, such as the A380 Airbus, where control systems play a very important role; a factory production line, etc. We can act upon these systems and these react to actions exerted on them.

Objectives of systems theory

Several objectives exist in control systems theory.

Modeling

The above-mentioned systems are part of the material world and are thus governed by the laws of physics. By putting the system in an equation based on these laws, we obtain a mathematical model which will greatly facilitate its understanding. Therefore, *modeling* is one of the important activities of the control systems analyst. The modeling process does not always start with the laws of physics; it can simply be based on a more or less empirical and qualitative observation of the behavior of the system. Within the framework of this book, we will however limit ourselves to cases where a mathematical description of the system behavior is possible. In general, the model obtained consists of a set of differential equations (sometimes of partial differential equations) or of difference equations.

Identification

The modeling process, as understood, involves determining the structure of the equations which govern the behavior of the system, and also in fixing *a priori* the values of certain system parameters: for example, the lengths, masses, resistance values, capacitance values, etc. But it is often impossible to come to an *a priori* complete and precise understanding of all the parameters of this model. In order to refine and complete this understanding, it becomes necessary to proceed to an *identification* of the system: from the reactions of the latter to known and given stimulations, we can, under certain conditions, identify the yet unknown parameters. Identification is one of the major aspects in control theory.

Analysis

Once the system is modeled and identified, it becomes possible to analyze its behavior. This *analysis* can be very complex. As an example, the analysis of the European and North American electric systems is difficult and requires powerful computing resources and enormous databases because we deal with very large systems. And understanding these systems well is essential for security reasons: points such as the possibility of "black-out" risk? In general, the analysis of a system makes it possible to determine its essential properties.

Control

Thus, systems theory is a "theoretical" science in which one of its objectives is *knowing* systems, through modeling, identification and analysis. But it is also (we may be tempted to say essentially) a "practical" science, a science of "action". We try to understand systems in order to be able to control them, and to regulate them in the best possible way. The last major chapter of systems theory is that of *control.*

Open-loop and closed-loop systems

A fundamental difference exists between "open-loop" and "closed-loop" systems. Controlling an open-loop system is doing it blindly, without taking into account the results of the action taken. We know the expression: "I could walk down this path blind-folded". Because I know at which moment I need to turn left, and then right. I know when I need to accelerate, when to slow down,... And it is true that if, as with Laplace's genius, we had a perfect knowledge of the world, we could control, in open-loop, every system belonging to it. However, our knowledge of things is incomplete. Even with a route we know by heart, unpredictable events may occur: a child unexpectedly crossing the street, a spell of rain making the road more slippery than usual, etc. It is thus necessary to note at all times, and to adjust actions, taking into account the reality that appears from moment to moment. Control theory is the art (or science) of making this unceasing adjustment, which we call a *loop* or a *feedback* or a *servo-mechanism.* It is a common concept which nevertheless poses numerous problems.

One of the major difficulties encountered with feedback systems is their possible instability. There are certainly unstable open-loop systems, but they are relatively rare: a helicopter, an alternator connected with a long power line, certain types of fighter jets, etc. On the other hand, nothing is more commonplace than making a system unstable with a poor-performing feedback loop. Let us think of a child striving to take a shower: when the water is cold, he wants to heat it up, turns the hot water tap on too quickly, and gets boiling hot water; then, over-doing it the same way with the cold water tap, he makes the water ice cold, and so on.

The notion of feedback control is thus very powerful, but cannot be applied without proper knowledge.

Presentation

This book is divided into chapters, and sections; for example, section 2.1 is the first section of Chapter 2, and section 2.1.3 is the third subsection of section 2.1.

It contains the basics that a non-specialized engineer must know in control engineering. I had the opportunity to teach this course at Conservatoire National des Arts et Métiers (Paris), several engineering institutes Grandes Ecoles, Ecole Normale Supérieure de Cachan, and, for the most difficult parts, at Paris XI University (Master 2 level). The course contained in this book is progressive, making it accessible to any reader with an L2 level in science, and includes many examples. Nevertheless, to make it coherent, passages (sections or groups of sentences) had to be included in the first chapters that call upon somewhat more difficult notions which may need a second or third reading. *These passages are preceded by asterisks for sections, and situated between two asterisks for groups of words or sentences.*

The purpose of this book is to study *system modeling, identification, analysis and control.* Among these four themes, *modeling* is perhaps the trickiest problem: to know how to model electrical, mechanical, thermal, hydraulic, or other systems – as they are complex – a person has to be an electrical, mechanical, thermal, hydraulic engineer. It is physics in general, all of physics, which is useful for modeling; and modeling, which is the subject matter of Chapter 1, is not exclusive to the control engineer. Some examples will serve as basic reminders of hydraulics and thermodynamics. Reminders relative to electricity are somewhat less succinct. With respect to mechanics, I thought that it would be useful to go into more details. However, the only objective of the presentation is to enable the reader to understand examples and resolve exercises. Obviously, it cannot replace a treatise on mechanics.

From Chapters 2 to 11, the control engineer finds himself/herself in his/her own private domain. These chapters deal with linear systems analysis, control, and identification. I have put together some *mathematical elements* which I deemed essential in the *appendices* included in Chapters 12 and 13. This will spare the reader from constantly referring to the bibliography: elements of analysis first, and then of algebra. Some of them (Smith form of a polynomial matrix, $\Re\mathcal{H}_\infty$ algebra, module theory etc.) are probably new to most readers. But I preferred not to discuss them in the main body of the text, to save the latter, to the extent possible, for what truly comes within the scope of control theory. Likewise, the Laplace transform and the z-transform, which are *mathematical tools* that classically appear high in textbooks on control theory, are included in these appendices. These are organized as a presentation of "mathematics for systems theory" with proofs when they are constructive or simply useful for the reader's understanding. The reader can thus choose to refer to the appendices if needed or to read them entirely as actual chapters. I have decided not to present the theory of measure and integration. Concerning the latter, I refer the reader to the second volume of [35] or to [103]. I achieved this with the help of some mathematical gymnastics here and there.

About 80 exercises are given to the readers to test and refine their understanding of the subject. Solutions to most of the exercises are provided (sometimes in brief) in Chapter 14.

MATLAB and/or SCILAB files to run examples and solve exercises can be found at www.iste.co.uk/bourles/LS.zip.

Course outline

Beginner's course

The beginner at L3 (i.e. third-year undergraduate) level can start with the easiest parts of Chapter 1. The part that might be difficult for the beginner is section 1.2 which is devoted to the modeling of a mechanical system. I advise the reader to approach this part in a pragmatic way, using the examples and exercises which show what results are essential and how to make good use of them.

In Chapter 2, readers should leave aside everything that relates to multi-input multi-output systems. It is better if they focus on "left-forms", "right-forms", and on methods which make it possible to obtain system transfer functions. They will eventually develop particular interest in treated examples and exercises.

The reader should study Chapters 3 to 6 in depth (leaving aside, of course, the starred passages "intended for a second reading"). From my point of view, the reader's course therefore ends with the RST controller ("usual case" section).

Advanced course

The reader at M1 (i.e. first-year graduate) level should begin to go deeper into what he/she only skimmed over when he/she was a beginner, i.e. section 1.2 of Chapter 1, and Chapter 2. Afterwards, the reader will resolutely move on to Chapter 7, which includes complementary material on systems theory, and will then be ready to study

Chapters 8 to 11. I advise the reader to systematically skip the passages related to module theory.

Advanced graduate level

The reader at M2 (i.e. second-year graduate) level still has many things to discover before launching into specialized courses. These are all the passages that he/she skipped during previous readings, in particular those presenting the theory of systems in the language of modules. It may also be in the interest of the readers to systematically re-read the elements of mathematics in Appendices 1 and 2 (Chapters 12 and 13).

Coherent course and "butterfly" course

This book should be read in a more systematic way to understand the contents. Another approach to this book is also possible for readers who wish to be systematic while not necessarily belonging to the above categories: these readers should read the chapters of the book, without skipping any passage, in the order I have presented them (with the exception of the appendices – Chapters 12 and 13 – which they should read first). I have often started reading a science book in that spirit, and only a lack of time has made me take an opposite turn, which is to "flit about", trying to go as fast as possible (a strategy, besides, which does not always pay off). I do not have any way to propose to the "butterfly" readers, but I have included a detailed index at the end of the book for them.

Choices and necessities

Since this book is about linear time-invariant systems, I have deemed it essential to present this concept correctly. Such a system can no longer be properly defined by its transfer matrix (too poor a representation), or by Kalman's approach, as one of its state realizations, or by Rosenbrock's approach, as a "system matrix" (too contingent a representation). Wonham's "geometric approach", which was proposed in the 1980s, in the end, only reformulates Kalman's approach in a little more modern mathematical language, and does not seem to be a good answer. Following the works of experts in differential algebra, it appeared that, what control engineers call "system" and "linear system", should be defined as an *extension of a differential field* and a *module defined over a ring of differential operators,* respectively. In the early 1990s, this property was revealed by M. Fliess to the community of control engineering (see the synthesis presented in [47]). Such a language may have discouraged a beginner, but it is this conception that I have attempted to make the reader gradually sensitive to, in Chapters 2 and 7. In reality, module theory, at the level used here (finitely generated modules over principal ideal domains) is very simple for a reader somewhat accustomed to abstraction. It does clarify manipulations that can be done with polynomial matrices, as well as the Jordan canonical form. For this reason, I have made it the main theme of "passages intended for a third reading". The reader wishing to go further into systems theory can refer to [47] on nonlinear aspects, and [22] on linear time-varying systems ([15] is a preliminary version of [22]). Module theory is helpful in presenting (in the case of linear systems) the notion of flatness (due to Fliess and his collaborators [48]), which has become essential to efficiently resolve the problem of motion planning (see section 7.5).

I wanted to show that control engineering is not magic but science. On that premise, I have insisted, in Chapter 4, on the *limitations due to certain characteristics of the system to be controlled* (unstable poles and zeros, specifically). For more details on this point, see [51].

The control engineer is permanently confronted with the problem of uncertainties in modeling. Thus, this book is constantly preoccupied with the *robustness of control laws*. However, *this is not a book on robust control*. The theory of robustness has become extremely sophisticated, and its recent development is not covered here. Besides, excellent books on this issue ([107] and [122], among others) do exist.

I have emphasized in this book on control (whether in polynomial or state formalism) that system stabilization is exceptionally the role of the feedback loop; the point is often to compel the output of such a system to follow a reference signal, in spite of various disturbances that it may be subjected to. This is why – in order to design an RST controller as well as a control by state feedback, possibly with an observer – *it is essential to first make the necessary model augmentations* (following what Wonham called the "internal model principle"). Without this prerequisite, "modern methods" only generate nonsense.

It also appeared to me that it was important to show that, in the multi-input multi-output case, *a control law with execrable qualities can correspond to a good pole placement*, because the latter does not determine the former. The worst method (even though it is quite elegant from a strictly theoretical point of view) is probably the one that consists of reverting to a cyclic state matrix through a first loop, and then proceeding as in the single-input single-output case.

I have abstained from presenting the linear-quadratic (LQ) optimal control and its "dual" version, the Kalman filter. For the conception of control by state feedback and observer, I have indeed preferred to propose essentially algebraic methods for several reasons. Once these methods are well assimilated, it becomes easy to implement an efficient "LQG control".[1] The theory of LQG control (which encompasses LQ control and Kalman filter theories) is now classic and, at a basic level, is well covered in treatises such as [1] and [2]. On the other hand, the minimization of a criterion is only a convenient intermediate step for the control engineer to obtain a control law with the desired properties [78]. Another element that kept me from including optimal control

1. "LQG" stands for "Linear Quadratic Gaussian".

is that it would have made this book longer (especially to correctly present continuous-time stochastic optimal control, with, in the background, Ito's differential calculus [68], [118]). I found the volume of this book to be large enough. Nevertheless, except for a few subtleties, the methods proposed in the multi-input multi-output case in Chapters 8 and 9 only differ from LQ control and Kalman filtering in their presentation style (the minimization of a criterion is not presented as a goal – see Remark 245 in section 8.1.4). I encourage the reader to continue the study of this book along with that of "LQG control with frequency-shaped weights" [2], which is a good complement.

Only Chapter 10 deals with discrete-time systems and control. All previous chapters (including the one on RST controller) are thus presented in the context of continuous-time, which is a bit unusual. Despite the ideas spread by some people, I have experienced that continuous-time formalism offers much more flexibility for the design of control laws, especially with respect to the choice of poles and the question of "roll-off", which are essential to robustness. Once the synthesis of continuous-time control laws is mastered, it is very simple to switch to the synthesis of discrete-time control laws – without any approximation (this point is of course discussed in detail in Chapter 10).

Chapter 11 discusses parametric identification of discrete-time systems by minimization of the l_2 norm of the prediction error. The presentation is relatively classic as far as open-loop identification is concerned; as for closed-loop identification, it essentially includes Ljung and Forsell's contribution [82] (as a complement, the reader can consult [75] and Chapter 10 of [110]).

Many other topics would have been worth presenting: for example, anti-windup methods in the presence of saturations (which is clearly expounded in [4]) or gain scheduling (see survey papers [104] and [80]), not to mention nonlinear control (for a general presentation of this theme, see [108] and [62], and begin with the first reference because it is simple; *robust* nonlinear control is discussed in [49]; see also [48] which deals with *flatness* from both theoretical and practical viewpoints, in addition to section 7.5 of this book). We can also cite the following subjects: adaptive control (which is the subject of a lot of literature, but [57], [3] and [108] can be recommended as a good initiation on this theme); fuzzy or neural control (the reader can find a *genuinely scientific presentation* of the last two types of control in [40]); the control of systems governed by delayed differential equations or partial differential equations (on this subject, see [55] and [33] in the linear context, as well as [58] and [59] where nonlinear problems are discussed). As already mentioned, the extension of many methods presented in this book to linear time-varying systems can be found in [22].

Note on the English edition

This English edition has given me the opportunity to correct many errors which, despite proof-reading, had been left in the original French edition, and also include additional information which might be useful. Furthermore, it has given me the opportunity to complete Chapter 7 with a section on flatness and to entirely revise Chapter 11 on identification: the mathematical bases included therein are given in Appendix 1, and a section on closed-loop identification has also been added. Finally, this edition contains additional exercises and one of them presents the basics of what needs to be known about the "delta transform".

Acknowledgments

I would first like to thank my "elders", especially E. Irving, I.D. Landau and P. de Larminat, who have shared their experience with me during informal, but always passionate, discussions. I also thank M. Fliess who has opened my eyes to the algebraic theory of systems – which I hope this book has benefited from – and P. Chantre for our exchanges on identification. I thank my son Nicolas who has helped me finalize figures. I also thank my translator G.K. Kwan for his efforts and his kindness. Last but not the least, I thank my wife Corinne, whose patience has often been severely tested during this work which has demanded a lot of time and energy from me.

Chapter 1

Physical Models

1.1. Electric system

Electric circuits can be modeled by applying Kirchhoff's two laws: Kirchhoff's Current Law and Kirchhoff's Voltage Law. They are also known as the nodal rule and the mesh rule. Following are the two examples in which these two laws have been applied.

1.1.1. *Mesh rule*

Consider the "RLC" circuit shown in Figure 1.1.

This circuit constitutes a mesh. *The mesh rule* states that between any two points on a mesh, points A and B in this circuit for example, the potential difference is independent of the path that is taken (which, in physics, is a fundamental property of a *potential*, in this case the electric potential). Therefore, we have $V_A - V_B = V = V_R + V_L + V_C$, where V_R, V_L, and V_C are the voltages across the resistance R, the inductance L, and the capacitance C, respectively. We have $V_R = Ri, V_L = L\frac{di}{dt}, V_C = \frac{1}{C}\int_{-\infty}^{t} i(\tau)d\tau$; therefore we obtain the integro-differential equation $V = Ri + L\frac{di}{dt} + \frac{1}{C}\int_{-\infty}^{t} i(\tau)d\tau$. Differentiating this equation we get the following linear differential equation:

$$LC\frac{d^2 i}{dt^2} + RC\frac{di}{dt} + i - C\frac{dV}{dt} = 0. \tag{1.1}$$

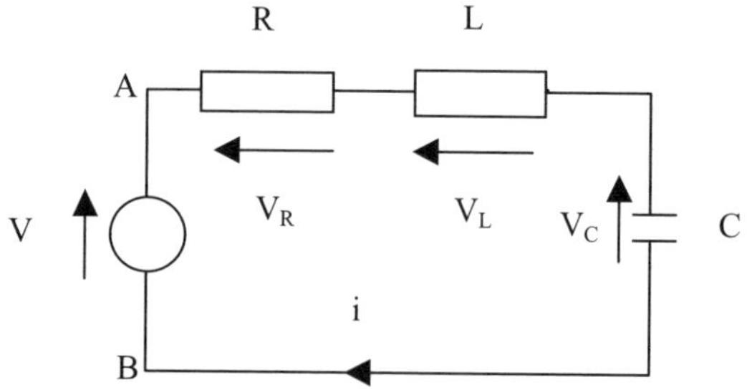

Figure 1.1. *RLC circuit*

1.1.2. *Nodal rule*

Consider the electric circuit given in Figure 1.2, which is a "pi circuit" used to represent a transmission line in the domain of electric networks.

According to *the mesh rule* described above, we have

$$V_A = \frac{1}{C}\int_{-\infty}^{t} i_1(\tau)d\tau, V_B = \frac{1}{C}\int_{-\infty}^{t} i_2(\tau)d\tau, V_A - V_B = Ri + L\frac{di}{dt}.$$

The nodal rule is based on the conservation of charge. According to this rule, at any node of a circuit, which can be any point within the circuit, the sum of current entering a node is equal to the sum of current leaving that node. Hence, we have the following two equations: $i_A = i + i_1$ and $i = i_B + i_2$. Eliminating the variable i, we can now re-arrange the equations as follows:

$$\left\{\begin{array}{c} C\frac{dV_A}{dt} - i_1 = 0, \\ C\frac{dV_B}{dt} - i_2 = 0, \\ V_A - V_B - R(i_A - i_1) - L\frac{d}{dt}(i_A - i_1) = 0, \\ i_A - i_1 - (i_B + i_2) = 0. \end{array}\right. \tag{1.2}$$

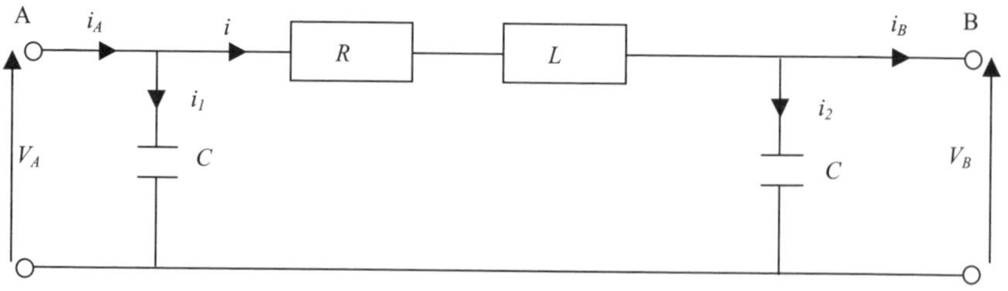

Figure 1.2. *Pi circuit*

1.2. Mechanical system

There are two main approaches to model a mechanical system. The first one uses the fundamental principle of dynamics, and is based on the *analysis of forces and moments* applied to the various elements of the system (particularly, internal forces, i.e. forces applied by subsystems on one another). The second approach uses the Lagrangian formalism and is based on the *energy of the system* (kinetic and potential energy). The analysis of internal forces is then unnecessary. These two approaches are succinctly expounded and applied to two examples.

1.2.1. *Fundamental principle of dynamics*

Torsor

A *torsor* is a field of antisymmetric vectors in a three-dimensional affine space. Let $\overrightarrow{M_\bullet}$ be a field of such vectors; this is a function $A \mapsto \overrightarrow{M_A}$, and $\overrightarrow{M_A}$ is called the *moment* of the torsor $\overrightarrow{M_\bullet}$ at point A. There exists a vector $\overrightarrow{V}$, called the *characteristic vector*, such that for any points A and B, we have the relation $\overrightarrow{M_A} = \overrightarrow{M_B} + \overrightarrow{AB} \wedge \overrightarrow{V}$. The torsor is thus entirely determined by $\overrightarrow{V}$ and the moment $\overrightarrow{M_A}$, for any point A under consideration. We are agreed on the notation $\left\{\overrightarrow{V}\right\}_A$ for the pair $\left(\overrightarrow{V}, \overrightarrow{M_A}\right)$ and $\left\{\overrightarrow{V}\right\}$ for the torsor $\overrightarrow{M_\bullet}$. The two vectors $\overrightarrow{V}$ and $\overrightarrow{M_A}$ are called the elements of reduction of torsor $\left\{\overrightarrow{V}\right\}$ at point A.

The *comoment of* $\left\{\overrightarrow{V}\right\}_A = \left(\overrightarrow{V}, \overrightarrow{M_A}\right)$ and $\left\{\overrightarrow{V'}\right\}_A = \left(\overrightarrow{V'}, \overrightarrow{M'_A}\right)$ is the scalar $\overrightarrow{V}.\overrightarrow{M'_A} + \overrightarrow{V'}.\overrightarrow{M_A}$. This comoment is invariant in the sense that it is independent of the point A considered (as can easily be verified by the reader). Therefore, the comoment of the *torsors* $\left\{\overrightarrow{V}\right\}$ and $\left\{\overrightarrow{V'}\right\}$ is well defined and is denoted by $\left\{\overrightarrow{V}\right\}.\left\{\overrightarrow{V'}\right\}$.

Kinematic torsor

An example of a torsor is the field of velocities of a *rigid body* S, called the *kinetic torsor* of S; this torsor is defined relative to an orthonormal frame of reference $\mathcal{R}$ (we also say the torsor is defined *in this frame of reference*). This is to say that the translational velocities and angular velocities are measured relative to the origin and along the axes of $\mathcal{R}$. Its characteristic vector is the vector of "instantaneous rotation" $\overrightarrow{\omega}$ (this torsor is thus denoted by $\{\overrightarrow{\omega}\}$). Indeed, as easily shown, the velocities at two points A and B on this rigid body are related by

$$\boxed{\overrightarrow{v_A} = \overrightarrow{v_B} + \overrightarrow{AB} \wedge \overrightarrow{\omega}}. \tag{1.3}$$

Now, let us consider a frame of reference $\mathcal{R}'$ attached to the rigid body S. This consists of an origin A and three basis vectors $\overrightarrow{e_i}, i = 1, 2, 3$. Put $\overrightarrow{e_i} = \overrightarrow{AA_i}$, where

A_i's are points on S. We have $\frac{d\overrightarrow{e_i}}{dt} = \overrightarrow{v_{A_i}} - \overrightarrow{v_A} = \overrightarrow{A_iA} \wedge \overrightarrow{\omega}$ according to (1.3), and thus $\frac{d\overrightarrow{e_i}}{dt} = \overrightarrow{\omega} \wedge \overrightarrow{e_i}$. Let $\overrightarrow{Y}$ be any vector; its coordinates within $\mathcal{R}'$ are those Y_i's such that $\overrightarrow{Y} = \sum_{i=1}^{3} Y_i \overrightarrow{e_i}$. Differentiating this expression, we get

$$\frac{d\overrightarrow{Y}}{dt} = \sum_{i=1}^{3} \frac{dY_i}{dt} \overrightarrow{e_i} + \sum_{i=1}^{3} Y_i \frac{d\overrightarrow{e_i}}{dt} = \sum_{i=1}^{3} \frac{dY_i}{dt} \overrightarrow{e_i} + \overrightarrow{\omega} \wedge \overrightarrow{Y}.$$

This quantity, denoting it by $\frac{d\overrightarrow{Y}}{dt/\mathcal{R}}$, is the derivative of $\overrightarrow{Y}$ in the reference frame $\mathcal{R}$. The vector $\sum_{i=1}^{3} \frac{dY_i}{dt} \overrightarrow{e_i}$ is the derivative of $\overrightarrow{Y}$ in this moving reference frame $\mathcal{R}'$ and is denoted by $\frac{d\overrightarrow{Y}}{dt/\mathcal{R}'}$. We have thus obtained the formula establishing the relation between the differentiation in a moving reference frame and the differentiation in a fixed reference frame as follows:

$$\boxed{\frac{d\overrightarrow{Y}}{dt/\mathcal{R}} = \frac{d\overrightarrow{Y}}{dt/\mathcal{R}'} + \overrightarrow{\omega} \wedge \overrightarrow{Y}}. \tag{1.4}$$

Kinetic torsor

Kinetic moment

The kinetic moment of a *material system* S (i.e. a collection of material points and rigid bodies – a material point can be considered a punctual rigid body) with respect to *any point* A in reference frame $\mathcal{R}$ is

$$\boxed{\overrightarrow{\sigma_A} \triangleq \int_S \overrightarrow{AM} \wedge \overrightarrow{v_M}\, dm}$$

where dm is the density of mass at point M; this density is assumed to be constant over time.[1]

Elements of reduction

The field of vectors $\overrightarrow{\sigma_\bullet}$ is a torsor, as will be shown below, and it is called the *kinetic torsor.*

Indeed, if B is any other point, we have

$$\overrightarrow{\sigma_A} = \int_S (\overrightarrow{AB} + \overrightarrow{BM}) \wedge \overrightarrow{v_M}\, dm = \overrightarrow{AB} \wedge \int_S \overrightarrow{v_M}\, dm + \int_S \overrightarrow{BM} \wedge \overrightarrow{v_M}\, dm.$$

Now let O be the origin of $\mathcal{R}$; we then get $\overrightarrow{v_M} = \frac{d\overrightarrow{OM}}{dt}$, thus $\int_S \overrightarrow{v_M}\, dm = \frac{d}{dt} \int_S \overrightarrow{OM}\, dm$. By definition of the *center of mass* G of S, we have

$$\int_S \overrightarrow{OM}\, dm = m\,\overrightarrow{OG}$$

1. The symbol $\triangleq$ means "equals by definition".

where m is the total mass of S. Therefore,

$$\boxed{\overrightarrow{\sigma_A} = \overrightarrow{\sigma_B} + \overrightarrow{AB} \wedge \overrightarrow{p}} \tag{1.5}$$

where $\overrightarrow{p}$ is the *momentum* of S, defined as

$$\boxed{\overrightarrow{p} \triangleq m\,\overrightarrow{v_G}}.$$

According to (1.5), $\overrightarrow{\sigma_\bullet}$ is actually a torsor $\{\overrightarrow{p}\}$ with characteristic vector $\overrightarrow{p}$.

Case of a rigid body

Suppose S is a *rigid body* and A is *any* point on this rigid body; according to (1.3) we get $\overrightarrow{v_M} = \overrightarrow{v_A} + \overrightarrow{MA} \wedge \overrightarrow{\omega} = \overrightarrow{v_A} + \overrightarrow{\omega} \wedge \overrightarrow{AM}$, thus

$$\begin{aligned}\overrightarrow{\sigma_A} &= \int_S \overrightarrow{AM} \wedge \left(\overrightarrow{v_A} + \overrightarrow{\omega} \wedge \overrightarrow{AM}\right)\, dm \\ &= \int_S \overrightarrow{AM}\, dm \wedge \overrightarrow{v_A} + \int_S \overrightarrow{AM} \wedge \overrightarrow{\omega} \wedge \overrightarrow{AM}\, dm.\end{aligned}$$

The linear mapping $\mathcal{I}_A\ :\ \overrightarrow{\omega} \mapsto \int_S \overrightarrow{AM} \wedge \overrightarrow{\omega} \wedge \overrightarrow{AM}\, dm$ is called the *inertia tensor* of S with respect to A, and hence we have

$$\boxed{\int_S \overrightarrow{AM} \wedge \overrightarrow{\omega} \wedge \overrightarrow{AM}\, dm = \mathcal{I}_A\, \overrightarrow{\omega}}. \tag{1.6}$$

We can express this tensor in terms of its coordinates in an orthonormal system of reference $\mathcal{R}'$. Let (x, y, z) be the components of $\overrightarrow{AM}$ in $\mathcal{R}'$. Developing expression (1.6), the linear mapping $\mathcal{I}_A$ is identified[2] with the *inertia matrix* of S with respect to A in $\mathcal{R}'$, given by

$$\mathcal{I}_A = \begin{bmatrix} \int_S \left(y^2 + z^2\right)\, dm & -\int_S xy\, dm & -\int_S xz\, dm \\ -\int_S xy\, dm & \int_S \left(x^2 + z^2\right)\, dm & -\int_S yz\, dm \\ -\int_S xz\, dm & -\int_S yz\, dm & \int_S \left(x^2 + y^2\right)\, dm \end{bmatrix}.$$

We thus obtain

$$\boxed{\overrightarrow{\sigma_A} = m\,\overrightarrow{AG} \wedge \overrightarrow{v_A} + \mathcal{I}_A\, \overrightarrow{\omega}}. \tag{1.7}$$

This expression can be simplified when $A = G$ or when A is a *fixed* point on S (if S rotates around this point); then (1.7) reduces to

$$\boxed{\overrightarrow{\sigma_A} = \mathcal{I}_A\, \overrightarrow{\omega}}. \tag{1.8}$$

---

2. Once the choice of bases is made, a linear mapping (also called a homomorphism) is represented by a matrix (see section 13.3.2). Abusing the language, we can identify this linear mapping with this matrix. And this is what we do here. "Abuse of language" (in the sense widely used in mathematics) and "abuse of notation" are considered to be synonymous in this book.

Inertia matrix

Matrix $\mathcal{I}_A$ is only constant when the reference frame $\mathcal{R}'$ is rigidly linked to rigid body S and it is therefore our interest to look at things in the context of this case. Of course, this reference frame is in motion, and the derivative of (1.8) is obtained by applying (1.4). On the other hand, the matrix $\mathcal{I}_A$ is symmetric real, and thus can be diagonalized in an orthonormal reference frame (see section 13.5.6), and whose axes are by definition the *principal axes of inertia* of S. The diagonal elements obtained in such a matrix are called the *principal moments of inertia* of S (with respect to A and relative to the principal axes in question).

Torque

In the case where $\overrightarrow{p} = 0$, the kinetic moment is independent of the point being considered, according to (1.5): for any points A and B, $\overrightarrow{\sigma_A} = \overrightarrow{\sigma_B}$. Such a kinetic moment is called a *torque*, which we shall denote by $\overrightarrow{C}$.

Kinetic energy

The kinetic energy T of a material system S is equal to the comoment $\{\overrightarrow{\omega}\}.\{\overrightarrow{p}\}$. In case of a rigid body, we obtain

$$\boxed{T = \frac{1}{2}\left(m\, v_G^2 + \overrightarrow{\omega}.\mathcal{I}_G\, \overrightarrow{\omega}\right)}. \qquad (1.9)$$

Force torsor

Now consider a set of external forces $\overrightarrow{f_1}, ..., \overrightarrow{f_n}$ being applied to points $A_1, ..., A_n$ of a material system S. The resultant force is given by $\overrightarrow{f} = \sum_{k=1}^{n} \overrightarrow{f_k}$, while the resultant moment of these forces with respect to a point O, arbitrary at this time, is $\overrightarrow{\mathcal{M}_O} = \sum_{k=1}^{n} \overrightarrow{OA_k} \wedge \overrightarrow{f_k}$. If A is another arbitrary point, we immediately obtain the equality

$$\boxed{\overrightarrow{\mathcal{M}_A} = \overrightarrow{\mathcal{M}_O} + \overrightarrow{AO} \wedge \overrightarrow{f}} \qquad (1.10)$$

which shows that the field $\overrightarrow{\mathcal{M}_\bullet}$ defines a torsor with characteristic vector $\overrightarrow{f}$; this torsor $\left\{\overrightarrow{f}\right\}$ is called the *force torsor* (or, more precisely, the *torsor of external forces*).

Fundamental principle of dynamics (Newton's law)

Expression in a Galilean reference frame

Let O be a *fixed point* in a *Galilean reference frame* (also known as *inertial reference frame*) $\mathcal{R}$, and S be a material system. The *fundamental principle of dynamics* or *Newton's law* is written as

$$\boxed{\left\{\overrightarrow{f}\right\}_O = \tfrac{d}{dt}\left\{\overrightarrow{p}\right\}_O}. \tag{1.11}$$

This equality between torsors shows that we have the following two relations

$$\overrightarrow{f} = \frac{d\overrightarrow{p}}{dt} \tag{1.12}$$

$$\overrightarrow{\mathcal{M}_O} = \frac{d\overrightarrow{\sigma_O}}{dt}. \tag{1.13}$$

It is reasonable to emphasize the fact that (1.11) is valid in the reference frame $\mathcal{R}$ (or in any other Galilean reference frame). In a non-Galilean frame of reference, it is necessary to take into account the torsor of inertial forces as well as Coriolis forces [76] – notions that can be derived from (1.4).

Expression in a moving frame of reference firmly fixed in the center of mass

Nevertheless, let us consider the center of mass G of the material system S; from (1.5) we get

$$\begin{aligned}\frac{d\overrightarrow{\sigma_G}}{dt} &= \frac{d\overrightarrow{\sigma_O}}{dt} + \frac{d}{dt}\left(\overrightarrow{GO}\wedge\overrightarrow{p}\right) = \overrightarrow{\mathcal{M}_O} - \overrightarrow{v_G}\wedge\overrightarrow{p} + \overrightarrow{GO}\wedge\frac{d\overrightarrow{p}}{dt}\\ &= \overrightarrow{\mathcal{M}_O} + \overrightarrow{f}\end{aligned}$$

according to (1.12). Therefore,

$$\frac{d\overrightarrow{\sigma_G}}{dt} = \overrightarrow{\mathcal{M}_G}$$

which shows that

$$\boxed{\left\{\overrightarrow{f}\right\}_G = \tfrac{d}{dt}\left\{\overrightarrow{p}\right\}_G}.$$

Expression in an arbitrary moving frame of reference

Let A be any *arbitrary* point. We have, according to (1.8), $\overrightarrow{\sigma_O} = \overrightarrow{\sigma_A} + \overrightarrow{OA}\wedge\overrightarrow{p}$, and thus

$$\frac{d\overrightarrow{\sigma_O}}{dt} = \frac{d\overrightarrow{\sigma_A}}{dt} + \overrightarrow{v_A}\wedge\overrightarrow{p} + \overrightarrow{OA}\wedge\frac{d\overrightarrow{p}}{dt}.$$

From (1.10) and (1.12) we also have

$$\overrightarrow{\mathcal{M}_A} = \overrightarrow{\mathcal{M}_O} + \overrightarrow{AO}\wedge\frac{d\overrightarrow{p}}{dt}. \tag{1.14}$$

From (1.13) we obtain

$$\boxed{\overrightarrow{\mathcal{M}_A} = \frac{d\overrightarrow{\sigma_A}}{dt} + \overrightarrow{v_A} \wedge \overrightarrow{p}}$$

which is a generalization of (1.13).

Moving frame of reference: case of a rigid body

Let A be a point on *rigid body* S. The kinetic moment $\overrightarrow{\sigma_A}$ is given by expression (1.7), therefore

$$\frac{d\overrightarrow{\sigma_A}}{dt} = m\,\overrightarrow{AG} \wedge \frac{d\overrightarrow{v_A}}{dt} + m\,\overrightarrow{v_G} \wedge \overrightarrow{v_A} + \mathcal{I}_A \frac{d\overrightarrow{\omega}}{dt} + \overrightarrow{\omega} \wedge \mathcal{I}_A\,\overrightarrow{\omega}.$$

According to (1.14) we obtain the following:

$$\boxed{\overrightarrow{\mathcal{M}_A} = \mathcal{I}_A \frac{d\,\overrightarrow{\omega}}{dt} + \overrightarrow{AG} \wedge m\frac{d\overrightarrow{v_A}}{dt} + \overrightarrow{\omega} \wedge \mathcal{I}_A\,\overrightarrow{\omega}}. \tag{1.15}$$

This expression can be simplified in the following cases:

i) A is fixed and S rotates around A, or $A = G$ (the second term on the right-hand side drops out);

ii) rotation is around one of the principal axes of inertia (the third term drops out).

Case of a rigid body rotating around an axis

Consider the case where S is a rigid body rotating around one of the principal axes of inertia, denoted by $\overrightarrow{Oz}$ for example. Let $\overrightarrow{C}$ be the resultant torque exerted on S (directed along $\overrightarrow{Oz}$) and write $\overrightarrow{C} = C\,\overrightarrow{k}$, where $\overrightarrow{k}$ is the unit vector of $\overrightarrow{Oz}$. The rotation vector of S is $\overrightarrow{\omega} = \omega\,\overrightarrow{k}$. We have $\mathcal{I}_O\,\overrightarrow{\omega} = J\,\overrightarrow{\omega} = J\,\omega\,\overrightarrow{k}$, where J is the principal moment of inertia of Σ with respect to the axis $\overrightarrow{Oz}$, i.e.

$$\boxed{J = \int_\Sigma \left(x^2 + y^2\right)\,dm}. \tag{1.16}$$

Equation (1.15) reduces to

$$\boxed{J\,\frac{d\omega}{dt} = C}. \tag{1.17}$$

Principle of action and reaction

In the case of a material system S that consists of N interacting rigid bodies S_i, the global equation (1.11) is not sufficient to determine the motion of each of the rigid bodies. We therefore are led to decompose the system into each of its elements S_i and to take into account the forces and moments these rigid bodies exert on one another.

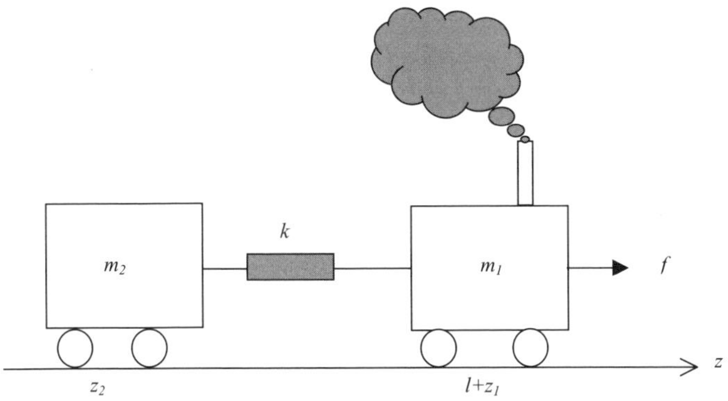

Figure 1.3. *Train*

According to the principle of action and reaction, if $\left\{\overrightarrow{f_{i\to j}}\right\}$ and $\left\{\overrightarrow{f_{j\to i}}\right\}$ represent the torsor of the forces exerted on S_j by S_i, and on S_i by S_j, respectively, we have

$$\boxed{\left\{\overrightarrow{f_{i\to j}}\right\} + \left\{\overrightarrow{f_{j\to i}}\right\} = 0}. \tag{1.18}$$

Example of a frictionless train

Consider a train composed of a locomotive and a wagon such as depicted in Figure 1.3. The equations of this system are formulated with the help of (1.12) and (1.18).

The coupling between the locomotive of mass m_1 and the wagon of mass m_2 is represented by a spring with constant k (according to Hooke's law). The positions of center of mass of the locomotive and that of the center of mass of the wagon are denoted by z_2 and $l+z_1$, where l is the distance between the two centers of mass when the train is at rest. The motional force put into action by the locomotive is denoted by f. The frictional forces are neglected along with the mass of the spring. The resultant force exerted on the locomotive is therefore $f-f_r$, where f_r is the force exerted by the spring on the locomotive. As a result, we have $m_1\frac{d^2z_1}{dt^2} = f - f_r$, and $f_r = k(z_1 - z_2)$, with $z_1 - z_2$ being the lengthening of the spring. The spring transmits the integral force f_r according to (1.12). The wagon is only subject to this force f_r and we thus have $m_2\frac{d^2z_2}{dt^2} = f_r$. Eliminating the variable f_r, we obtain the equations

$$\left\{\begin{array}{l} m_1\frac{d^2z_1}{dt^2} + k\left(z_1 - z_2\right) - f = 0 \\ m_2\frac{d^2z_2}{dt^2} + k\left(z_2 - z_1\right) = 0. \end{array}\right. \tag{1.19}$$

Example of the inverted pendulum

Consider the inverted pendulum in Figure 1.4. It consists of a carriage with which a rod of length l terminated by a mass m is articulated. The frictional forces as well

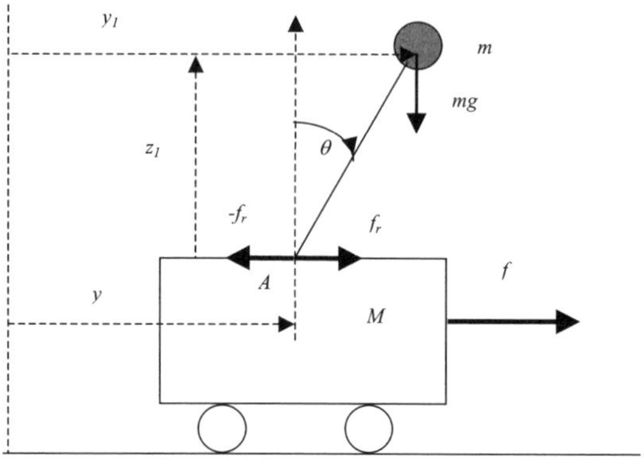

Figure 1.4. *Inverted pendulum*

as the mass of the rod are considered negligible; in addition, mass m is considered punctual.

The inverted pendulum is acted upon by a horizontal force f that is exerted onto the carriage.

First, let us examine the forces that are exerted on the carriage. Since the carriage moves horizontally, only the horizontal components of the forces have to be considered. The horizontal resultant force on the carriage is $f - f_r$, where f_r is the horizontal component of the force exerted by the rigid body (rod + mass m) onto the carriage. According to (1.12) we thus have $M\frac{d^2y}{dt^2} = f - f_r$.

Now, let us examine the rigid body (rod + mass), whose center of mass G has coordinates (y_1, z_1). The only horizontal force exerted on this rigid body is f_r, hence $m\frac{d^2y_1}{dt^2} = f_r$.

We have the geometrical relations

$$\begin{cases} y_1 = y + l\sin\theta \\ z_1 = l\cos\theta. \end{cases} \tag{1.20}$$

By way of the first of these equations and by eliminating f_r, we obtain

$$(M+m)\frac{d^2y}{dt^2} - m\,l\sin\theta\left(\frac{d\theta}{dt}\right)^2 + m\,l\cos\theta\frac{d^2\theta}{dt^2} = f.$$

We can write equation (1.15) at the point A. The term $\vec{\omega} \wedge \mathcal{I}_A\,\vec{\omega}$ is zero, since the rotation is effected around one of the principal axes of inertia. Therefore, (1.15) is written as $\mathcal{I}_A\frac{d\vec{\omega}}{dt} = \overrightarrow{\mathcal{M}_A} - \overrightarrow{AG} \wedge m\frac{d\vec{v_A}}{dt}$ and the second term on the right-hand side can be interpreted as the moment with respect to A of the inertial force $-m\frac{d\vec{v_A}}{dt}$

applied at G. The other moment to be accounted for is that of the gravitational force. We thus obtain $J\frac{d^2\theta}{dt^2} = mgl\sin\theta - ml\frac{d^2y}{dt^2}\cos\theta$, where J is the moment of inertia of m with respect to A, i.e. ml^2.

Finally, we obtain

$$\begin{cases} (M+m)\frac{d^2y}{dt^2} - m\,l\,\sin\theta\left(\frac{d\theta}{dt}\right)^2 + m\,l\,\cos\theta\,\frac{d^2\theta}{dt^2} - f = 0 \\ l\,\frac{d^2\theta}{dt^2} - g\,\sin\theta + \frac{d^2y}{dt^2}\,\cos\theta = 0. \end{cases} \tag{1.21}$$

1.2.2. *Lagrangian formalism*

For certain types of mechanical systems, it may be more convenient to use the Lagrangian formalism instead of the fundamental principle of dynamics, for the formulation of the equations. As mentioned above, one of the advantages is that the internal forces of the system do not have to be considered.

Holonomic system

A mechanical system S is said to be *holonomic* if the position of its different parts can be characterized by n independent variables $q_1, ..., q_n$, called the *generalized coordinates* of the system. We then say that S is a holonomic system with n *degrees of freedom*.

For example, the train above is a holonomic system with generalized coordinates $q_1 = y_1$ and $q_2 = y_2$. The inverted pendulum is also a holonomic system with generalized coordinates $q_1 = y$ and $q_2 = \theta$.

The vector space of dimension n in which these coordinates are defined is called the *configuration space*.

Holonomic relations

The variables $q_1, ..., q_n$ are only independent if all existing relations between the subsystems have been taken into account and all redundant coordinates have been eliminated.* For example, suppose that, when all relations are removed, the system S only depends on N coordinates $q_1, ..., q_N$. Let us also assume that the only relations by which S is constrained are p relations of the form

$$g_k(q_1, ..., q_N) = 0, \quad k = 1, ..., p \tag{1.22}$$

where g_k are scalar functions. Relations of the form (1.22), involving only the positions q_i, are called the *holonomic relations*. These relations are of a pure geometrical nature. Let $n = N - p$ and suppose that the Jacobian $\frac{\partial(g_1,...,g_p)}{\partial(q_{n+1},...,q_N)}$ is non-zero. Then, according to the Implicit Mapping Theorem, we can express (locally) $q_{n+1}, ..., q_N$ as a function of $q_1, ..., q_n$ which are independent variables. In this case, S is a holonomic system and $q_1, ..., q_n$ are its generalized coordinates.*

Non-holonomic relations

The situation becomes more complicated in the case where the relations involve the velocities $\dot{q}_1, ..., \dot{q}_N$, according to $h_j(q_1, ..., q_N, \dot{q}_1, ..., \dot{q}_N) = 0$. These relations, of kinematic (not geometric) nature, are called *non-holonomic*. Then, the system S itself is also called *non-holonomic*. This situation will not be considered in the sequel.

Kinetic and potential energies; external forces and torques

Let S be a holonomic system with n degrees of freedom. Its kinetic energy is the sum of the kinetic energies of its subsystems; it depends on the time derivatives of the generalized coordinates and very often depends on these coordinates themselves. For each part of the rigid body S, it is given by (1.9). The kinetic energy of S is denoted by $T(q_1, ..., q_n, \dot{q}_1, ..., \dot{q}_n)$.

In the same manner, the *potential energy* of S is the sum of the potential energies of its subsystems. This energy only depends on the generalized coordinates; it is denoted by $U(q_1, ..., q_n)$.

Moreover, S can also be subjected to *external* forces and torques, i.e. which are not derived from the potential U (whether we have chosen not to take them into account in U, or whether they cannot be derived from any potential). So, the force f has to be treated as an external force in the example of the train and that of the inverted pendulum.

Generalized forces

Consider the external forces and torques exerting on a holonomic system S. While one of the generalized coordinates q_i varies with an infinitesimal quantity δq_i, and while the other generalized coordinates remain constant, an infinitesimal work $\delta W_i = Q_i \, \delta q_i$ is done as a result of these external forces and torques. Then Q_i is said to be the (external) generalized force in the ith direction of the configuration space – or, in short, the generalized external force along q_i. In the example of the train, f is the generalized external force along y_1, whereas in the example of the inverted pendulum, f is the generalized external force along y.

Lagrangian equations

The Lagrangian of the system is

$$\boxed{L = T - U} \tag{1.23}$$

and the *Lagrangian equations* are

$$\boxed{\frac{d}{dt}\left(\frac{\partial L}{\partial \dot{q}_i}\right) - \frac{\partial L}{\partial q_i} = Q_i, \; i = 1, ..., n} \tag{1.24}$$

where $\dot{q}_i \triangleq \frac{d\, q_i}{dt}$.

Example of the train

We have $T = \frac{1}{2}\left(m_1\,\dot{y}_1^2 + m_2\,\dot{y}_2^2\right)$. The potential energy is that of what is stored inside the spring, and is given by $U = \frac{1}{2}\,k\,\left(y_1 - y_2\right)^2$. We can derive (1.19) from (1.24).

Example of the inverted pendulum

The kinetic energy of the carriage is $T_1 = \frac{1}{2}M\,\dot{y}^2$; and that of the assembly (rod + mass) is $T_2 = \frac{1}{2}\,m\,\left(\dot{y}_1^2 + \dot{z}_1^2\right)$. By way of the holonomic relations (1.20), we obtain

$$T = T_1 + T_2 = \frac{1}{2}\left[M\,\dot{y}^2 + m\left(\dot{y}^2 + 2l\,\dot{y}\,\dot{\theta}\cos\theta + l^2\,\dot{\theta}^2\right)\right].$$

The potential energy is that of the weight, which is given by $U = m\,g\,l\,\cos\theta$. Applying (1.24), we immediately obtain (1.21).

1.3. Electromechanical system

A lot of systems consist of an electrical and a mechanical part. This is the case, for example, of an alternator connected to an electric network [70]. This is also the case of a synchronous motor [112] or that of an asynchronous motor [24]. The modeling of these systems requires too much circuit theory and techniques to be covered in this work. We will content ourselves with considering only the case of a *DC motor* [112], even though there seems to be a trend where the DC motor is being replaced by the motors we just mentioned.

The windings of the rotor are connected to the armature by rings and brush. The stator generates a fixed field (either using permanent magnets or using windings in which a DC current is flowing) wherein the turns of the rotor are rotated with angular speed ω. With a voltage V applied to the armature, a current i passes through these turns. The charge is characterized by its moment of inertia J (see Figure 1.5).

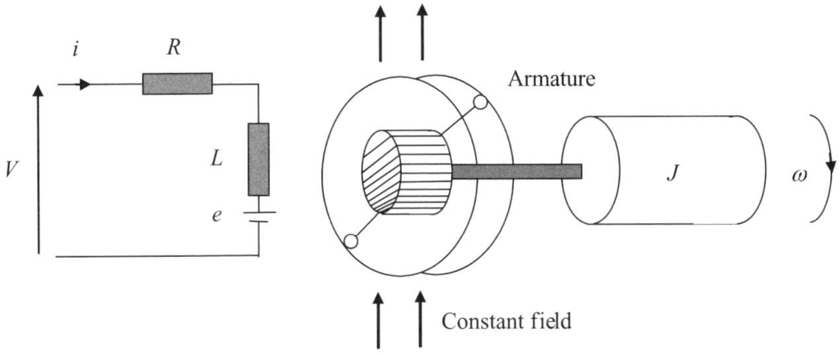

Figure 1.5. *DC motor*

According to the mesh rule,

$$V = R\,i + L\,\frac{di}{dt} + e$$

where e is the back electromotive force. According to *Faraday's law*, $e = \frac{d\varphi}{dt}$, where φ is the magnetic flux in the windings of the rotor. This flux is constant with respect to the stator, and varies in proportion to ω and with respect to the rotor. We can then let $e = K\,\omega$; the coefficient $K \neq 0$ is called the *motor constant*. The power exchanged between the stator and the rotor is $P = e\,i$ (in electrical terms) or $P = C_m\,\omega$ (in mechanical terms). Therefore, $C_m\,\omega = K\,\omega\,i$, from which we deduce that $C_m = K\,i$. Finally we get, according to (1.17),

$$J\,\frac{d\omega}{dt} = C_m - C_r$$

where C_r is the resistive torque, due to friction. This resistive torque is of the form $C_r = \lambda\omega$, where λ is the frictional coefficient. In addition, the angular position θ of the rotor satisfies the equation $\frac{d\theta}{dt} = \omega$. We can now regroup the equations in the following form:

$$\begin{cases} R\,i + L\,\frac{di}{dt} + K\,\omega - V = 0 \\ J\,\frac{d\omega}{dt} + \lambda\,\omega - K\,i = 0 \\ \frac{d\theta}{dt} - \omega = 0. \end{cases} \tag{1.25}$$

1.4. Thermal hydraulic system

Thermal or hydraulic systems are largely formulated using relations of balance: conservation of mass, conservation of energy. "Matter can neither be created nor destroyed, only transformed". This formula of Lavoisier is undoubtedly the most universal principle of science.

Consider, for example, the heated tank in Figure 1.6.

This tank consists of a reservoir with section S. It contains water which is heated by a resistance R. Q_e is the rate of discharge of water (at a constant temperature T_e) entering the reservoir. The tank is emptying through an opening of cross-section σ located at the bottom of the reservoir; rate of water discharging from the tank is denoted by Q_s. The temperature of the water inside the tank is assumed to be uniform and denoted by T_s. The heating power is P.

1.4.1. *Balance in volume*

For an infinitesimal interval of time dt, let dV be the increase in volume of water inside the reservoir. We have $dV = S\ dh$, where dh is the increase in height

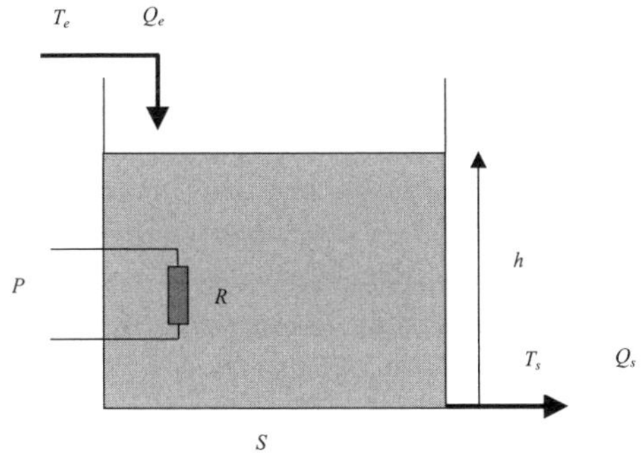

Figure 1.6. *Heated tank*

of water. The volume of water entering the reservoir during this time interval is $Q_e\ dt$. And the volume exiting is $Q_s dt$. We have therefore (under the assumption of incompressibility)

$$S\,\frac{dh}{dt} = Q_e - Q_s.$$

1.4.2. *Exit rate: Torricelli's formula*

This formula gives velocity v_s of water exiting the reservoir under height of water h:

$$\boxed{v_s = \sqrt{2\,g\,h}}$$

where g is the gravitational acceleration. Since $Q_s = \sigma\, v_s$, we obtain $Q_s = \sigma\,\sqrt{2\,g\,h}$.

1.4.3. *Energy balance*

Let dW be the amount of electric energy furnished during the infinitesimal time interval dt. Let also dW_e be the calorific energy (i.e. heat energy) furnished by the liquid entering the tank, and dW_i be the increase in calorific energy in the interior of the tank that produces variation of the temperature T_s. According to the *first principle* of thermodynamics, or *principle of conservation of energy*:

$$dW + dW_e = dW_i$$

On the other hand, $dW = P\,dt$. Denoting by μ the mass of the volume of liquid and by c the specific heat, we have $dW_e = Q_e\,\mu\,c\,(T_e - T_s)\,dt$ and $dW_i = S\,h\,\mu\,c\,dT_s$.

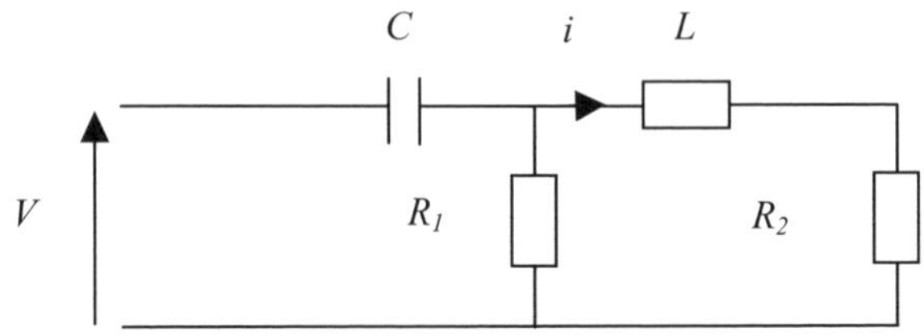

Figure 1.7. *Electric circuit*

Re-arranging the equations, we obtain

$$\left\{ \begin{array}{c} S\,\frac{dh}{dt} + \sigma\,\sqrt{2\,g\,h} - Q_e = 0 \\ S\,h\,\frac{dT_s}{dt} + Q_e\,(T_s - T_e) - \frac{P}{\mu\,c} = 0. \end{array} \right. \tag{1.26}$$

1.5. Exercises

EXERCISE 1.– *Consider the electric circuit in Figure 1.7. Determine the differential equation relating current i to voltage V.*

EXERCISE 2.– *Consider the double inverted pendulum as shown in Figure 1.8. Determine the equations of its motion.*

EXERCISE 3.– *[72] Consider the mixer given in Figure 1.9.*

This mixer consists of a reservoir that receives two liquids of constant concentrations c_1 and c_2 and whose discharge rates Q_1 and Q_2 are regulated by means of valves. The liquid is stirred within the reservoir, for concentration c_s to remain uniform. The mixture exits through an opening of cross-section σ located at the bottom of the tank. Determine the equations of this system.

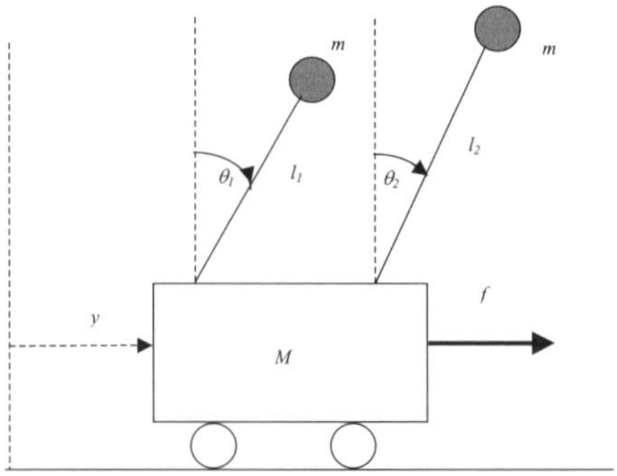

Figure 1.8. *Double inverted pendulum*

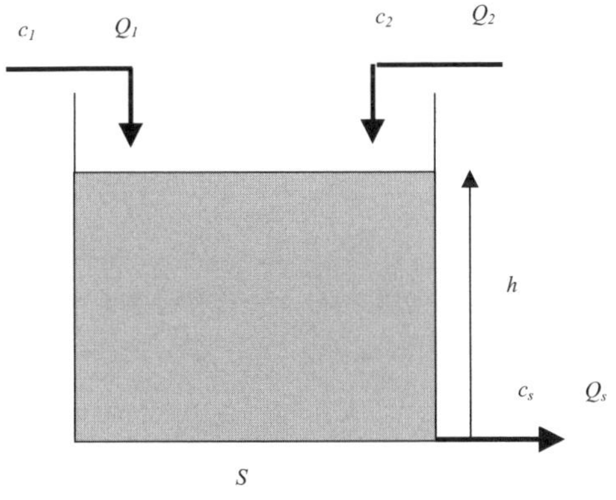

Figure 1.9. *Mixer*

EXERCISE 4.– *We consider another mixer, but this time the hot water at temperature T_1 enters the tank with rate Q_1, and the cold water with temperature T_2 enters with rate Q_2. (It is a bathtub.) The temperatures T_1 and T_2 are fixed. Water flows to the bottom of the tub where we have omitted to plug the opening having cross-section σ; temperature of water inside the tub is T_s. Determine the system equations. What are the connections between this exercise and Exercise 3?*

Chapter 2

Systems Theory (I)

The objective of this chapter is to specify the notion of system and to introduce a few basic concepts. Chapter 7 provides the complementary material that is necessary for a good understanding of the notions and methods presented in Chapters 3 to 6. In this present work, we are mainly interested in "linear time-invariant" systems (see section 2.2.5). The other types of systems will only be briefly mentioned.

2.1. Introductory example

Consider the RLC circuit in Figure 1.1 (section 1.1.1). It is governed by equation (1.1). To simplify the calculations, let us assume $R = 0$. Equation (1.1) can be written as

$$\frac{d^2\, i}{dt^2} + \omega_0^2\, i - k\,\frac{dV}{dt} = 0 \tag{2.1}$$

with

$$\omega_0^2 = \frac{1}{LC},\ k = \frac{1}{L}.$$

Representation by a second-order scalar differential equation

The system variables that are used in representation (2.1) are $w_1 = i$ and $w_2 = V$. Equation (2.1) takes the form

$$\begin{bmatrix} \partial^2 + \omega_0^2 & -k\,\partial \end{bmatrix} \begin{bmatrix} w_1 \\ w_2 \end{bmatrix} = 0 \tag{2.2}$$

where ∂ is the differential operator $\frac{d}{dt}$.

This representation is a second-order scalar differential equation, with two variables w_1 and w_2. These variables are physical quantities, and as we will see in the sequel, they are of particular interest to us.

Representation by a system of first-order differential equations

The above representation is not the only possible one. Indeed (2.2) can be written as

$$\partial\left(\partial w_1 - k\, w_2\right) + \omega_0^2\, w_1 = 0.$$

Introducing the variable $w_3 = \partial\, w_1 - k\, w_2$, we obtain the representation

$$\begin{bmatrix} \partial & -k & -1 \\ \omega_0^2 & 0 & \partial \end{bmatrix} \begin{bmatrix} w_1 \\ w_2 \\ w_3 \end{bmatrix} = 0. \tag{2.3}$$

This representation is a system of first-order differential equations (composed of two scalar differential equations). It is equivalent to (2.2). This time, three variables are involved: w_1, w_2 and w_3. Variable w_3 has no physical meaning: it is simply an intermediate variable. This is what is called a *latent variable* [96].

Therefore, *system variables and equations are not unique*.

2.2. General representation and properties

The systems studied in Chapter 1 have been modeled using variables and equations. These are differential (or algebraic-differential) equations involving time. From a mathematical point of view, time is a real variable which we call *continuous time*. Later (in Chapter 10), we will study computer-controlled systems. A computer only "knows" "discrete time", analogous to the set of integers. For the moment, we are only concerned with the context of continuous time.

2.2.1. *Variables*

In a general fashion, we make use of variables $w_1, \ldots, w_k$.

Case of the RLC circuit in Figure 1.1 (section 1.1.1): $w_1 = i, w_2 = V$.

Case of the train in Figure 1.3 (section 1.2.1): $w_1 = y_1, w_2 = y_2, w_3 = f$.

Case of the inverted pendulum in Figure 1.4 (section 1.2.1): $w_1 = y, w_2 = \theta$, $w_3 = f$.

Case of the heated tank in Figure 1.6 (section 1.4): $w_1 = h, w_2 = T_s, w_3 = P$.

Etc.

We write $w = \left[w_1, \ldots, w_k\right]^T$.

2.2.2. *Equations*

The algebraic and differential equations relating the variables can be gathered in the form

$$\boxed{F(w, \dot{w}, ...w^{(\alpha)}) = 0}. \tag{2.4}$$

The above function F is very often a vector-valued function, having q components when q is the number of scalar equations. Equation (2.4) is the general representation of a nonlinear system (in the sense of: *possibly* nonlinear). (Here, we exclude "infinite-dimensional systems", although such systems can be encountered in practice: for example time-delay systems.) We denote the system defined by (2.4) as $\mathbf{\Sigma}$.

2.2.3. *Time-invariant systems*

The function F involves a number of coefficients (in case of the RLC circuit, these coefficients are R, L and C; in case of the inverted pendulum, these coefficients are M, m, l, g, etc.). If all these coefficients are constant (i.e. real numbers), $\mathbf{\Sigma}$ is said to be *time-invariant,* and it is said to be *time-varying* in the opposite case, that is when the coefficients depend on time (*e.g. when they are elements of a differential field*).

2.2.4. *Linear systems*

If the function F is linear, $\mathbf{\Sigma}$ is said to be *linear*. Equation (2.4) can then be written in the form

$$\sum_{i=0}^{\alpha} E_i\, w^{(i)} = 0 \tag{2.5}$$

where the matrices E_i are of size $q \times k$ (q = number of equations and k = number of variables).

REMARK 5.– *Quite often a system is said to be linear when it satisfies the* principle of superposition. *Let us clarify this. (i) Suppose* $\mathbf{\Sigma}$ *is time-invariant and, to clarify ideas, that the function* F *of* (2.4) *is rational with real coefficients. Then,* $\mathbf{\Sigma}$ *is linear if and only if, for any variables* w_A *and* w_B *solutions of* (2.4) *and for any scalars* $\lambda_A, \lambda_B \in \mathbb{R}$, $\lambda_A w_A + \lambda_B w_B$ *is again a solution of* (2.4). *This signifies that the set of solutions is an* $\mathbb{R}$*-vector space. (ii)* Consider now the case of a time-varying system,* f *being a rational function with coefficients in a differential field* $\mathbf{K}$*. Then,* $\mathbf{\Sigma}$ *is linear if and only if the set of solutions of* (2.4) *is a* $\mathbf{k}$*-vector space where* $\mathbf{k}$ *is the subfield of constants of* $\mathbf{K}$ *(see section 13.1.1). This set is not a* $\mathbf{K}$*-vector space [22].**

2.2.5. *Linear time-invariant systems*

Let $E(\partial)$ be the *polynomial matrix* (see section 13.1.4, Example 490) defined by

$$E(\partial) = \sum_{i=0}^{\alpha} E_i \, \partial^i .$$

If $\mathbf{\Sigma}$ is assumed to be linear and time-invariant, the entries of $E(\partial)$ are from the ring $\mathbf{R} = \mathbb{R}[\partial]$ of polynomials in ∂ with real coefficients; $E(\partial)$ is of size $q \times k$. Equation (2.5) can be put in the equivalent form

$$\boxed{E(\partial)\, w = 0}. \tag{2.6}$$

The q equations of (2.6) are said to be *linearly independent* if the rank over $\mathbf{R}$ of $E(\partial)$ (denoted by $r = \mathrm{rk}_{\mathbf{R}}\ E(\partial)$: see section 13.1.4) is equal to q. This is of course only possible if k, the number of variables, is such that $k \geq r$.

It follows that the two electric circuits of section 1.1 as well as the train and the DC motor of sections 1.2 and 1.3 are all linear systems.

In reality, however, a system is never linear in the rigorous sense. Take an extremely simple example: according to Ohm's law, the potential difference V across a resistance R is related to the current i traversing it by the linear relation $V = Ri$. But when the voltage V, and thus the current intensity i, takes on too important values (even without getting to the point of burning off the resistance!), the relation between V and i becomes nonlinear and needs to be written in the form $V = \rho(i)$. This function ρ is the *nonlinear characteristic* of the resistance. But under the circumstances of "normal" functioning, we can write with good precision that $f(i) = Ri$. Therefore, linear systems are only purely theoretical views and results of simplifications. But the material world is immensely complex; we can only claim to understand and master it, at least partially, if we make adequate simplifications. This constitutes, *par excellence*, the know-how of engineers and scientists. It is true that linear systems do not exist in the absolute, but as notions, they play a very important role in control theory as well as in other sciences.

DEFINITION 6.– *For a linear time-invariant system* $\mathbf{\Sigma}$*, we call the* $m = k - r$ *the* rank *of the system where* $r = \mathrm{rk}_{\mathbf{R}}\ E(\partial)$.

DEFINITION 7.– *A linear time-invariant system* $\mathbf{\Sigma}$ *is said to be* determined *if its rank is zero. It is said to be* underdetermined *if its rank is positive.*

REMARK 8.– * *(i) Equation* (2.6) *is that of the module* $M = [w]_{\mathbf{R}}$ *(see section 13.4.1, Theorem 540) so that we can associate the finitely presented* $\mathbf{R}$*-module* M *with the linear system* $\mathbf{\Sigma}$*. From a more abstract point of view, we can identify* $\mathbf{\Sigma}$ *and* M *in such* *a way that a linear system is defined (from a mathematical point of view) as being a finitely presented* **R***-module ([42], [22]). (ii) The rank of* Σ*, according to Definition 6, coincides with the rank of the module* M*, according to Definition 552 of section 13.4.2.**

Among the examples of Chapter 1, there are nonlinear systems and they are modeled as such: the inverted pendulum and the heated tank of sections 1.2 and 1.4. By linearizing a system around an equilibrium point, we come back to the linear case. This linearization is a first-order Taylor expansion. The representation obtained is of course only valid for tiny variations around the chosen equilibrium point. We are now going to specify these notions.

2.2.6. *Equilibrium point*

Consider the time-invariant system Σ defined by (2.4).

DEFINITION 9.– *The point* $w^* = [w_1^*, ..., w_k^*]^T$ *is an* equilibrium point *of* Σ *if* $F(w^*, 0, ..., 0) = 0$.

In other words, we locate the equilibrium point(s) of a system by zeroing the derivatives of all orders of all variables.

Case of the inverted pendulum

By zeroing the derivatives in (1.21) (section 1.2.1), we obtain

$$\begin{cases} f^* = 0 \\ \sin\theta^* = 0. \end{cases}$$

This system has infinite equilibrium points: y^* is arbitrary, $\theta^* = 0$ or $\pi \pmod{2\pi}$, $f^* = 0$. The physical interpretation is clear: the pendulum can be at equilibrium in any region of the y axis. For it to be at equilibrium, the force applied onto the carriage must be zero (otherwise, a movement of acceleration will be given to the latter). Finally, there are only two possible positions of equilibrium for the pendulum itself: the rod being vertical, with the mass on top, or at bottom; we will see later that in the first situation, the equilibrium point is unstable (which is the case of the pendulum said to be *inverted*) and that it is stable in the second case.

Case of the heated tank

Proceeding similarly with (1.26) (section 1.4.3), we obtain

$$\begin{cases} \sigma\sqrt{2\,g\,h^*} - Q_e^* = 0 \\ Q_e^*\,(T_s^* - T_e) - \frac{1}{\mu\,c} P^* = 0. \end{cases}$$

The first of these equations expresses the equilibrium in volume: the rate of flow into the tank is equal to the rate of flow out of it. The second equation expresses the equilibrium in energy.

2.2.7. *Linearization around an equilibrium point*

Consider the system $\boldsymbol{\Sigma}$ defined by equation (2.4); assume that it is time-invariant and admits an equilibrium point w^*. This system is *linearizable* at the point w^* if F is differentiable in a neighborhood of $(w^*, 0, ...0)$.

Let $w = w^* + \Delta w$, where Δw is a "small increment" in the sense that Δw and its derivatives up to order α are "small increments".[1] We have, neglecting second-order terms,

$$\begin{aligned} F(w, \dot{w}, ...w^{(\alpha)}) &\simeq F(w^*, 0, ..., 0) + \sum_{i=0}^{\alpha} \frac{\partial F}{\partial w^{(i)}}(w^*, 0, ..., 0)\, \Delta w^{(i)} \\ &= \sum_{i=0}^{\alpha} \frac{\partial F}{\partial w^{(i)}}(w^*, 0, ..., 0)\, \Delta w^{(i)} \end{aligned}$$

and therefore

$$\sum_{i=0}^{\alpha} \frac{\partial F}{\partial w^{(i)}}(w^*, 0, ..., 0)\, \Delta w^{(i)} \simeq 0.$$

DEFINITION 10.– *The* linear approximation *of the system* $\boldsymbol{\Sigma}$ *around the equilibrium point* w^* *is the linear system* $\boldsymbol{\Sigma}_l$ *with equation* $\sum_{i=0}^{\alpha} \frac{\partial F}{\partial w^{(i)}}(w^*, 0, ...0)\, \Delta w^{(i)} = 0$.

The equation of $\boldsymbol{\Sigma}_l$ is of the form (2.5) with $E_i = \frac{\partial F}{\partial w^{(i)}}(w^*, 0, ..., 0)$, the variable w being replaced by the increment $\Delta w = w - w^*$.

Case of the inverted pendulum (section 1.2.1)

Linearizing (1.21) in the vicinity of $(y^* = 0, \theta^* = 0, f^* = 0)$, we obtain the equations of the linearized inverted pendulum:

$$\left\{ \begin{array}{r} (M+m)\, \frac{d^2 y}{dt^2} + m\, l\, \frac{d^2 \theta}{dt^2} - f = 0 \\ l\, \frac{d^2 \theta}{dt^2} - g\, \theta + \frac{d^2 y}{dt^2} = 0. \end{array} \right. \qquad (2.7)$$

Case of the heated tank (section 1.4.3)

The linearization of (1.26) leads to

$$\left\{ \begin{array}{r} S\, \frac{d\Delta h}{dt} + \sigma \sqrt{2\, g\, h^*}\, \Delta h - \Delta Q_e = 0 \\ S\, h^*\, \frac{d\Delta T_s}{dt} + Q_e^*\, \Delta T_s + (T_s^* - T_e)\, \Delta Q_e - \frac{1}{\mu\, c}\, \Delta P = 0. \end{array} \right.$$

1. *From the point of view of functional analysis, we can, for example, endow the codomain of the function w with the norm $\|w\| = \sum_{i=0}^{\alpha} \left\| w^{(i)} \right\|_\infty$ where $\|.\|_\infty$ denotes the norm in the space L_∞.*

Set $\tau = \frac{S}{\sigma}\sqrt{\frac{2\,h^*}{g}}$ and $V^* = S\,h^*$. Using the equilibrium relations, we obtain

$$\left\{\begin{array}{c} \frac{d\Delta h}{dt} + \frac{1}{\tau}\,\Delta h - \frac{1}{S}\,\Delta Q_e = 0 \\ \frac{d\Delta T_s}{dt} + \frac{2}{\tau}\,\Delta T_s - (T_s^* - T_e)\left(-\frac{\Delta Q_e}{V^*} + \frac{2}{\tau}\frac{\Delta P}{P^*}\right) = 0. \end{array}\right.$$

To simplify and improve condition equations, it is now in our interest to normalize variables, or, to be more precise, to "reduce" them. Set $w_1 = \frac{\Delta h}{h^*}$, $w_2 = \frac{\Delta T_s}{T_s^* - T_e}$, $w_3 = \frac{\tau\,\Delta Q_e}{V^*}$, and $w_4 = \frac{\Delta P}{P^*}$ (on condition, of course, that none of the denominators of these fractions is zero); these variables $w_i = i = 1, ..., 4$, which are unitless, are called the *reduced variables*. We then obtain the *reduced equations*:

$$\left\{\begin{array}{c} \frac{d\,w_1}{d\,t} + \frac{1}{\tau}\,w_1 - \frac{1}{\tau}\,w_3 = 0 \\ \frac{d\,w_2}{d\,t} + \frac{2}{\tau}\,w_2 - \frac{1}{\tau}\,(-w_3 + 2\,w_4) = 0. \end{array}\right. \tag{2.8}$$

It equally makes sense to choose the *unit of time* properly in such a way that τ takes a numerical value close to 1.

2.3. Control systems

2.3.1. *Inputs*

General properties of input variables

As already mentioned in the preface of this book, the systems we are interested in are the systems a user can act upon. This signifies that by means of actuators, we can impose the evolution of certain variables, which we call the *control variables*.

The environment can also act upon the considered system Σ (the environment is itself a system). This imposes upon the evolution of other variables: which are the *disturbances* (as opposed to the controls, the disturbances are variables over which the user has no mastery).

The control variables and the disturbances constitute the *input variables* of Σ. Let $u_1, ..., u_m$ be those variables; we put $u = [u_1, ..., u_m]^T$, and we succinctly call it the *input* of Σ.

We will make the following three hypotheses:

i) Once the input is fixed, the evolution of the remaining variables $\xi_1, ..., \xi_{k-m}$ of Σ depends only on their initial conditions.

ii) The input u is an "independent" variable, in the sense that there does not exist a non-zero function[2] f and an order of differentiation β such that $f\left(u, \dot{u}, ..., u^{(\beta)}\right) = 0$.

---

2. To be rigorous, it is of course necessary to make some hypotheses about the regularity of function f. *In the framework of differential algebra, for example [47], it is (and so are also the components of function F of equation (2.4)) a rational function of several variables, with coefficients in $\mathbb{R}$ or in a differential field.*

It is important to understand this hypothesis well. Consider, for example, a system Σ whose only input is a disturbance u. For the modeling of Σ, we assume that u is independent. Now, this disturbance can be a constant, and then $\dot{u} = 0$. This last equation is a model of the *environment* of Σ, and is therefore not part of the equations of this system. Let $\breve{\Sigma}$ be the system that consists of Σ and its environment, whose equations are those of Σ together with the equation $\dot{u} = 0$. The variable u (which is "dependent") is *not* an input of $\breve{\Sigma}$.

iii) When the initial conditions of system Σ are fixed, the evolution of its variables depends only on the input u.

Linear time-invariant case

The above can only further be specified (except if notions outside of the scope of this text are called upon) in the linear time-invariant case. Consider system (2.6), where the matrix $E(\partial)$ is of size $r \times k$ and of rank r. It follows that there exists a submatrix $D(\partial)$ of $E(\partial)$, of size $r \times r$ and non-singular, i.e. the determinant of $D(\partial)$ is non-zero and is an element of $\mathbf{R} = \mathbb{R}[\partial]$ (see section 13.1.4). It is clear that $D(\partial)$ is composed of columns of $E(\partial)$. After renumbering these columns if necessary, we can assume that these are the first r columns, corresponding to the first r variables $\xi_1 = w_1, ..., \xi_r = w_r$. Let $w_{r+1} = u_1, ..., w_k = u_m$. In addition, let

$$E(\partial) = \begin{bmatrix} D(\partial) & -N(\partial) \end{bmatrix}. \tag{2.9}$$

Then, setting $\xi = [\xi_1 \dots \xi_r]^T$ and $u = [u_1 \dots u_m]^T$, (2.6) is written as

$$\boxed{D(\partial)\, \xi = N(\partial)\, u}. \tag{2.10}$$

Let $\bar{\xi}$ be the variable ξ when $u = 0$. We thus have

$$D(\partial)\, \bar{\xi} = 0. \tag{2.11}$$

The system defined by (2.11) is *determined* (see Definition 7, section 2.2.5). As shown in section 2.3.8, the evolution of $\bar{\xi}$ is determined by the initial conditions, therefore hypothesis *i)* is satisfied.

On the other hand, equation (2.10) does not involve any relation between the components of u and its successive derivatives; this variable u is therefore "independent". As a result, hypothesis *ii)* is satisfied.

Finally, we will see from section 2.5.1 that hypothesis *iii)* is also satisfied.

The variable u, constructed as mentioned above, can therefore be the input of the system. This is not the only possibility, since there are several ways to extract from $E(\partial)$ a square submatrix $D(\partial)$ of rank r. But whatever method we take, we will always find an independent variable u having m components. We therefore arrive at the following result:

THEOREM 11.– *(i) The input of a system has a number of components equal to its rank. (ii) It is always possible to select a finite sequence* $(u_1, ..., u_m)$ *from the system variables that satisfies hypotheses i), ii), iii) given above (in such a way that* $u = [u_1, ..., u_m]^T$ *can be the system input).*

REMARK 12.– * *Consider the module* $M = [w]_{\mathbf{R}}$ *defined by (2.6) (see Remark 8, section 2.2.5). The module* $[u]_{\mathbf{R}}$ *is a submodule of* M *(since the input variables belong to* M*). (i) The independence of the input means that* $[u]_{\mathbf{R}}$ *is a free* **R**-*module of rank* m. *(ii) We have* $M = [\xi]_{\mathbf{R}} + [u]_{\mathbf{R}}$ *and according to* (2.10)*, equation* (2.11) *is that of the quotient module* $M/[u]_{\mathbf{R}}$ *(where* $\bar{\xi}$ *is the column matrix* $(\bar{\xi}_i)_{1\leq i\leq r}$ *such that* $\bar{\xi}_i$ *is the canonical image of* ξ_i *in* $M/[u]_{\mathbf{R}}$*). The fact that the system* $(\bar{2.11})$ *is determined means that the module* $M/[u]_{\mathbf{R}}$ *is torsion (see section 13.4.2, Corollary 555(ii)).**

Example

Let there be a system of equations

$$\begin{cases} \dot{w}_1 = w_2 - w_3 \\ w_2 + w_3 = 0. \end{cases}$$

In view of the first equation, we may perhaps try to choose input variables: $u_1 = w_2$ and $u_2 = w_3$. But this is not possible from the second equation where we need $u_1 + u_2 = 0$. Theorem 11 confirms this impossibility. We have

$$E(\partial) = \begin{bmatrix} \partial & -1 & 1 \\ 0 & 1 & 1 \end{bmatrix}.$$

This matrix is of rank $r = 2$ over $\mathbb{R}[\partial]$; the number of variables is $k = 3$; thus the rank of the system is $m = k - r = 1$. There is therefore only one independent input.

2.3.2. *Outputs*

The user wishes to keep watch over the evolution of the system, and therefore of a certain number of its variables (or of functions of such variables as well as their derivatives, their second-order derivatives, etc.). These quantities are measured by means of *sensors* and, from then on, constitute the *measurements* of the system.

In many cases, the user would like to more precisely regulate certain variables to some reference values, fixed by the former. For that to be possible, these variables

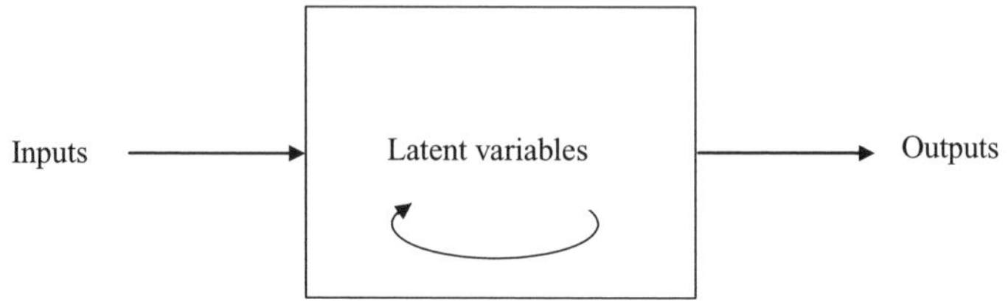

Figure 2.1. *Control system*

need to be measured (unless some complex procedures of estimation are developed). The *controlled variables* are in general part of the measurements.

The *output* variables consist of those measured variables and possibly other manifestations of the system, its (non-measured) actions on other systems for example.

2.3.3. *Latent variables*

The system variables often do not reduce to input and output variables. The supplementary variables are called the *latent variables* [117]. These variables are not "intrinsic", as we will show below using the example of the DC motor.

By distinguishing its input and output variables from all the other variables, we endow a system with a supplementary structure. This is what is expressed by the following definition:

DEFINITION 13.– *A* control system[3] *is a triple* (Σ, u, y)*, where* Σ *is a* system*,* $u = [u_1, ..., u_m]^T$ *is the* input*, and* $y = [y_1, ..., y_p]^T$ *is the* output.

A control system can be represented as in Figure 2.1.

Examples

Case of RLC circuit

In the case of the RLC circuit given in Figure 1.1 (section 1.1.1), we can impress a voltage V by means of a potentiometer. After this, equation (1.1) can be written as

$$LC\frac{d^2 i}{dt^2} + RC\frac{di}{dt} + i = C\frac{dV}{dt}. \tag{2.12}$$

3. In what follows, we will call a "control system" simply a "system" where there is no ambiguity. This will allow us to return to the terminology commonly used in the literature.

The right-hand member of this equation being fixed, the current i is governed by a second-order linear differential equation and its evolution only depends on $i(0)$ and $\frac{di}{dt}(0)$ ($t = 0$ taken as the initial instant); in other words, the initial condition of the system at initial instant 0 is $\left\{i(0), \frac{di}{dt}(0)\right\}$. We can monitor the evolution of i using an ammeter. Let $u = V$ and $y = i$: u is the control and y is the measurement of the control system as defined.

We can also do the reverse, fixing the value of i using a current generator; and hence now the voltage V depends only on $V(0)$. We obtain another control system, with input i and output V. We will see later that this approach creates some inconvenience, even though it is theoretically possible.

In both cases, there are no latent variables.

Case of the train

Very naturally, the control variable of the train (section 1.2.1) is the force f. The conductor of the train, using this variable, can decide how to position the wagon, hence controlling z_2. If all he monitors is z_2, this variable is the measurement and z_1 is the latent variable. If it uses the information of the position of the wagon as well as that of the locomotive (which makes the task easier), the measurement becomes $y = [z_1, z_2]^T$ and there will no longer be any latent variable.

Let us verify that the control can be chosen as the force f: putting $\xi_1 = z_1$, $\xi_2 = z_2$, and $u = f$, equations (1.19) can be written in the form (2.10) with

$$D(\partial) = \begin{bmatrix} m_1\,\partial^2 + k & -k \\ -k & m_2\,\partial^2 + k \end{bmatrix}, \quad N(\partial) = \begin{bmatrix} 1 \\ 0 \end{bmatrix}.$$

Therefore,

$$\det D(\partial) = m_1\, m_2\, \partial^2 \left(\partial^2 + k\frac{m_1 + m_2}{m_1\, m_2}\right) \neq 0 \qquad (2.13)$$

and the rank condition is satisfied.

Recall that writing $\det D(\partial) \neq 0$ means that $\det D(\partial)$ *is not the zero polynomial*. Later, we will have to consider values of the complex variable s for which $\det D(s)$ is zero: that is a completely different question: we then calculate the *roots* of $\det D(s)$, and that only makes sense if $\det D(s)$ is not the zero polynomial.

Case of the DC motor

This system (see section 1.3) has $r = 3$ linearly independent equations and $k = 4$ variables (including among them the velocity ω), hence the number of inputs is

$m = k - r = 1$. Let us show that we can choose the voltage V to be the control variable. Equations (1.25) can be written in the form

$$\begin{bmatrix} R + L\,\partial & K & 0 \\ -K & J\,\partial + \lambda & 0 \\ 0 & -1 & \partial \end{bmatrix} \begin{bmatrix} i \\ \omega \\ \theta \end{bmatrix} = \begin{bmatrix} 1 \\ 0 \\ 0 \end{bmatrix} V. \tag{2.14}$$

The determinant of the matrix $D(\partial)$ on the left of this equation is: $\det D(\partial) = \partial\left[(J\,\partial + \lambda)(R + L\,\partial) + K^2\right] \neq 0$. It is therefore quite possible to choose voltage V to be the control variable.

There exist two modes of control: "position control", where the controlled variable is θ, and "velocity control", where the controlled variable is ω. In the first case, a position sensor is always available to us, rarely a velocity sensor and in general the current i is not measured. We can therefore consider $y = \theta$ as the output, i and ω are then latent variables. But note that we can rewrite (1.25) by replacing in the first two equations ω by $\frac{d\theta}{dt}$, and by suppressing the last equation. In this case, there are only $r = 2$ linearly independent equations and $k = 3$ variables. The rank $m = k - r$ remains unchanged: it is an *intrinsic* quantity, that is it is independent of the representation chosen (*see Remark 8(ii), section 2.2.5*). However, the latent variables depend on the representation chosen, as seen in this example.

Case of the heated tank

Equations (2.8) (section 2.2.5) can be written in the form

$$\begin{bmatrix} \partial + \frac{1}{\tau} & 0 & -\frac{1}{\tau} & 0 \\ 0 & \partial + \frac{2}{\tau} & \frac{1}{\tau} & -\frac{2}{\tau} \end{bmatrix} \begin{bmatrix} w_1 \\ w_2 \\ w_3 \\ w_4 \end{bmatrix} = 0.$$

We can take $y_1 = w_1$ and $y_2 = w_2$ as output variables, and $u_1 = w_3$ and $u_2 = w$ as input variables. Physically, this is logical as well: the two means of action are actually the input discharge and the heating, i.e. w_3 and w_4 up to "reduction factors"; and the two variables to be regulated are the height of the water inside the tank and the temperature of water of the same, i.e. w_1 and w_2 after reduction. The above equations can then be written as

$$\begin{bmatrix} \partial + \frac{1}{\tau} & 0 \\ 0 & \partial + \frac{2}{\tau} \end{bmatrix} y = \begin{bmatrix} \frac{1}{\tau} & 0 \\ -\frac{1}{\tau} & \frac{2}{\tau} \end{bmatrix} u. \tag{2.15}$$

2.3.4. *Classification of systems*

We refer to the systems having one input and one output variable as *Single-Input Single-Output* (SISO) systems, and those with more than one input- and output-variable as *Multi-Input Multi-Output* (MIMO) systems. There are of course systems that are intermediate in nature (SIMO and MISO systems, but these expressions are less often used in the literature).

2.3.5. *Rosenbrock representation*

General formulation

The output variables $y_1, ..., y_p$, as already seen, are linear combinations of the variables $w_1, ..., w_k$, of their derivatives, of their second-order derivatives, etc. Thus y can be written in the form

$$y = H(\partial)\, w$$

where $H(\partial)$ is a polynomial matrix of size $p \times k$. By decomposing $H(\partial)$ in the same way as in (2.9) (section 2.3.1), which is $H(\partial) = \begin{bmatrix} Q(\partial) & W(\partial) \end{bmatrix}$, where $Q(\partial)$ and $W(\partial)$ are polynomial matrices of size $p \times r$ and $p \times m$, respectively, we finally obtain

$$\boxed{\begin{cases} D(\partial)\, \xi = N(\partial)\, u \\ y = Q(\partial)\, \xi + W(\partial)\, u \end{cases}} \tag{2.16}$$

where $D(\partial)$ is a non-singular matrix of size $r \times r$ over $\mathbf{R} = \mathbb{R}[\partial]$; (2.16) is a very general representation of a control system [100].

DEFINITION 14.– *The set of equations* (2.16) *is called a* Rosenbrock representation *and is noted as* $\{D, N, Q, W\}$. *The vector* ξ *of such a representation is called the* partial state *(or the* pseudo-state*).*

REMARK 15.– **Consider the system* $\mathbf{\Sigma}$ *as an* $\mathbf{R}$*-module, denoted by* M *(see Remark 8, section 2.2.5). The first equation of* (2.16) *has already been interpreted in Remark 12 (section 2.3.1). The second one simply means that the variables* y_i $(1 \leq i \leq p)$ *belong to* M. *See [19] for more details.**

Left form

The partial state generally consists of latent variables, except, in particular, when the second equation of (2.16) is $y = \xi$ (with $Q(\partial) = I_p$, $W(\partial) = 0$); in this case, indeed, the Rosenbrock representation reduces to one equation whose latent variables are eliminated: this equation, called a *left form*, is written as

$$\boxed{D(\partial)\, y = N(\partial)\, u}. \tag{2.17}$$

The representation (2.15) of the heated tank is of this form.

It is also the case in representation (2.12) of the RLC circuit with $u = V$, $y = i$ and

$$\begin{cases} D(\partial) = \partial^2 + \frac{R}{L}\,\partial + \frac{1}{LC} \\ N(\partial) = \frac{1}{L}\,\partial \end{cases}$$

after expression (2.12) has been divided by LC, for $D(\partial)$ to be a monic polynomial.

Right form

Equations of the RLC circuit in Figure 1.1 can also be obtained by using the charge q of the capacitor as a latent variable. We then obtain

$$\begin{cases} V = \left(L\partial^2 + R\,\partial + \frac{1}{C}\right) q. \\ i = \partial q. \end{cases}$$

In order to get a monic polynomial of second degree in the first equation, put $\xi = L\,q$; we obtain, with $u = V$ and $y = i$,

$$\begin{cases} u = \left(\partial^2 + \frac{R}{L}\,\partial + \frac{1}{L\,C}\right) \xi. \\ y = \frac{1}{L}\partial\xi. \end{cases} \tag{2.18}$$

This is a Rosenbrock representation such as (2.16) where $N(\partial) = 1$ and $W(\partial) = 0$:

$$\boxed{\begin{cases} u = D(\partial)\,\xi \\ y = Q(\partial)\,\xi \end{cases}}. \tag{2.19}$$

A representation of this form is called, in general, a *right form*.

In the present case, we have $D(\partial) = \partial^2 + \frac{R}{L}\,\partial + \frac{1}{L\,C}$ and $Q(\partial) = \frac{1}{L}\,\partial$.

Left form ↔ right form duality

For the left form (2.17) and the right form (2.19) to represent the same control system, it is necessary to have the following correspondences (assuming that $D(\partial)$ is a monic polynomial):

$$\begin{cases} D(\partial) \longleftrightarrow D(\partial) \\ N(\partial) \longleftrightarrow Q(\partial) \end{cases}$$

(see e.g. the RLC circuit case).

This is what we call the *left form ↔ right form duality* (for a *linear time-invariant SISO system*).

Soon, we will see from section 7.1 that a left form and its dual (which satisfies the duality relation above) *may not always be representations of the same control system.* The same applies for a right form and its dual.

2.3.6. *State-space representation*

A state-space representation is a particular type of Rosenbrock representation, where $D(\partial)$ is of the form $\partial I_n - A$, $A \in \mathbb{R}^{n \times n}$, $N(\partial) = B \in \mathbb{R}^{n \times m}$ and $Q(\partial) = C \in \mathbb{R}^{n \times p}$. Such a representation is of the form

$$\boxed{\left\{ \begin{array}{c} \dot{x} = A\,x + B\,u \\ y = C\,x + W(\partial)\,u \end{array} \right.}. \tag{2.20}$$

This type of representation (called a *state-pace system*) will be studied in detail from Chapter 7 onwards.

REMARK 16.– *The codomain of the state x is $\mathbb{R}^n$, i.e. the* state-space *is of dimension n. Abusing the language, we can also say that the* state-space representation *considered, or its* state vector, *is of dimension n.*

2.3.7. *Poles and order of a system*

Consider a control system Σ described by the Rosenbrock representation (2.16).

DEFINITION 17.– *(i) The* poles *of Σ are the roots of* $\det D(s)$ *(taking into account multiplicities). (ii) The* order *of Σ is the degree of the polynomial* $\det D(s)$.

The order of a system is therefore equal to the number of its poles. The following result is obvious:

THEOREM 18.– *Suppose Σ is given by the state representation* (2.20). *Then, the poles of Σ are the eigenvalues of A and its order is the dimension of its state vector.*

REMARK 19.– * *Consider the system Σ and the associated* $\mathbf{R}$*-module M (see Remarks 8, 12, and 15). Then, the poles of Σ are the Smith zeros of the torsion module $M/[u]_{\mathbf{R}}$ (see section 13.4.2).* *

2.3.8. *Free response and behavior*

Let Σ be a linear time-invariant system of order n.

DEFINITION 20.– *We call* free behavior *of Σ the set of its variables when the system input is maintained at zero (for any initial conditions). The* free response *of this system is its output in the same situation.*

Suppose again that Σ is described by the Rosenbrock representation (2.16). Let $\bar{\Sigma}$ be the system obtained from Σ by forcing the input u to 0. Its partial state $\bar{\xi}$ is a solution of

$$D(\partial)\ \bar{\xi} = 0. \tag{2.21}$$

Case of a scalar partial state

One of the fundamental results in the classic theory of linear differential equations with constant coefficients is the following [36] (see also section 12.5.2, Theorem 449):

THEOREM 21.– *Let $p_1, ..., p_q$ be the distinct roots of $D(s)$ (also called the* distinct poles *of Σ or of $\bar{\Sigma}$), with order of multiplicity $\rho_1, ..., \rho_q$, respectively. Then any solution of equation (2.21) is of the form*

$$\boxed{\bar{\xi}(t) = \sum_{j=1}^{q} \sum_{k=1}^{\rho_j} \alpha_{jk} t^{k-1} e^{p_j t}}$$

where $\alpha_{jk} \in \mathbb{C}$.

The set of these solutions (which is the free behavior of the system) is a $\mathbb{C}$-vector space of dimension n; a basis of this space consists of the n functions

$$\epsilon_{jk} : t \mapsto t^{k-1} e^{p_j t},\ 1 \leq k \leq \rho_j,\ 1 \leq j \leq q.$$

Examples

– Consider first the doubly integrating system

$$\ddot{y} = u. \tag{2.22}$$

This equation is a left form (2.17) (section 2.3.5) with $D(\partial) = \partial^2$ and $N(\partial) = 1$. The polynomial $\det(D(s)) = D(s) = s^2$ has double root $s = 0$. The set of system poles is therefore $\{0, 0\}$. This is a second-order system.

– Consider now the system

$$\ddot{y} = \dot{u}. \tag{2.23}$$

It is again a left form with $D(\partial) = \partial^2$ and $N(\partial) = \partial$. The set of system poles is, like previously, $\{0, 0\}$, and this system is of the second order. It presents a particularity which will be studied in Chapter 7.

**General case*

Generalization of Theorem 21

Consider now the general case where the partial state has $r > 1$ components. There exist matrices $P(\partial)$ and $R(\partial)$, invertible over $\mathbf{R} = \mathbb{R}[\partial]$, such that

$$P(\partial)\,D(\partial)\,R(\partial) = S(\partial)$$

where

$$S(\partial) = \operatorname{diag}(\alpha_1(\partial), ..., \alpha_r(\partial))$$

is the Smith form of $D(\partial)$ (see section 13.2.3).

Let $\tilde{\xi} = R^{-1}(\partial)\,\xi$ (which is licit since $R^{-1}(\partial)$ is a polynomial matrix in ∂). Then, the matrix differential equation (2.21) is equivalent to r scalar differential equations

$$\begin{cases} \alpha_1(\partial)\,\tilde{\xi}_1 = 0 \\ \vdots \\ \alpha_r(\partial)\,\tilde{\xi}_r = 0. \end{cases}$$

For each of these equations, we can apply Theorem 21. We can therefore factorize the invariant factors $\alpha_k(\partial), 1 \leq k \leq r$, into primes and thus make apparent the elementary divisors of $D(\partial)$. These divisors determine the Smith zeros of the matrix $D(s)$ as well as their structural indices (see section 13.2.5).

DEFINITION 22.– *A* pole *of* Σ *is a Smith zero of* $D(s)$ **(or, more intrinsically, of* $M/[u]_{\mathbf{R}}$ *: see Remark 19, section 2.3.7)*. The* structural indices, the order and the degree *of such a pole are the structural indices, the order and the degree of this Smith zero.*

One obtains the following theorem ([12], IV.2.9):

THEOREM 23.– *Let* $p_1, ..., p_q$ *be the distinct poles of* Σ*, and let* $\sigma(p_j)$ *be the set of structural indices of the pole* p_j $(1 \leq j \leq q)$*. Any solution of (2.21) is of the form*

$$\boxed{\bar{\xi}(t) = \sum_{j=1}^{q} \sum_{k \in \sigma(p_j)} a_{jk} t^{k-1} e^{p_j t}} \qquad (2.24)$$

where $a_{jk} \in \mathbb{C}^r$.

The set of these solutions (that is the free behavior of the system) is a $\mathbb{C}$*-vector space of dimension* n.

Example

Let us take a simple case where the matrix $D(\partial)$ is already in Smith form:

$$D(\partial) = \begin{bmatrix} \partial - 1 & 0 & 0 \\ 0 & (\partial-1)^3(\partial-2) & \\ 0 & 0 & (\partial-1)^3(\partial-2)^2 \end{bmatrix}.$$

The order of the system is $n = \deg \det D(\partial) = 10$.

On the other hand, (2.21) can be written in the form of three scalar equations:

$$\begin{cases} (\partial - 1)\ \bar{\xi}_1 = 0 \\ (\partial - 1)^3\ (\partial - 2)\ \bar{\xi}_2 = 0 \\ (\partial - 1)^3\ (\partial - 2)^2\ \bar{\xi}_3 = 0. \end{cases}$$

We thus obtain the following solutions (in the space $\mathcal{E}(\mathbb{R})$ consisting of all indefinitely differentiable functions $\mathbb{R} \to \mathbb{C}$)

$$\begin{cases} \bar{\xi}_1\ (t) = c_1\ e^t, \\ \bar{\xi}_2\ (t) = c_2\ e^t + c_3\ t\ e^t + c_4\ t^2\ e^t + c_5\ e^{2t}, \\ \bar{\xi}_3\ (t) = c_6\ e^t + c_7\ t\ e^t + c_8\ t^2\ e^t + c_9\ e^{2t} + c_{10}\ t\ e^{2t}, \end{cases}$$

where the 10 constants $c_1, ..., c_{10}$ are arbitrary.

We obtain the form (2.24) in the following manner:

$$\bar{\xi}\ (t) = \begin{bmatrix} c_1 \\ c_2 \\ c_6 \end{bmatrix} e^t + \begin{bmatrix} 0 \\ c_3 \\ c_7 \end{bmatrix} t\,e^t + \begin{bmatrix} 0 \\ c_4 \\ c_8 \end{bmatrix} t^2\,e^t + \begin{bmatrix} 0 \\ c_5 \\ c_9 \end{bmatrix} e^{2t} + \begin{bmatrix} 0 \\ 0 \\ c_{10} \end{bmatrix} t\,e^{2t}.$$

As expected, the solutions form a $\mathbb{C}$-vector space of dimension 10.

2.4. Transfer matrix

2.4.1. *Laplace transforms*

The Laplace transform is detailed in section 12.3.4. As far as we are concerned here, the essential points are as follows:

Let $u : \mathbb{R} \to \mathbb{R}^m$ be a function with positive support, i.e. which is zero for $t < 0$. Its Laplace transform is a function of the complex variable $\hat{u}$, from $\mathbb{C}$ to $\mathbb{C}^m$, defined for $\mathrm{Re}(s) > \gamma$ (where γ is the abscissa of convergence) by

$$\boxed{\hat{u}(s) = \int_{0^-}^{+\infty} u(t)e^{-st}\,dt} \qquad (2.25)$$

(the definition in section 12.3.4 is extended here to the case of vector-valued functions).

If $\hat{u}\ (s)$ is a rational function, the abscissa of convergence is given by

$$\gamma = \max_{p \in P} \mathrm{Re}\, p$$

where P is the set of poles of $\hat{u}(s)$.

The Laplace transformation $\mathcal{L} : u \mapsto \hat{u}$ is linear. It transforms the convolution product into an ordinary product.

If u and all its derivatives are zero at $t = 0$, then $\left(\mathcal{L}\, u^{(n)}\right)(s) = s^n\, \hat{u}(s)$, irrespective of what the order n of the derivative is. In other words, with zero initial conditions, Laplace transformation transforms differential operator ∂ into the multiplication by the "Laplace variable" s, which is a complex variable.

2.4.2. *Transfer matrix: definition*

THEOREM 24.– *Let (Σ, u, y) be a control system. There exists a unique matrix of rational functions $G(s)$ such that for zero initial conditions,*

$$\boxed{\hat{y}(s) = G(s)\hat{u}(s)}\,. \tag{2.26}$$

PROOF. We can assume that the system is described by a Rosenbrock representation, because, as already said, this is the most general type of representation of a linear control system. By applying the Laplace transform with zero initial conditions to (2.16), we obtain

$$\begin{cases} D(s)\,\hat{\xi}(s) = N(s)\,\hat{u}(s) \\ \hat{y}(s) = Q(s)\,\hat{\xi}(s) + W(s)\,\hat{u}(s)\,. \end{cases}$$

The matrix $D(s)$ is a square, non-singular, thus invertible over the field $\mathbb{R}(s)$, and we have $\hat{\xi}(s) = D^{-1}(s)\,N(s)\,\hat{u}(s)$; from (2.26) we obtain

$$\boxed{G(s) = Q(s)\,D^{-1}(s)\,N(s) + W(s)}\,. \tag{2.27}$$

This proves the existence of $G(s)$. Its uniqueness is also ensured, since the equality $G_1\hat{u} = G_2\hat{u}$ for every $\hat{u}$ (where the input variables are independent) implies $G_1 = G_2$. ■

DEFINITION 25.– *The matrix $G(s) \in \mathbb{R}(s)^{p\times m}$ (where $\mathbb{R}(s)$ denotes the field of rational functions with real coefficients) is called the* transfer matrix *of the control system. In the case $m = p = 1$ (SISO), $G(s)$ is called the* transfer function *of the control system.*

From an algebraic point of view, the major difference between working with temporal signals or with their Laplace transforms is that, in the case of temporal signals, the operators belong to the ring $\mathbb{R}[\partial]$ of polynomials in ∂; these elements are not invertible in general. In the case of Laplace transforms, on the contrary, the operators are rational functions, i.e. belonging to the field $\mathbb{R}(s)$, and so are invertible whenever they are non-zero. This simplification brought along by the Laplace transformation with zero initial conditions is of course paid for by a loss of information which will be studied in Chapter 7.

Note that a left form and a right form which are dual have the same transfer matrix.

2.4.3. *Examples*

Let us re-examine some of the examples studied in Chapter 1.

RLC circuit

Consider equation (2.12) (section 2.3.3).

– First, suppose the input chosen is $u = V$, and the output is $y = i$. The transfer function is then

$$G(s) = \frac{C\,s}{L\,C\,s^2 + R\,C\,s + 1}. \tag{2.28}$$

– If on the contrary we choose $u = i$ as input and $y = V$ as output, the transfer function is the inverse of the previous one, which is

$$G(s) = \frac{L\,C\,s^2 + R\,C\,s + 1}{C\,s}. \tag{2.29}$$

DC motor

Equation (2.14) (section 2.3.3) is of the form $D\left(\partial\right)\xi = N\left(\partial\right)u$. Suppose the output is $y = \theta$ (position control). We then have the relation $y = Q\left(\partial\right)\xi + W\left(\partial\right)u$ with $Q\left(\partial\right) = \begin{bmatrix} 0 & 0 & 1 \end{bmatrix}$ and $W\left(\partial\right) = 0$. Since we have a Rosenbrock representation, we can make use of the relation (2.27) to calculate the transfer function $G(s)$.

In the present case, anyhow, it is more effective to return to equations (1.25) (section 1.3) and then pass into the Laplace domain. We have

$$\begin{cases} (R + L\,s)\,\hat{\imath}(s) + K\,\hat{\omega}\left(s\right) = \hat{V}\left(s\right) \\ (J\,s + \lambda)\,\hat{\omega}\left(s\right) = K\,\hat{\imath}(s) \\ s\,\hat{\theta}\left(s\right) = \hat{\omega}\left(s\right). \end{cases}$$

From the first two equations, we get

$$\left[(R + L\,s)\,(J\,s + \lambda) + K^2\right]\,\hat{\omega}\left(s\right) = K\,\hat{V}\left(s\right)$$

from which we finally obtain

$$\hat{\theta}(s) = \frac{K}{s\left[(R + L\,s)(J\,s + \lambda) + K^2\right]}\hat{V}(s).$$

The transfer function of the control system is therefore

$$G(s) = \frac{K}{s\left[(R + L\,s)(J\,s + \lambda) + K^2\right]}. \tag{2.30}$$

Heated tank

This is an MIMO system; it therefore has a transfer *matrix*. According to (2.15) (section 2.3.3), it is

$$G(s) = \begin{bmatrix} \frac{1}{1+\tau\,s} & 0 \\ \frac{-\frac{1}{2}}{1+\frac{\tau}{2}\,s} & \frac{1}{1+\frac{\tau}{2}\,s} \end{bmatrix}. \tag{2.31}$$

2.4.4. *Transmission poles and zeros*

SISO case

Consider a SISO system $\mathbf{\Sigma}$, with rational transfer function $G(s)$. We will see later that the behavior of $\mathbf{\Sigma}$ is largely characterized by the poles and zeros of $G(s)$ (defined in section 13.6.1).

DEFINITION 26.– *The poles and zeros of $G(s)$ are called the* transmission poles and zeros *of* Σ.

Examples

Case of RLC circuit

Consider the case of the RLC circuit with transfer function (2.28). This transfer function can be put in the form

$$G(s) = \frac{k\omega_0\,s}{s^2 + 2\varsigma\omega_0\,s + \omega_0^2} \tag{2.32}$$

with $k = \sqrt{\frac{C}{L}}$, $\omega_0 = \frac{1}{\sqrt{L\,C}}$, $\varsigma = \frac{R}{2}\sqrt{\frac{C}{L}}$.

If $\varsigma \geq 1$, the system has two real transmission poles, forming the set

$$\left\{-\omega_0\left(\varsigma - \sqrt{\varsigma^2 - 1}\right), -\omega_0\left(\varsigma + \sqrt{\varsigma^2 - 1}\right)\right\}.$$

If $0 \leq \varsigma < 1$, then the system has two complex conjugate transmission poles:

$$\left\{-\omega_0\left(\varsigma - i\sqrt{1 - \varsigma^2}\right), -\omega_0\left(\varsigma + i\sqrt{1 - \varsigma^2}\right)\right\}.$$

Moreover, the set of transmission zeros of the system is $\{0\}$.

Case of the DC motor

The transfer function (2.30) can be put in the form

$$G(s) = \frac{k}{s\left(s^2 + 2\varsigma\,\omega_0\,s + \omega_0^2\right)}. \tag{2.33}$$

The system therefore has three transmission poles: $\{0, p_1, p_2\}$, where p_1 and p_2 are the roots of the polynomial $s^2 + 2\varsigma\,\omega_0\,s + \omega_0^2$, and which are calculated above.

This system has no transmission zero.

MIMO case

Definition 26 remains valid in the MIMO case. We are led to define what we mean by a pole and a zero of a transfer *matrix*.

Naive definition

Let us first look at a naive definition of these notions and its limitations. In a second step, we will follow a more rigorous approach, but a more complicated one from a mathematical point of view.

Transmission poles

A pole of a transfer *function* $G(s)$ is a complex number p such that $\lim_{s \to p} |G(s)| = +\infty$.

Let us examine the case of a transfer *matrix*. Elements of such a matrix $G(s)$ are denoted by $G_{ij}(s)$. A pole of $G(s)$ is a complex number p such that at least one of the elements $G_{ij}(s)$ satisfies $\lim_{s \to p} |G_{ij}(s)| = +\infty$.

Consider, for example, the heated tank, whose transfer matrix is given by (2.31). From the above definition, we get $-\frac{1}{\tau}$ and $-\frac{2}{\tau}$ as the transmission poles of this system. The difficulty here (resulting from the lack of precision of the definition) is that we know nothing about the multiplicity (or order) of the pole $-\frac{2}{\tau}$. Indeed, two elements of $G(s)$ have an absolute value that tends to $+\infty$ when $s \to -\frac{2}{\tau}$. Is the set of poles of $G(s)$ (repeating a pole a number of times equal to its multiplicity) $\left\{-\frac{1}{\tau}, -\frac{2}{\tau}\right\}$ or $\left\{-\frac{1}{\tau}, -\frac{2}{\tau}, -\frac{2}{\tau}\right\}$? We will answer this question in section 2.4.5.

Transmission zeros and blocking zeros

A zero of a *transfer function* $G(s)$ is a complex number z such that $G(z) = 0$.

In the case of MIMO systems, a straightforward extension of this definition allows us to obtain the notion of "blocking zero":

DEFINITION 27.– *A* blocking zero *of system* Σ *(or of its transfer matrix* $G(s)$*) is a complex number* z *such that* $G(z) = 0$.

We can nevertheless envisage a second type of extension to the classic notion of zero of a rational function, proceeding in the following manner:

The rank of the transfer *function* $G(s)$ over the field $\mathbb{R}(s)$ of rational functions (this rank is denoted by $\mathrm{rk}_{\mathbb{R}(s)}\, G(s)$ and is sometimes called the "normal rank" of $G(s)$) is equal to 1. So, a zero of $G(s)$ is a complex number z such that $\mathrm{rk}_{\mathbb{C}}\, G(z) < \mathrm{rk}_{\mathbb{R}(s)}\, G(s)$. We can now envisage to extend this definition to the case of a transfer matrix.

For example, if we consider the transfer matrix $G(s)$ of the heated tank, $\mathrm{rk}_{\mathbb{R}(s)}\, G(s) = 2$ and $\mathrm{rk}_{\mathbb{C}}\, G(z) = 2$ for all values of z at which $G(z)$ is defined, i.e. for any z that is not a pole of $G(s)$. "Therefore", this system has no transmission zero.

But there are problematic cases. Consider, for example,

$$G(s) = \begin{bmatrix} \frac{1}{s} & 0 \\ 0 & s \end{bmatrix}.$$

The second column becomes zero when $s = 0$, "therefore" 0 is a zero of $G(s)$, i.e. a transmission zero of the system. But, on the other hand, $s = 0$ is a pole of $G(s)$, which renders $G(0)$ undefined. In view of such an incoherence, one must use a more precise approach, leading us to define the "MacMillan poles and zeros".

Definition 26 extends to the MIMO case in the following manner:

DEFINITION 28.– *The transmission poles and transmission zeros of a system are the MacMillan poles and MacMillan zeros of its transfer matrix.*

2.4.5. **MacMillan poles and zeros*

Smith–MacMillan form of a transfer matrix

Let $G(s)$ be a transfer matrix, the elements $G_{ij}(s)$ of which are rational functions. Let $d(s)$ be the least common denominator of these rational functions (i.e the lcm of its denominators). Then we have

$$G(s) = \frac{N(s)}{d(s)}$$

where $N(s)$ is a polynomial matrix. We know there exist polynomial matrices $U(s)$ and $V(s)$, invertible over $\mathbf{R} = \mathbb{R}[\partial]$, such that

$$N(s) = U^{-1}(s)\, S(s)\, V(s)$$

where

$$S(s) = \operatorname{diag}(\alpha_1(s), ..., \alpha_r(s), 0, ..., 0)$$

is the Smith form of $N(s)$ (see section 13.2.3). The polynomials $\alpha_1(s), ..., \alpha_r(s)$ are the invariant factors of $N(s)$; they are non-zero and such that $\alpha_1(s) \mid ... \mid \alpha_r(s)$.

As a result, we have

$$G(s) = U^{-1}(s) \operatorname{diag}\left(\frac{\alpha_1(s)}{d(s)}, ..., \frac{\alpha_r(s)}{d(s)}, 0, ..., 0\right) V(s).$$

Let

$$\frac{\alpha_i(s)}{d(s)} = \frac{\varepsilon_i(s)}{\psi_i(s)} \quad (1 \leq i \leq r)$$

where the rational function on the right-hand side is irreducible. We obtain

$$G(s) = U^{-1}(s) \Sigma(s) V(s), \tag{2.34}$$

$$\Sigma(s) = \operatorname{diag}\left(\frac{\varepsilon_1(s)}{\psi_1(s)}, ..., \frac{\varepsilon_r(s)}{\psi_r(s)}, 0, ..., 0\right). \tag{2.35}$$

It is easy to show that

$$\varepsilon_1(s) \mid ... \mid \varepsilon_r(s), \quad \psi_r(s) \mid ... \mid \psi_1(s). \tag{2.36}$$

On the other hand, the uniqueness of the Smith form $S(s)$ of $N(s)$ implies the *uniqueness* of $\Sigma(s)$ satisfying (2.34), (2.35) with the divisibility conditions (2.36).

DEFINITION 29.– *The matrix* $\Sigma(s)$ *is called the* Smith–MacMillan form *of* $G(s)$.

Definition of the MacMillan poles and zeros

Let

$$\boldsymbol{\epsilon}(s) = \operatorname{diag}(\varepsilon_1(s), ..., \varepsilon_r(s), 0, ..., 0)$$

$$\boldsymbol{\Psi}(s) = \operatorname{diag}(\psi_r(s), ..., \psi_1(s), 1, ..., 1)$$

so that

$$\Sigma(s) = \boldsymbol{\epsilon}(s) \boldsymbol{\Psi}^{-1}(s) = \boldsymbol{\Psi}^{-1}(s) \boldsymbol{\epsilon}(s)$$

where the polynomial matrices $\boldsymbol{\epsilon}(s)$ and $\boldsymbol{\Psi}(s)$ are left- and right-coprime (see section 13.2.6).

DEFINITION 30.– *(i) The MacMillan poles (resp., the MacMillan zeros) of $G(s)$ are the Smith zeros of $\mathbf{\Psi}(s)$ (resp., of $\boldsymbol{\epsilon}(s)$). (ii) The structural indices, the orders and the degrees of these MacMillan poles (or of these MacMillan zeros) are the structural indices, the orders and the degrees, respectively, of these Smith zeros. (iii) If $G(s)$ is a* proper *transfer matrix (see section 13.6.1), its* MacMillan degree *is the sum of the degrees of the polynomials $\psi_i(s)$ $(1 \leq i \leq r)$.*

REMARK 31.– *(i) According to (2.34) and (2.35), $G(s) = D_l^{-1}(s) N_l(s)$ where $D_l(s) = \mathbf{\Psi}(s) U(s)$ and $N_l(s) = \boldsymbol{\epsilon}(s) V(s)$. We have the equality*

$$\begin{bmatrix} D_l(s) & N_l(s) \end{bmatrix} = \begin{bmatrix} \mathbf{\Psi}(s) & \boldsymbol{\epsilon}(s) \end{bmatrix} \operatorname{diag}\{U(s), V(s)\}.$$

The matrices $U(s)$ and $V(s)$ are invertible over $\mathbf{R} = \mathbb{R}[\partial]$, so is also $\operatorname{diag}\{U(s), V(s)\}$*, therefore the matrices $\{D_l(s), N_l(s)\}$ are left-coprime. Therefore, $(D_l(s), N_l(s))$ is, like $(\mathbf{\Psi}(s), \boldsymbol{\epsilon}(s))$, a left-coprime factorization of $G(s)$ over $\mathbf{R}$. According to Theorem 509 (section 13.2.7), the MacMillan poles (resp., the MacMillan zeros) of $G(s)$ are the Smith zeros of $D_l(s)$ (resp., $N_l(s)$). (ii) We can equally write $G(s) = N_r(s) D_r^{-1}(s)$ where $D_r(s) = V^{-1}(s) \mathbf{\Psi}(s)$ and $N_r(s) = U^{-1}(s) \boldsymbol{\epsilon}(s)$. Since*

$$\begin{bmatrix} D_r \\ N_r \end{bmatrix} = \begin{bmatrix} V^{-1} & 0 \\ 0 & U^{-1} \end{bmatrix} \begin{bmatrix} \mathbf{\Psi} \\ \boldsymbol{\epsilon} \end{bmatrix},$$

$(N_r(s), D_r(s))$ is a right-coprime factorization of $G(s)$ over $\mathbf{R}$. The MacMillan poles (resp., zeros) of $G(s)$ are the Smith zeros of $D_r(s)$ (resp., $N_r(s)$).

DEFINITION 32.– *A control system which has a transmission pole (resp., zero) at $s = 0$ is called an* integrator system *(resp., a* derivator system*).*

Example of the heated tank

The transfer matrix (2.31) of the heated tank can be written as

$$G(s) = a \begin{bmatrix} \frac{1}{s+a} & 0 \\ \frac{-1}{s+2a} & \frac{1}{s+2a} \end{bmatrix}$$

with $a = \frac{1}{\tau}$. It is clear that the transmission poles and zeros of $G(s)$ are identical to those of $\tilde{G}(s) = \frac{G(s)}{a}$; it is therefore this last matrix that we will now be interested in. We have

$$\tilde{G}(s) = \frac{1}{(s+a)(s+2a)} \begin{bmatrix} s+2a & 0 \\ -(s+a) & s+a \end{bmatrix} = \frac{1}{(s+a)(s+2a)} N(s).$$

The Smith form of $N(s)$ is

$$S(s) = \begin{bmatrix} 1 & 0 \\ 0 & (s+2a)(s+a) \end{bmatrix}.$$

As a result, the Smith–MacMillan form of $\tilde{G}(s)$ is

$$\Sigma(s) = \begin{bmatrix} \frac{1}{(s+2a)(s+a)} & 0 \\ 0 & 1 \end{bmatrix}.$$

The heated tank therefore has two transmission poles $\{-a, -2a\}$. These are simple poles (meaning that their order and their degree are equal to 1). On the other hand, this system has no transmission zero.

2.4.6. *Minimal systems*

Relation between the poles of a system and its transmission poles

Consider a system described by the Rosenbrock representation (2.16). According to (2.27), its transfer matrix is

$$\begin{aligned} G(s) &= Q(s)\, D^{-1}(s)\, N(s) + W(s) \\ &= \frac{1}{\det D(s)} \left[Q(s)\, \mathrm{adj}(D(s))\, N(s) + \det(D(s))\, W(s)\right] \end{aligned} \quad (2.37)$$

where $\mathrm{adj}(D(s))$ designates the classical adjoint of $D(s)$ (see section 13.1.4). Now, the roots of $\det D(s)$ are the poles of the system being considered (see section 2.3.7, Definition 17). It follows from (2.37) that

$$\boxed{\{\text{ transmission poles}\} \subset \{\text{system poles}\}}. \quad (2.38)$$

This inclusion is an equality, except when the fraction (2.37) is not irreducible, i.e. when

$$\det D(s) \quad \text{and}$$

$$\left[Q(s)\, \mathrm{adj}(D(s))\, N(s) + \det(D(s))\, W(s)\right]$$

have common factors. Indeed, once these common factors are cancelled, the (MacMillan) poles of $G(s)$ become no more than just a strict subset of the roots of $\det(D(s))$.

This case, which in a way is "pathological", is studied in Chapter 7.

Minimal control system

DEFINITION 33.– *A control system is said to be* minimal *if the inclusion* (2.38) *is an equality.*[4]

The reader can verify that all systems having been envisaged up till now in the examples and exercises are minimal, with the exception of system (2.23) in section 2.3.8. Indeed, this one here has poles $\{0, 0\}$, while it has a unique transmission pole $\{0\}$.

Transmission order

The *order of a system* is defined at section 2.3.7 (see Definition 17(ii), which we may reformulate by saying that the order of a system is the number of its poles, counting multiplicities).

DEFINITION 34.– *We call the number of transmission poles of a system the* transmission order *of this system (counting multiplicities).*

According to (2.38),

$$\boxed{(\text{transmission order}) \leq (\text{system order})}\,. \tag{2.39}$$

For example, the system (2.23) is of order 2, but its transmission order is 1.

It is clear that a control system is minimal if and only if,

$$(\text{transmission order}) = (\text{system order})\,.$$

When a control system is minimal, we can talk about its "order" without ambiguity.

As a result:

The RLC circuit with transfer function (2.28) is of order 2.

The DC motor with transfer function (2.30) is of order 3.

The heated tank with transfer matrix (2.31) is of order 2.

4. Note that in [42] and [96], the minimality corresponds to a different situation. Our definition corresponds to the notion of minimal (or irreducible) realization introduced by Kalman [65].

2.4.7. *Transmission poles and zeros at infinity*

SISO case

Consider a transfer function $G(s) = \frac{b(s)}{a(s)}$, where $b(s)$ and $a(s)$ belong to $\mathbb{R}[s]$, are both non-zero, and of degree n and m, respectively; write

$$\begin{aligned} b(s) &= b_0 s^m + b_1 s^{m-1} ... + b_m, \\ a(s) &= a_0 s^n + a_1 s^{n-1} ... + a_n, \end{aligned}$$

where $a_0 b_0 \neq 0$. The *relative degree* of the rational function $G(s)$ is therefore $\delta(G) = n - m$ (see section 13.6.1).

We can embed the ring $\mathbb{R}[s]$ in the field $\mathbb{R}((\sigma))$ of the Laurent series with indeterminate $\sigma = 1/s$ (see section 13.1.1). Indeed, any element $c(\sigma) \neq 0$ of $\mathbb{R}((\sigma))$ is of the form

$$c(\sigma) = \sum_{i \geq \nu} c_i \sigma^i, \quad c_\nu \neq 0;$$

this element belongs to $\mathbb{R}[s]$ if and only if $c_i = 0$ for $i > 0$. As a result, we can embed the field $\mathbb{R}(s)$ in $\mathbb{R}((\sigma))$. We obtain

$$c(\sigma) = c_\nu \sigma^\nu \left(1 + \frac{c_{\nu+1}}{c_\nu} \sigma + \cdots\right) = c_\nu \sigma^\nu \upsilon(\sigma)$$

where $\upsilon(\sigma)$ is a unit of the ring $\mathbb{R}[[\sigma]]$. Applying this to the rational function $G(s)$, we obtain

$$G(s) = \frac{b_0}{a_0} \sigma^{n-m} \upsilon(\sigma).$$

Remember now that $\sigma = 0$ if and only if $s = \infty$. As a result:

– if $n - m > 0$, we say that $G(s)$ has a *zero at infinity* of order $n - m$ (or that $G(s)$ has $n - m$ zeros at infinity);

– if $m - n > 0$, we say that $G(s)$ has a *pole at infinity* of order $m - n$ (or that $G(s)$ has $m - n$ poles at infinity).

These *poles* (resp., *zeros*) at infinity are the *transmission poles* (resp., *zeros*) *at infinity* of the system with transfer function $G(s)$.

The transfer function $G(s)$ has m *finite* zeros (in $\mathbb{C}$, for example), which are the roots of $b(s)$, and n *finite* poles – the roots of $a(s)$. Therefore:

– if $n - m > 0$, $G(s)$ has m finite zeros and $n - m$ zeros at infinity, say n zeros in the set $\mathbb{C}_e$ consisting of the complex plane to which we added the "point infinity"; on the other hand, $G(s)$ has n poles in $\mathbb{C}_e$;

– if $m - n > 0$, $G(s)$ has n finite poles and $m - n$ poles at infinity, say m poles in $\mathbb{C}_e$; and $G(s)$ has m zeros in $\mathbb{C}_e$.

From which we get the following result:

PROPOSITION 35.– *Let $G(s)$ be a rational function. It has as many poles as it has zeros,* if we also account for the point at infinity.

On the other hand, the following result is a simple consequence of the definitions (see section 13.6.1):

PROPOSITION 36.– *A transfer function is proper if and only if it has no pole at infinity.*

* *MIMO case*

The rationale behind section 2.4.4 for defining *finite* MacMillan poles and zeros of a transfer matrix can also be used for the poles and zeros at *infinity*. Let $G(s) \in \mathbb{R}(s)^{p\times m}$ be a transfer matrix and consider $G(s)$ as an element of $\mathbb{R}((\sigma))^{p\times m}$, where $\sigma = 1/s$. We know that $\mathbb{R}((\sigma))$ is the field of fractions of $\mathbf{S} = \mathbb{R}[[\sigma]]$, the ring of formal series in σ, and that the ring $\mathbf{S}$ is a principal ideal domain (see section 13.1.1). Thus let σ^κ $(\kappa \geq 0)$ be the least common denominator of the elements of $G(s)$, when the elements are considered the way we have just indicated. We obtain

$$G(s) = \frac{N_\infty(\sigma)}{\sigma^\kappa}$$

where $N_\infty(\sigma) \in \mathbf{S}^{p\times m}$. The matrix $N_\infty(\sigma)$ admits a Smith form $S_\infty(\sigma)$, and there exist matrices $U_\infty(\sigma)$ and $V_\infty(\sigma)$, invertible over the ring $\mathbf{S}$, such that $N_\infty(\sigma) = U_\infty^{-1}(\sigma)\, S_\infty(\sigma)\, V_\infty(\sigma)$ where

$$S_\infty(\sigma) = \operatorname{diag}\left(\sigma^{\mu_1}, ..., \sigma^{\mu_r}, 0, ..., 0\right),$$

$0 \leq \mu_1 \leq ... \leq \mu_r$. Therefore, $G(s) = U_\infty^{-1}(\sigma)\, \Sigma_\infty \sigma\, V_\infty(\sigma)$ where

$$\Sigma_\infty(\sigma) = \frac{1}{\sigma^\kappa} S_\infty(\sigma) = \operatorname{diag}\left(\sigma^{\nu_1}, ..., \sigma^{\nu_r}, 0, ..., 0\right)$$

with $\nu_i = \mu_i - \kappa$, and thus $\nu_1 \leq \cdots \leq \nu_r$.

DEFINITION 37.– *(i) The matrix $\Sigma_\infty(\sigma)$ is called the* Smith–MacMillan form *of $G(s)$ at infinity. (ii) Let $(\bar{\varsigma}_i)_{1\leq i\leq r}$ and $(\bar{\pi}_i)_{1\leq i\leq r}$ be the finite sequence of natural numbers defined by: $\bar{\varsigma}_i = \max(0, \nu_i)$ and $\bar{\pi}_i = \max(0, -\nu_i)$. Among the natural numbers $\bar{\varsigma}_i$ (resp., $\bar{\pi}_i$), those that are non-zero (if any) are called the structural indices of the zeros at infinity (resp., the poles at infinity) of the transfer matrix $G(s)$; they are arranged in increasing (resp., decreasing) order and are denoted by $\varsigma_i, 1 \leq i \leq \rho$ (resp.,*

$\pi_i, 1 \leq i \leq \varpi$). (iii) If $\rho \geq 1$ (resp., $\varpi \geq 1$), $G(s)$ is said to have ρ zeros (resp., ϖ poles) at infinity, the ith one with order ς_i (resp., π_i). (iii) The integer $\sum_{1 \leq i \leq \varpi} \pi_i$ (resp., $\sum_{1 \leq i \leq \rho} \varsigma_i$) is called the degree of the poles (resp., the zeros) of $G(s)$ at infinity.

REMARK 38.– *We can write $G(s)$ (as an element of $\mathbb{R}((\sigma))^{p \times m}$) in the form*

$$G(s) = \sum_{i \geq \nu_1} \Theta_i \, \sigma^i$$

where $\Theta_i \in \mathbb{R}^{p \times m}$. Therefore, $G(s)$ is proper if and only if $\nu_1 \geq 0$, i.e. if $G(s)$ has no poles at infinity; $G(s)$ is strictly proper if and only if $\nu_1 \geq 1$, and then $G(s)$ is said to have a blocking zero at infinity *of order ν_1.*

We can now generalize Definition 30(iii):

DEFINITION 39.– *The MacMillan degree of a transfer matrix (not necessarily proper) is the sum of the degrees of all its poles (including its poles at infinity).*

REMARK 40.– *We naturally will ask the question of whether Proposition 36 will generalize to the MIMO case. The answer is* negative*: see ([64], section 6.5). The number of poles (finite and infinite) of a transfer matrix can exceed its number of zeros (finite and infinite) by a quantity called the* defect *of the transfer matrix $G(s)$. The defect of an invertible square transfer matrix is zero.*

2.5. Responses of a control system

2.5.1. *Input–output operator*

**General case*

Consider a time-invariant control system $\mathbf{\Sigma}$, possibly nonlinear, defined by an equation such as (2.4) (section 2.2.2). In distinguishing the inputs from the other variables of the system, we can suppose that the system equation is of the form

$$F\left(\xi, ..., \xi^{(\beta)}, u, ..., u^{(\gamma)}\right) = 0.$$

According to section 2.2.6, an equilibrium point $w^* = (\xi^*, u^*)$ is such that

$$F(\xi^*, 0, ..., 0, u^*, 0..., 0) = 0.$$

Let t_0 be an initial instant and u be an input such that

$$\begin{cases} u(t_0) = u^*, \\ u^{(i)}(t_0) = 0, \; 1 \leq i \leq \gamma. \end{cases} \quad (2.40)$$

According to Condition *iii)* of section 2.3.1, there exists one and only one solution ξ such that

$$\left\{ \begin{array}{c} \xi\left(t_0\right) = \xi^*, \\ \xi^{(j)}\left(t_0\right) = 0, \ 1 \leq j \leq \beta. \end{array} \right. \tag{2.41}$$

In particular, once the initial conditions (2.40) and (2.41) are imposed at time t_0, the output y of $\mathbf{\Sigma}$ depends only on u. Therefore, there exists a unique nonlinear operator $\tilde{\Sigma}_{t_0,w^*}$ such that

$$y = \left(\tilde{\Sigma}_{t_0,w^*}\right)(u). \tag{2.42}$$

DEFINITION 41.– *The operator* $\tilde{\Sigma}_{t_0,w^*}$ *is called the* input–output operator *associated with* $\mathbf{\Sigma}$ *for the initial condition* (t_0, w^*).

The function y defined by (2.42) is called the *response* of Σ for the input u and the initial condition (t_0, w^*).

* *Case of a linear or time-invariant system*

If $\mathbf{\Sigma}$ is a *linear* system, then $w^* = 0$ is an equilibrium point. It is not necessarily the only one – as the reader can show, as part of an exercise, using the example of the system (2.23) (section 2.3.8) – but the problem can always come down to the case where $w^* = 0$ by translating the origin of the space where w is "living". The input–output operator associated with $\mathbf{\Sigma}$ for the initial condition $(t_0, 0)$ can then be denoted more concisely by $\tilde{\Sigma}_{t_0}$.

Likewise, if $\mathbf{\Sigma}$ is linear and time-invariant, the problem comes down to the case where $t_0 = 0$ by a translation of the origin of time. The input–output operator associated with $\mathbf{\Sigma}$ for the initial condition $(0, w^*)$ can then be denoted by $\tilde{\Sigma}_{w^*}$.

Last, if $\mathbf{\Sigma}$ is *linear and time-invariant*, we denote by $\tilde{\Sigma}$ the input–output operator associated with $\mathbf{\Sigma}$ for the initial condition $(0, 0)$. This operator is explored in what follows next.

Case of a linear time-invariant SISO system

Consider a *linear time-invariant SISO system* $\mathbf{\Sigma}$, with transfer function $G(s)$ assumed to be a rational function. *With zero initial conditions*, we have relation (2.26) between the Laplace transform of the input and that of the output (see section 2.4.2). Let g be the inverse Laplace transform of $G(s)$. In time domain, we have

$$\boxed{y(t) = (g * u)(t), \ t \geq 0}. \tag{2.43}$$

Consequently, the input–output operator associated with $\mathbf{\Sigma}$ is the convolution operator $\tilde{\Sigma} : u \mapsto g * u$.

Case of a linear time-invariant MIMO system

Let $\mathbf{\Sigma}$ be a *linear time-invariant MIMO* system with transfer matrix $G(s) = (G_{ij}(s))$ of size $p \times m$. Suppose that each element $G_{ij}(s)$ is a rational function. At zero initial conditions, we have again relation (2.26). Let g_{ij} be the inverse Laplace transform of $G_{ij}(s)$ and let g be the matrix with entries of the elements g_{ij}. By extension, we can call G the Laplace transform of g and we can easily verify that

$$\begin{aligned} G(s) &= \int_{0^-}^{+\infty} g(t) e^{-st} dt \text{ for } \operatorname{Re}(s) > \gamma, \\ \gamma &= \max_{p \in P} \operatorname{Re} p, \end{aligned} \tag{2.44}$$

where P is the set of poles of $G(s)$. This expression is identical to (2.25) (section 2.4.1). We have therefore

$$g = \mathcal{L}^{-1}(G(s)).$$

Let u be an input, assumed to be a locally integrable function with values in $\mathbb{R}^m$. The initial conditions are assumed to be zero, and the input is assumed to be positively supported: $u(t) = 0$ for $t < 0$. The convolution product $g * u$ can then be defined: it is a function y with values in $\mathbb{R}^p$ the ith component y_i of which is given by

$$y_i = \sum_{j=1}^{m} g_{ij} * u_j \quad (1 \leq i \leq p).$$

With this definition of the convolution product, the expression (2.26) in Laplace domain corresponds to expression (2.43) in time domain, which is thus extended to the MIMO case. We have thus obtained the following general result:

THEOREM 42.– *The input–output operator associated with the control system* $\mathbf{\Sigma}$ *is the convolution operator* $\tilde{\Sigma} : u \mapsto g * u$.

Abusing the language, we can say that $G(s)$ is the transfer matrix of the convolution operator $\tilde{\Sigma}$ (as well as of system $\mathbf{\Sigma}$).

Control system and associated input–output operator

The knowledge of the transfer matrix $G(s)$ is equivalent to that of the convolution operator $\tilde{\Sigma}$. But it does not determine the control system $\mathbf{\Sigma}$ unless it is *minimal*. If that is not the case, indeed, the inclusion (2.38) is not an equality, so there exist poles of the system $\mathbf{\Sigma}$ which are not represented in the transfer matrix. We will see in Chapter 7 that the minimality of a linear time-invariant system is a *necessary and sufficient condition* for such system to be entirely characterized by its transfer matrix (or by its associated convolution operator).

2.5.2. *Impulse and step responses*

Impulse response

In mathematics, g is called the *kernel* of the convolution operator $\tilde{\Sigma}$.

From the point of view of systems theory, let us first consider the case of a SISO system: g is the output of the system $\mathbf{\Sigma}$ when the input u is the "Dirac impulse" (which, in mathematical language, is the Dirac distribution) δ, and the initial conditions are zero. That is why we call g the *impulse response* of $\mathbf{\Sigma}$.

Therefore, we can *identify* a system by subjecting it to a Dirac impulse: we only have to measure the corresponding response, which is the impulse response, and it remains to calculate its Laplace transform to get the transfer function $G(s)$.

Step response

The method just described is, however, not very practical because the Dirac distribution is not an input which is physically realizable. It is preferable to replace it by the unit step $\mathbf{1}(t)$, defined by

$$\boxed{\mathbf{1}(t) = \begin{cases} 0, & t < 0 \\ 1, & t \geq 0 \end{cases}}. \tag{2.45}$$

The corresponding response (with zero initial conditions) is the *unit step response* (or the *step response*). The Laplace transform of the step response is $\frac{G(s)}{s}$. The derivative of the step response (in the sense of distributions) is the impulse response (for the derivative of the unit step is the Dirac distribution): see equation (12.23), section 12.2.3.

Extension to MIMO case

In the MIMO case, recall that g is a matrix (of size $p \times m$). It is again called the impulse response of the system (or sometimes the matrix of impulse responses). The entry g_{ij} is the ith component of the output when the jth component of the input is δ and the other components are zero.

The extension of the notion of step response to the MIMO case (sometimes called, in this case, the *matrix of step responses*) is done in a similar manner.

2.5.3. *Proper, biproper and strictly proper systems*

Proper systems

THEOREM 43.– *Let $\mathbf{\Sigma}$ be a linear time-invariant control system. The following conditions are equivalent: (i) its step response is a piecewise continuous function (having no discontinuity except at the origin); (ii) its transfer matrix $G(s)$ is proper.*

PROOF. The proof of this theorem makes use of the results of section 13.6.1; it is detailed in the SISO case, and its extension to the MIMO case is left to the reader. (1) If $G(s)$ is proper, we have

$$G(s) = q_0 + H(s)$$

where $q_0 = \lim_{|s|\to+\infty} G(s)$ and where $H(s)$ is a strictly proper rational function. The inverse Laplace transform of $H(s)$ is a locally integrable function h. Therefore, the impulse response of $\mathbf{\Sigma}$ is $g = q_0\,\delta + h$, and its response to a locally integrable input u is

$$y(t) = q_0 u(t) + \int_0^t h(t-\tau)\,u(\tau)\,d\tau$$

which is again a locally integrable function. If u is the unit step, we obtain

$$y(t) = q_0 \mathbf{1}(t) + \int_0^t h(\tau)\,d\tau \tag{2.46}$$

which is a continuous function, except at $t = 0$ if $q_0 \neq 0$. (2) Conversely, suppose $G(s)$ is improper. Then, there exists a polynomial $Q(s)$, with positive degree, such that

$$G(s) = Q(s) + H(s)$$

where $H(s)$, just as before, admits an inverse Laplace transform h which is a locally integrable function. Put $Q(s) = \sum_{k=0}^{n} q_k s^k$, where $n = d^\circ(Q) \geq 1$. The step response is

$$y(t) = \sum_{k=1}^{n} q_k \delta^{(k-1)}(t) + q_0 \mathbf{1}(t) + \int_0^t h(\tau)\,d\tau \tag{2.47}$$

which is a singular distribution, since $q_n \neq 0$. ∎

DEFINITION 44.– *The control system* $\mathbf{\Sigma}$ *is said to be* proper *if it satisfies one of the equivalent conditions in Theorem 43.*

REMARK 45.– *As was shown in the proof of Theorem 43, a linear time-invariant control system is proper if and only if its response to a locally integrable function is again locally integrable.*

The control engineer must always choose the inputs and outputs of a system in such a way that the resulting control system be proper. Indeed, the unit step is an input commonly used and if the system is not proper (we say that it is *improper*), the resulting output is a "singular distribution" from a mathematical point of view (see the proof of Theorem 43). Such a signal does not exist in reality and, in more concrete terms, what actually happens is the destruction of the system (or the saturation of its components). An *improper* system is therefore a badly designed system. Such a system is considered not physically realizable in general. This point may be contested; but it is true that an improper system is not *practically* realizable.

Examples

Consider again the RLC circuit in section 1.1.1. With input $u = V$ and output $y = i$, its transfer function is (2.28), and thus it is a proper control system. On the other hand, with input $u = i$ and output $y = V$, its transfer function is (2.29) and this time it is an improper control system.

With the conventions of section 2.3.3, the DC motor in section 1.3 and the heated tank in section 1.4 are proper, since they have as transfer functions (2.30) and (2.31), respectively.

Strictly proper System

The proof of the following theorem is similar to that of Theorem 43:

THEOREM 46.– *Let Σ be a linear time-invariant control system. The following conditions are equivalent: (i) its step response is a continuous function; (ii) its transfer matrix $G(s)$ is strictly proper.*

DEFINITION 47.– *The control system Σ is said to be* strictly proper *if it satisfies one of the conditions in Theorem 46.*

THEOREM 48.– *The system in state-space form* (2.20) *(section 2.3.6) is* proper *(resp.,* strictly proper*) if and only if $W(\partial) \subset \mathbb{R}^{p \times m}$ (resp., $W(\partial) = 0$).*

PROOF. Notice that the transfer matrix of this system is

$$G(s) = C\,(sI_n - A)^{-1}\,B + W(s)\,. \tag{2.48}$$

■

One can prove the following property: the output of a proper system is at least "as regular as" its input; for example, its response for an input having a certain number of discontinuities can have the same discontinuities, but not more. While the output of a strictly proper system is "more regular" than its input, the output of such a system is continuous even if its input is a discontinuous function. A strictly proper system therefore has a "regularizing" effect.

The RLC circuit having input $u = V$ and output $y = i$ is a strictly proper control system, so is the case of DC motor with transfer function (2.30), and the heated tank with transfer matrix (2.31).

We insisted above on the fact that the inputs and outputs of a system have to be chosen in such a manner that the resulting control system be proper. Generally, this system is strictly proper for this system as it does not instantaneously react to a solicitation because of its "inertia". To clarify ideas, consider an SISO system at rest at initial time $t = 0$ (zero initial conditions). Take the unit step as input, the resulting output is therefore the step response (2.47). For a system with a certain inertia, like the DC motor (where its output is $y = \theta$ or $y = \omega$) for example, this step response cannot go from the value $y = 0$ (at instant $t = 0^-$) to a value $\geq \alpha$, $\alpha > 0$, at $t = 0^+$: the angular speed of the motor cannot "jump" from zero to a quantity $\geq \alpha$ when we apply a voltage step as the input, and this is equally true, *a fortiori*, for the angular position. *Physical* control systems are, in general, designed in such a way that they are *strictly proper*.

REMARK 49.– *We have seen from Theorem 11 that it is possible to choose a finite sequence* $(u_1, ..., u_m)$ *among the variables of a system such that* $u = [u_1, ..., u_m]^T$ *can be an input of this system. Another question is whether the input* u *can be chosen in such a way that the resulting control system be* proper. *This question is studied in [22]. Obviously, the properness of a control system depends not only on the choice of its inputs, but also on the choice of its outputs. The detailed analysis made in the cited reference, where an algorithm is also proposed to realize in a systematic manner the choice of* u*, showed the importance of the* non-controllable poles at infinity, *also called the* input-decoupling zeros at infinity *(a notion that is beyond the scope of this work: see [16] or [22]). Consider a proper (or even strictly proper) control system; if that system has uncontrollable poles at infinity, it can still generate – in response to a step input for example – "impulsive motions", consisting of linear combinations of the Dirac distribution and its derivatives, which will lead to the system destruction.*

Biproper system

Some purely electrical or electronic systems are proper but not strictly proper, the simplest example being an electric circuit consisting of a resistance R and a voltage generator. The input $u = V$ (voltage across the resistance) is related to the output $y = i$ (electric current intensity through the resistance) by the relation $V = Ri$. The system transfer function is therefore $G(s) = \frac{1}{R}$: which is proper, but not strictly proper. Its inverse $G^{-1}(s) = R$ is defined and is proper, thus $G(s)$ is *biproper* (see section 13.6.1). For this reason, the corresponding control system is also qualified as *biproper*. It is quite rare having to control a physical system of this nature; on the other hand, it is quite common for a *controller* to be biproper. In general:

DEFINITION 50.– *A linear time-invariant system* Σ *is* biproper *if its transfer matrix has this property.*

2.5.4. *Frequency response*

Definition

We consider only the SISO case here, the extension to the MIMO case is trivial (see, in section 2.5.2, how the notion of impulse response is extended to the MIMO case).

Let $u(t) = A \cos(\omega_0 t)$, $t \in (-\infty, +\infty)$, a sinusoidal signal, be the input of system Σ. To calculate the output of the system, it is convenient to write $u(t) = \operatorname{Re} u_c(t)$, where $u_c(t) = A\, e^{i\,\omega_0\, t}$. Let us examine why this is so.

The response y of Σ to the input u is equal to the convolution product $g * u$, where g denotes the impulse response of Σ. Likewise, the (fictitious) response y_c of Σ to the (fictitious) input u_c is equal to the convolution product $g * u_c$. We have therefore $\operatorname{Re}(g * u_c) = \operatorname{Re} g * \operatorname{Re} u_c$, and since g is real, $y = \operatorname{Re} y_c$.

1° First of all, consider the case where $G(s)$ has no poles with positive real part.

The impulse response g is therefore a tempered distribution and it results according to section 12.3.1 in: $\mathcal{F}(g * u_c) = \mathcal{F} g\, \mathcal{F} u_c$. Now,

$$(\mathcal{F}\, u_c)(\omega) = A \int_{-\infty}^{+\infty} e^{i\,(\omega - \omega_0)\, t} dt = 2\pi\, A\, \delta(\omega - \omega_0)$$

(see (12.34), section 12.3.1). Therefore,

$$\mathcal{F}\, g\, \mathcal{F}\, u_c = (\mathcal{F}\, g)(\omega_0)\; \mathcal{F}\, u_c$$

and by taking the inverse Fourier transform, we obtain $y_c(t) = (\mathcal{F}\, g)(\omega_0)\; u_c(t)$, and finally

$$\boxed{y_c(t) = G(i\,\omega_0)\, u_c(t)}. \tag{2.49}$$

By definition, the function (or distribution) $\omega \mapsto G(i\omega)$ is called the *frequency response* of the system Σ. This is the Fourier transform of the impulse response.

It is now easy to calculate the output $y(t)$ by using the polar decomposition of $G(i\,\omega_0)$: $G(i\omega_0) = |G(i\omega_0)|\, e^{i \arg G(i\,\omega_0)}$. We get: $y_c(t) = A\,|G(i\omega_0)|\, e^{[i\omega_0 t + \arg G(i\omega_0)]}$, from which

$$y(t) = A\,|G(i\omega_0)| \cos\left[\omega_0 t + \arg G(i\omega_0)\right]. \tag{2.50}$$

Expression (2.49), using complex signals, is simpler than (2.50), which uses real signals. That is why it is interesting to work with complex numbers.

REMARK 51.– *The frequency response $\mathcal{F} g$ is also defined in the case where $G(s)$ has poles on the imaginary axis, but it is a singular distribution (not a function).*

Nevertheless, the output (2.50) is only defined if $s = i\omega_0$ *is not a pole of* $G(s)$*. This output is thus defined for all finite values of* ω_0 *if and only if* $G(s)$ *has no poles on the imaginary axis and thus only has poles that have a negative real part (i.e. all poles belong to the left half-plane*[5]*). Moreover, the set of these responses is bounded when* ω_0 *varies (and in particular tends to* $+\infty$*) if and only if* $G(s)$ *is a proper rational function, then belonging to* $\Re\mathcal{H}_\infty$ *: see section 13.6.2.*

2° The above definition can be extended to any transfer function $G(s)$ having no poles on the imaginary axis: its *frequency response* is the function $\omega \mapsto G(i\omega)$. On the other hand, this frequency response can no longer be defined as the Fourier transform of the impulse response, since this Fourier transform does not exist. With this generalized definition, the frequency response is still the quantity which multiplies a sinusoidal input $e^{i\omega t}$ to produce the output signal, provided that the system is stabilized by feedback.

Experimental determination of the frequency response of a system

The frequency response of a system can be determined experimentally if the transfer function $G(s)$ is proper and has only poles with negative real part (this double condition is essential, as shown in Remark 51 ; but as it turns out from n°2 above, the procedure may be extended to the case where $G(s)$ also has poles with positive real part provided that the system is stabilized by feedback. It is appropriate to operate in the following manner:

– For a representative set of frequencies ω, put a sinusoidal signal $u_\omega(t) = A\cos(\omega t)$ at the system input. Wait for a sufficiently long time for the steady state to be reached (in theory, as we have seen above, we have to wait an infinitely long time).

– For each of the frequencies ω above, measure the output $y_\omega(t)$. According to (2.50), the ratio between the amplitude of y_ω and that of u_ω is equal to $|G(i\omega)|$, and the phase difference between y_ω and u_ω is given by $\arg G(i\omega)$.

– We thus obtain, point by point, the frequency response $G(i\omega)$.

2.6. Diagrams and their algebra

2.6.1. *Diagram of a control system*

Let $\mathbf{\Sigma}$ be a system with input u and output y. A common usage in control theory is to represent $\mathbf{\Sigma}$ by a box and some arrows, as shown in Figure 2.2. We call this the *system diagram*.

---

5. This expression refers to the complex plane. In what follows, we call the "left half-plane" the *open* set $\mathbb{C}_- = \{s \in \mathbb{C} : \mathrm{Re}(s) < 0\}$. Its closure $\bar{\mathbb{C}}_- = \{s \in \mathbb{C} : \mathrm{Re}(s) \leq 0\}$ is called "closed left half-plane". Similarly, we call the *open* set $\mathbb{C}_+ = \{s \in \mathbb{C} : \mathrm{Re}(s) > 0\}$ the "right half-plane" and its closure $\bar{\mathbb{C}}_+ = \{s \in \mathbb{C} : \mathrm{Re}(s) \geq 0\}$ the "closed right half-plane".

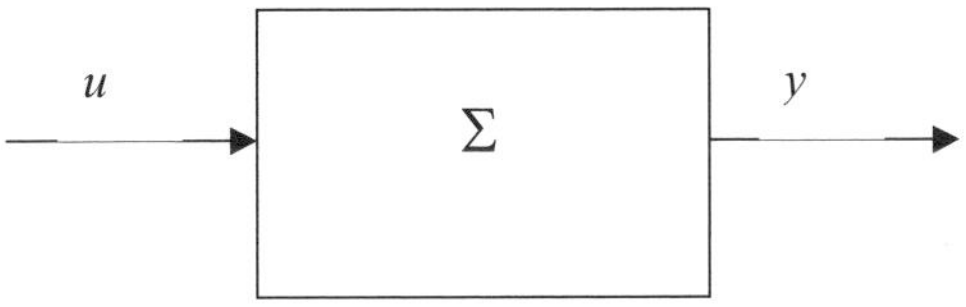

Figure 2.2. *System diagram*

Such a representation can be used in very general cases and in particular even for a system $\mathbf{\Sigma}$ that is nonlinear and/or time-varying. Nevertheless, this representation suggests that we are only interested in the input–output operator $\tilde{\Sigma}$ associated with $\mathbf{\Sigma}$. In the linear time-invariant case, it does not result in a loss of information if and only if $\mathbf{\Sigma}$ is minimal (see section 2.5.1). This remains true in the nonlinear and/or time-varying case [18]. In the linear time-invariant case, we can (see Figure 2.2) replace $\mathbf{\Sigma}$ by its transfer matrix $G(s)$.

2.6.2. *General algebra of diagrams*

There can be extremely diverse connections among systems [46]. We consider below connections in parallel, in series, and with unit feedback. The first two connections correspond to an addition and to a multiplication, respectively; the third is specific to systems theory and is a fundamental operation of this science.

Systems in parallel

The systems $\mathbf{\Sigma}_1$ and $\mathbf{\Sigma}_2$ in Figure 2.3 are said to be "in parallel". Let $\mathbf{\Sigma}$ be the resulting system, with input u and output y.

Let y_1 and y_2 be the outputs of $\mathbf{\Sigma}_1$ and $\mathbf{\Sigma}_2$, respectively. We have

$$y_1 = \tilde{\Sigma}_1\, u, \quad y_2 = \tilde{\Sigma}_2\, u, \quad y = y_1 + y_2,$$

as a result

$$y = \tilde{\Sigma}\, u, \quad \tilde{\Sigma} = \tilde{\Sigma}_1 + \tilde{\Sigma}_2. \tag{2.51}$$

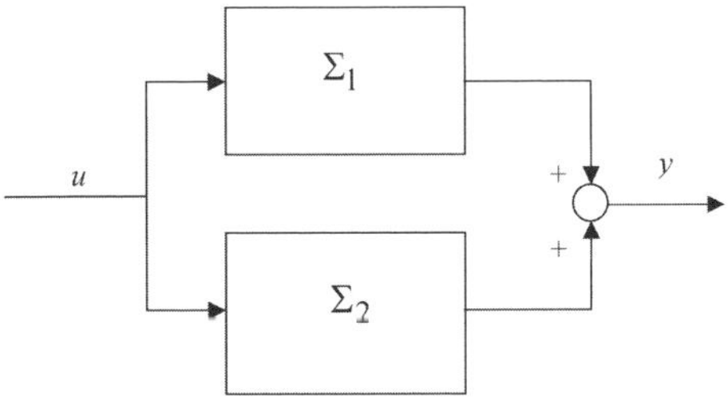

Figure 2.3. *Systems in parallel*

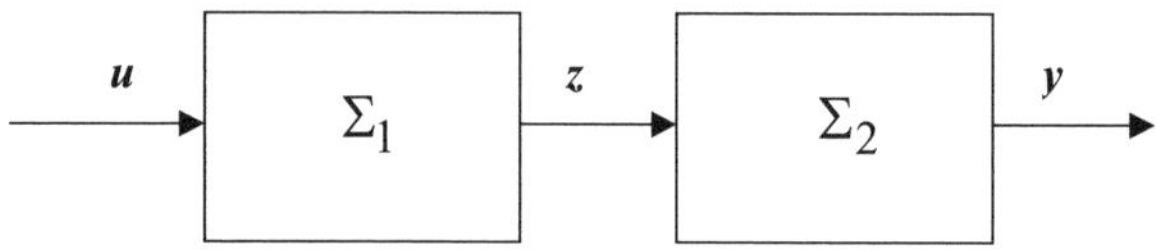

Figure 2.4. *Systems in series*

Note that $\mathbf{\Sigma}$ is not necessarily minimal, even if $\mathbf{\Sigma}_1$ and $\mathbf{\Sigma}_2$ are so [46].

Expression (2.51) is valid even in the nonlinear and/or time-varying case. If $\mathbf{\Sigma}_1$ and $\mathbf{\Sigma}_2$ are linear time-invariant with transfer matrices $G_1(s)$ and $G_2(s)$, respectively, then $\mathbf{\Sigma}$ is again linear time-invariant and its transfer matrix $G(s)$ is given by

$$\boxed{G(s) = G_1(s) + G_2(s)}. \tag{2.52}$$

Systems in series

The systems $\mathbf{\Sigma}_1$ and $\mathbf{\Sigma}_2$ in Figure 2.4 are said to be "in series". Consider the resulting system $\mathbf{\Sigma}$, with input u and output y.

The output z of $\mathbf{\Sigma}_1$ is equal to the input of $\mathbf{\Sigma}_2$. We have

$$z = \tilde{\Sigma}_1 u, \quad y = \tilde{\Sigma}_2 z,$$

therefore

$$y = \tilde{\Sigma} u, \quad \text{with } \tilde{\Sigma} = \tilde{\Sigma}_2 \tilde{\Sigma}_1. \tag{2.53}$$

The composition of the operators is denoted by a multiplication.

The system $\mathbf{\Sigma}$ is not necessarily minimal, even if $\mathbf{\Sigma}_1$ and $\mathbf{\Sigma}_2$ are so [46].

Expression (2.53) is valid even in the nonlinear and/or time-varying case. If $\mathbf{\Sigma}_1$ and $\mathbf{\Sigma}_2$ are linear time-invariant with transfer matrices $G_1(s)$ and $G_2(s)$, respectively, then $\mathbf{\Sigma}$ is again linear time-invariant and its transfer matrix $G(s)$ is such that

$$\boxed{G(s) = G_2(s)\, G_1(s)}. \tag{2.54}$$

It is important to be aware that the product (2.54) is not commutative in the MIMO case. Regarding the composition (2.53), it is not commutative in the nonlinear case, even if $\mathbf{\Sigma}_1$ and $\mathbf{\Sigma}_2$ are SISO.

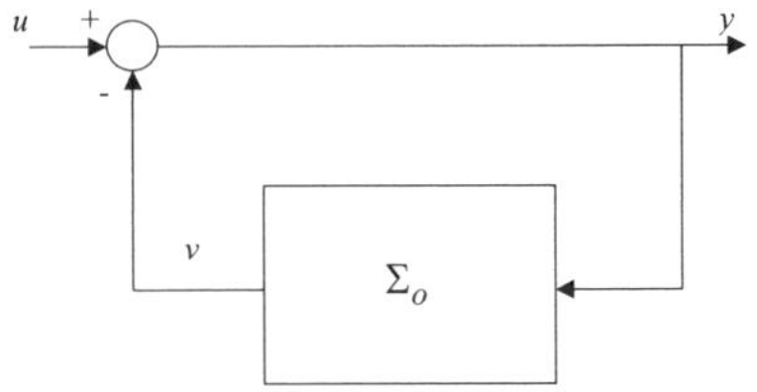

Figure 2.5. *Elementary feedback*

Elementary feedback

Consider the "elementary feedback" in Figure 2.5. The system $\mathbf{\Sigma}_0$, with input y and output v, is called the "open-loop system". The "closed-loop system", with input u and output y, is denoted by $\mathbf{\Sigma}$.

We have the following relations:

$$v = \tilde{\Sigma}_0 y, \quad y = u - v,$$

thus

$$\left(I_p + \tilde{\Sigma}_0\right) y = u \tag{2.55}$$

where p is the number of components of y.

We can express y as a function of u only if the operator $\left(I_p + \tilde{\Sigma}_0\right)$ is invertible.[6] Then the feedback system is said to be "well-defined".

In the case where $\mathbf{\Sigma}_0$ is linear time-invariant, with transfer matrix $G_0(s)$, the feedback system is well-defined if and only if $I_p + G_0(s)$ is invertible in the algebra $\mathbb{R}(s)^{p \times p}$. We will return to this in section 4.1.2.

Assuming that the feedback system is well-defined, we obtain from (2.55) $y = \tilde{\Sigma}\, u$ where

$$\tilde{\Sigma} = \left(I_p + \tilde{\Sigma}_0\right)^{-1}. \tag{2.56}$$

Expression (2.56) is valid even in the case where $\mathbf{\Sigma}_0$ is nonlinear and/or time-varying. If it is linear time-invariant, so it is also $\mathbf{\Sigma}$ and its transfer matrix $G(s)$ can be expressed as a function of the transfer matrix $G_0(s)$ of Σ_0 as follows:

$$\boxed{G(s) = \left[I_p + G_0(s)\right]^{-1}}.$$

6. We will not explain the meaning of this invertibility in the general case (see [116] for further reading on this subject).

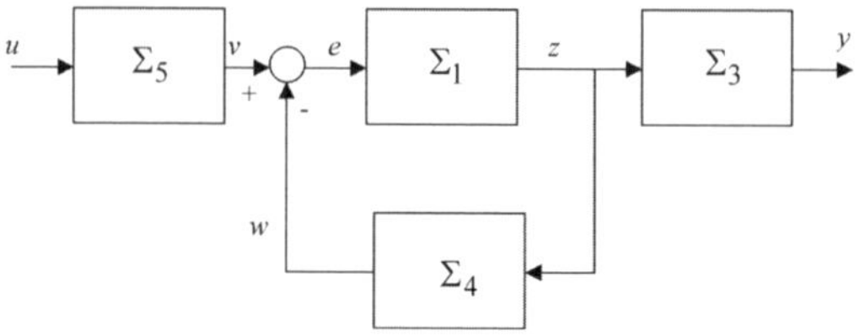

Figure 2.6. *Diagram example*

Application example

Consider Figure 2.6.

The three rules developed above can be applied to determine the input–output operator relating u to y. The system $\mathbf{\Sigma}$ defined by this diagram is constituted by putting four systems in series:

– the system $\mathbf{\Sigma}_5$;

– the loop with input v and output e, which is an elementary feedback such as in Figure 2.5 with $\tilde{\Sigma}_0 = \tilde{\Sigma}_4\tilde{\Sigma}_1$ (pay attention to the order!);

– the system $\mathbf{\Sigma}_1$;

– the system $\mathbf{\Sigma}_3$.

We have, therefore, with the assumption that the elementary feedback is well-defined,

$$y = \tilde{\Sigma}\, u$$

with

$$\tilde{\Sigma} = \tilde{\Sigma}_3\tilde{\Sigma}_1 \left(I + \tilde{\Sigma}_4\tilde{\Sigma}_1\right)^{-1} \tilde{\Sigma}_5. \tag{2.57}$$

2.6.3. *Specificity of linear systems*

"Multiplication" of operators, defined in section 2.6.2, is always *right-distributive* with respect to addition; this means (as shown immediately)

$$\left(\tilde{\Sigma}_1 + \tilde{\Sigma}_2\right)\tilde{\Sigma}_3 = \tilde{\Sigma}_1\tilde{\Sigma}_3 + \tilde{\Sigma}_2\tilde{\Sigma}_3.$$

On the other hand, multiplication is *left*-distributive with respect to addition only in the case of *linear* systems (time-invariant or not). More precisely, we have the relation

$$\tilde{\Sigma}_3\left(\tilde{\Sigma}_1 + \tilde{\Sigma}_2\right) = \tilde{\Sigma}_3\tilde{\Sigma}_1 + \tilde{\Sigma}_3\tilde{\Sigma}_2$$

if $\mathbf{\Sigma}_3$ is a *linear* system.

Take again the example in Figure 2.6 with, to simplify, $\tilde{\Sigma}_3 = I$ and $\tilde{\Sigma}_5 = I$. We have according to (2.57), no matter of what nature are the systems $\mathbf{\Sigma}_1$ and $\mathbf{\Sigma}_4$

$$\tilde{\Sigma} = \tilde{\Sigma}_1 \left(I + \tilde{\Sigma}_4 \tilde{\Sigma}_1 \right)^{-1}.$$

Suppose now that $\mathbf{\Sigma}_1$ is a linear system. We can write

$$z = \tilde{\Sigma}_1 \left(v - \tilde{\Sigma}_4 z \right)$$

from which we have

$$\left(I + \tilde{\Sigma}_1 \tilde{\Sigma}_4 \right) z = \tilde{\Sigma}_1 v$$

and finally (assuming that the loop is well-defined)

$$z = \left(I + \tilde{\Sigma}_1 \tilde{\Sigma}_4 \right)^{-1} \tilde{\Sigma}_1.$$

We have thus obtained the following result, which is similar to Lemma 520 (section 13.3.3):

THEOREM 52.– *If $\mathbf{\Sigma}_1$ is a linear system, we have the equality*

$$\boxed{\tilde{\Sigma}_1 \left(I + \tilde{\Sigma}_4 \tilde{\Sigma}_1 \right)^{-1} = \left(I + \tilde{\Sigma}_1 \tilde{\Sigma}_4 \right)^{-1} \tilde{\Sigma}_1}.$$

The algebra of diagrams is therefore richer in the context of linear systems than in that of nonlinear systems. It is possible to derive rules that allow the simplification of linear diagrams with small calculations, in particular the "Mason rule" (see e.g. [77]).

2.7. Exercises

EXERCISE 53.– *Determine the transfer function of the train in Figure 1.3, with input the force f and output the position $y = z_2$. What are the transmission poles and zeros of this system?*

EXERCISE 54.– *Same questions as above for the case of the DC motor, where the output chosen is $y = \omega$ (speed control).*

EXERCISE 55.– * *Let $G(s)$ have the following expression:*

$$(a) \quad \begin{bmatrix} \frac{s+1}{(s-1)^2} & \frac{1}{(s+1)(s-1)} \\ 0 & \frac{(s+1)^2}{(s-1)^3} \end{bmatrix}; \quad (b) \quad \begin{bmatrix} \frac{1}{(s+1)^2} & \frac{1}{(s+1)(s+2)} \\ \frac{1}{(s+1)(s+2)} & \frac{s+3}{(s+2)^2} \end{bmatrix};$$

$$(c) \quad \begin{bmatrix} \frac{1}{(s+1)^2} & \frac{1}{(s+1)(s+2)} \\ \frac{1}{(s+1)(s+2)} & \frac{s+1}{(s+2)^2} \end{bmatrix}.$$

In these three cases, determine the MacMillan poles and zeros of the transfer matrix. What is the transmission order of a system that has such a transfer matrix?

EXERCISE 56.– *Determine the equilibrium points of the double inverted pendulum of Exercise 2 (section 1.5) and linearize this system around the points* $y^* = 0$, $\theta_1^* = \theta_2^* = 0$.

EXERCISE 57.– *Same problem but for the mixer of Exercise 3.*

EXERCISE 58.– *In the case of the inverted pendulum in Figure 1.4, linearized as above, what is the number of input variables? Which one can be reasonably chosen as the control variable? The two measures being* z *and* θ*, are there any latent variables? Will a representation other than (2.7) have latent variables?*

EXERCISE 59.– *We consider the mixer linearized in Exercise 57. What is the number of input variables? Which ones are to be chosen? How many variables will need to be regulated? And what are they?*

EXERCISE 60.– *Many nonlinear control systems are described by a* nonlinear state representation *of the following form:*

$$\begin{aligned} \dot{x} &= f(x, u), \\ y &= g(x, u), \end{aligned}$$

where f *and* g *are of class* C^1. *(i) Let* (x^*, u^*, y^*) *be an equilibrium point. What are the relations that these quantities must satisfy? (ii) Write down the state equations of the linearized system about this equilibrium point.*

Chapter 3

Open-Loop Systems

In this chapter, we only consider linear time-invariant SISO systems unless otherwise stated.

3.1. Stability and static gain

3.1.1. *Stability*

DEFINITION 61.– *The linear system* Σ *(assumed to be* minimal*) is said to be* stable *if its transfer function* $G(s)$ *(assumed to be rational) is* proper *and* if all its poles belong to the left half-plane.[1] *This definition is also valid in the case of an MIMO system; then* $G(s)$ *is the transfer* matrix *of the minimal system* Σ *and the poles of this system coincide with the MacMillan poles of* $G(s)$ *(see section 2.4.4).*

We have already seen this double condition (see Remark 51 in section 2.5.4). As a result, stability is the condition that allows for an "open-loop" experimental determination of the frequency response of a system.

The set of transfer functions that satisfies the above double condition is written as $\Re\mathcal{H}_\infty$ (and $\Re\mathcal{H}_\infty^{p\times m}$ in the MIMO case with m inputs and p outputs) (see section 13.6.2). We call this set: *the set of stable transfer functions*.

1. See section 2.5.4, footnote 5.

As a result of Theorem 589 (section 13.6.2), a system is stable if, in ways that are equivalent:

– its output is so long as its input,

– its output has finite energy as long as its input has finite energy (with, in addition, a continuity condition that is specified in the previously cited theorem).

3.1.2. *Static gain*

Case of a stable system

Let Σ be a stable linear system, and f be its step response. It is given as in (2.46), where $h \in L_1$. It has a limit as t tends to $+\infty$, given by

$$\lim_{t \to +\infty} f(t) = g_0 + \int_0^{+\infty} h(\tau)\, d\tau.$$

We can apply the final value theorem (see section 12.3.4): the Laplace transform of f is $\hat{f}(s) = \frac{\hat{g}(s)}{s}$, so that

$$\boxed{\lim_{t \to +\infty} f(t) = \lim_{s \in \mathbb{R}, s \to 0^+} s \hat{f}(s) = \lim_{s \in \mathbb{R}, s \to 0^+} \hat{g}(s) = \hat{g}(0)}. \tag{3.1}$$

This quantity is called the *static gain* of the system (or of its transfer function).

According to (3.1), the static gain has several interpretations, which of course are equivalent:

(i) In the time domain, it is the final value of the step response.

(ii) In the frequency domain, it is $\hat{g}(0) = \mathcal{F}g(0)$: frequency response of a system at frequency zero.

(iii) We can also say that it is the gain multiplying a constant signal u producing an output y (also constant).

Case of an unstable system

Let Σ be an unstable system having no transmission pole at $s = 0$, so that the quantity $\hat{g}(0)$ is defined. We can also call this quantity the static gain of the system. Interpretation *(i)* is no longer valid (the final value of the step response is no longer defined); in *(ii)* the equality $\hat{g}(0) = \mathcal{F}g(0)$ is no longer true, since the Fourier transform $\mathcal{F}g$ is not defined; and, finally, *(iii)* remains valid if the system is stabilized by feedback.

3.2. First-order systems

3.2.1. *Transfer function*

The transfer function of a first-order system, assumed to be strictly proper, is of the form

$$G(s) = \frac{b}{s-a}. \tag{3.2}$$

As a result, this system has a unique pole $s = a$; and therefore, it is stable if and only if $a < 0$. Let us consider such a case (the case of an unstable system will be treated at section 3.2.5).

Let $k = -\frac{b}{a}$ and $\tau = -\frac{1}{a}$. Then, the system transfer function is written as

$$\boxed{G(s) = \frac{k}{1+\tau s}}. \tag{3.3}$$

The quantity $\tau > 0$, which represents a period of time, is called the *time constant* of the system.

It is immediately clear that k is the static gain.

3.2.2. *Time domain responses*

Impulse response

The impulse response g of this system is the inverse Laplace transform of $G(s)$, i.e.

$$g(t) = be^{a\,t}\,\mathbf{1}(t) = \frac{k}{\tau}e^{-\frac{t}{\tau}}\,\mathbf{1}(t).$$

In particular, $g(0^+) = \frac{k}{\tau}$.

The graph of this function is represented in Figure 3.1 (solid line) where $k = 1$ (unit static gain) and $\tau = 10\,s$.

Step response

The step response is the function h defined by

$$h(t) = \int_0^t g(\tau)\,d\tau = \frac{b}{a}\left(e^{a\,t} - 1\right)\mathbf{1}(t) = k\left(e^{-\frac{t}{\tau}} - 1\right)\mathbf{1}(t). \tag{3.4}$$

The graph of this function is represented in Figure 3.2 (solid line) for $k = 1$ (unit static gain) and $\tau = 10\,s$.

Note that the step response is continuous at the origin (which is due to the fact that the transfer function is proper) contrary to the impulse response (the derivative of the step response). We see in Figures 3.1 and 3.2 how to construct the slope at the origin (dashed lines) from the time constant τ, for the impulse response and the step response, respectively.

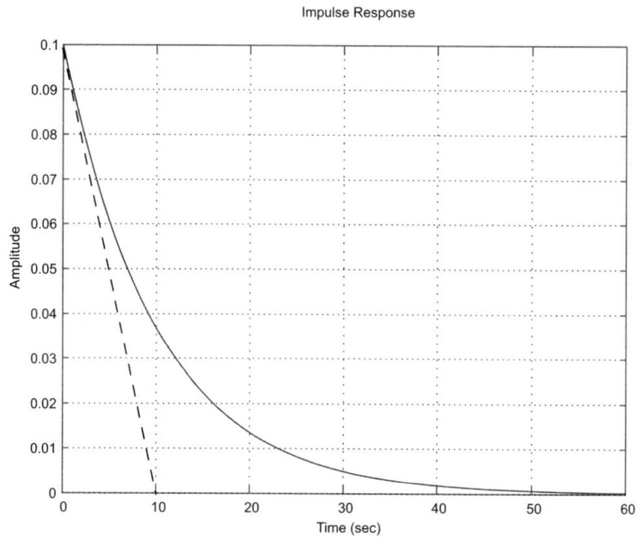

Figure 3.1. *Impulse response of a first-order system*

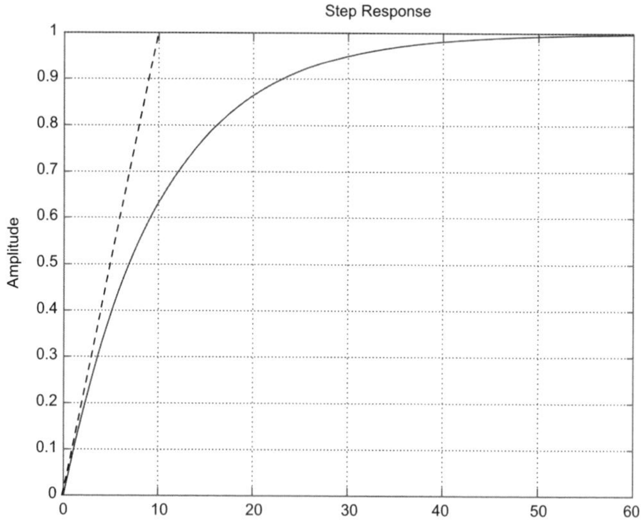

Figure 3.2. *Step response of a first-order system*

3.2.3. *Frequency response*

The frequency response is the (complex-valued) function defined by

$$\mathcal{F}g(\omega) = G(i\omega) = \frac{k}{1 + i\tau\omega}.$$

There are several ways to represent this frequency response:

– Amplitude and phase as a function of the frequency: this is the *Bode plot* representation.

– Curves that trace the evolution of the complex number $\mathcal{F}g(\omega) = G(i\omega)$ in the complex plane as a function of the parameter ω : this is the *Nyquist plot* representation.

– Curves that trace the evolution of the coordinates

$$(|\mathcal{F}g(\omega)|, \arg \mathcal{F}g(\omega))$$

as a function of the parameter ω : this is the *Black plot*.

Nyquist and Black plots are only interesting for the study of closed-loop systems, which is done later. At this moment, we will study the frequency response using Bode plots only.

3.2.4. *Bode plot*

Magnitude

We have, assuming $k > 0$,

$$|G(i\omega)| = \frac{k}{\sqrt{1 + \tau^2\omega^2}}. \tag{3.5}$$

This number, expressed in decibels, is called the *magnitude* of $G(i\omega)$. We have by definition[2]

$$\boxed{|G(i\omega)|_{dB} = 20 \log |G(i\omega)|}. \tag{3.6}$$

On the other hand, we will only consider positive frequencies, without generating any loss of information since

$$\mathcal{F}g(-\omega) = \overline{\mathcal{F}g(\omega)}$$

and we will plot these frequencies on a logarithmic scale.

2. Throughout this text, we denote by log the decimal logarithm, whereas the natural logarithm is denoted by ln.

For $0 < \tau\omega \ll 1$:

$$|G(i\omega)|_{dB} \cong 20 \log k \tag{3.7}$$

(with an error that tends to 0 while $\omega \to 0^+$).

For $\tau\omega \gg 1$:

$$\begin{aligned} |G(i\omega)|_{dB} &= 20 \log k - 10\ |G(i\omega)|_{dB} = 20\ \log k - 20\ \log\left(1 + \tau^2\omega^2\right) \\ &= 20 \log k - 20 \log(\tau\,\omega) - 10\ \log\left(1 + \frac{1}{\tau^2\omega^2}\right) \\ &\cong 20 \log k - 20\ \log(\tau\omega) \end{aligned} \tag{3.8}$$

(with an error that tends to 0 while $\omega \to +\infty$).

This last expression is a line that has a slope of -20 dB per decade (the frequency changes by a *decade* when it is multiplied by 10, an *octave* when multiplied by 2; a rate of change of -20 dB per decade corresponds to approximately a rate of change of -6 dB per octave).

As a result, the Bode magnitude plot has two asymptotes:

– the horizontal asymptote (3.7), for low frequencies;

– the oblique asymptote (3.8), for high frequencies.

These two asymptotes intersect at the "corner point" at abscissa $\omega = \frac{1}{\tau}$, and form what we call the "asymptotic Bode magnitude curve".

For $\omega = \frac{1}{\tau}$,

$$|G(i\omega)|_{dB} = 20 \log k - 20 \log 2 \simeq 20 \log k - 3.$$

Thus, at the "corner frequency" point, the Bode magnitude curve is 3 dB below its asymptotic curve.

Phase

On the other hand, we have

$$\arg G(i\omega) = -\arctan(\tau\omega)\,. \tag{3.9}$$

This argument, expressed in degrees, is called the *phase* of $G(i\omega)$. It is equally traced as a function of ω, this last variable is represented in a logarithmic scale, and we call this graph the "Bode phase plot".

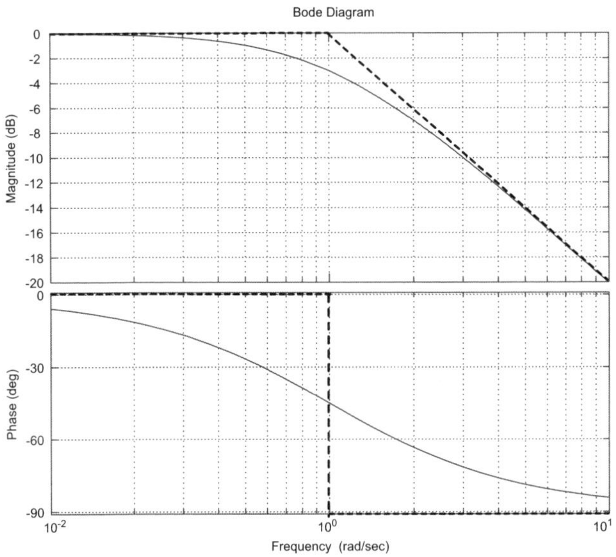

Figure 3.3. *Bode plot of a first-order system*

It is clear that

$$\begin{aligned} \lim_{\omega \to 0} \arg G(i\omega) &= 0^\circ \\ \lim_{\omega \to +\infty} \arg G(i\omega) &= -90^\circ. \end{aligned}$$

The Bode phase plot has two horizontal asymptotes, at 0° for low frequencies and at -90° for high frequencies. At the corner frequency as defined before, we have

$$\arg G\left(\frac{i}{\tau}\right) = -\arctan 1 = -45^\circ$$

Bode plot

The set of the two Bode curves, that of the magnitude and the phase, form the "Bode plot", on which we often add the "asymptotic diagram", as is shown in Figure 3.3 (corresponding to $k = 1$ and $\tau = 10$); the Bode plot is a thin rigid body line while the asymptotic plot is a thicker dashed line.

3.2.5. *Case of an unstable first-order system*

A first-order unstable system has a transfer function of the form (3.2) with $a \geq 0$.

Integrator system

First, consider the case where $a = 0$, i.e. $G(s) = \frac{b}{s}$. Assuming that $b > 0$ (the case $b = 0$ is trivial and the case $b < 0$ is only different by a change of sign), we can put $b = \frac{1}{\tau}$, $\tau > 0$. Therefore,

$$G(s) = \frac{1}{\tau s}.$$

Consequently:

- $|G(i\omega)| = \frac{1}{\tau\omega}$, from which we have for $\omega > 0$: $20 \log |G(i\omega)| = -20 \log(\tau\omega)$.

The Bode magnitude plot is therefore a straight line, with a slope of -20 dB per decade. We have $|G(i\omega)| = 1$ (which is 0 dB) for $\omega = \frac{1}{\tau}$.

- $\arg G(i\omega) = -90°$.

System with a positive pole

Consider now the case where $a > 0$. We can write $G(s)$ in the form

$$G(s) = \frac{k}{1 - \tau s}.$$

As a result, $G(i\omega) = \frac{k}{1-\tau i\omega}$ and so

$$|G(i\omega)| = \frac{k}{\sqrt{1 + \tau^2\omega^2}},$$

which is identical to (3.5): *if the pole is changed to its opposite, the amplitude remains unchanged.*

On the other hand,

$$\arg G(i\omega) = \arctan(\tau\omega),$$

an expression that is of opposite sign to (3.9): *if the pole is changed to its opposite, the phase is changed in the same way.*

3.3. Second-order systems

3.3.1. *Transfer function*

Conventions

The transfer function of a second-order system, assumed to be strictly proper, is of the form

$$G(s) = \frac{b_1 s + b_2}{s^2 + a_1 s + a_2}.$$

If $b_1 \neq 0$, this system has a zero (of transmission) $z = -\frac{b_2}{b_1}$; if $b_1 = 0$, this system has no zero.

In what follows in this section, we will only be interested in a system without zero $(b_1 = 0)$; see section 3.4 for a study of the general case.

Poles

The discriminant of the denominator is $\Delta = a_1^2 - 4a_2$.

– If $\Delta \geq 0$, $G(s)$ has two real poles (indistinct for $\Delta = 0$) and is thus the product of two first-order transfer functions. We are therefore led to the same study as before (see section 3.4 for more details).

– If $\Delta < 0$, we have necessarily $a_2 > 0$, and we can put

$$a_2 = \omega_0^2, \quad \omega_0 > 0.$$

To clarify the calculations, we can further put

$$a_1 = 2\varsigma\omega_0, \quad b_2 = k\omega_0^2.$$

With these conventions, the system transfer function can be written as

$$\boxed{G(s) = k\frac{\omega_0^2}{s^2 + 2\varsigma\,\omega_0\, s + \omega_0^2}}.$$

We have $k = G(0)$, thus k is the static gain.

The reduced discriminant of the denominator is $\Delta' = \omega_0^2\left(\varsigma^2 - 1\right)$. Since $\Delta' < 0$, we have $|\varsigma| < 1$ and the poles are

$$\left\{-\varsigma\,\omega_0 - i\omega_0\sqrt{1-\varsigma^2}, -\varsigma\,\omega_0 + i\omega_0\sqrt{1-\varsigma^2}\right\}.$$

The system is stable if and only if these two complex conjugate poles have negative real parts, i.e. if $\varsigma > 0$. These two poles then lie in the complex plane as shown in Figure 3.4.

The quantity ω_0 is the absolute value of both poles considered, i.e. the distance in the complex plane between any of these poles and the origin. We call this the *undamped natural frequency*.

The angle Ψ in Figure 3.4 is related to ς by the relation $\varsigma = \sin\Psi$. This ς term is called *the damping coefficient*.

3.3.2. *Time domain responses*

The two typical time domain responses (impulse and step) are now looked at in the case of unity static gain $(k = 1)$.

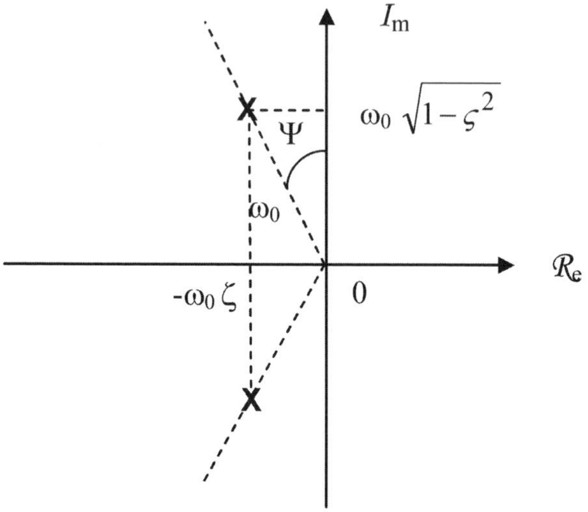

Figure 3.4. *Complex conjugate poles*

Impulse response

The impulse response g is given by

$$\begin{aligned} g(t) &= \mathcal{L}^{-1} G(s) = \mathcal{L}^{-1} \left\{ \frac{\omega_0^2}{s^2 + 2\varsigma\,\omega_0\, s + \omega_0^2} \right\} \\ &= \frac{\omega_0}{\sqrt{1-\varsigma^2}}\, e^{-\varsigma\,\omega_0\, t} \sin\left(\omega_0 \sqrt{1-\varsigma^2}\, t\right) \mathbf{1}(t). \end{aligned} \tag{3.10}$$

This response oscillates at the *natural frequency*

$$\boxed{\omega_p = \omega_0 \sqrt{1-\varsigma^2}}\,.$$

The impulse response (3.10) is plotted in Figure 3.5 for $\omega_0 = 10$ rad/s and for different positive values of the damping coefficient ς : 0.1 ('-'), 0.3 ('- -'), 0.5 ('-.'), and 0.7 (':').[3] We observe that the smaller the value of the damping coefficient, the more oscillatory the impulse response. For $\varsigma = 0$, the impulse response becomes purely sinusoidal with frequency ω_0 (from which comes the name of this oscillation).

3. That is rigid body, dashed, dash-dotted, dotted lines, respectively.

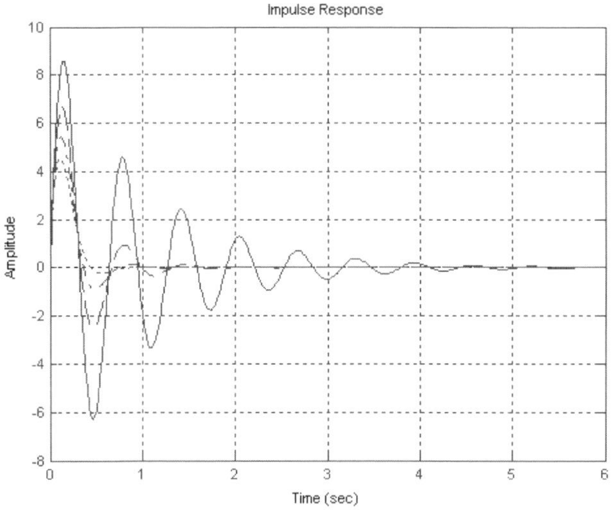

Figure 3.5. *Impulse response of a second-order system*

Step response

The step response is the function h defined by

$$\begin{aligned} h(t) &= \int_0^t g(\tau)\, d\tau \\ &= \left[1 - e^{-\varsigma\omega_0 t}\left(\cos\omega_p\, t + \frac{\varsigma}{\sqrt{1-\varsigma^2}}\sin\omega_p\, t\right)\right] \mathbb{1}(t)\,. \end{aligned} \tag{3.11}$$

This step response is plotted in Figure 3.6 for $\omega_0 = 10$ rad/s and for $\varsigma = 0.1$, 0.3, 0.5, and 0.7 (with the same conventions). The main difference with the impulse response is the final value: 1 (because the static gain of $G(s)$ is 1) instead of 0 (of course, this final value is only defined for $\varsigma > 0$, i.e. when the system is stable).

The maximum of the step response h is obtained at the first time that its derivative g is zero, that is $t = \frac{\pi}{\omega_p}$. This maximum is therefore

$$h\left(\frac{\pi}{\omega_p}\right) = 1 + e^{-\varsigma\frac{\pi}{\sqrt{1-\varsigma^2}}} \triangleq M(\varsigma)\,.$$

The graph of the function $\varsigma \mapsto M(\varsigma)$ is represented in Figure 3.7 (for $\varsigma \in [0, 1)$). We observe that the overshoot is zero for $\varsigma = 1$, is of the order of 5% for $\varsigma = 0.7$, and of the order of 10% for $\varsigma = 0.6$; it tends to 100% when ς tends to 0.

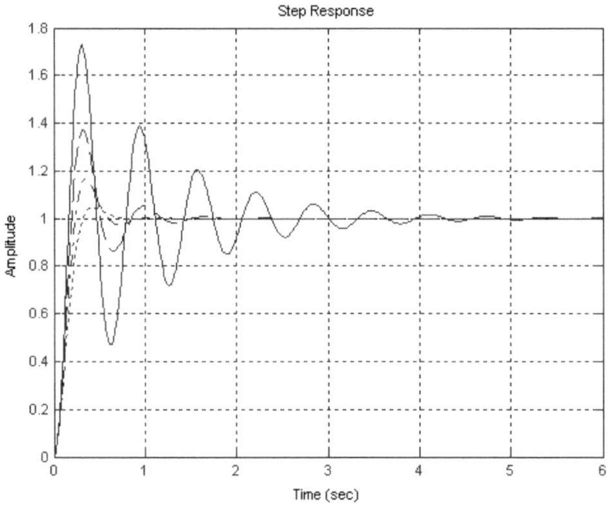

Figure 3.6. *Step response of a second-order system*

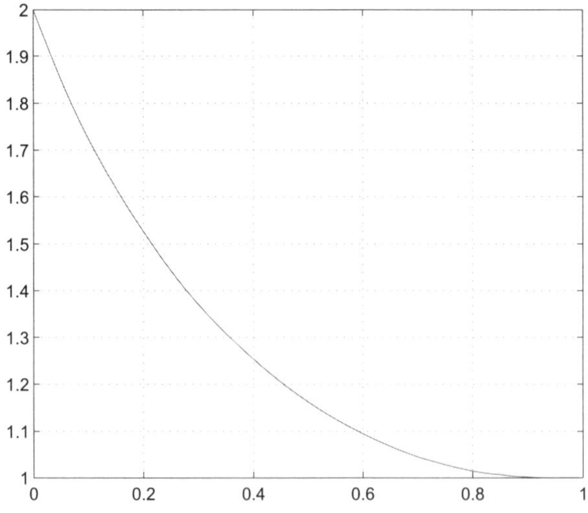

Figure 3.7. *Overshoot as a function of the damping coefficient*

3.3.3. *Bode plot*

The frequency response of the system is given by

$$G(i\omega) = \frac{\omega_0^2}{\omega_0^2 - \omega^2 + 2\,i\,\varsigma\,\omega_0\,\omega} = \frac{1}{1 - \upsilon^2 + 2\,i\,\varsigma\,\upsilon}$$

where $\upsilon \triangleq \frac{\omega}{\omega_0}$ is the *normalized frequency*.

Study of the amplitude

The square of the amplitude of $G(i\omega)$ is

$$|G(i\omega)|^2 = \frac{1}{\left(1 - \upsilon^2\right)^2 + 4\,\varsigma^2 \upsilon^2}.$$

This quantity is independent of the sign of the damping coefficient ς; without loss of generality, let $\varsigma > 0$ (case of a stable system).

This value becomes a maximum when its denominator D is minimum. On the other hand, this is a function of only $\varpi = \upsilon^2$. We have

$$\frac{d\,D}{d\,\varpi} = 2\,\left(\upsilon^2 - 1\right) + 4\,\varsigma^2 = 2\left[\upsilon^2 - \left(1 - 2\,\varsigma^2\right)\right]. \tag{3.12}$$

Therefore

i) For $\varsigma > \frac{\sqrt{2}}{2}$:

The derivative (3.12) will not be zero, and so the function $\omega \mapsto |G(i\,\omega)|$ does not have a maximum on $(0, +\infty)$. We say that the system does not have a *resonance*.

ii) For $0 < \varsigma < \frac{\sqrt{2}}{2}$:

The derivative (3.12) becomes zero for $\upsilon = \sqrt{1 - 2\,\varsigma^2}$, i.e. for $\omega = \omega_r$ with

$$\boxed{\omega_r \triangleq \omega_0\,\sqrt{1 - 2\,\varsigma^2}}.$$

This frequency ω_r is that at which $|G(i\,\omega)|$ attains its maximum on $(0, +\infty)$: it is the *resonance frequency*.

We can easily verify that

$$\boxed{|G(i\omega_r)| = \frac{1}{2\varsigma\,\sqrt{1-\varsigma^2}}}.$$

This quantity is the ratio between the maximum value of $|G(i\omega)|$ and the static gain (which is 1 in this case). This value is called the *resonance factor* and is denoted by λ. It is sometimes expressed in decibels and denoted by Q : $Q = 20\,\log\lambda$.

Note that for the small values of ς, we have $\omega_r \simeq \omega_0$ and $\lambda \simeq \frac{1}{2\,\varsigma}$.

As a result, when the damping coefficient ς tends to 0, the resonance factor tends to $+\infty$. In addition, the resonance frequency ω_r as well as the natural frequency ω_p tend towards the undamped natural frequency ω_0.

Let $u(t) = A\cos(\omega t)$, a sinusoidal signal of amplitude A, be the input of the system. As shown in section 2.5.4, the output is a sinusoidal signal of amplitude $A\,|G(i\,\omega)|$. This amplitude is maximum when $\omega = \omega_r$, and this maximum amplitude tends to $+\infty$ when ς tends to 0. This shows that the system becomes unstable when its damping coefficient tends to 0, since we can find a bounded input to which corresponds an unbounded output (which contradicts Property *v*) of Theorem 589, given in section 13.6.1.

iii) For $\varsigma = \frac{\sqrt{2}}{2}$:

This is the value that makes the transition between the absence or presence of resonance. This number $\left(\frac{\sqrt{2}}{2} \simeq 0.7\right)$ is called the *critical value* of the damping coefficient (or of the damping).

Recall that according to (3.12)

$$\frac{dD}{d\varpi}(0) = -2\left(1 - 2\varsigma^2\right)\omega_0^2.$$

This quantity is zero if and only if the system is critically damped. It is obvious that it is also true for $\frac{d\,|G|}{d\,\omega}(0)$. The critical damping therefore happens when the graph of the amplitude, $|G|$ as a function of the frequency ω, has a a horizontal tangent at $\omega = 0$ (the amplitude and the frequency are both plotted in a linear – and not logarithmic – scale in this case).

Study of the phase

We have

$$\arg G(i\,\omega) \triangleq \varphi(\upsilon) = \begin{cases} -\arctan\frac{2\,\varsigma}{1-\upsilon^2} & \text{if } \upsilon < 1 \Leftrightarrow \omega < \omega_0, \\ -\arctan\frac{2\,\varsigma}{1-\upsilon^2} - \pi & \text{if } \upsilon > 1 \Leftrightarrow \omega > \omega_0. \end{cases}$$

We see that if we change ς to $-\varsigma$, $\varphi(\upsilon)$ is changed to $-\varphi(\upsilon)$.

Case of a stable system

Assume that $\varsigma > 0$. We have in particular

– For $\upsilon \to 0$, $\varphi(\upsilon) \to 0$.

– For $\upsilon \to +\infty$, $\varphi(\upsilon) \to -\pi$.

– For $\upsilon \to 1$, $\varphi(\upsilon) \to -\frac{\pi}{2}$, and this function $\varphi(\upsilon)$ is continuous at $\upsilon = 1$.

On the other hand,

$$\frac{d\varphi}{dv} = -\frac{1}{1+\left(\frac{2\varsigma}{1-v^2}\right)^2}\,(2\varsigma)\,\frac{2v}{(1-v^2)^2}$$

$$= -\frac{4\varsigma v}{(1-v^2)^2+4\varsigma^2}$$

therefore

$$\boxed{\frac{d\varphi}{dv}(1) = -\frac{1}{\varsigma}}.$$

The variation of the phase in the vicinity of $v = 1$ is therefore all the more rapid as the damping coefficient is weakened.

Case of an unstable system

As already mentioned, a sign change in the damping coefficient implies a sign change in phase. The above study remains valid, apart from a difference in sign.

Bode plot in logarithmic scale

As for the first-order system already studied, we can now plot the amplitude and the phase as a function of frequency (to provide a more intrinsic character to these curves, we have chosen to use the normalized frequency on the abscissa); the amplitude is expressed in decibels, the phase is expressed in degrees, and a logarithmic scale is used for the frequency. This representation makes apparent the Bode asymptotic curves:

– For $v \to 0$, $|G(i\omega)| \to 1$ (i.e. 0 dB).

– For $v \to +\infty$, $|G(i\omega)| \sim \frac{1}{v^2}$, and therefore

$$20\,\log|G(i\omega)| = -40\,\log v + \text{(terms that tend to 0 when } v \to +\infty).$$

The asymptotic diagram of the magnitude thus consists of two line segments:

– A horizontal line segment, at 0 dB (or, more generally, the value of the static gain expressed in decibels).

– A line segment of slope -40 dB per decade.

These two line segments intersect at the abscissa point $v = 1$ (i.e. $\omega = \omega_0$).

The asymptotic diagram of the phase follows immediately from the study of the phase made above.

We finally obtain the Bode plots in Figure 3.8, corresponding to damping ratios $0.1, 0.3, 0.5$ and 0.7, with the same conventions as above. The asymptotic diagram is also plotted.

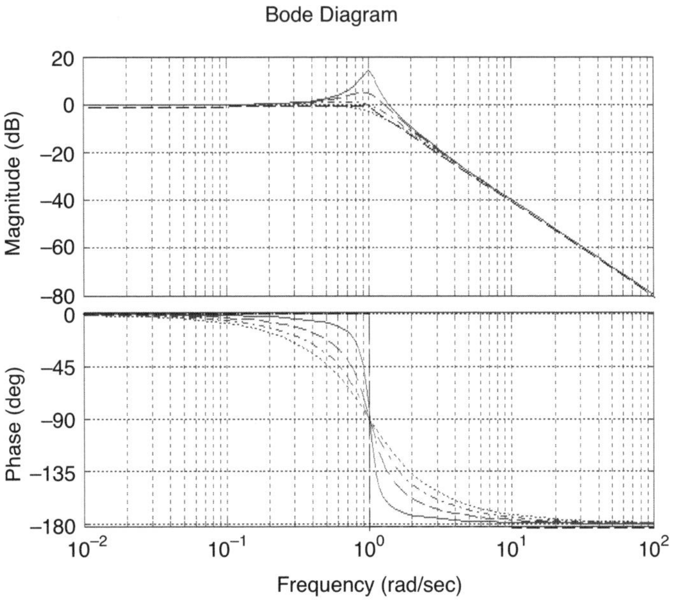

Figure 3.8. *Bode plot of a second-order system*

3.4. Systems of any order

3.4.1. *Stability*

Consider the system of order $n \geq 1$ represented by the left form (2.17). Its transfer function $G(s) = N(s)/D(s)$ is proper if and only if (i): $d^{\circ}(N) \leq d^{\circ}(D)$ (see section 13.6.1). This system is minimal if and only if (ii): the polynomials $N(s)$ and $D(s)$ are coprime (according to Definition 33 of section 2.4.6). Suppose Conditions (i) and (ii) are satisfied. Therefore, according to Definition 61 (section 3.1.1), the system considered is stable if and only if Condition (iii), given below, is satisfied:

(iii) All roots of the polynomial $D(s)$ belong to left half-plane.

DEFINITION 62.– *A polynomial $D(s) \in \mathbb{R}[s]$ is said to be* Hurwitz *if Condition (iii) above is satisfied.*

Write

$$D(s) = \sum_{j=0}^{n} a_j s^{n-j}, \quad a_0 \neq 0. \tag{3.13}$$

PROPOSITION 63.– *For a polynomial $D(s)$ to be Hurwitz, it is necessary that all the coefficients a_j $(0 \leq j \leq n)$ be of the same sign.*

PROOF. The polynomial $D(s)$ can be factorized into the following form

$$\begin{aligned} D(s) &= a_0 \prod_k (s+\alpha_k) \prod_l (s+\beta_l+i\gamma_l)(s+\beta_l-i\gamma_l) \\ &= a_0 \prod_k (s+\alpha_k) \prod_l \left(s^2+\gamma_l^2+2\beta_l\right) \end{aligned}$$

where $\alpha_k > 0$, $\beta_l > 0$ if $D(s)$ is Hurwitz. In developing the above expression, we obtain (3.13) with all the coefficients a_j $(0 \leq j \leq n)$ of the same sign as a_0. ■

REMARK 64.– *(i) The condition in Proposition 63 is* necessary, but not sufficient, *for the polynomial $D(s)$ to be Hurwitz, except if $n \leq 2$. In the general case, a necessary and sufficient condition is provided by the* Routh criterion, *which we will not elaborate (see e.g. [26]). (ii) It is equally possible, thanks to an extension of the Routh criterion, to determine the number of roots of $D(s)$ that belong to the right half-plane [6]. (iii) At last, suppose the coefficients a_j of $D(s)$ are uncertain in the sense that each of them satisfies the bounds $\underline{a_j} \leq a_j \leq \overline{a_j}$, and only the lower and upper bounds, $\underline{a_j}$ and $\overline{a_j}$, are known. Let $\mathcal{D}$ be the set of polynomials $D(s)$ whose coefficients a_j $(0 \leq j \leq n)$ satisfy the above conditions. A criterion owed to Kharitonov provides a necessary and sufficient condition, depending only on the bounds $\underline{a_j}$ and $\overline{a_j}$ $(0 \leq j \leq n)$, for all polynomials belonging to $\mathcal{D}$ to be Hurwitz (see e.g. [7]). When $\underline{a_j} = \overline{a_j}$ $(0 \leq j \leq n)$, the* Kharitonov criterion *reduces to the Routh criterion.*

3.4.2. *Decomposition of the transfer function*

The transfer function $G(s)$ of a system of order $n \geq 1$ is of the form

$$G(s) = \frac{\prod_k N_k(s)}{\prod_k D_k(s)}$$

where $\frac{1}{N_k(s)}$ and $\frac{1}{D_k(s)}$ are transfer functions of the first order, or of the second order with two complex conjugate poles. The study of these "elementary transfer functions" has been done above.

We have

$$\log|G(i\omega)| = \sum_k \log\left|\frac{1}{D_k(i\omega)}\right| - \sum_k \log\left|\frac{1}{N_k(i\omega)}\right|, \tag{3.14}$$

$$\arg G(i\omega) = \sum_k \arg\left(\frac{1}{D_k(i\omega)}\right) - \sum_k \arg\left(\frac{1}{N_k(i\omega)}\right).$$

Therefore, we easily obtain the curves of $G(s)$ (amplitude and phase) from the Bode plots of the elementary transfer functions $\frac{1}{N_k(s)}$ and $\frac{1}{D_k(s)}$.

REMARK 65.– *From* (3.14), *it is easy to show that the Bode magnitude asymptotic diagram has a slope of* $-20\,\delta\,(G)$ *dB/decade in the high frequencies, where* $\delta\,(G)$ *is the relative degree of the transfer function* $G(s)$ *(see section 13.6.1).*

3.4.3. *Asymptotic Bode plot*

Construction of asymptotic plot

The asymptotic Bode plot of $G(s)$ is particularly simple to construct. The method is as follows :

i) Determine the abscissae of the "corner frequencies" of the asymptotic diagram, which are the absolute values of the poles and zeros. Place them on the axis of the abscissae.

ii) For each of the corner frequencies, apply the following rule to the variation of slope of the magnitude and the variation of phase, a rule which follows from previous results:

	Stable pole	Unstable pole	Stable zero	Unstable zero
Δ (slope)	-20 dB/dec.	-20 dB/dec.	20 dB/dec.	20 dB/dec.
Δ (phase)	$-90°$	$90°$	$90°$	$-90°$

Table of the rule of corner frequencies

We call a stable (resp., unstable) pole a pole which has negative (resp., non-negative) real part, and similarly for a zero.

Example 1

Let

$$G\,(s) = \frac{100\,(s-1)}{s\,(s+10)}.$$

– The corner frequencies are: $\omega = 0$ (corresponding to the pole $s = 0$), $\omega = 1$ (corresponding to the unstable zero $s = 1$) and $\omega = 10$ (corresponding to the stable pole $s = -10$).

– Construct the asymptotic Bode plot at low frequencies: for $\omega \rightarrow 0$, $G\,(i\omega) \sim \frac{-10}{i\omega}$. Amplitude: slope -20 dB/decade passing through the point $\left(\omega = 1, |G|_{\text{dB}} = 20\right)$. Phase: $90°$ (modulo $360°$).

– The rest of the diagram can be obtained by applying the "corner frequency" rule from the above table.

The asymptotic plot obtained is represented in Figure 3.9 (dashed lines), along with the "true" Bode plot (solid lines). The phase varies from $90°$ to $-90°$.

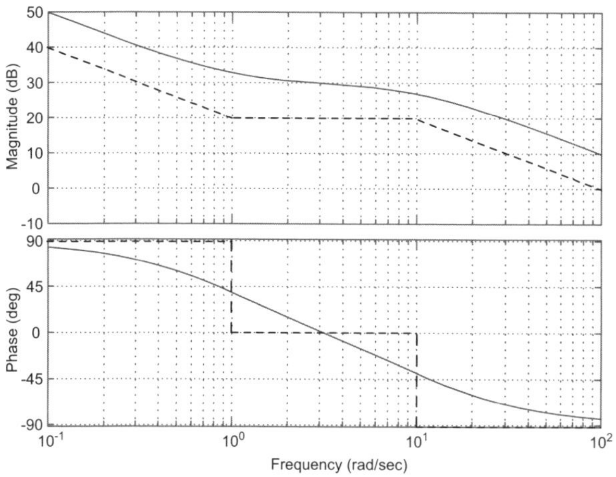

Figure 3.9. *Bode plot – Example 1*

Example 2

Let

$$G(s) = \frac{s(s+10)}{(s^2 - s + 100)(s+1)}.$$

– The corner points are $\omega = 0$ (corresponding to the zero $s = 0$), $\omega = 1$ (corresponding to the stable pole $s = -1$), and $\omega = 10$ (corresponding to the stable zero $s = -10$ as well as to two unstable complex conjugate poles with absolute value 10).

– Construct the asymptotic plot in the low frequencies: for $\omega \to 0$, $G(i\omega) \sim \frac{i\omega}{10}$. Amplitude: slope of 20 dB/decade passing through the point $\left(\omega = 1, |G|_{\text{dB}} = -20\right)$. Phase: $90°$.

– The corner frequency at $\omega = 1$ poses no problem.

– At the corner point $\omega = 10$, we must account for:

- the effect of the zero, which induces a variation of slope of magnitude of $+20$ dB/decade and a variation of phase of $+90°$;

- the effect of the two complex conjugate poles, which induces a variation of slope of magnitude of -40 dB/decade and a variation of phase of $+180°$;

- i.e. in total: a variation of slope of magnitude of -20 dB/decade and a variation of phase of $+270°$.

The asymptotic plot obtained is depicted in Figure 3.10, together with the “true” Bode plot (with the same conventions as for Figure 3.9). Note that the true plot separates significantly from the asymptotic plot at the vicinity of resonance.

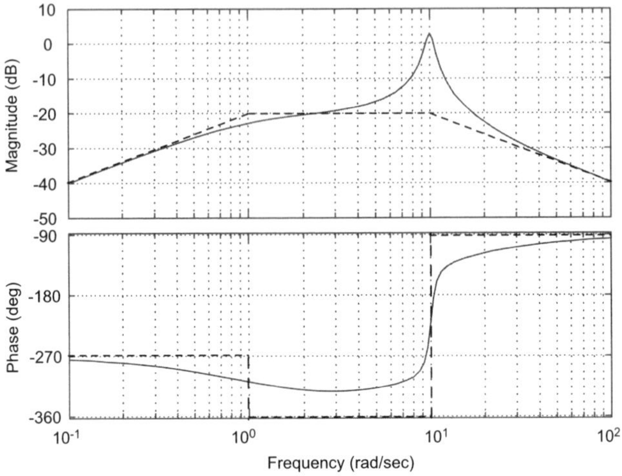

Figure 3.10. *Bode plot – Example 2*

3.4.4. *Amplitude/phase relation*

Amplitude/phase relation in the asymptotic Bode plot

The table of the "corner frequency rule" shows that the complete asymptotic Bode plot of a system is entirely determined by its asymptotic diagram of magnitude if we assume that this system only has stable poles and zeros. Such a system is stable (a property that follows from the stability of its poles); we are now going to identify which property follows from the stability of its zeros.

Minimum phase systems

The opposite of its phase, that is $-\arg G(i\omega)$, is called the *phase shift* of the system (or of its transfer function).

Let z be a complex number with negative real part, and let $G(s)$ and $G^*(s)$ be two stable transfer functions such that

$$G^*(s) = \frac{s+z}{s-z}G(s).$$

We have

$$\left|\frac{i\omega + z}{i\omega - z}\right| = 1$$

and according to the "table of the corner frequency rule", the phase of the transfer function $\frac{s+z}{s-z}$ decreases from $0°$ to $-180°$ when the frequency goes from 0 to $+\infty$.

As a result, the amplitude $A(\omega)$ of $G^*(s)$ is equal to that of $G(s)$ at all frequencies, whereas the phase shift of $G^*(s)$ is higher than that of $G(s)$.

With the same rationale, we immediately establish the following result:

THEOREM 66.– *Given the amplitude $\omega \mapsto A(\omega)$ of the frequency response of a stable transfer function, there exists an infinite number of stable transfer functions having a frequency response with the same amplitude. Among these transfer functions, there exists a unique one $G(s)$ whose phase shift is minimum: all zeros of this have a negative real part. All the others are of the form*

$$G^*(s) = G(s) \prod_k \frac{s + z_k}{s - z_k}$$

where the product is finite and the z_k are complex numbers in the left half-plane (provided that $\bar{z}_k$ appears in the product if z_k does, for the rational function $G^*(s)$ to have real coefficients).

DEFINITION 67.– *A stable system is said to have* minimum phase *if all its transmission zeros belong to the left half-plane (in other words, if its transfer function has a stable inverse), and* non-minimum phase *otherwise.*

REMARK 68.– *By extension, a minimal control system, possibly unstable and/or MIMO, is said to have* minimum phase *if all its transmission zeros lie in the left half-plane.*

Bode' amplitude/phase relations

For a minimum phase system, we have seen above that the asymptotic diagram of the phase is completely determined by the asymptotic diagram of the magnitude. We can go a little further, and show that the phase $\arg G(i\omega)$ itself is entirely determined by the magnitude $A(\omega) = \ln |G(i\omega)|$.[4] Indeed, we have the following result: for any frequency $\omega_c \geq 0$,

$$\begin{aligned} \arg G(i\omega_c) - \arg G(0) &= \frac{2}{\pi} \int_0^{+\infty} \frac{A(\omega) - A(0)}{\omega^2 - \omega_c^2} d\omega \\ &= \frac{1}{\pi} \int_{-\infty}^{+\infty} \frac{dA}{d\nu} \left[\ln \coth \frac{|\nu|}{2} \right] d\nu \end{aligned} \quad (3.15)$$

where $\nu = \ln\left(\frac{\omega}{\omega_c}\right)$. (For a complete proof, see [9], Chapter 14, or [39], section 7.2).

4. For convenience, we consider $\ln |G(i\omega)|$, the natural logarithm of $|G(i\omega)|$, rather than $20 \log |G(i\omega)|$, the expression of magnitude in decibels.

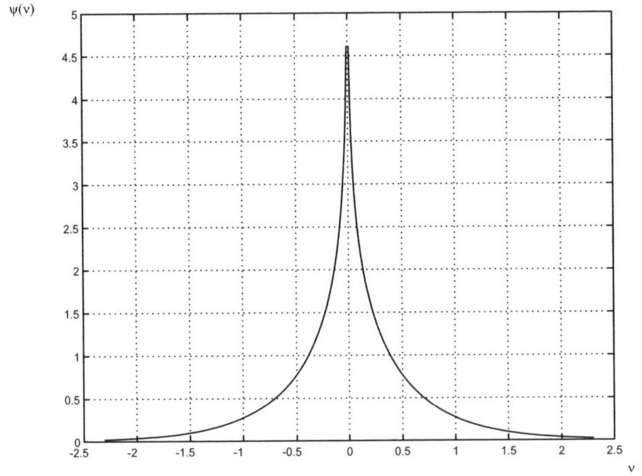

Figure 3.11. *Density function psi*

This relation, often called *Bode's amplitude/phase relation*, is also called (more justly) the *Bayard–Bode relation.*

Put

$$\psi(\nu) = \ln \coth \frac{|\nu|}{2}$$

in a way that according to (3.15),

$$\boxed{\arg G(i\omega_c) - \arg G(0) = \tfrac{1}{\pi} \int_{-\infty}^{+\infty} \tfrac{dA}{d\nu} \psi(\nu)\, d\nu}. \tag{3.16}$$

The graph of the "density function" ψ is plotted in Figure 3.11.

Note that ψ is a positive even function and that $\psi(\nu)$ tends to $+\infty$ as ν tends to 0. In addition, one can show that

$$\int_{-\infty}^{+\infty} \psi(\nu)\, d\nu = \frac{\pi^2}{2}. \tag{3.17}$$

Approximation of phase from magnitude

Suppose that the slope of the magnitude (i.e. an element of $\mathbb{Z}$ is $-20n$ dB/decade in the vicinity of the frequency ω_c, where n is a rational integer). Assume also that this neighborhood is an interval of k decades centered around ω_c (frequencies are represented in a logarithmic scale). Getting back to variable ν, this means that the slope of magnitude is $-20\,n$ dB/decade for $\nu \in \left[-\frac{k}{2}\ln 10, \frac{k}{2}\ln 10\right]$.

– For $k \to +\infty$, we have exactly, according to (3.16) and (3.17),

$$\arg G(i\omega_c) = \arg G(0) - n\frac{\pi}{2}. \tag{3.18}$$

This is also the value we get by using the "table of corner frequency rule" of the Bode asymptotic diagram.

– For finite k, the value (3.18) is obviously only an approximation of the phase, the precision of which depends not only on k, but also on the slope of magnitude outside the interval $\left[-\frac{k}{2}\log 10, \frac{k}{2}\log 10\right]$ of values of ν. To give an idea of the error that can be made if we use (3.18) to calculate the phase, let

$$I(k) = \frac{2}{\pi^2}\int_{-\frac{k}{2}\ln 10}^{\frac{k}{2}\ln 10} \psi(\nu)\, d\nu.$$

We have

$$I(0,5) = 0.31,\ I(1) = 0.74,\ I(2) = 0.92,\ I(4) = 0.99. \tag{3.19}$$

This just means that when $k = 1$, for example, the error incurred on the phase cannot exceed 25%, but this percentage can constitute an order of magnitude and most often a lower bound. Therefore, if the slope of magnitude is approximately -20 dB/decade over one decade centered around ω_c, the estimated phase is between $-112°$ and $-68°$ at frequency ω_c.

3.5. Time-delay systems

3.5.1. *Left form time-delay systems*

The general theory of time-delay systems is beyond the scope of this book (see e.g. [55] and [58]). We will limit ourselves, here, to giving elements which are among the simplest but still very useful ones.

Consider a linear time-invariant SISO system defined by the left form

$$D(\partial)\, y(t) = N(\partial)\, u(t).$$

Now suppose that the input u is delayed because, for example, of a time lapse between the moment the order of the action is transmitted and when the actuator starts to act (this delay can be the time of information propagation or the time of transport in the medium). The above equation becomes

$$\boxed{D(\partial)\, y(t) = N(\partial)\, u(t-\tau)} \tag{3.20}$$

where $\tau > 0$ denotes the time lapse in question.

Below, we will only study time-delay systems governed by an equation of this form.

3.5.2. *Transfer function*

We know that the delayed input $u_{(\tau)} : t \mapsto u(t-\tau)$ is equal to the convolution product $\delta_{(\tau)} * u$, where $\delta_{(\tau)}$ is the delayed Dirac distribution of the time τ (section 12.2.3). Therefore, according to the exchange theorem and the fact that $\mathcal{L}\left(\delta_{(\tau)}\right)(s) = e^{-\tau s}$, we have

$$\mathcal{L}\left(u_{(\tau)}\right)(s) = e^{-\tau s}\hat{u}(s)$$

where $\hat{u}(s)$ is the Laplace transform of u.

Therefore, the transfer function of system (3.20) is

$$\boxed{G(s) = e^{-\tau s} H(s)} \tag{3.21}$$

where

$$H(s) = \frac{N(s)}{D(s)}.$$

3.5.3. *Bode plot*

Magnitude

We have according to (3.21)

$$G(i\omega) = e^{-i\tau\omega} H(i\omega).$$

As a result,

$$|G(i\omega)| = |H(i\omega)|. \tag{3.22}$$

Phase

On the other hand,

$$\arg G(i\omega) = \arg H(i\omega) - \tau\omega. \tag{3.23}$$

3.5.4. *Example: first-order time-delay system*

The simplest case is a first-order time-delay system, with transfer function

$$G(s) = \frac{k\,e^{-\tau\,s}}{1 + T\,s}. \tag{3.24}$$

In the following, we assume that the static gain $G(0) = k$ is equal to 1.

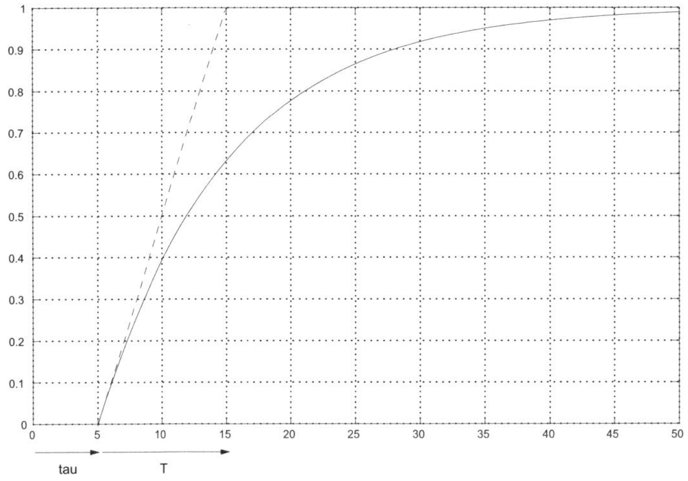

Figure 3.12. *Step response of a first order time-delay system*

Step response

The step response of this system is represented in Figure 3.12 for $T = 10\,\mathrm{s}$ and $\tau = 5$ s. We see that for the first five seconds, which is the duration of the delay, the output remains at zero.

Bode plot

The Bode plot of this system is shown in Figure 3.13.

Note that the magnitude plot is identical to that of the non-delayed system (conforming to (3.22)); on the other hand, the phase plot has a very novel allure in relation to what we have seen so far. When ω tends to $+\infty$, we have indeed, according to (3.23),

$$\arg G(i\omega) = -\frac{\pi}{2} - \tau\omega + (\text{terms tending to } 0).$$

In other words, $\arg G(i\omega)$ decreases linearly with respect to ω, and exponentially with respect to $\log \omega$ (recall that the frequency is in logarithmic scale), from which there is a very rapid decrement of the phase in high frequencies

3.5.5. *Approximations of a time-delay system*

The transfer function (3.21) is not rational and that is what is particular with time-delay systems. It is however still possible to approach this kind of transfer functions in a different manner, by way of rational functions, leading to the study of these

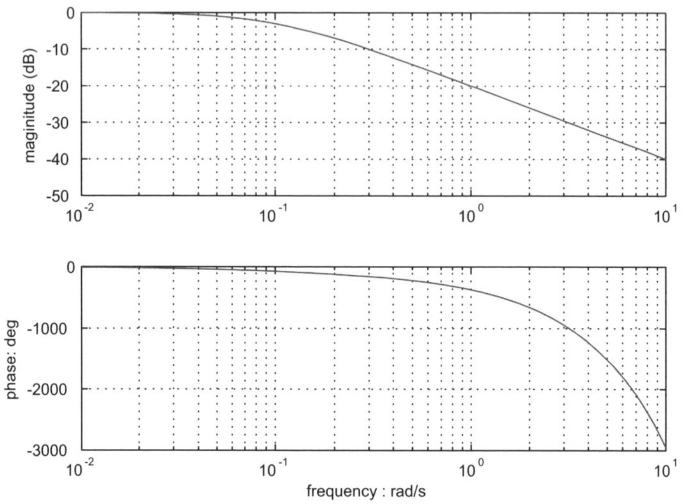

Figure 3.13. *Bode plot of a first-order time-delay system*

time-delay systems in the classic framework of systems governed by ordinary differential equations.

A-approximations

Let $e_n(s)$ $(n \geq 1)$ be the rational function defined by

$$\boxed{e_n(s) = \left(\frac{1-(\tau s)/(2n)}{1+(\tau s)/(2n)}\right)^n}. \tag{3.25}$$

THEOREM 69.– *We have, for every $n \geq 1$ and every ω, $|e_n(i\omega)| = 1$, and*

$$\lim_{n\to+\infty} e_n(s) = e^{-\tau s};$$

this convergence is uniform on every compact set $|s| \leq \sigma$ starting from any rank n such that $n > \frac{\tau\sigma}{2}$.

PROOF. We can write $e_n(s) = e^{n\left(\ln\left(1-\frac{\tau s}{2n}\right)-\ln\left(1+\frac{\tau s}{2n}\right)\right)}$ with

$$n\left(\ln\left(1-\frac{\tau s}{2n}\right)-\ln\left(1+\frac{\tau s}{2n}\right)\right) = n\left[-\frac{\tau s}{n}+\frac{1}{n}\varepsilon\left(\frac{1}{n}\right)\right] = -\tau s+\varepsilon\left(\frac{1}{n}\right)$$

where $\varepsilon\left(\frac{1}{n}\right)$ is a term that tends to 0 when n tends to $+\infty$. As a result, $e_n(s) = e^{-\tau s}e^{-\varepsilon\left(\frac{1}{n}\right)}$, and therefore $e_n(s)$ converges to $e^{-\tau s}$ because $e^{-\varepsilon\left(\frac{1}{n}\right)}$ tends to 1. To prove the uniform convergence, we make use of the first-order Taylor expansion with the Lagrange remainder. Let k be a positive integer; the reader can verify that inside the disc $|s| \leq \frac{k}{\tau}$, we have, for $n > \frac{k}{2}$ the upper bound

$$\left|\frac{e_n(s) - e^{-\tau s}}{e^{-\tau s}}\right| \leq \frac{k^2}{2n}$$

which shows how fast the convergence of the relative error is. Writing $e_n(i\omega) = e^{-i\varphi_n(\omega)}$ and setting $\Delta\varphi_n(\omega) = \omega\tau - \varphi_n(\omega)$ (which is the phase error, while the amplitude error is zero), we easily obtain, from the upper bound stated above, $|\Delta\varphi_n(\omega)| \leq \arcsin\left(\frac{k^2}{2n}\right)$ for $|\omega| \leq \frac{k}{\tau}$. ■

Theorem 69 shows that a time-delay system can be viewed as an "infinite order system". Indeed, the transfer function $e_n(s)$ has n poles equal to $-\frac{2n}{\tau}$. As n tends to infinity, the number of these poles also becomes infinite. They tend to $-\infty$ while remaining on the real axis. As far as the function of the complex variable $s \mapsto e^{-\tau s}$ is concerned, it has no pole in the complex plane (it is an entire function). Infinite-order systems are the subject of extensive research and literature [85].

Example

Consider again the system with transfer function (3.24), with the same values as k, τ, and T considered above. Replace $e^{-\tau s}$ by $e_n(s)$.

The step responses are shown in Figure 3.14 for $n = 1$ (-), 2 (- -) and 3 (-.). These responses are to be compared with the step response of the exact delay system (Figure 3.12). We see that the accuracy of the approximation gets much better with n. The approximation, already satisfied when $n = 1$, is good when $n = 2$ and excellent when $n = 3$.

Padé approximation

General case

The Padé approximations are more classic than the previous ones that we have called "A-approximations". Let

$$\epsilon_n(s) = \frac{\sum_{k=0}^{n} h_k(-\tau s)^k}{\sum_{k=0}^{n} h_k(\tau s)^k} \tag{3.26}$$

where the coefficients h_k are defined by the following recurrent formula:

$$\begin{aligned} h_{k+1} &= \frac{n-k}{(2n-k)(k+1)} h_k,\ 0 \leq k \leq n-1 \\ h(0) &= 1. \end{aligned}$$

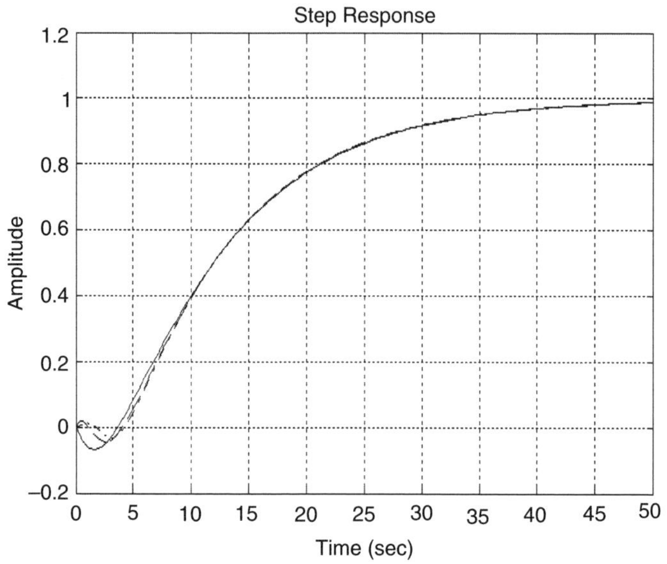

Figure 3.14. *Step responses of A-approximations*

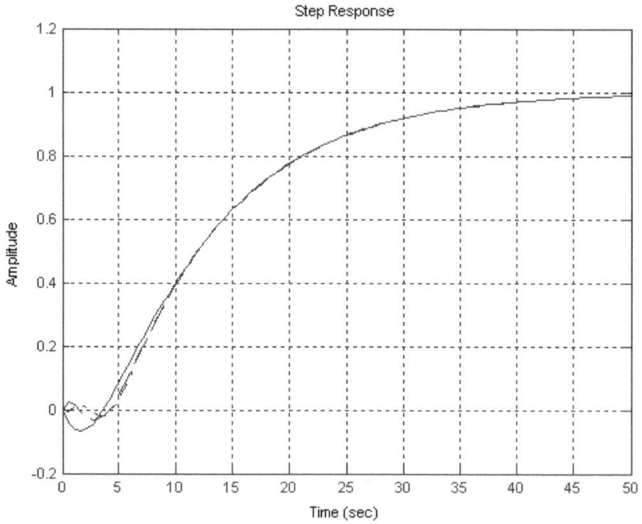

Figure 3.15. *Step responses of Padé approximations*

One can show that the Taylor power series expansions of $\epsilon_n(s)$ and of $e^{-\tau s}$ in s coincide up to the order $2n$. In addition,

$$|\epsilon_n(i\omega)| = 1 = \left|e^{-i\tau\omega}\right|. \tag{3.27}$$

The rational function $\epsilon_n(s)$ is called the Padé estimate of order n of $e^{-\tau s}$. According to (3.27), $\epsilon_n(s)$ is a perfect approximation of $e^{-\tau s}$ as far as the *amplitude* is concerned; the quality of the approximation of the phase depends on the order n (and increases with it, up till the occurrence of numerical errors).

Padé estimations of the first and second order

We have

$$\epsilon_1(s) = \frac{1 - \frac{\tau s}{2}}{1 + \frac{\tau s}{2}}$$

$$\epsilon_2(s) = \frac{1 - \frac{\tau s}{2} + \frac{(\tau s)^2}{12}}{1 + \frac{\tau s}{2} + \frac{(\tau s)^2}{12}}.$$

The reader can show that the Taylor expansion of $\epsilon_1(s)$ and of $\epsilon_2(s)$ coincide with that of $e^{-\tau s}$ up to the orders 2 and 4, respectively.

Example

Consider one more time the system with transfer function (3.24) using the previous values of k, τ, and T. Replace $e^{-\tau s}$ by $\epsilon_n(s)$.

The step responses are shown in Figure 3.15 for $n = 1$, 2, and 3 with the same conventions as above.

A-approximations (3.25) and Padé approximations (3.26) are of a quality that is quasi-equivalent. The advantage of (3.25) lies in the quantification that has been done for the approximation error.

3.6. Exercises

EXERCISE 70.– *We consider the RLC circuit with transfer function (2.32).*

i) Determine the static gain of this system.

ii) Determine the frequency ω_r at which $|G(i\omega)|$ is maximum.

iii) Determine the maximum of $|G(i\omega)|$.

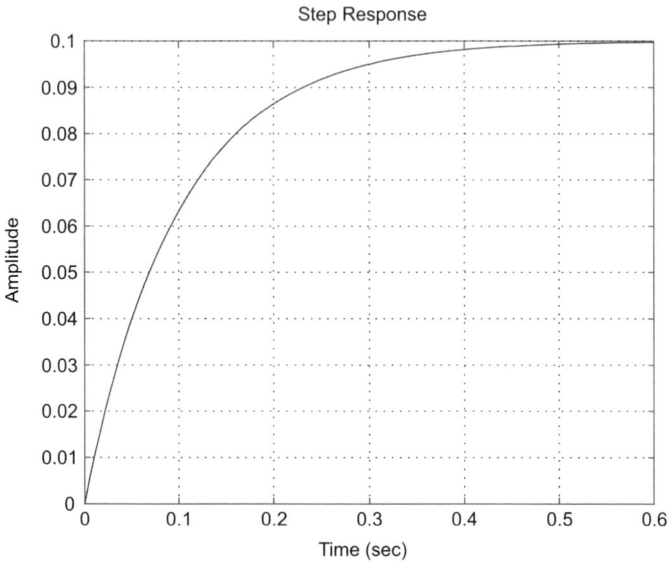

Figure 3.16. *Step response – Question (a)*

EXERCISE 71.– *In this exercise, "identifying a system" is synonymous with "determining its transfer function experimentally". We can identify a system by a graphical method, either from the time domain response (e.g. step response) or from the frequency domain response. (a) Determine the system which has a step response as shown in Figure 3.16. (b) Same question for the step response shown in Figure 3.17. (c) A point by point list (see section 2.5.4) allows us to obtain the frequency response on the Bode plot in Figure 3.18. Identify the corresponding system.*

EXERCISE 72.– "To have a negative start". *We hereby study under what condition the step response of a stable system* Σ *"goes in the wrong direction". That is to say, it heads, during the first moments, toward a direction that is opposite to the final value. We usually refer to such a system as having a "negative start" or an undershoot (and "having a positive start" in the opposite case). Intuition tells us that a system with negative start is hard to control (if the reader imagines a car which, when we turn the steering wheel clockwise, goes for a brief instant toward the left before going toward the right ...). Let* Σ *be a stable system, with transfer function* $G(s)$ *of relative degree* $r > 0$ *(see section 13.6.1). (a) Write down the expression of* $G(s)$ *as a function of its zeros, its poles, and a gain. (b) Let* $y^{(i)}(0^+) = 0$, $0 \leq i \leq r-1$, *and* $y^{(r)}(0^+) \neq 0$. *What is the necessary and sufficient condition involving the ratio* $\rho = \frac{y^{(r)}(0^+)}{G(0)}$ *for the*

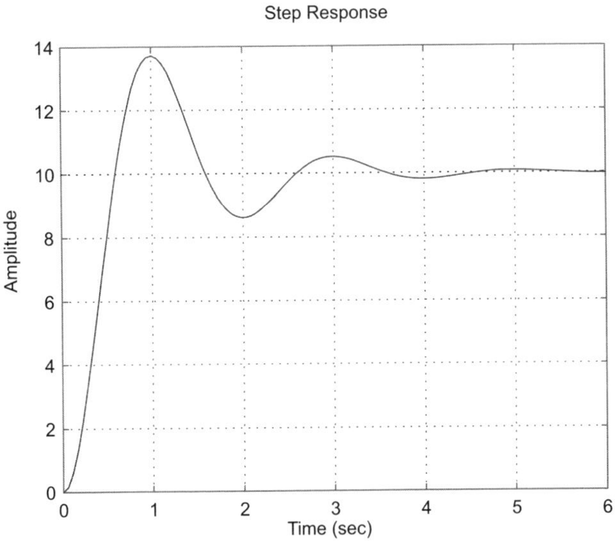

Figure 3.17. *Step response – Question (b)*

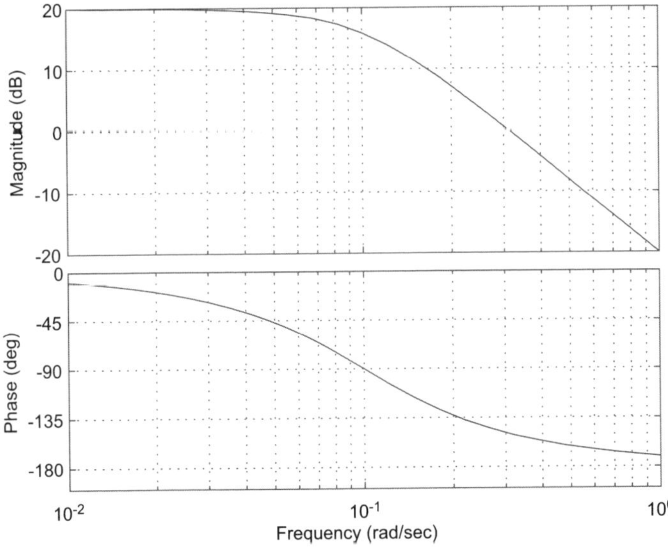

Figure 3.18. *Bode plot – Question (c)*

system to have a negative start? (c) Calculate ρ *as a rational function of the poles and zeros of* $G(s)$. *(d) Show that the denominator of this rational function is positive. Can we say the same thing about its numerator? (e) Show that the system* Σ *has a negative start if and only if the number of its positive real zeros is odd. (f) Is there agreement between this result and A-approximations as well as Padé approximations of the above-studied time-delay system?*

Chapter 4

Closed-Loop Systems

4.1. Closed-loop stability

4.1.1. *Standard feedback system*

We have emphasized in the preface to this book the importance of stability for closed-loop systems. We call such a system as that depicted in Figure 4.1 a "standard feedback system".

The system to be controlled $\mathbf{P}$, with transfer matrix $P(s)$, is fed back through a regulator $\mathbf{K}$ with transfer matrix $K(s)$. Systems $\mathbf{P}$ and $\mathbf{K}$ are assumed to be minimal.[1] Recall that the Laplace transformation of a signal x is denoted by $\hat{x}$.

4.1.2. *Closed-loop equations*

Let $y = \begin{bmatrix} y_1 \\ y_2 \end{bmatrix}$, $u = \begin{bmatrix} u_1 \\ u_2 \end{bmatrix}$, and $v = \begin{bmatrix} v_1 \\ -v_2 \end{bmatrix}$. Also let $M = \begin{bmatrix} P & 0 \\ 0 & K \end{bmatrix}$ and $J = \begin{bmatrix} 0 & -I \\ I & 0 \end{bmatrix}$ (where I is an identity matrix of a suitable size). Imposing zero initial conditions, we obtain

$$\hat{y} = M\,\hat{u}, \quad \hat{u} = \hat{v} + J\,\hat{y} \tag{4.1}$$

1. This hypothesis is missing in numerous works on stabilization of infinite-dimensional systems. The definition of a "minimal system" is indeed, in this case, not a trivial problem. See [14] for more details.

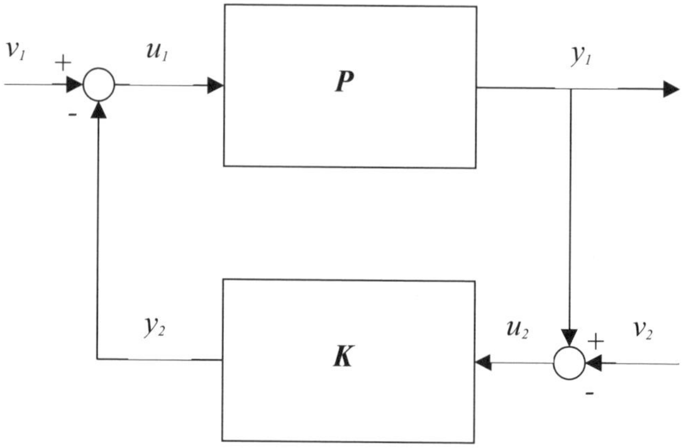

Figure 4.1. *Standard feedback system*

from which

$$[I - M\,J]\,\hat{y} = M\,\hat{v}.$$

Note that $I - M\,J \in \mathbb{R}(s)^{(p+m)\times(p+m)}$, where $p = \dim(y)$ and $m = \dim(u)$ (see section 13.6.1). What is in question now is whether $\hat{y}$ can be expressed in a unique manner as a function of $\hat{v}$; then it will also be the same for $\hat{u}$ according to (4.1). The definition below specifies that given in section 2.6.2.

DEFINITION 73.– *The closed-loop system in Figure 4.1 is* well-posed *if the matrix of rational functions $I - M\,J$ is invertible in the algebra $\mathbb{R}(s)^{(p+m)\times(p+m)}$.*

Recall that $I - J\,M = \begin{bmatrix} I & P \\ -K & I \end{bmatrix}$, thus $\det(I - J\,M) = \det(I + P\,K)$. Therefore, the closed-loop system is well-posed if and only if $\det(I + P\,K)$ is not identically zero.

From this point onwards, we will consider only a well-posed closed-loop system.[2] Let

$$\begin{aligned} L_o &= P\,K, \quad L_i = K\,P, \\ S_o &= (I_p + L_o)^{-1}, \quad T_o = I_p - S_o, \\ S_i &= (I_m + L_i)^{-1}, \quad T_i = I_m - S_i. \end{aligned}$$

2. A detailed analysis of those phenomena which arise in the case of an *ill-posed* closed-loop system is given in [46].

The transfer matrices L_o, S_o, and T_o are called the *open-loop transfer matrix*, the *sensitivity function*, and the *complementary sensitivity function*, respectively, at the output of System $\mathbf{P}$. The transfer matrices L_i, S_i, and T_i are called the *open-loop transfer matrix*, the *sensitivity function*, and the *complementary sensitivity function*, respectively, at the input of System $\mathbf{P}$. We can easily establish the relations

$$\begin{bmatrix} \hat{y}_1 \\ \hat{y}_2 \end{bmatrix} = \begin{bmatrix} S_o P & T_o \\ T_i & -S_i K \end{bmatrix} \begin{bmatrix} \hat{v}_1 \\ \hat{v}_2 \end{bmatrix}, \tag{4.2}$$

$$\begin{bmatrix} \hat{u}_1 \\ \hat{u}_2 \end{bmatrix} = \begin{bmatrix} S_i & S_i K \\ S_o P & -S_o \end{bmatrix} \begin{bmatrix} \hat{v}_1 \\ \hat{v}_2 \end{bmatrix}, \tag{4.3}$$

which are the closed-loop equations.

4.1.3. *Stability of a closed-loop system*

DEFINITION 74.– *The closed-loop system $\mathbf{P}_c$ in Figure 4.1 is* stable *if: (i) it is well-posed, and (ii) the system with input $v = \begin{bmatrix} v_1 \\ v_2 \end{bmatrix}$ and output $y = \begin{bmatrix} y_1 \\ y_2 \end{bmatrix}$ is stable, in the sense of Definition 61 (section 3.1.1).*[3]

REMARK 75.– *According to (4.1), Condition (ii) given above is equivalent to the following : (ii') the system with input $v = \begin{bmatrix} v_1 \\ v_2 \end{bmatrix}$ and output $u = \begin{bmatrix} u_1 \\ u_2 \end{bmatrix}$ is stable (in the sense of Definition 61).*

In what follows in this paragraph, the systems $\mathbf{P}$ and $\mathbf{K}$ are supposed to be SISO. The transfer functions $P(s)$ and $K(s)$ are thus rational functions, written in the form

$$P(s) = \frac{B(s)}{A(s)}, \quad K(s) = \frac{R(s)}{S(s)} \tag{4.4}$$

where $A(s)$, $B(s)$, $R(s)$, $S(s)$ are polynomials with real coefficients, i.e. are elements of the ring $\mathbb{R}[s]$ (see section 13.1.1). The rational functions in (4.4) are assumed to be irreducible. The *open-loop transfer function*

$$L(s) = P(s) K(s) \tag{4.5}$$

is identical at the input and the output of $\mathbf{P}$ ($L_i = L_o = L$).

3. According to some authors, what is defined here is the "internal stability" of a feedback system.

THEOREM 76.– *(i) The feedback system* $\mathbf{P}_c$ *is well-posed if and only if* $A_{cl}(s) = A(s)\,S(s) + B(s)\,R(s) \neq 0$ *in* $\mathbb{R}[s]$. *(ii) The transmission poles of* $\mathbf{P}_c$ *are the roots of* $A_{cl}(s)$. *(iii) If the systems* $\mathbf{P}$ *and* $\mathbf{K}$ *are proper and at least one of them is strictly proper, then the feedback system* $\mathbf{P}_c$ *is well-posed and proper. (iv) Suppose the feedback system* $\mathbf{P}_c$ *is proper; then it is stable if and only if the roots of* $A_{cl}(s)$ *all lie in the left half-plane.*

PROOF. * (i) We have $1 + P\,K = 1 + \frac{B}{A}\frac{R}{S} = \frac{A\,S + B\,R}{A\,S}$. (ii) According to (4.2),

$$P_c = \begin{bmatrix} \frac{B\,S}{A\,S+B\,R} & \frac{B\,R}{A\,S+B\,R} \\ \frac{B\,R}{A\,S+B\,R} & \frac{-A\,R}{A\,S+B\,R} \end{bmatrix} = \frac{1}{AS+BR}\begin{bmatrix} BS & BR \\ BR & -AR \end{bmatrix}.$$

Let $\begin{bmatrix} \alpha_1 & 0 \\ 0 & \alpha_2 \end{bmatrix}$ be the Smith form of the matrix $\begin{bmatrix} BS & BR \\ BR & -AR \end{bmatrix}$. The first invariant factor α_1 is the greatest common divisor of the polynomials $B\,S$, $B\,R$, and $A\,R$ (see section 13.2.3, Proposition 502); it divides $B\,R$ and $A\,R$ and so it divides R since B and A are coprime; it divides $B\,S$ and $B\,R$ and so it divides B since S and R are coprime. This invariant factor α_1 is therefore prime with $A\,S + B\,R$ (since it divides $B\,R$ and is prime with $A\,S$). Moreover, there exist polynomials B_1 and R_1 such that $B = B_1\,\alpha_1$ and $R = R_1\,\alpha_1$. The second invariant factor α_2 is a multiple of α_1, and is of the form $\beta\,\alpha_1$ where $\beta \in \mathbb{R}[s]$. The product $\alpha_1\,\alpha_2 = \beta\,\alpha_1^2$ is equal to the determinant

$$\begin{aligned} -\left(A\,R\,B\,S + B^2\,R^2\right) &= -B\,R\,(A\,S + B\,R) \\ &= -\alpha_1^2\,B_1\,R_1\,(A\,S + B\,R)\,. \end{aligned}$$

As a result, $\beta = -B_1 R_1\,(A\,S + B\,R)$ and $\alpha_2 = -\alpha_1 B_1 R_1\,(A\,S + B\,R)$. The Smith–MacMillan form of P_c is therefore

$$\frac{1}{AS+BR}\begin{bmatrix} \alpha_1 & 0 \\ 0 & \alpha_2 \end{bmatrix} = \begin{bmatrix} \frac{\alpha_1}{A\,S+B\,R} & 0 \\ 0 & -B_1 R_1 \alpha_1 \end{bmatrix}.$$

Since α_1 is prime with $A\,S + B\,R$, the MacMillan poles of P_c (i.e. the transmission poles of $\mathbf{P}_c$) are the roots of $A_{cl}(s)$. (iii): The verification is left to the reader. (iv) This is an obvious consequence of Definition 61.* ■

REMARK 77.– *The two systems* $\mathbf{P}$ *and* $\mathbf{K}$ *can be proper while* $\mathbf{P}_c$ *is not, as shown by the example where* $P(s) = \frac{-s+1}{s}$ *and* $K(s) = 1$. *It can also happen that* P_c *is ill-posed, such as in the case* $P(s) = -1$ *and* $K(s) = 1$.

DEFINITION 78.– $A_{cl}(s) = A(s)\,S(s) + B(s)\,R(s)$ *is the* characteristic polynomial *of the feedback system* $\mathbf{P}_c$ *(that is of the closed-loop).*

4.1.4. *Nyquist criterion*

Assume that the open-loop transfer function $L(s)$ is strictly proper (which implies that $\mathbf{P}_c$ is well-posed and proper according to Theorem 76). Let

$$f(s) = \frac{A_{cl}(s)}{(s+1)^{\nu}} \tag{4.6}$$

where $\nu = d^{\circ}(A_{cl})$, so that $0 \neq \lim_{|s|\to+\infty} f(s) < +\infty$. By (4.5),

$$f(s) = \frac{A(s)\, S(s)}{(s+1)^{\nu}} \,(1 + L(s)).$$

Let n_P (resp., n_K) be the number of poles of $P(s)$ (resp., $K(s)$) (counting multiplicities) located in the *closed* right half-plane $\bar{\mathbb{C}}_+ = \{s \in \mathbb{C} : \mathrm{Re}(s) \geq 0\}$.

Case where $P(s)$ and $K(s)$ have no poles on the imaginary axis

DEFINITION 79.– *The path $\omega \mapsto L(i\omega)$, $\omega \in (-\infty, +\infty)$ is called the* Nyquist plot *of $L(s)$ (see section 12.4.3).*

Since $L(-i\omega) = \bar{L}(i\omega)$ (where $\bar{L}(i\omega)$ is the complex conjugate of $L(i\omega)$), the Nyquist plot of $L(s)$ is entirely determined by the path $[0, +\infty) \ni \omega \mapsto L(i\omega)$. If the Nyquist plot passes through the point -1 (called the *critical point*), the closed-loop system is unstable because it has an imaginary pole. If not, let N be the number of turns the Nyquist plot $L(s)$ goes around the point -1 in the direct sense (which is anti-clockwise). The following theorem is fundamental (and is generalized to the case of linear systems with commensurate delays in [14]; see also [84], [85]).

THEOREM 80.– *(Nyquist criterion). The closed-loop system is stable if and only if the Nyquist plot does not pass through the point -1 and $n_P + n_K = N$.*

PROOF. Let $A > 0$ and consider the closed path λ formed by the concatenation of the following paths: (i) $\gamma_1 : \omega \mapsto i\omega$, $\omega \in (-A, A)$; (ii) $\gamma_2 : \theta \mapsto A\,e^{\,i\theta}$, $\theta \in \left(-\frac{\pi}{2}, \frac{\pi}{2}\right)$. (The reader is requested to draw a diagram.) The closed path λ is called the "Bromwich contour". Let A be large enough for λ to encircle one time clockwise all the poles and zeros of $P(s)$ and $K(s)$ that are located in $\bar{\mathbb{C}}_+$, if any. Let

$$f_1(s) = \frac{A(s)\, S(s)}{(s+1)^{\nu}}, \quad f_2(s) = 1 + L(s).$$

According to (12.76) (see section 12.4.3), $j(0; f \circ \lambda) = j(0; f_1 \circ \lambda) + j(0; f_2 \circ \lambda)$. On the other hand, according to Cauchy's argument principle (Theorem 447, section 12.4.5), $j(0; f_1 \circ \lambda) = n_P + n_K$ and $j(0; f_2 \circ \lambda) = -N$. As a result, $j(0; f \circ \lambda) = n_P + n_K - N$. According to (4.6), $j(0; f \circ \lambda)$ is the number of roots of $A_{cl}(s)$ located in $\bar{\mathbb{C}}_+$, and so, by Theorem 76, a necessary and sufficient condition for stability of the closed-loop system $\mathbf{P}_c$ is $j(0; f \circ \lambda) = 0$, i.e., $n_P + n_K = N$. Since $L(s)$ is strictly proper, as $A \to +\infty$, the image of $L \circ \gamma_2$ reduces to the origin. ∎

EXAMPLE 81.– *Let* $P(s) = \frac{1-s}{(1+s)^3}$. *The Nyquist plot of this transfer function is shown in Figure 4.3. The real axis is cut at* $\frac{-1}{k_0}$ *where* $k_0 = 2.01$. *Suppose the system* $\mathbf{P}$ *is fed back by a proportional regulator* $K(s) = k > 0$. *We have* $n_P = n_K = 0$, *therefore, according to the Nyquist criterion, the closed-loop system is stable if and only if* $k < k_0$. *We refer to* k_0 *as the* critical gain.

EXAMPLE 82.– *Let* $P(s) = \frac{2s+2}{s^2-s-2}$. *The Nyquist plot of* $P(s)$ *is shown in Figure 4.4. The poles of* $P(s)$ *are* -1 *and* 2, *and so* $n_P = 1$. *If, as in Example 81, we feed back the system through a gain* $k > 0$, *we have* $n_K = 0$. *The Nyquist plot of* $P(s)$ *cuts the real axis at the point* -1. *The necessary and sufficient condition for stability of the closed-loop system as provided by the Nyquist criterion is* $N = 1$, *i.e.* $k > 1$.

Case where $P(s)$ *and* $K(s)$ *have poles on the imaginary axis*

Suppose now that $A(s)$ or $S(s)$ has roots on the imaginary axis. Then, on one hand, n_P and n_K do not have a clear meaning, and, on the other hand, the Argument principle is no longer applicable if the closed path λ is defined as in the proof of Theorem 80. The solution consists in replacing the imaginary axis by an "indented imaginary axis" as defined below.

Let $p_k = i\omega_k$ $(k = 1, ..., r)$ be the distinct roots of $A(s)\,S(s)$ located on the imaginary axis, arranged in such a way that $\omega_{k+1} > \omega_k$. Let $I_k(\varepsilon)$ be a segment on the imaginary axis, centered at p_k, and of length 2ε, where $\varepsilon > 0$ is a real number sufficiently small in order that $I_k(\varepsilon) \cap I_{k+1}(\varepsilon) = \varnothing$ $(k = 1, ..., r-1)$. Replace each segment $I_k(\varepsilon)$ by the half circle $J_k(\varepsilon)$, of radius ε and centered at p_k and going round that point to the left (see Figure 4.2).

The indented imaginary axis is the set $I(\varepsilon)$ obtained after replacing all the segments $I_k(\varepsilon)$ $(k = 1, ..., r)$ by the semi-circles $J_k(\varepsilon)$. Let γ_ε denote the path traversed by the complex variable s when it travels over $I(\varepsilon)$ with a strictly increasing imaginary part (we can call γ_ε the "directed indented imaginary axis"). The definition of the Bromwich contour can clearly be extended to this case.

We are now going to replace Definition 79 by the following :

DEFINITION 83.– *The Nyquist plot of* $L(s)$ *is* $L \circ \gamma_\varepsilon$.

Let n_P and n_K be the integers as defined above and N be the number of turns the Nyquist plot of $L(s)$ made around point -1 anti-clockwise. We obtain the following:

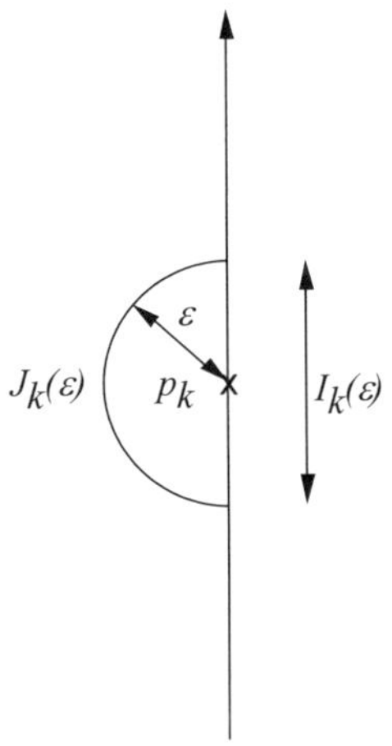

Figure 4.2. *Construction of the indented imaginary axis*

THEOREM 84.– *With N as defined above, the statement of Theorem 80 remains valid.*

EXAMPLE 85.– *Let*

$$P(s) = \frac{1}{s\left(s^2 + s + \frac{1}{2}\right)}.$$

The system with such a transfer function is fed back by a proportional regulator with gain $k > 0$. Therefore, $A(s) = s\left(s^2 + s + \frac{1}{2}\right)$. The product $A(s)\, S(s)$ has a root on the imaginary axis, at the origin. The directed indented imaginary axis consists of the following parts: (a) The quarter circle parameterized by $\varepsilon e^{-i\theta}$, where θ increases from $-\pi$ to $-\frac{\pi}{2}$; (b) The set of all $i\omega$, when ω increases from $\varepsilon > 0$ to $+\infty$; (c) The symmetric with respect to the real axis of the concatenated paths (a) and (b). For $\varepsilon \to 0^+$, $P\left(\varepsilon e^{-i\theta}\right) \sim \frac{2}{\varepsilon} e^{i\theta}$. Therefore, for a "very small" ε, the image of (a) by $P(s)$ is the quarter circle of ("very large") radius $\frac{2}{\varepsilon}$ covered with an angle θ increasing from $-\pi$ to $-\frac{\pi}{2}$.
The complete Nyquist plot is shown in Figure 4.5 (where we have to imagine the semi-circle at infinity; it is traversed in the direct sense). The poles of $P(s)$ are 0, $\frac{1}{2}(-1+i)$, $\frac{1}{2}(-1-i)$. As a result, $n_P = 1$. We have $n_K = 0$, therefore a necessary and sufficient condition for stability of the closed-loop system is $N = 1$. Now, the Nyquist plot cuts the real axis at around -2. Thus, closed-loop stability holds when $0 < k < \frac{1}{2}$.

* *Nyquist criterion in the MIMO case*

In the MIMO case, it is necessary to distinguish the open-loop transfer matrices at the *input* and *output* of the system $\mathbf{P}$, as we have seen in section 4.1.1. Anyhow,

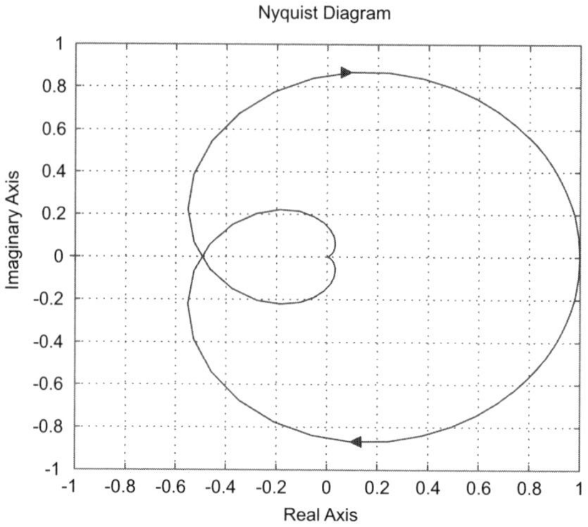

Figure 4.3. *Nyquist plot of* $P(s)$ *– Example 81*

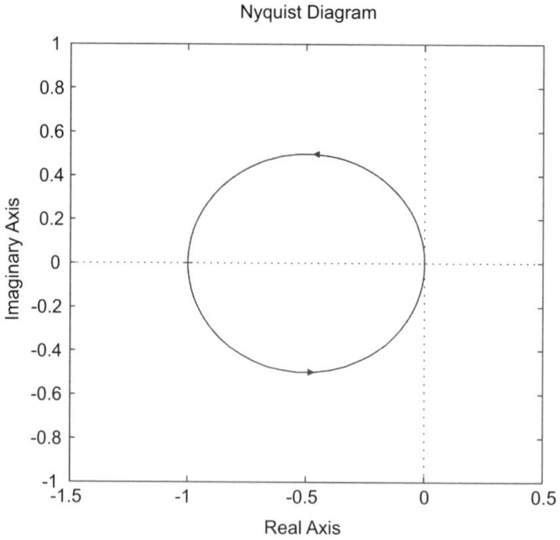

Figure 4.4. *Nyquist plot of* $P(s)$ *– Example 82*

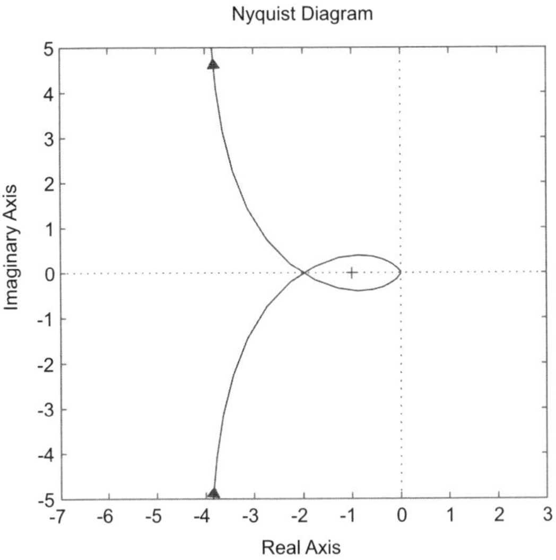

Figure 4.5. *Nyquist plot of $P(s)$ – Example 85*

$\det(I_p + L_o(s))$ is zero if and only if $\det(I_p + L_i(s))$ is also zero according to Lemma 520 (section 13.3.3). Therefore, if we put

$$\begin{aligned} g_o(s) &= \det(I_p + L_o(s)) - 1, \\ g_i(s) &= \det(I_m + L_i(s)) - 1, \end{aligned}$$

we have $g_o(s) \neq -1$ if and only if $g_i(s) \neq -1$. Let N be the number of turns effected by the Nyquist plot of $g_o(s)$ (also called the "MIMO Nyquist plot") around point -1 anti-clockwise. The following theorem is proved in ([114], section 7.4):

THEOREM 86.– *With N as defined above, the statement of Theorem 80 remains valid.*

4.1.5. *Small gain theorem*

The small gain theorem is often used in robustness theory. Consider the standard feedback system in Figure 4.1; suppose $P(s) \in \Re\mathcal{H}_\infty^{p\times m}$ and $K(s) \in \Re\mathcal{H}_\infty^{m\times p}$. In the linear time-invariant case considered here, the small gain theorem is expressed as follows:

THEOREM 87.– *If $\|P\|_\infty \|K\|_\infty < 1$, then the feedback system is stable.*

PROOF. We have $L_i(s) \in \Re\mathcal{H}_\infty^{m\times m}$ and $\|L_i\|_\infty \leq \|P\|_\infty \|K\|_\infty < 1$. Let $s \in \bar{\mathbb{C}}_+$ and $\lambda(s)$ be an eigenvalue of $L_i(s)$. According to Proposition 578 (section 13.5.7), we have $|1 + \lambda(s)| \geq 1 - |\lambda(s)|$ and $|\lambda(s)| \leq \|L_i\|_\infty$. Therefore, $|1 + \lambda(s)| > 0$, and so the matrix $I_m + L_i(s)$ is invertible in $\mathbb{C}^{m \times m}$. It follows that all the transfer matrices of the expression (4.2) have their entries belonging to $\Re\mathcal{H}_\infty$; thus, according to Definition 74, the proof is complete. ■

4.2. Robustness and performance

4.2.1. *Generalities*

The *robustness of a feedback system* is the capacity of such a system to preserve stability in the presence of model errors (robustness of stability) or even to maintain a certain performance in the presence of such errors (robustness of performance). Robustness of performance implies robustness of stability, but converse does not hold. In most cases, "robust control" is synonymous with "prudent control". The control engineer designs the control of a system based on a model (more or less explicit). The reality being infinitely complex; the more precise the model is, the more complicated it becomes (of the higher order, with nonlinearities, etc.) and the less usable it is to synthesize a control law.

Recall what was stated in section 2.2.5: an apparently very simple system, such as a voltage generator supplying a current through a resistance; is, with a good approximation, represented by the elementary equation $V = Ri$. However, this equation does not take into account the fact that, if we consider important variations of the voltage, the resistance becomes a nonlinear element; it is therefore more precise to write $V = \rho(i)$ where ρ is the characteristic function of the resistance and ρ is linear only in the vicinity of 0. Moreover, if we study the behavior of such a system, we have to take into account the propagation velocity of the electric field. The model obtained is then governed by the wave equation (which is a partial differential equation) and is of an "infinite order"; since it is nonlinear according to what has been said above, and, even though it was simple at the beginning, it has become very complicated. If such big complication (which becomes infinite in the last analysis) is most often useless, it is because, in practice, it is not necessary to represent reality in the most accurate way.

For a control engineer, a "good model" has just the necessary complexity that enables him to design the control of his/her system in such a way as to obtain the performance which corresponds to his/her specifications sheet. The more stringent the performance is required, the more precise and the more complex the model must be. Conversely, if the only model available is very imprecise, there will be no other solution but to run the system with a poor performance, and so with caution. However, caution is a virtue that is not sufficient in itself: skill is also necessary. Robustness theory is not limited to recommending servo-control adjustment flabbiness. It even happens that the robustness of a control law requires the use of large gains; it all depends on the nature of model errors or uncertainties.

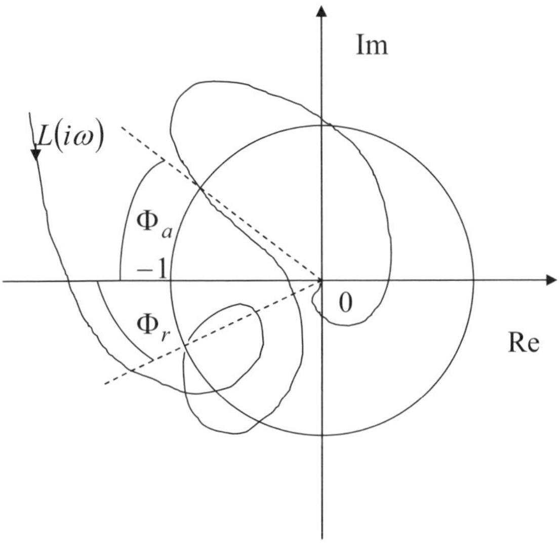

Figure 4.6. *Nyquist plot of L(s)*

Consider, for example, the system with input u and output y, governed by the differential equation

$$\dot{y} = \theta\, y + u \tag{4.7}$$

where we only know that the coefficient θ is inside an interval $[\theta_{min}, \theta_{max}]$, $0 \leq \theta_{min} < \theta_{max}$. Using a proportional control law $u = -k\, y$, we can only be sure to stabilize the system by using $k > \theta_{max}$, thus choosing a sufficiently large gain.

In reality, robustness is not an absolute notion: it is always relative to a class of modeling errors. In most of the cases encountered in practice, these errors are such that robustness against these errors is synonymous with prudence, but example (4.7) shows that it is important not to generalize this rule improperly.

4.2.2. *Robustness margins*

Suppose that the Nyquist plot of $L(s) = P(s)\, K(s)$ is as shown in Figure 4.6. The nominal feedback system (i.e. with no modeling error) is assumed to be stable.

Gain margin

Let $-1/g_1$ and $-1/g_2$ be the two numbers closest to -1 and situated on either side of that point, and where the Nyquist plot cuts the real axis. The *gain margin* is the interval $Mg = (g_1, g_2)$ $(0 \leq g_1 < 1 < g_2 \leq +\infty)$. According to the Nyquist criterion, the significance of this margin is as follows: suppose that the transfer function $P(s)$ is replaced (due to the model error) by $gP(s)$, where g is positive real number, the feedback system remains stable if and only if $g \in Mg$. The values g_1 and g_2 are often expressed in decibels. We can estimate that a gain margin is correct if it includes $[-3\text{ dB}, 6\text{ dB}]$. One may call $g_{1\text{ dB}}$ the *gain reduction margin*, and $g_{2\text{ dB}}$ the *gain augmentation margin*. But note that many control engineers refer to what we call above the *gain augmentation margin* as the *gain margin*; this usage is discouraged as it is a source of ambiguity.

Phase margin

The *phase lag margin* is the angle $\Phi_r \in [0, 2\pi)$ as shown in Figure 4.6. It has the following significance: suppose that, due the to modeling error, the transfer function $P(s)$ is replaced by $e^{-i\phi}P(s)$ $(0 \leq \phi < 2\pi)$. Then the closed-loop system remains stable if and only if $\phi < \Phi_r$.

The *phase lead margin* is the angle $\Phi_a \in [0, 2\pi)$ as shown in Figure 4.6, for analogous reasons. However, instead of considering a phase lag $-\phi$, we consider this time a phase lead ϕ.

The *phase margin* is the interval $Mp = (-\Phi_a, \Phi_r)$. The angles are often expressed in degrees. One can consider that a phase margin is correct if it includes $[-30°, 30°]$. Many control engineers refer to what we have called above the *phase lag margin* as the *phase margin*.

Delay margin

The notions of gain margin and phase margin are very classic and important. They are, however, very insufficient in characterizing robustness [20]. In numerous situations, the modeling error is analogous to a delay in series with the system. This delay may be due to sampling (in the case of a computer-controlled system); it may also be due, for various reasons, to delays between the sensor and the regulator and between this and the actuator. In the end, we will see that this delay can approximate certain kind of neglected dynamics.

Suppose now that we introduce a parasitic delay $\tau > 0$ into the loop, with transfer function $e^{-\tau s}$. We call *unity gain frequency* a frequency ω such that $|L(i\omega)| = 1$. For $s = i\omega$, where ω is a unity gain frequency, the delay τ introduces a phase shift $-\omega\tau$, thus a phase delay that has a linear dependency on ω. This phase delay is thus higher than the phase lag margin if the unity gain frequency considered is too large. This explains why the phase lag margin is not of much significance in such a situation and leads us to the following definition [20]: the *delay margin* Mr is the least upper bound of the parasitic delays for which the closed-loop system remains stable.

Let us see now how we can calculate the delay margin based on the Nyquist plot. (i) If $|L(i\omega)| < 1$ for all ω, the delay margin is infinite. (ii) Suppose that such is not the case, then the set of frequencies ω for which $|L(i\omega)| = 1$ non-empty; let $\{\omega_1, ...\omega_n\}$ be this set.[4] For every $k \in \{1, ...n\}$, let Φ_k be the angle in $[0, 2\pi)$ having a value $\arg L(i\omega_k) - \pi$ (modulo 2π). The delay margin is defined as

$$\boxed{Mr = \min\left\{\frac{\Phi_k}{\omega_k}, \quad k = 1, ..., n\right\}}. \tag{4.8}$$

In the case where the unity gain frequency is unique and is denoted by ω_0, we obtain

$$Mr = \frac{\Phi_r}{\omega_0}.$$

We will take great care in expressing the angles Φ_k and the frequencies ω_k in coherent units. For example, if Φ_k is expressed in radians, and ω_k in rad/s, Mr is obtained in seconds.

We cannot in the absolute provide a value with which the delay margin can be considered sufficient: such a value (which is one of the interests in such a notion) depends on the system to be controlled, the expected performance, the sampling period, etc. One of the rules we can formulate is that, in the case where control is discretized, the delay margin has to be at least the sampling period [74].

Modulus margin

Consider the Nyquist plot as shown in Figure 4.7. In the case considered, the gain augmentation margin g_2 is large (of the order of 3, i.e. close to 10 dB) whereas the gain reduction margin is $g_1 = 0$ (or, in dB, $-\infty$). Similarly, the phase lag margin is large (of the order of 60°) while the phase lead margin has the same value, by symmetry. However, the Nyquist plot of $L(s)$ passes very close to the critical point -1, consequently a small error in the model can destabilize the feedback system. This shows that the phase and gain margins need to be completed by another criterion which is the distance between the Nyquist plot and point -1. This distance is called the *modulus margin*. We can estimate that the modulus margin is correct if it is at least 0.5.

Using the modulus margin, we can obtain a lower bound for the gain and phase margins. Indeed, elementary geometry shows that

$$Mg \supset \left(\frac{1}{1+Mm}, \frac{1}{1-Mm}\right) \tag{4.9}$$

$$Mp \supset \left(-2\arcsin\frac{Mm}{2}, 2\arcsin\frac{Mm}{2}\right). \tag{4.10}$$

4. This set is necessarily finite when the open-loop system is finite-dimensional, i.e. when $L(s)$ is a rational function.

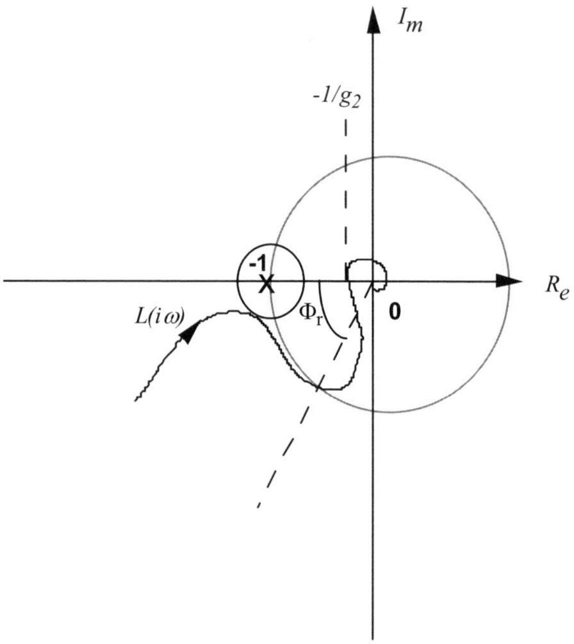

Figure 4.7. *Modulus margin*

To clarify ideas, with $Mm = 0.5$, the above lower bounds become: $Mg \supset (-3.5 \text{ dB}, 6 \text{ dB})$, $Mp \supset (-29°, 29°)$.

The modulus margin can easily be expressed using the sensitivity function $S_i = \frac{1}{1+L}$. First note that this sensitivity function belongs to $\Re\mathcal{H}_\infty$ due to the fact that the nominal feedback system is stable (see sections 3.1.1 and 4.1.3). We have

$$Mm = \inf_{\omega \geq 0} |1 + L(i\omega)| = \frac{1}{\sup_{\omega \geq 0} \left| \frac{1}{1+L(i\omega)} \right|}$$

thus according to section 13.6.2,

$$\boxed{Mm = \frac{1}{\|S_i\|_\infty}}. \tag{4.11}$$

Since $L(s)$ is a strictly proper transfer function, we have $\lim_{\omega \to +\infty} |S_i(i\omega)| = 1$. As a result, $\|S_i\|_\infty$ can be interpreted as the *resonance factor* of the sensitivity function S_i. The expression (4.11) shows that the modulus margin is the inverse of this resonance factor.

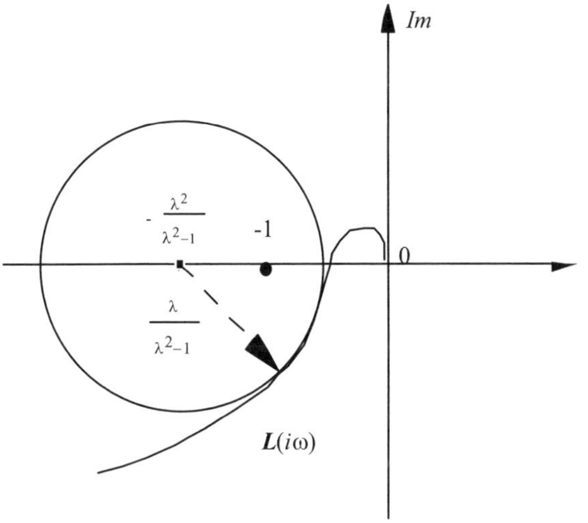

Figure 4.8. *M-circle*

Hall chart

It is common to trace in the Nyquist plane the "λ-circles" (also called "M-circles") making it possible to geometrically determine the resonance factor of the "complementary sensitivity function" $T_o = T_i = 1 - S_i$. We then obtain the *Hall chart*. The "λ-circle" C_λ is the locus of the points M such that $\frac{OM}{AM} = \lambda \ (\lambda \geq 0)$ where A represents the critical point -1. Let $M = x + i\,y$. The reader can easily verify that the coordinates x and y of M in the complex plane are solutions of the algebraic equation

$$\left(\lambda^2 - 1\right) x^2 + 2\lambda^2 x + \left(\lambda^2 - 1\right) y^2 + \lambda^2 = 0,$$

and C_λ is therefore the circle with center $-\frac{\lambda^2}{\lambda^2-1}$ and radius $\frac{\lambda}{|\lambda^2-1|}$. One such circle C_λ is shown in Figure 4.8 in the case $\lambda > 1$ (the most common in practice). The Nyquist plot of $L(s)$ is also plotted in this figure (at least the part of the plot that corresponds to frequencies $\omega \geq 0$). The norm $\|T_o\|_\infty$ is λ because this Nyquist plot is tangential to C_λ and does not penetrate into the interior of this circle. If $\lim_{s \to 0} |T_o(s)| = 1$, $\|T_o\|_\infty = \lambda$ is the the *resonance factor* of $T_o(s)$.

Complementary modulus margin

The *complementary modulus margin* Mmc is the least upper bound of the quantities $1/\lambda$ such that the Nyquist plot of $L(s)$ does not penetrate into the circle C_λ (with $\lambda > 1$). It thus follows that

$$\boxed{Mmc = \frac{1}{\|T_o\|_\infty}}.$$

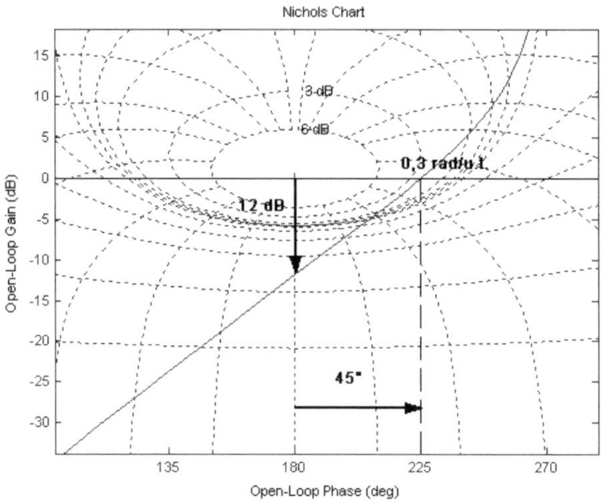

Figure 4.9. *Black plot of L(s)*

4.2.3. *Use of the Nichols chart*

The Black plot of a transfer function $L(s)$ has already been defined (see section 3.2.3): it is the curve that represents the evolution of the coordinates $(|L(i\omega)|_{\text{dB}}, \arg L(i\omega))$ as a function of the parameter ω. It is convenient to represent this curve on the *Nichols chart* where the curves $|T_o| = const$ and $\arg T_o = const$ (where $T_o = \frac{L}{1+L}$) are plotted.

Consider, for example, the system **P** with transfer function

$$P(s) = \frac{1 - s/3}{s(1 + s/2)(1 + 2s)},$$

fed back by the regulator **K** with transfer function $K(s) = 0.35$ (this transfer function being a constant, the regulator – or the control it generates – is said to be "proportional"). The transfer function $L(s) = P(s)\,K(s)$ is represented in the Nichols chart in Figure 4.9. The phase lag margin is $225° - 180° = 45°$ (obtained at the frequency of 0.3 rad/t.u., where t.u. is the time unit used). The gain margin is about $(-\infty, 12\,\text{dB})$. The quantity $\|T_0\|_\infty$ (dB) is seen on the chart: it is slightly less than 3 dB (a precise calculation shows that $\|T_0\|_\infty \simeq 1.31$, i.e. about 2.39 dB). The Nichols chart can be used to determine the gain of an adequate proportional regulator. The gain $k = 0.35$ is that of a proportional regulator which, when feeding back the system **P**, provides a phase lag margin of 45°.

4.2.4. *Robustness against neglected dynamics*

Neglected dynamics without resonance

The transfer function of the *model* **P** is again denoted by $P(s)$. We make the hypothesis that the *actual control system* $\tilde{\mathbf{P}}$ (assumed to be linear time-invariant) has the following transfer function:

$$\tilde{P}(s) = P(s)\,\frac{1}{1+\tau_1 s}\cdots\frac{1}{1+\tau_n s}$$

where the $\tau_k > 0$ $(1 \le k \le n)$ denote the neglected time constants. Let $s = i\omega$ and suppose $|\tau_k \omega| \ll 1$ $(1 \le k \le n)$, that is to say $|\tau_k s| \ll 1$. It follows that ([17], section 2.2.1)

$$\begin{aligned}\tilde{P}(s) &\cong P(s)(1-\tau s) \cong P(s)\,e^{-\tau s},\\ \tau &= \sum_{k=1}^{n} \tau_k.\end{aligned}$$

In other words, the "neglected time constants" have an effect analogous to a neglected delay. The *delay margin* therefore correctly represents the robustness of the feedback system against this kind of neglected dynamics.

Neglected dynamics of any kind

In a more general case, the delay margin is insufficient as a criterion for robustness.

Consider the case of a "multiplicative modeling error". The transfer function $\tilde{P}(s)$ of the *actual control system* $\tilde{\mathbf{P}}$ is expressed as a function of the transfer function $P(s)$ of the *model* **P** according to a relation of the form

$$\tilde{P}(s) = (1 + E(s))\,P(s)\,.$$

The transfer function $E(s)$ (or the associated minimal system **E**) is called a "multiplicative modeling error".

Consider the following conditions:

(i) The regulator with transfer function $K(s)$ stabilizes $P(s)$;

(ii) $P(s)$ and $\tilde{P}(s)$ have the same number of poles in the closed right half-plane.

(iii) $P(s)$ and $\tilde{P}(s)$ have the same poles on the imaginary axis (taking into account multiplicities) – the set of these poles can of course be empty.

(iv) There exists a transfer function $W_1(s) \in \Re\mathcal{H}_\infty$ (see section 13.6) such that $|E(i\omega)| < |W_1(i\omega)|$, for any frequency $\omega \ge 0$.

Let $\mathcal{E}$ be the set of transfer functions $E(s)$ satisfying Conditions (ii)–(iv) above, assuming that Condition (i) is satisfied.

THEOREM 88.– *A necessary and sufficient condition for the feedback system (consisting of the* control system $\tilde{\mathbf{P}}$ *fed back by the regulator* $\mathbf{K}$*) to be stable for any modeling error* $E(s) \in \mathcal{E}$ *is* $\|W_1 T_o\|_\infty \leq 1$.

PROOF. (A) According to Condition (iii), the Bromwich contours of $\tilde{L}(s) = \tilde{P}(s)\,K(s)$ and that of $L(s) = P(s)\,K(s)$ are identical. (B) Let N be the number of turns the Nyquist plot of $L(s)$ goes around the point -1 in the direct sense. According to Hypothesis (i) and the Nyquist criterion (Theorem 84), we have $N = n_P + n_K$, where n_P and n_K denote, respectively, the number of poles of $P(s)$ and $K(s)$ that belong to the closed right half-plane. According to this same theorem and observation (A) here above, a necessary and sufficient condition for $\mathbf{K}$ to stabilize $\tilde{\mathbf{P}}$ is $\tilde{N} = n_{\tilde{P}} + n_K$, where $\tilde{N}$ is the number of turns the Nyquist plot goes around point -1 in the direct sense and $n_{\tilde{P}}$ denotes the number of poles of $\tilde{P}(s)$ belonging to the closed right half-plane. According to Conditions (ii) and (iii), we have $n_{\tilde{P}} = n_P$, and thus the necessary and sufficient condition above is equivalent to $\tilde{N} = N$. (C) The only supplementary information we have on $\tilde{P}(s)$ is Condition (iv). It is equivalent to $\left|\tilde{L}(i\omega) - L(i\omega)\right| < |W_1(i\omega)\,L(i\omega)|$ for all $\omega \geq 0$, which means that for every frequency ω, $\tilde{L}(i\omega)$ belongs to the open disk centered at $L(i\omega)$ and with radius $|W_1(i\omega)\,L(i\omega)|$. The necessary and sufficient condition determined at (B) is therefore satisfied if and only if the radius $|W_1(i\omega)\,L(i\omega)|$ does not exceed the distance $|1 + L(i\omega)|$ from the point $L(i\omega)$ to the point -1, and this for any frequency ω (see Figure 4.10). This condition is written as $|W_1(i\omega)\,L(i\omega)| \leq |1 + L(i\omega)|$, and is equivalent to $|W_1(i\omega)\,T_o(i\omega)| \leq 1$. This inequality is satisfied for every frequency ω if and only if $\|W_1\,T_o\|_\infty \leq 1$. ■

Suppose that ω is a frequency at which the uncertainty or model error is large, that is $|W_1(i\omega)| \gg 1$. The necessary and sufficient condition in Theorem 88 implies

$$|T_o(i\omega)| \leq \frac{1}{|W_1(i\omega)|} \ll 1. \tag{4.12}$$

We thus have $T_o(i\omega) \sim L(i\omega)$, and Condition (4.12) leads to

$$|L(i\omega)| \preceq \frac{1}{|W_1(i\omega)|} \ll 1 \tag{4.13}$$

where the symbol $\preceq$ signifies "is dominated by" ([12], section V.1). As a result, a good amount of robustness against a multiplicative error is obtained using a "small loop gain".

In general, the neglected dynamics are essentially located in high frequencies. And thus the condition $|L(i\omega)| \ll 1$ needs to be satisfied for large values of ω.

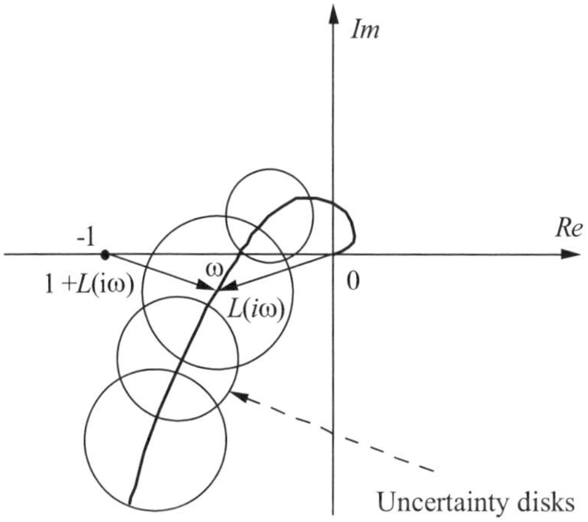

Figure 4.10. *Uncertainty disks*

4.2.5. *Performance*

Section 4.2.1 already suggested that the analysis of robustness is not separable from that of performance. One of the essential tasks of the control engineer is indeed to manage the best compromise between robustness and performance. As we will see below, these two properties are in general antagonistic.

Suppose that, in Figure 4.1, the variable being controlled (or regulated)[5] is $z = u_2$, and that $-v_2 = d$ is a disturbance. With $v_1 = 0$, we obtain from (4.3) the expression

$$\hat{z}(s) = S_o(s)\,\hat{d}(s).$$

In the absence of any feedback, we have $\hat{z}(s) = \hat{d}(s)$ under the same conditions and the transfer function between the disturbance d and the variable to be regulated z is therefore equal to 1. As a result, the feedback is efficient (in terms of performance) at frequency ω if and only if

$$|S_o(i\omega)| < 1.$$

This is indeed the condition under which the feedback will *attenuate the disturbance* that affects the controlled variable (at frequency ω). We say that there is a *rejection* of this disturbance (at this same same frequency ω) if $S_o(i\omega) = 0$.

5. A regulation problem is a special case of tracking problem, where the controlled variable is to be maintained constant. Here this constant is zero.

Note that, because $L(s)$ is a strictly proper transfer function, we have $\lim_{\omega \to +\infty} |S_o(i\omega)| = 1$, therefore the feedback will necessarily no longer be efficient at very high frequencies.

We can quantify the expected performance (in accordance with the specifications) by choosing a transfer function $W_2(s)$ in $\Re\mathcal{H}_\infty$, such that the performance is considered to be satisfactory if $|W_2(i\omega)\ S_o(i\omega)| \leq 1$, for every frequency ω. This condition is equivalent to

$$\boxed{\|W_2 S_o\|_\infty \leq 1}. \tag{4.14}$$

An advantage of such a formulation is that Condition (4.14) is in the same form as the one stated in Theorem 88. This allows us to transform the problem of managing the compromise robustness/performance to a problem of optimization in the $\Re\mathcal{H}_\infty$ space – which is outside the scope of this book (see e.g. [122]).

Consider a frequency ω at which a heavy attenuation of the disturbance is required. We have therefore $|W_2(i\omega)| \gg 1$. Condition (4.14) leads to

$$|S_o(i\omega)| \leq \frac{1}{|W_2(i\omega)|} \ll 1. \tag{4.15}$$

We thus have $L(i\omega) \sim 1/S_o(i\omega)$, and (4.15) yields

$$|L(i\omega)| \gtrsim |W_2(i\omega)| \gg 1. \tag{4.16}$$

It is clear that Conditions (4.13) and (4.16) are contradictory if they are required at the same frequency ω. *We therefore cannot obtain from a closed-loop system a good performance at a frequency at which a large modeling error is present.* In general, performance is required at low frequencies. And that is the reason why the condition $|L(i\omega)| \gg 1$ must be satisfied at small values of ω.

4.2.6. *Sensitivity to measurement noise*

Now suppose that (see Figure 4.1) v_2 represents the *measurement noise* adding to the controlled variable (especially because of imperfections of the sensor). It is very important that the control u_1 of the system should not be too sensitive to this noise. If not, indeed, the control will be very agitated and it will result in premature wear on the actuators (especially if these are made of mechanical parts). According to (4.3), the transfer function between v_2 and u_1 is $S_i(s)K(s) = S_o(s)K(s)$ (because, at this moment, only the *SISO* case is considered). The agitation due to this measurement noise is harmful essentially at high frequencies, for which we have the relation (4.13), and thus $|S_o(i\omega)| \sim 1$ and $|S_o(i\omega)K(i\omega)| \sim |K(i\omega)|$. It is therefore important to *limit the gain of the regulator in high frequencies.*

In the absence of feedback, the transfer function between v_2 and the control variable y_1 is obviously zero. A perverse effect of the feedback can make y_1 agitated because of the noise v_2. In closed-loop, the transfer function between v_2 and y_1 is (according to (4.2)) $T_o(s)$. To limit the pernicious effect examined here, we are led to impose a condition such as (4.12), which implies (4.13) on the open-loop transfer function.

4.2.7. *Loopshaping of* $L(s)$

Conditions (4.13) and (4.16), uniquely imposed on the open-loop transfer function $L(s)$, must be satisfied: the first one in high frequencies and the second one in low frequencies.

On the other hand, it is necessary to have sufficient phase margin. Assuming that $L(0) > 0$, let us consider the *most favorable case* where $L(s)$ is stable and minimum phase (we will elaborate later why this will be the most favorable case) and assume that $L(0) > 0$. Let ω_0 be the unity gain frequency (which we will assume to be unique for simplicity). Let $-20\,n$ dB/decade ($n \in \mathbb{N}$) be the slope of the magnitude of $L(i\omega)$ in a neighborhood of $\omega = \omega_0$. The asymptotic Bode plot shows that, for $n = 2$, the phase is about $-180°$, which corresponds to a zero phase margin. If $n > 2$, it is not hard to see that the Nyquist plot of $L(s)$ encircles point -1 clockwise, and so the closed-loop system is unstable. The only two possible values for n are 1 and 0. In the case $n = 1$, the Bayard–Bode relation (see section 3.4.4, (3.19)) shows that, if the corresponding slope is maintained over 1 decade centered logarithmically at ω_0, the phase of $L(s)$ at that frequency cannot be smaller than $-90\,(1 + 0,26) \cong -113°$, which guarantees a phase lag margin of $180 - 113 = 67°$, largely sufficient. On the other hand, maintaining the same slope over only half a decade guarantees an insufficient phase lag margin.

Having gotten these results, we can deduce the typical shape a Bode plot of $L(s)$ has to have, i.e. the *loopshaping*[6] we need to achieve for this transfer function (see Figure 4.11).

The difficult problems are those where performance and robustness requirements are such that $L(s)$ must be the transfer function of a very selective filter in the two transition bands: the region to the immediate left of the line segment of slope -20 dB/decade, and the region located to the immediate right of that same segment. This large selectivity necessitates a controller of high order.

The value of the cutoff frequency ω_0 is limited by the necessary delay margin.

6. This loopshaping does take into account the necessity, previously pointed out, of limiting $|K(i\omega)|$ in high frequencies.

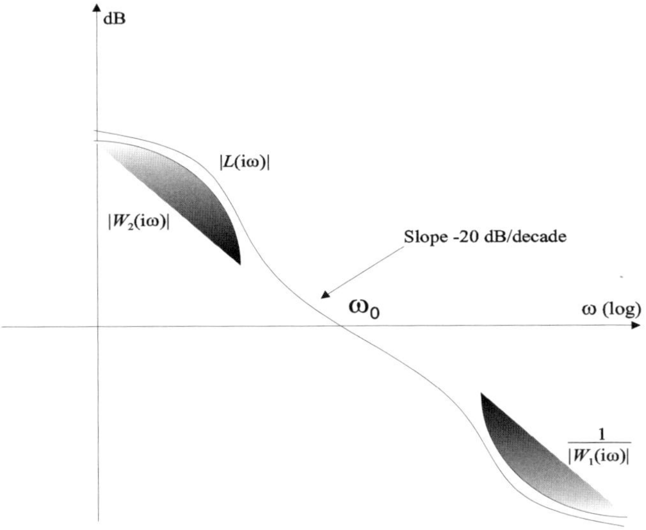

Figure 4.11. *Loopshaping of L(s)*

4.2.8. *Degradation of robustness/performance trade-off*

Unstable poles

Intuitively, an unstable system is harder to control than a stable system. We can support this intuitive judgment by reasoning on the sensitivity function $S_o(s)$. The closed-loop system possesses both a good modulus margin and a good capacity of attenuation of a disturbance adding to the controlled variable, under the condition that the quantity $\|S_o\|_\infty$ should not be too large compared to 1. In particular, a modulus margin $Mm \geq 0.5$ is obtained when $\|S_o\|_\infty \leq 2$ and the feedback in terms of disturbance attenuation is efficient at frequency ω if $|S_o(i\omega)| \leq 1$ (see sections 4.2.2 and 4.2.5).

The following result was obtained by Bode [9] in the case of a stable open-loop and generalized by Freudenberg and Looze [50] (see also [51], section 3.4) in the case of an arbitrary open-loop. Suppose the transfer function $L(s)$ has the following two properties:

(i) The set of poles of $L(s)$ with positive real part is $\{p_j : j \in J\}$, where J is a finite set of indices (possibly empty);

(ii) $L(s)$ has a relative degree $\delta(L) \geq 2$.

Then the following relation (called the *Bode relation*, or the *Bode–Freudenberg–Looze relation*) holds:

$$\boxed{\int_{-\infty}^{+\infty} \ln |S_o(i\omega)|\, d\omega = 2\pi \sum_{j \in J} \operatorname{Re} p_j}. \tag{4.17}$$

The quantity on the right-hand side of (4.17) is non-negative, and is zero if and only if $L(s)$ only has stable poles or poles at the limit of instability (meaning they are on the imaginary axis). And we deduce the following result:

PROPOSITION 89.– *If there exists an interval of frequencies I_b, with non-empty interior, on which $|S_o(i\omega)| < 1$ (i.e. $\ln |S_o(i\omega)| < 0$), there must exist another interval of frequencies I_m, with non-empty interior, on which $|S_o(i\omega)| > 1$, so that the integral on the left-hand side of* (4.17) *is non-negative (and positive if $L(s)$ has poles in the right half-plane).*

In other words, the efficiency of a control system (in terms of disturbance attenuation) at certain frequencies is obtained at the price of a harmful effect of such a control system (relative to the same point) at other frequencies. This is like a "communicating vessels" effect.

When $L(s)$ has unstable poles, the right-hand side of (4.17) is positive, which, on the whole, makes the situation more unfavorable.

Unstable zeros

Contrary to what we might think, "unstable zeros" are more penalizing than "unstable poles". This is what we will show below. Suppose the open-loop transfer function $L(s)$ has the following two properties:

(i) the set of poles of $L(s)$ with positive real part is $\{p_j : j \in J\}$, where J is a finite set of indices (possibly empty);

(ii) $L(s)$ has at least one zero z with *positive* real part.

The *Blaschke product* [103] associated with the set of poles $P = \{p_j, j \in J\}$ is the rational function

$$\begin{aligned} B_P(s) &= \prod_{j \in J} \frac{p_j - s}{\bar{p}_j + s} \quad \text{if } J \text{ is non-empty,} \\ &= 1 \quad \text{if } J \text{ is empty.} \end{aligned}$$

This transfer function $B_P(s)$ belongs to $\Re\mathcal{H}_\infty$. Assume that J is non-empty. Since for any index $j \in J$, $\left|\frac{p_j - s}{\bar{p}_j + s}\right| \leq 1$ if and only if $\mathrm{Re}(s) \geq 0$, we have $|B_P(s)| \leq 1$ if and only if $\mathrm{Re}(s) \geq 0$.

Let $z = x + iy$ be a zero of $L(s)$ with real part $x > 0$, and let θ_z be the function defined by

$$\boxed{\theta_z(\omega) = \arctan \frac{\omega - y}{x}}.$$

We can prove the following result [50]:

$$\boxed{\int_{-\infty}^{+\infty} \ln |S_o(i\omega)| \, d\theta_z(\omega) = -\pi \ln |B_P(z)|}. \tag{4.18}$$

The right-hand side of (4.18) is non-negative. In addition, $d\theta_z(\omega) = W_z(\omega)\, d\omega$ with $W_z(\omega) = \frac{d\theta_z}{d\omega}(\omega) = \frac{x}{x^2 + (y-\omega)^2} > 0$, which leads to the following conclusion:

PROPOSITION 90.– *The property stated in Proposition 89 still holds.*

Proposition 90 is not a redundant version of Proposition 89 because it is based on different hypotheses: while hypotheses (i) and (i') are similar, (ii) and (ii') are very different. If we compare the left-hand side of (4.17) to that of (4.18), the integral in the last quantity includes, as an additional term, "weighting function" $W_z(\omega)$ which, being maximum at $\omega = y$, "focalizes" this integral on the frequencies in the neighborhood of y. The reader can check the following using Theorem 432 (section 12.2.3): if (x_n) is a sequence of positive numbers tending to zero, then $\left(\frac{1}{\pi} d\theta_{z_n}/d\omega\right)$ (where $z_n = x_n + iy$) tends to $\delta_y : t \mapsto \delta(t-y)$ in $\mathcal{D}'$; therefore, $\frac{1}{\pi}\int_{-\infty}^{+\infty} \ln|S_o(i\omega)|\, d\theta_{z_n}(\omega) \to \ln|S_o(iy)|$.

As a result, when the open-loop transfer function of $L(s)$ has an "unstable zero" z, it is penalizing for the sensitivity function and thus for the *modulus margin*. Let us see what remedy we can find in such a situation. Let

$$L'(s) = \left(\frac{z+s}{z-s}\right)^k L(s)$$

where k is the multiplicity of the zero z. The transfer function $L(s)$ does not have a zero at $s = z$ any more, but at $s = -z$ (with a multiplicity equal to k). Proceeding in this manner for each unstable zero of $L(s)$, we arrive at an open-loop transfer function which has no unstable zero any more. To simplify the rationale, suppose $L(s)$ has a unique unstable zero z. We have

$$L(s) - L'(s) = L'(s)\, E(s) \tag{4.19}$$

with $E(s) \sim 2k\frac{s}{z}$ as $s \to 0$. As a result, the "multiplicative error" that is caused by replacing $L(s)$ with $L'(s)$ is such that $|E(s)| \ll 1$, if and only if $|s| \ll \frac{|z|}{2k}$. Let ω_0 be the largest unity gain frequency of $L(s)$, that is

$$\omega_0 = \sup\{\omega : |L(i\omega)| = 1\}.$$

According to the above, unstable zero z brings an insignificant deterioration on the modulus margin on the condition that

$$\boxed{\omega_0 \ll \frac{|z|}{2k}}. \tag{4.20}$$

Therefore, an unstable but "rapid" zero (which means the absolute value $|z|$ is large) is at the end only a little penalizing. If it is "slow", it is necessary to "re-shape" the closed-loop system until condition (4.20) is satisfied, and, therefore, to give up obtaining good performance.

REMARK 91.– *The above is only valid when* $\mathrm{Re}(z) > 0$. *Indeed, if* $L(s)$ *has a zero* $z = iy$ *(on the imaginary axis),* $-z$ *is again on the imaginary axis and the transfer function* $L'(s)$ *as constructed above is unstable. To simplify the analysis of this situation, suppose that* $z = iy$ *is a simple zero (i.e. with multiplicity* 1*) of* $L(s)$*. Put*

$$L'(s) = \frac{z + \alpha + s}{z - s} L(s), \quad \alpha > 0. \tag{4.21}$$

The transfer function $L'(s)$ *no longer has a zero at* $s = z$ *but at* $s = -z - \alpha$*, the real part of which is* $-\alpha < 0$*. We then have relation* (4.19) *with* $|E(i\omega_0)| \ll 1$ *if*

$$\alpha \ll |\omega_0| \ll \frac{|y|}{2}. \tag{4.22}$$

Condition (4.22) *is coherent with* (4.20)*, even though it corresponds to a quite different situation.*

4.2.9. * *Extension to the MIMO case*

Robustness margins

To extend the above results to the case of MIMO systems, it is necessary to carefully distinguish S_i and T_i, which are the sensitivity function and the complementary sensitivity function, respectively, at the *input* of system **P**, from S_o and T_o, which are at the *output* of **P** (see section 4.1.2). This leads us to define two *modulus margins*: that at the *input* of **P**, denoted by Mm_i, and that at the *output* of **P**, denoted by Mm_o. These quantities (called the *input modulus margin* and the *output modulus margin*) are defined by the following relations:

$$Mm_i = \frac{1}{\|S_i\|_\infty}, \quad Mm_o = \frac{1}{\|S_o\|_\infty}$$

(see section 13.6.2 for the definition of the "$\mathcal{H}_\infty$ norm" of a transfer *matrix*). We can similarly define two *complementary modulus margins* – at the *input* of **P**, denoted by Mmc_i, and at the *output* of **P**, denoted by Mmc_o – by the relations

$$Mmc_i = \frac{1}{\|T_i\|_\infty}, \quad Mmc_o = \frac{1}{\|T_o\|_\infty}.$$

The MIMO Nyquist plot cannot provide a geometrical interpretation for these robustness margins (see Exercise 105).

We can also define a gain margin Mg_i and a phase margin Mp_i at the *input* of the MIMO system **P**, and a gain margin Mg_o and a phase margin Mp_o at the *output* of same system, proceeding in the following manner [13]:

DEFINITION 92.– *(i) The input (resp., output) gain margin is the largest interval* Mg_i *(resp.,* Mg_o*), containing* 1*, such that the feedback system remains stable when the transfer matrix* $P(s)$ *is replaced by* $P(s)\Lambda$ *(resp.,* $\Lambda P(s)$*) for every non-negative symmetric real matrix* Λ *whose eigenvalues belong to such an interval.*
(ii) The input (resp., output) phase margin is the largest interval Mp_i *(resp.,* Mp_o*), containing* 0*, such that the feedback system remains stable when the transfer matrix* $P(s)$ *is replaced by* $P(s)\,e^{i\Phi}$ *(resp.,* $e^{i\Phi}\,P(s)$*) for every symmetric real matrix* Φ *whose eigenvalues belong to such an interval.*

Note that Λ and Φ represent the uncertainties, the first is on the "gain" and the second is on the "phase".

THEOREM 93.– *Relations* (4.9) *and* (4.10) *extend to the MIMO case, that is*

$$Mg_* \supset \left(\frac{1}{1+Mm_*}, \frac{1}{1-Mm_*}\right) \tag{4.23}$$

$$Mp_* \supset \left(-2\arcsin\frac{Mm_*}{2}, 2\arcsin\frac{Mm_*}{2}\right) \tag{4.24}$$

with $* = i$ *or* o.

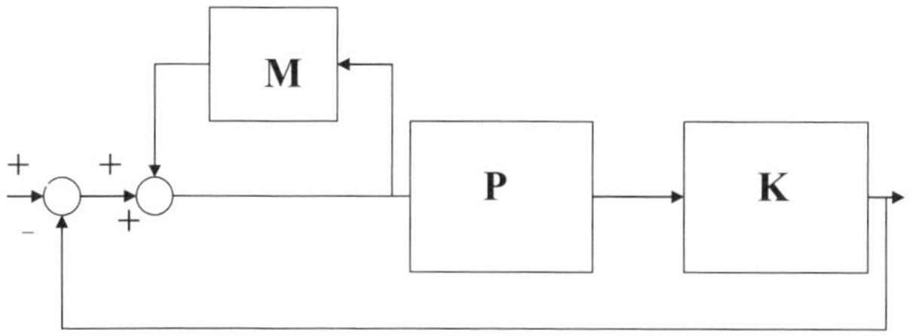

Figure 4.12. *Uncertain feedback system*

PROOF. The proof is made for the gain and phase margins at system input; the reasoning is similar for the same margins at system output. (i) Let $\Lambda = (I_m - M)^{-1}$. Then the block diagram of the feedback system, with $P(s)$ replaced by $P(s)\,\Lambda$, is represented as shown in Figure 4.12.

The matrix M is fed back to the transfer matrix $(I_m + K\,P)^{-1} = S_i$. According to the small gain theorem (Theorem 87, section 4.1.5), a sufficient condition for the closed-loop to remain stable is therefore

$$\bar{\sigma}(M)\,\|S_i\|_\infty < 1$$

and according to (4.23) this yields

$$\bar{\sigma}(M) < Mm_i. \tag{4.25}$$

Since $M = I_m - \Lambda^{-1}$, condition (4.25) holds if and only if for every eigenvalue λ of Λ, we have $\lambda \in \left(\frac{1}{1+Mm_i}, \frac{1}{1-Mm_i}\right)$. (ii) Let $e^{i\Phi} = (I_m - M)^{-1}$. The sufficient condition of stability (4.25) can now be written as[7]

$$\bar{\sigma}\left(I_m - e^{-i\Phi}\right) < Mm_i. \tag{4.26}$$

We have

$$I_m - e^{-i\Phi} = 2ie^{-i\Phi/2}\sin\Phi/2$$

([52], section 9.12), and therefore $\bar{\sigma}\left(I_m - e^{-i\Phi}\right) = 2\bar{\sigma}\left(\sin\Phi/2\right)$. Condition (4.26) is therefore satisfied if and only if $|\varphi| < 2\arcsin\left(Mm_i/2\right)$ for every eigenvalue φ of Φ. ■

7. The matrix $M = I_m - e^{-i\,\Phi}$ has complex entries, nevertheless the small gain theorem is still valid in that case. On the other hand, $e^{i\,\Phi}$ is the "phase", at a given frequency, of a transfer matrix belonging to $\Re\mathcal{H}_\infty^{m\times m}$.

MIMO delay margin

DEFINITION 94.– *The input (resp., output) delay margin is the least upper bound* MR_i *(resp.,* MR_o*) of the real numbers* r *for which the closed-loop system remains stable when the transfer matrix* $P(s)$ *is replaced by* $P(s)\, e^{-R\,s}$ *(resp.,* $e^{-R\,s}P(s)$*) for any symmetric real matrix* $R \geq 0$ *whose* r *is the largest eigenvalue.*

THEOREM 95.– *An upper bound of the delay margin* MR_* *is* $\|sT_*(s)\|_\infty^{-1}$ *with* $* = i$ *or* o*.*

PROOF.* The proof is done for the delay margin at input of $\mathbf{P}$. Since $\tilde{P}(s) = P(s)(I_m + E(s))$, replacing $P(s)$ by $\tilde{P}(s) = P(s)\, e^{-Rs}$ introduces a multiplicative error $E(s) = e^{-R\,s} - I_m$ at the input. The ratio $E(s)/s$ is fed back to $sT_i(s)$, in the sense where the closed-loop has an equivalent diagram as depicted in Figure 4.1 with $P(s)$ and $K(s)$ replaced by $E(s)/s$ and $sT_i(s)$, respectively (see sections 2.6.2 and 2.6.3). Now,

$$\frac{E(s)}{s} = e^{-R\,s/2}\left(e^{-Rs/2} - e^{R\,s/2}\right)/s$$

and thus for $s = i\omega \neq 0$,

$$\bar{\sigma}\left(\frac{E(i\omega)}{i\omega}\right) \leq 2\,\bar{\sigma}\left(e^{-R\,i\omega/2}\right)\bar{\sigma}\left(\frac{\sin(R\omega/2)}{\omega/2}\right).$$

We immediately get

$$\sup_{\omega>0}\bar{\sigma}\left(\frac{\sin(R\omega/2)}{\omega/2}\right) = r. \tag{4.27}$$

On the other hand, $\frac{\sin(R\,s/2)}{s/2}$ does not have a pole in the closed right half-plane, and so according to (4.27), this transfer matrix belongs to the Hardy space $\mathcal{H}_\infty^{m\times m}$ (see section 13.6.3). According to the small gain theorem (Theorem 87, section 4.1.5), the statement of which is still valid when $\Re\mathcal{H}_\infty$ is replaced by $\mathcal{H}_\infty$, a sufficient condition for stability of the closed-loop system is $r\,\|s\,T_i(s)\|_\infty < 1$. This proves the theorem.* ■

Robust stability theorem

Let us generalize Theorem 88. Consider the so-called *standard diagram* in Figure 4.13.

The system $\mathbf{P}_a$ is called the *augmented system* and $\mathbf{\Delta}$ represents the model uncertainty. The systems $\mathbf{P}_a$, $\mathbf{\Delta}$ and $\mathbf{K}$ are assumed to be linear time-invariant and minimal. Therefore, they can be identified with their respective transfer matrices $P_a(s)$, $\Delta(s)$ and $K(s)$.

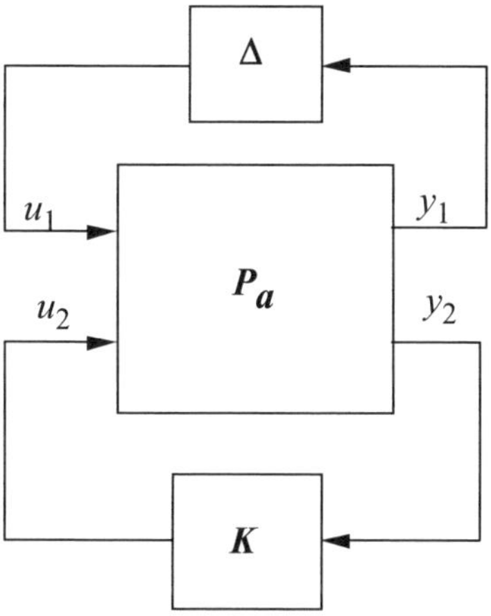

Figure 4.13. *Standard diagram*

Denote by $F_l(P_a, K)$ (resp., $F_u(P_a, \Delta)$) the linear time-invariant system – or the corresponding transfer matrix – with input u_1 and output y_1 when $\Delta = 0$ (resp., with input u_2 and output y_2 when $K = 0$); F_l and F_u are called *linear fractional transformations* (LFTs) and, as easily seen, they are given by

$$\begin{aligned} F_l(P_a, K) &= P_{a11} + P_{a12}K(I - P_{a22}K)^{-1} P_{a21}, \\ F_u(P_a, \Delta) &= P_{a21}\Delta(I - P_{a11}\Delta)^{-1} P_{a12} + P_{a22} \end{aligned}$$

where $P_a = \begin{bmatrix} P_{a11} & P_{a12} \\ P_{a21} & P_{a22} \end{bmatrix}$. Let $B_1(P_a)$ be the set of all rational transfer matrices Δ for which the two following conditions hold:

(i) $\overline{\sigma}(\Delta(i\omega)) < 1, \forall\omega \geq 0$;

(ii) $F_u(P_a, \Delta)$ and $F_u(P_a, 0)$ have the same number of poles in the closed right-half plane, and have the same poles on the imaginary axis (if any), taking into account multiplicities.

One can prove the following [89]:

THEOREM 96.– *Assume that K stabilizes $F_u(P_a, 0)$. Then K stabilizes $F_u(P_a, \Delta)$ for every $\Delta \in B_1(P_a)$ if and only if $\|F_l(P_a, K)\|_\infty \leq 1$.*

To see how Theorem 96 can be used, consider, e.g. the case of a multiplicative model error. Suppose the transfer matrix $\tilde{P}(s)$ of the *system to be controlled* $\tilde{\mathbf{P}}$ can be expressed as a function of the transfer matrix $P(s)$ of the *model* $\mathbf{P}$ following $\tilde{P}(s) = (I_p + E(s))P(s)$. The diagram of the closed-loop system can be represented as in

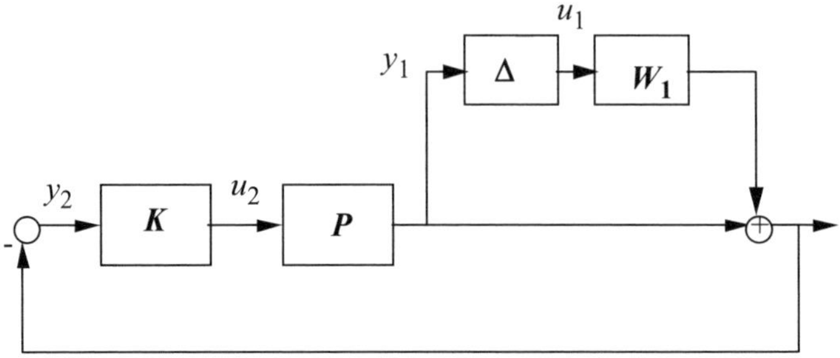

Figure 4.14. *Closed-loop with multiplicative error*

Figure 4.14 where $\overline{\sigma}\left(\Delta\left(i\omega\right)\right) < 1, \forall\omega \geq 0$. This diagram can then be put in the form given in Figure 4.13 with $\begin{bmatrix} 0 & P \\ -W_1 I & -P \end{bmatrix}$.

Therefore, $F_u\left(P_a, \Delta\right) = -\left(I + W_1\Delta\right)P = -\left(I + E\right)P = -\tilde{P}$ where $E = W_1\Delta$, and $F_l\left(P_a, K\right) = -PK\left(I + PK\right)^{-1}W_1$. Consider the following conditions:

(i) The controller with transfer matrix $K\left(s\right)$ stabilizes $P\left(s\right)$;

(ii) $P\left(s\right)$ and $\tilde{P}\left(s\right)$ have the same number of poles belonging to the closed right half-plane, taking into account multiplicities;

(iii) $P\left(s\right)$ and $\tilde{P}\left(s\right)$ have the same poles on the imaginary axis, taking into account multiplicities;

(iv) There exists a transfer function $W_1\left(s\right) \in \Re\mathcal{H}_\infty$ such that $\bar{\sigma}\left(E\left(i\omega\right)\right) < \left|W_1\left(i\omega\right)\right|$, for any frequency $\omega \geq 0$.

Let $\mathcal{E}$ be the set of transfer matrices $E(s)$ satisfying conditions (ii)–(iv), assuming that condition (i) holds. The following, which was proved in [114] (section 7.4), is now an obvious consequence of Theorem 96 (compare with Theorem 88):

COROLLARY 97.– *A necessary and sufficient condition for the closed-loop system to be stable for any modeling error $E\left(s\right) \in \mathcal{E}$ is $\left\|W_1 T_o\right\|_\infty \leq 1$.*

REMARK 98.– *(i) Theorem 96 is proved using the MIMO Nyquist criterion (Theorem 86), not the small gain theorem. Indeed, it is* not *assumed that Δ is a transfer matrix with entries in $\Re\mathcal{H}_\infty$ (and Δ can actually have poles in the closed right half-plane).*
(ii) In Corollary 97, the multiplicative model error $\mathbf{E}$ *is assumed to be at the output of* $\mathbf{P}$. *If this error is at the input of* $\mathbf{P}$, *the necessary and sufficient condition of closed-loop stability is changed to $\left\|W_1 T_i\right\|_\infty \leq 1$.*

Regarding performance, an inequality such as (4.14) can also characterize a satisfactory attenuation of a disturbance at the *output* of $\mathbf{P}$. The compromise between robustness and performance remains clear if the multiplicative uncertainty is located at the *output* of **P**, since $S_o + T_o = I_p$, but this is no longer the case when the multiplicative uncertainty is at the *input* of **P**, since then we must compare S_o and T_i, and these two quantities no longer have a simple relationship.

Shaping of singular values of $L_o(s)$

The loopshaping studied in section 4.2.7 can also be extended to the MIMO case. The transfer matrices $L_i(s) = K(s)P(s)$ and $L_o(s) = P(s)K(s)$ have to be distinguished. If we are interested in robustness against uncertainties located at the *output* of **P**, as well as in the attenuation of a disturbance affecting the *output* of **P**, the transfer matrix $L_o(s)$ of the open loop at *output* of **P** has to be considered. Suppose we have $\tilde{P}(s) = (I_p + E(s))P(s)$ and that the multiplicative error $E(s)$ at output of **P** is only large in high frequencies, in other words the above transfer function $W_1(s)$ satisfies $|W_1(i\omega)| \gg 1$ only for large values of ω. Then, the necessary and sufficient condition in Corollary 97, which is $\|W_1 T_o\|_\infty \leq 1$, only implies $\bar{\sigma}(T_o(i\omega)) \ll 1$ for large values of ω. Since $T_o = (I_p + L_o)^{-1} L_o$, we have $\bar{\sigma}(T_o(i\omega)) \sim \bar{\sigma}(L_o(i\omega))$ and the condition $\bar{\sigma}(T_o(i\omega))\, |W_1(i\omega)| \leq 1$ implies

$$\bar{\sigma}(L_o(i\omega)) \precsim \frac{1}{|W_1(i\omega)|} \ll 1. \qquad (4.28)$$

This replaces condition (4.13). Suppose also that heavy attenuation of the disturbance is required by specifications sheet only at output of **P** for small values of ω. This requirement translates into an inequality such as (4.14) where the transfer function $W_2(s)$ only satisfies $|W_2(i\omega)| \gg 1$ for small values of ω. Since $S_o = (I_p + L_o)^{-1}$, we have according to Proposition 581 (section 13.5.7)

$$\bar{\sigma}(S_o(i\omega)) = \frac{1}{\underline{\sigma}(I_p + L_o(i\omega))}$$

and this quantity cannot be larger than $\frac{1}{|W_2(i\omega)|}$, from which we get the condition

$$\underline{\sigma}(L_o(i\omega)) \gtrsim |W_2(i\omega)| \gg 1 \qquad (4.29)$$

which replaces (4.16). Conditions (4.29) (at high frequencies) and (4.29) (at low frequencies), plus the considerations on the modulus margin, contribute to shape of the singular values of $L_o(i\omega)$ as in Figure 4.15.

Robustness/performance compromise

The robustness/performance compromise seems more difficult to achieve in the case of MIMO systems (as compared to SISO ones) when the singular values of $L_o(i\omega)$ are very different,that is to say when the matrix $L_o(i\omega)$ is ill-conditioned (see

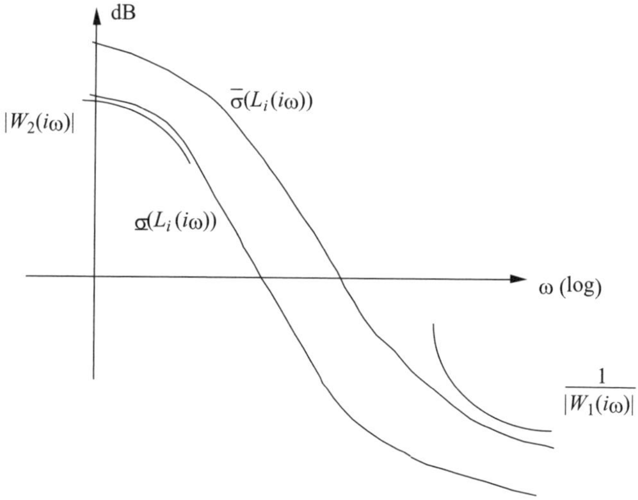

Figure 4.15. *Loopshaping in the MIMO case*

section 13.5.7). This is nonetheless inevitable when the system being controlled **P** is itself "ill-conditioned" (which is to say that its transfer matrix $P(i\omega)$ has non-zero singular values of a very different magnitude), a situation which is very unfavorable. The most important point is that $L_o(i\omega)$ then must have a condition number close to 1 at frequencies ω such that $\bar{\sigma}(L_o(i\omega)) \simeq 1$ (we then have $\bar{\sigma}(L_o(i\omega)) \simeq \underline{\sigma}(L_o(i\omega)) \simeq 1$): this is the so-called "singular value balancing" method (which is visibly not used in the case of Figure 4.15). We cannot go into much further details on this subject (see e.g. [90]).

The results presented in section 4.2.8, relative to the degradation of the robustness/performance trade-off due to the presence of unstable poles and zeros, can be generalized for the most part to the case of MIMO systems (see [51], [27]).

System conditioning

In the previous discussion, we have pointed out a major difficulty presented by "ill-conditioned" MIMO systems. We are now going to examine a second difficulty. The shaping of singular values, as shown in Figure 4.15, relates to the transfer matrix $L_o(s)$. Assuming that this shaping is correctly done, can we deduce that the closed-loop presents good robustness and performance at the *input* of **P**? In other words, does that also imply correct loopshaping with the singular values of $L_i(s)$? We have the following result for the case where the transfer matrices $P(s)$ of **P** and $K(s)$ of **K** are square (of dimension $m \times m$) and invertible (in the algebra $\mathbb{R}(s)^{m \times m}$), denoting by $\kappa(.)$ the condition number of the matrix in parentheses (section 13.5.7):

PROPOSITION 99.– *We have*

$$\frac{1}{\kappa(P)}\underline{\sigma}(L_o) \leq \underline{\sigma}(L_i) \leq \bar{\sigma}(L_i) \leq \kappa(P)\,\bar{\sigma}(L_o) \quad (4.30)$$

$$\frac{1}{\kappa(P)}\underline{\sigma}(L_i) \leq \underline{\sigma}(L_o) \leq \bar{\sigma}(L_o) \leq \kappa(P)\,\bar{\sigma}(L_i)\,. \quad (4.31)$$

PROOF. We have $L_i = K\,P = P^{-1}P\,K\,P = P^{-1}L_o\,P$, and therefore $\bar{\sigma}(L_i) \leq \bar{\sigma}(P^{-1})\,\bar{\sigma}(P)\,\bar{\sigma}(L_o) = \kappa(P)\,\bar{\sigma}(L_o)$. The other inequalities are proved by similar lines of reasoning. ■

A good approach for robust control of MIMO systems has to lead to a correct shaping of both the singular values of $L_o(s)$ and those of $L_i(s)$. The $\mathcal{H}_\infty$ synthesis method based on the representation of $P(s)$ by a "normalized coprime factorization", due to MacFarlane and Glover [89], is supposed to lead to such a result (however, their work is outside the scope of this book).

From Proposition 99, one can conclude that the regulation of *well-conditioned* MIMO systems poses no major problem. However, problems associated with *ill-conditioned* MIMO systems seem almost intractable, whatever the method used to synthesize the control law.

4.3. Exercises

EXERCISE 100.– *Consider the transfer function* $P(s) = \frac{1}{s(1+s)(1+0{,}25\,s)}$ *of a minimal system* $\mathbf{P}$. *(i) Trace the asymptotic Bode diagram of* $P(s)$ *and sketch from there the shape of its Bode plot. (ii) Trace the shape of the Nyquist diagram of* $P(s)$. *(iii) Trace the shape of its Black plot. (iv) Given below is a table of values of the Bode plot of* $P(s)$:

ω *(rad/s)*	*0,44*	*1*	*2*
Gain (dB)	*6,3*	*-3,3*	*-14*
Phase (°)	*-120*	*-149*	*-180*

Determine the proportional controller for which the closed-loop system has a phase margin of 60°. What are then the values of the delay margin and gain margin?

EXERCISE 101.– *Let* $P(s) = \frac{1}{s^2}$ *be the transfer function of a minimal system* $\mathbf{P}$ *fed back by a minimal controller* $\mathbf{K}$ *with transfer function* $K(s) = \frac{s}{s+1}$. *(i) According to Theorem 76, is the feedback system stable? (ii) Determine the transfer matrix of relation* (4.2). *Is that consistent with (i)? (iii) Trace the Nyquist plot of* $L(s)$. *(iv) By applying the Nyquist criterion (Theorem 84), study the stability of the closed-loop* *system. (v) Some authors state the Nyquist criterion with an indented imaginary axis that is different from that constructed for Theorem 84: the semi-circles $J_k(\varepsilon)$ go round the points p_k located on the imaginary axis* to the right, *instead of to the left (see Figure 4.2), n_P (resp., n_K) is therefore the number of poles of $P(s)$ (resp., $K(s)$) (accounting for multiplicities) that belong to the* open *right half-plane. According to this example, is the criterion obtained correct?*

EXERCISE 102.– *Let $P(s) = \frac{1}{s-1}$ be the transfer function of a minimal system $\mathbf{P}$ fed back by the minimal controller $\mathbf{K}$ with transfer function $K(s) = \frac{s-1}{s+1}$. (i) Answer again questions (i)–(iv) of Exercise 101 (using Theorem 80 instead of Theorem 84). (ii) Some authors replace the statement of Theorem 80 by the following: "The closed-loop system is stable if and only if $n_L = N$, where n_L is the number of poles of $L(s)$ (accounting for multiplicities) located in the closed right half-plane". Is this statement correct?*

EXERCISE 103.– *The hypotheses are those of Theorem 80. If $P(s)$ and $K(s)$ both belong to $\Re\mathcal{H}_\infty$, show that, according to the Nyquist criterion, we have the following sufficient condition of stability: "The closed-loop system is stable if all points of intersection between the curve $\{L(i\omega), \omega \geq 0\}$ and the real axis are located to the right of point -1. Is this a necessary condition?*

EXERCISE 104.– *A transfer function $P(s) \in \Re\mathcal{H}_\infty$ is said to be positive real (PR) if $\operatorname{Re} P(i\omega) \geq 0$ for any $\omega \geq 0$, and is quasi-strictly positive real (QSPR) if $\operatorname{Re} P(i\omega) > 0$ for any $\omega \geq 0$ [61]. The hypotheses are those of Theorem 80. Using the Nyquist criterion, show that if both $P(s)$ and $K(s)$ are PR and at least one of them is QSPR, then the closed-loop system is stable.*

EXERCISE 105.– **Let $\mathbf{P}_\gamma$ be a system defined by the differential equation $\dot{y} = A_0\, y + B_\gamma\, u$ with $A_0 = \begin{bmatrix} -1 & 0 \\ 0 & -1 \end{bmatrix}$ and $B_\gamma = \begin{bmatrix} 1 & -\gamma \\ 0 & 1 \end{bmatrix}$, $\gamma \geq 0$, fed back by the controller $\mathbf{K} = I_2$. (i) Trace the MIMO Nyquist plot of $L_o(s) = P(s)\, K(s) = K(s)\, P(s) = L_i(s)$. What is the distance from this plot to critical point -1? (ii) Assume that the modeling error can be effected by replacing the above matrix A_0 by the matrix $A_\varepsilon = \begin{bmatrix} -1 & 0 \\ \varepsilon & -1 \end{bmatrix}$, $\varepsilon > 0$. Show that for any $\varepsilon > 0$, the closed-loop system is unstable for a sufficiently large γ. (iii) Show that for $\gamma \to +\infty$, $Mm_{i,\gamma} = Mm_{o,\gamma}$ tends toward 0, where $Mm_{i,\gamma}$ (resp., $Mm_{o,\gamma}$) is the input (resp., output) modulus margin of the system $\mathbf{P}_\gamma$ fed back by the controller $\mathbf{K}$. (iv) What can be concluded?*

Chapter 5

Compensation and PID Controller

5.1. One degree of freedom controller

5.1.1. *Closed-loop system*

We have already seen in the previous chapter that in some cases, a proportional controller could be used to stabilize a system by feedback, or even to obtain a given phase margin (see Example 85, section 4.1.4, and Exercise 100, section 4.3). Nevertheless, it is clear that the class of proportional controllers is far too restricted to allow for a proper shaping of the open-loop transfer function (see section 4.2.7). This chapter is dedicated to the study of classic controllers that can improve the robustness/performance compromise when a controller with one degree of freedom (1-DOF) is used. The 1-DOF controller $\mathbf{K}$ is shown in Figure 5.1. In this figure, u and y denote, respectively, the control and the output of $\mathbf{P}$, n is the measurement noise (due to the presence of the sensor), d_u and d_y are additive disturbances acting on u and y, respectively, and r is a reference signal. The objective of the regulation is to render the error $e = y - r$ as small as possible in steady state, assuming that the spectrum of the disturbances and that of the reference signal is only located at low frequencies, and thus these signals can be modeled as constants. The spectrum of the measurement noise is assumed to be located at high frequencies.

5.1.2. *Closed-loop equations and static error*

The system $\mathbf{P}$ is SISO, therefore there is no need to distinguish the sensitivity functions at input and at output of $\mathbf{P}$; both of them are denoted by S_o. We thus have $S_o = \frac{1}{1+L}$, where $L = PK = KP$ is the open-loop transfer function. The complementary sensitivity function is denoted as $T_o = 1 - S_o$. It is easy to show

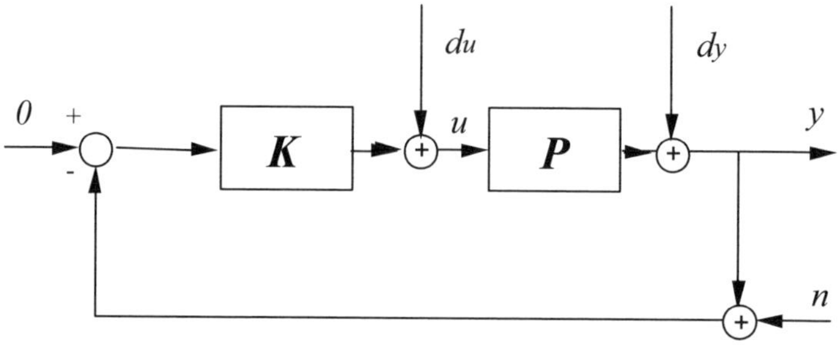

Figure 5.1. *1-DOF controller*

that the closed-loop system is governed by the following relations (assuming that the initial conditions are zero):

$$\hat{e}(s) = S_o(s)\left(\hat{d}_y(s) - \hat{r}(s) + P(s)\,\hat{d}_u(s)\right) - T_o(s)\,\hat{n}(s) \tag{5.1}$$

$$\hat{u}(s) = S_o(s)\,\hat{d}_u(s) + K(s)\,S_o(s)\,(\hat{r}(s) - \hat{n}(s)). \tag{5.2}$$

All the analysis done in Chapter 4 remains valid. We still need to precisely establish under what condition the static error is zero (i.e. $e(t) = 0$ in steady state); it is understood that the closed-loop system is stable. This only makes sense if the influence of n on e is neglected, which means one has to assume that condition (4.13) holds with an appropriate weighting function $W_1(s)$ (see section 4.2.6). Let us consider the following two cases:

(i) $d_u = 0$. Then the static error is zero if and only if the static gain of $S_o(s)$ is zero (see secton 3.1.2), i.e. $S_o(0) = 0$.

(ii) $d_u \neq 0$. Then the static error is zero if and only if the static gain of both $S_o(s)$ and $S_o(s)\,P(s)$ is zero.

To specify the above-given conditions, let us express the transfer functions $P(s)$ and $K(s)$ in the form of irreducible rational functions $\frac{B(s)}{A(s)}$ and $\frac{R(s)}{S(s)}$, respectively, where $A(s)$, $B(s)$, $R(s)$, $S(s)$ belong to $\mathbb{R}[s]$. We have

$$S_o(s) = \frac{A(s)\,S(s)}{A(s)\,S(s) + B(s)\,R(s)},$$

$$S_o(s)\,P(s) = \frac{B(s)\,S(s)}{A(s)\,S(s) + B(s)\,R(s)}.$$

As a result:

(a) In situation (i), the static error is zero if and only if $A(0) = 0$ or $S(0) = 0$. Since $L(s) = \frac{B(s)\,R(s)}{A(s)\,S(s)}$, this means that the open-loop transfer function has a zero pole (i.e. the open-loop is an integrator system: see Definition 32). *If P is already an integrator system, it is unnecessary that controller K be one too.*

(b) In situation (ii), the static error is zero if and only if the two following conditions hold: $A(0)\ S(0) = 0$ and $B(0)\ S(0) = 0$. Suppose the first equality holds, then $B(0)$ cannot be zero, for otherwise the characteristic polynomial of the closed-loop, $A_{cl}(s) = A(s)\ S(s) + B(s)\ R(s)$, would have a root equal to 0, and the closed-loop system would be unstable (see Theorem 76, section 4.1.3). *Therefore, it is necessary (and sufficient) that $S(0) = 0$, i.e. that the controller K be integrator.*

5.2. Lead compensator

5.2.1. *Characteristics of a lead compensator*

A *lead compensator* is a minimal system with transfer function form

$$H_{\alpha,\tau}(s) = \frac{1 + \alpha\,\tau\,s}{1 + \tau\,s}$$

where $\tau > 0$ is a time constant and α is a real number greater than 1. The Bode plot of $H_{\alpha,\tau}(s)$ (with its asymptotic diagram) is represented in Figure 5.2 for the case of $\tau = 1$ and $\alpha = 5$.

The maximum value φ_d of the phase (denoted as ϕ in the figure) is given by

$$\varphi_d = \arcsin\frac{\alpha - 1}{\alpha + 1} \tag{5.3}$$

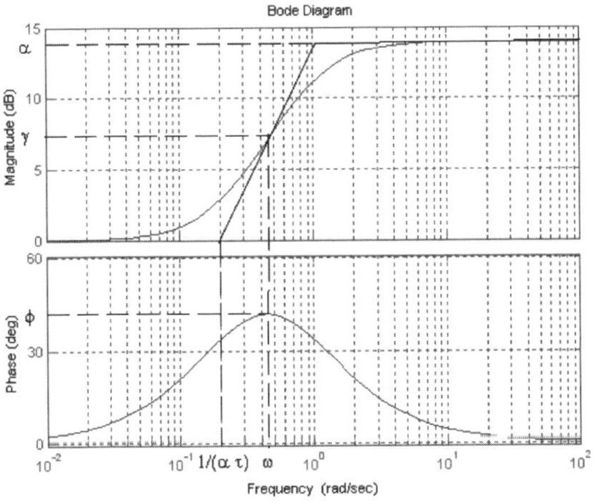

Figure 5.2. *Bode plot of a lead compensator*

and it is attained at the angular frequency

$$\omega_0 = \frac{1}{\tau\sqrt{\alpha}} \tag{5.4}$$

which is the logarithmic mean of $\frac{1}{\alpha\,\tau}$ and $\frac{1}{\tau}$ (this frequency is denoted as ω in Figure 5.2). We can easily verify that

$$\gamma = |H_{\alpha,\tau}(i\,\omega_0)| = \sqrt{\alpha} \tag{5.5}$$

(see Exercise 106(i)).

5.2.2. *Principles of a lead compensator*

Now, we will determine a compensator that:

(i) produces a maximum phase lead φ_d at a given frequency $\omega_0 > 0$ (with $0 < \varphi_d < 90°$);

(ii) has unity gain at that frequency.

According to section 5.2.1, the minimal system with transfer function $H(s) = \frac{1}{\gamma}H_{\alpha,\tau}(s)$ answers the question, with, according to (5.3) and (5.4),

$$\alpha = \frac{1+\sin\varphi_d}{1-\sin\varphi_d}, \qquad \tau = \frac{1}{\omega_0\sqrt{\alpha}}, \tag{5.6}$$

and finally we get

$$\boxed{H(s) = \frac{1}{\sqrt{\alpha}}\frac{1+\frac{\sqrt{\alpha}}{\omega_0}s}{1+\frac{1}{\omega_0\sqrt{\alpha}}s}}. \tag{5.7}$$

Consider again the example in section 4.2.3. Suppose we wish to obtain a phase lag margin of 45°, no longer at frequency 0.3 rad/t.u., but at 0.5 rad/t.u. It is thus necessary to "accelerate" the feedback system and one can for this purpose make use of a lead compensator. The Black plot of $P(s) = \frac{1-s/3}{s(1+s/2)(1+2s)}$ is represented in Figure 5.3, where the time unit is in seconds.

To bring the point at 0.5 rad/s to a gain of 0 dB and a phase of $180° + 45° = 225°$, two operations are necessary:

(i) to add a gain of -2.91 dB, which corresponds to a ratio of 0.72;

(ii) to add a phase of $225° - 202° = 23°$ at a frequency $\omega_0 = 0.5$ rad/s.

Operation (ii) is done using a lead compensator having characteristics, according to (5.6) and (5.7), $\alpha = 2.28$ and $H(s) = 0.66\,\frac{1+3.0\,s}{1+1.324\,s}$. The appropriate controller thus has a transfer function $K(s) = 0.72\,H(s) = 0.47\,\frac{1+3.0\,s}{1+1.32\,s}$. The Black plot of the compensated system (i.e. $L(s) = P(s)\,K(s)$) is represented in Figure 5.4.

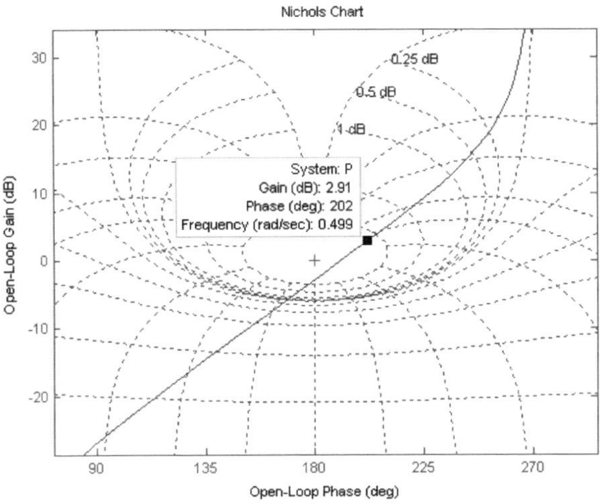

Figure 5.3. *Black plot of P(s)*

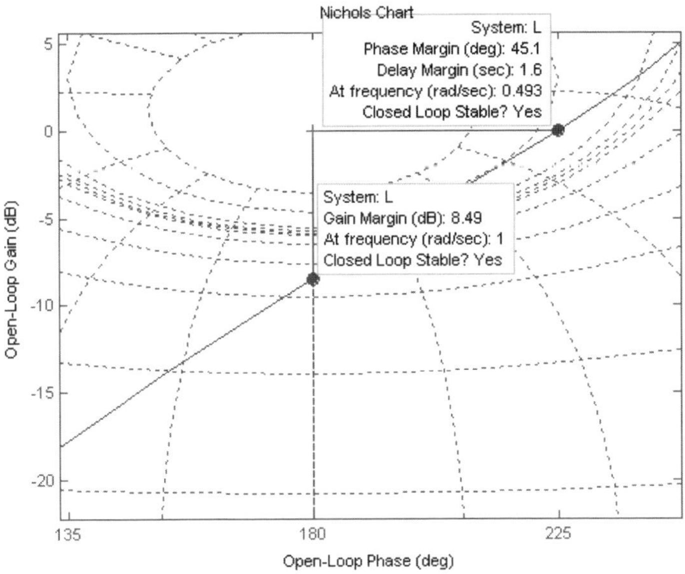

Figure 5.4. *Black plot of L(s)*

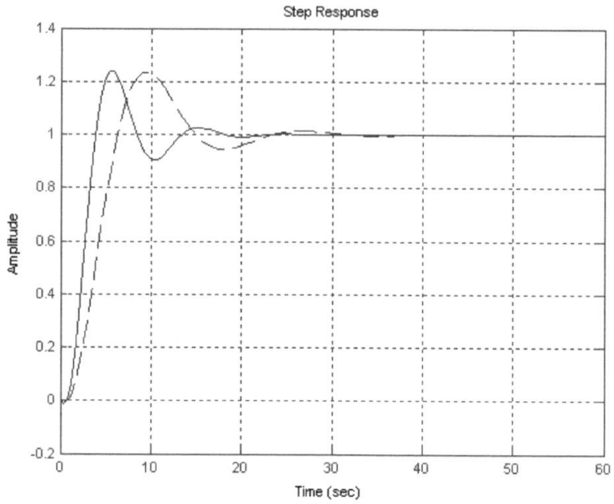

Figure 5.5. *Step responses*

In Figure 5.5, we see a comparison of two step responses of the feedback system: that obtained with the proportional controller determined at section 4.2.3 (with gain $k = 0.35$) (- -) and that obtained with the controller $K(s)$ determined here above (-). We clearly see the *acceleration due to the lead compensator.*

5.2.3. *PD controller*

The theoretical transfer function of a proportional and derivative (PD) controller is of the form

$$K_{PD}(s) = k(1 + T_d s) \tag{5.8}$$

where $k > 0$ is the "gain of the proportional term" and $T_d > 0$ is the "time constant of the derivative term". Such a controller has an improper transfer function, and therefore is not practically realizable (see section 2.5.3) and does not abide by the constraint as highlighted in section 4.2.6. The "pure derivative" of the transfer function $T_d s$ is therefore replaced by a "filtered derivative" with transfer function $\frac{T_d\, s}{1+\frac{T_d}{N}\, s}$, where $N > 0$ is a real number called the "filtering coefficient of the derivative term". The transfer function (5.8) is thus replaced by $K_{PDF}(s) = k\left(1 + \frac{T_d\, s}{1+\frac{T_d}{N}\, s}\right) = \lambda\frac{1+\alpha\,\tau\, s}{1+\tau\, s}$ with $\lambda = k$, $\tau = \frac{T_d}{N}$ and $\alpha = N + 1$. As a result, $K_{PDF}(s)$ is the transfer function of a lead compensator in series with a proportional controller.

5.3. PI controller

5.3.1. *Principle*

In the example considered in section 5.2.2, a step response without static error is obtained because the system **P** considered is integrator. Nonetheless, a constant disturbance d_u added at the input of **P** will introduce a static error, which is also inevitable in another example where **P** is not an integrator (see section 5.1.2). That is why the most widely used controllers have an integral action.[1] Among these controllers, those that have the simplest structure are of the "proportional and integral" type (a controller that is only of integral type cannot ensure the stability of the closed-loop for a class of control systems **P** that is sufficiently large).

The transfer function of a proportional and integral (PI) controller is of the form

$$K_{PI}(s) = k\left(1 + \frac{1}{T_I\, s}\right) \tag{5.9}$$

where $k > 0$ is the "gain of the proportional term" and $T_I > 0$ is the "time constant of the integral action".

We have for $\omega > 0$

$$\arg K_{PI}(i\omega) = \arg(1 + i\,T_I\,\omega) - \frac{\pi}{2} = \arctan(T_I\,\omega) - \frac{\pi}{2}. \tag{5.10}$$

We can therefore determine a PI controller, having "high-frequency gain" $\gamma > 0$, and such that for a given $\omega_0 > 0$,

$$\begin{aligned} |K_{PI}(i\,\omega_0)| &= 1 \\ -\arg K_{PI}(i\omega_0) &= \varphi_I, \quad 0 < \varphi_I < 90^\circ \end{aligned} \tag{5.11}$$

where φ_I is the phase lag produced by the PI controller at frequency ω_0. From (5.10), the two above inequalities hold if and only if

$$T_I = \frac{1}{\omega_0 \tan\varphi_I} \tag{5.12}$$

$$\gamma = \frac{T_I\omega_0}{\sqrt{1 + (T_I\omega_0)^2}} \tag{5.13}$$

(see Exercise 106(ii)).

1. One case where such an action must be avoided is that where **P** is a derivator, i.e. $P(0) = 0$ (see Definition 32).

With the above compensation, we obtain a phase lag margin equal to Mp on the condition that $\arg P(i\,\omega_0) - \varphi_I = -180° + Mp$. According to (5.11), this is realizable if and only if

$$\boxed{Mp - 180° < \arg P(i\,\omega_0) < Mp - 90°}. \tag{5.14}$$

5.3.2. *Example*

Consider the minimal system $\mathbf{P}$ with transfer function

$$P(s) = 2\,\frac{1 - 0.1\,s}{(1 + 0.5\,s)\,(1 + s)}.$$

The time unit is assumed to be in seconds. We wish to make use of this system in a way that the phase lag margin is 60°. The Black plot of $P(s)$ is represented in Figure 5.6.

Start with a proportional control. By way of vertical translation, we move point A_P with a phase of $-120°$ (obtained for an angular frequency of 2 rad/s) to unity gain by the addition of a gain of 3.79 dB, which corresponds to a ratio of 1.55. The open-loop transfer function thus becomes

$$L_P(s) = 3.1\,\frac{1 - 0.1\,s}{(1 + 0.5\,s)\,(1 + s)}.$$

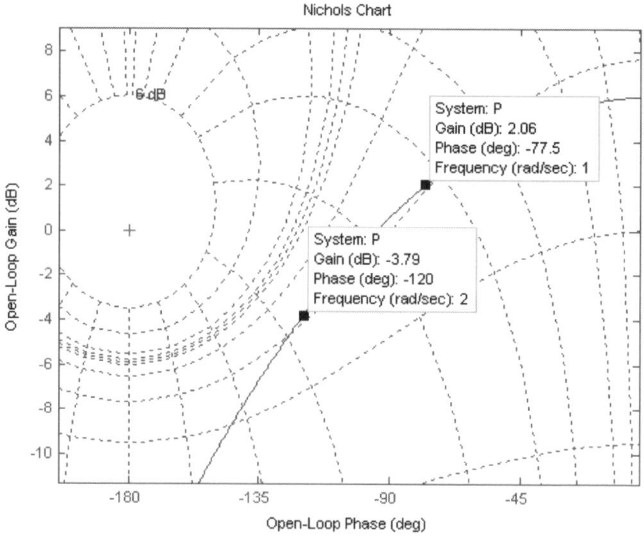

Figure 5.6. *Black plot of P(s)*

The unity gain angular frequency is then $\omega_{0P} = 2$ rad/s.

Now add an integral action. Since this produces a phase lag, it is necessary to first consider a point A_{PI} in the Black plot that has a phase larger than A_P; this point is obtained for an angular frequency smaller than 2 rad/s, $\omega_{0PI} = 1$ rad/s for example. We will bring this point to unity gain and a phase of $-120°$ by two successive operations:

(i) adding a gain of -2.06 dB, which corresponds to a ratio of 0.79;

(ii) a phase lag of $120° - 77.5° = 42.5°$.

Operation (ii) can be obtained by a PI controller that has integral action time constant $T_I = 1.09$ s (according to (5.12)) and whose high-frequency gain is $\gamma = 0.74$ (according to $(5.1.2)$).

The PI controller found thus has a transfer function $K_{PI}(s)$ given by (5.9) with $k = 0.79 \times 0.74 = 0.58$ and $T_I = 1.09$ s.

The Black plots of $L_P(s)$ (-) and $L_{PI}(s) = K_{PI}(s)\, P(s)$ (- -) are shown in Figure 5.7. Finally, the step responses of the two feedback systems are shown in Figure 5.8 (lines with the same conventions). One can observe the *"slowing-down" of the response because of the addition of the integral action* (but also the suppression of the static error).

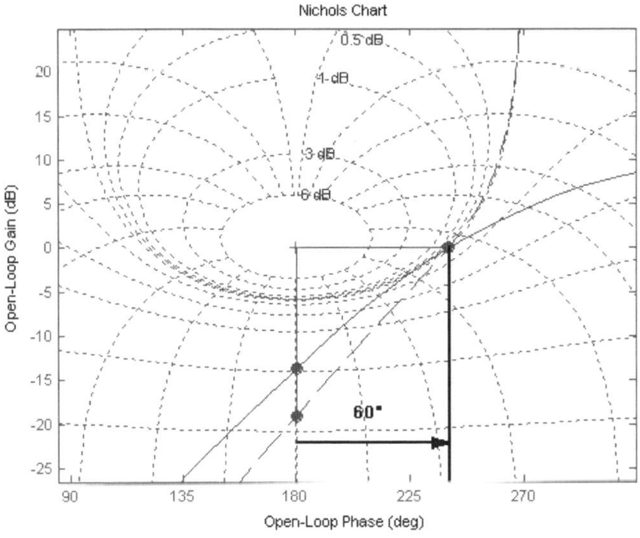

Figure 5.7. *Black plots of compensated systems*

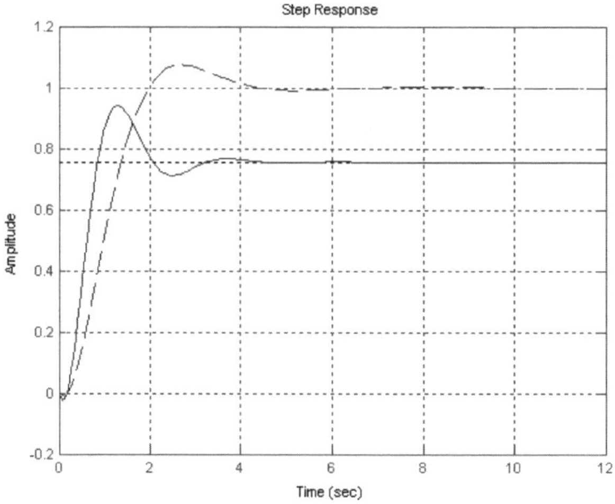

Figure 5.8. *Step responses*

5.4. PID controller

5.4.1. *Integral action and lead compensator*

As seen above, an integral action makes it possible to eliminate the static error at the price of slowing-down of the closed-loop system (section 5.1.2) whereas a lead compensator accelerates the closed-loop system (section 5.2.2). We thus understand that, by combining these two effects, one can eliminate the static error while maintaining a satisfactory rapidity on the closed-loop system.

If we use a simple PI controller, the frequency ω_0 at unity gain is restricted by condition (5.14). Suppose we want to obtain a higher unity gain frequency ω_0, not satisfying this condition. We can proceed in the following manner:

(i) realize a phase lead φ_d of unity gain at angular frequency ω_0 $(0 < \varphi_d < 90°)$.

(ii) put in series with the above phase lead a PI controller having unity gain at frequency ω_0 and phase lag φ_I at this same frequency $(0 < \varphi_I < 90°)$, where

$$\arg P\left(i\omega_0\right) + \varphi_d - \varphi_I = -180° + Mp. \tag{5.15}$$

(iii) put in series with the two above compensators a gain $\frac{1}{|P(i\,\omega_0)|}$ (if expressed in dB, add a gain of $-\left|P\left(i\,\omega_0\right)\right|_{\mathrm{dB}}$).

Relation (5.15) is realizable with the above constraints over φ_d and φ_I if and only if

$$\boxed{Mp - 270° < \arg P\left(i\omega_0\right) < Mp - 90°} \tag{5.16}$$

a condition that is less restrictive than (5.14). The controller finally obtained is of the PID type, as we will see next.

Note that relation (5.15) involves only the difference $\varphi_d - \varphi_I$; we can thus arbitrarily fix the sum $S = \varphi_d + \varphi_I$ or determine S in such a way so as to minimize an appropriate criterion (e.g. the overshoot of the step response). A reasonable choice is $S = 90°$.

Let us illustrate what has been said here above through the example in section 5.3.2. Suppose we would like the closed-loop system to have a phase margin $Mp = 60°$, but with unity gain angular frequency $\omega_0 = 2$ rad/s. We have $\arg P(i\omega_0) = -120°$, thus condition (5.14) is not satisfied, whereas (5.16) is. According to (5.15), we have $\varphi_d = \varphi_I$. Taking $S = 90°$, we thus obtain $\varphi_d = \varphi_I = 45°$.

According to (5.12), $T_I = 0.5$ s, and $\gamma = \frac{1}{\sqrt{2}}$ from (5.13); the PI controller having unity gain at angular frequency ω_0 has a transfer function $K_{PI}(s) = \frac{1}{\sqrt{2}}\left(1 + \frac{2}{s}\right) = 0.707\left(1 + \frac{2}{s}\right)$.

From (5.6), $\alpha = 3 + 2\sqrt{2}$ and the appropriate lead compensator has a transfer function given by (5.7), i.e. $H(s) = 0.414\,\frac{1+1.21\,s}{1+0.21\,s}$.

Finally, the last compensation to realize is a gain of 3.79 dB, corresponding to a ratio of 1.55.

The transfer function of the desired PID controller is therefore

$$K_{PID}(s) = 0.453\,\frac{1 + 1.21\,s}{1 + 0.21\,s}\left(1 + \frac{2}{s}\right).$$

The Black plot of the compensated system is shown in Figure 5.9. The Bode plots of the compensated systems are shown in Figure 5.10: with the proportional controller (-), the PI controller (- -), and the PID controller (-.). The step responses of the feedback systems are shown in Figure 5.11, and the Bode plots of the three controllers obtained are shown in Figure 5.12 (lines with the same conventions).

5.4.2. *Classic form of a PID controller*

The most classic form of the transfer function of a PID controller is

$$K(s) = k\left(1 + \frac{1}{T_I\,s} + \frac{T_d\,s}{1 + \frac{T_d}{N}\,s}\right) \tag{5.17}$$

where $k > 0$ is the "gain of the proportional term", $T_I > 0$ is the "time constant of the integral action", $T_d \geq 0$ is the "time constant of the derivative term", and where $N > 0$ is the "filtering coefficient of the derivative term".

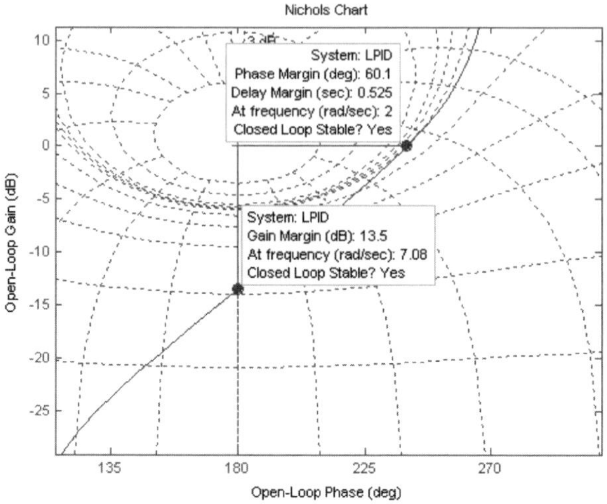

Figure 5.9. *Black plot of the compensated system with PID*

We can write (5.17) in the form

$$K(s) = \frac{k\,(1+N)\,s^2 + k\left(\frac{N}{T_d} + \frac{1}{T_I}\right)\,s + \frac{k\,N}{T_I\,T_d}}{s^2 + \frac{N}{T_d}s}.$$

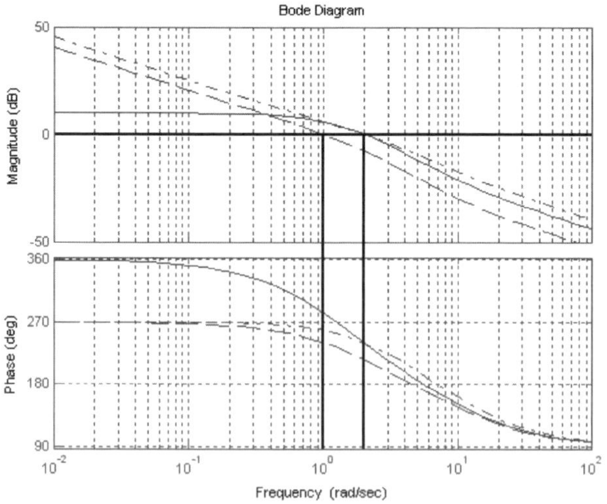

Figure 5.10. *Bode plots of compensated systems*

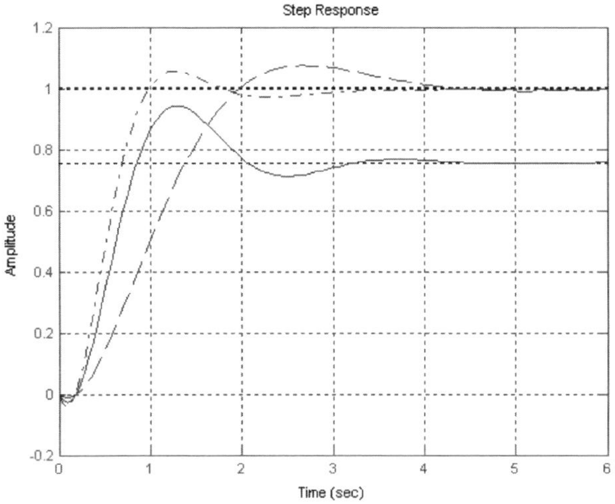

Figure 5.11. *Step responses (with P, PI, PID)*

The transfer function of the controller obtained in section 5.4.1 is of the form

$$K'(s) = \kappa \frac{1 + \alpha \tau_1 s}{1 + \tau_1 s} \left(1 + \frac{1}{\tau_2 s}\right). \tag{5.18}$$

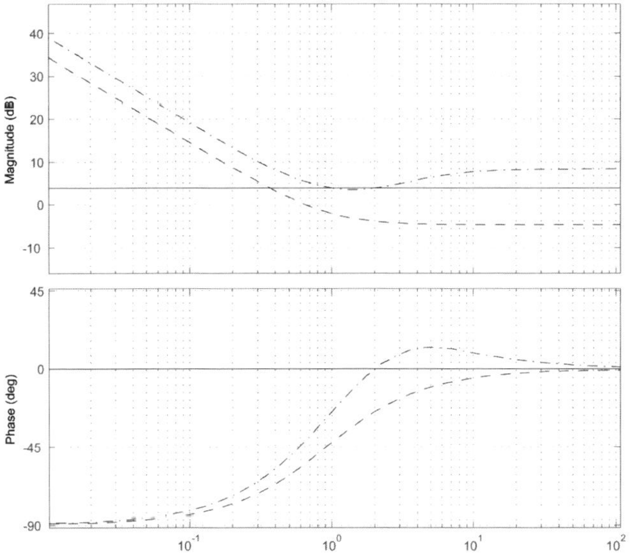

Figure 5.12. *Bode plots of P, PI and PID controllers*

As a result, we have $K(s) = K'(s)$ if and only if $\delta = 1 + (\alpha - 1)\frac{\tau_1}{\tau_2}$:

$$N = \frac{\alpha}{\delta} - 1, \quad T_d = N\,\tau_1, \quad k = \kappa\,\delta, \quad T_I = \delta\,\tau_2 \tag{5.19}$$

(see Exercise 106(iii)).

The controller obtained by cascading a lead compensator and a PI controller is thus a classic PID controller if and only if $N > 0$, i.e.

$$\boxed{\tau_1 < \tau_2} \tag{5.20}$$

a relation which is satisfied if the lead compensator acts upon frequencies that are higher than the integral action, which is often (but not always) the case.

Reconsider the example in given in section 5.4.1: we obtained a controller $K'(s)$ of the form (5.18) with $\kappa = 0.453$, $\tau_1 = 0.21$, $\alpha = 5.83$, $\tau_2 = 0.5$. Condition (5.20) is satisfied, and therefore the controller is called a classic PID. The relations (5.19) yield $\delta = 3.02$; $N = 0.92$; $T_d = 0.19$ s; $k = 1.37$; $T_I = 1.50$ s.

5.5. Exercises

EXERCISE 106.– *(i) Prove expressions* (5.3), (5.4), *and* (5.5). *(ii) Prove expressions* (5.12) *and* (5.13). *(iii) Prove the relations* (5.19).

EXERCISE 107.– *Let* **P** *be the system considered in Exercise 100 (section 4.3). (i) Suppose there is a constant disturbance that adds to the* output *of* **P** *(prior to the controlled variable). Does the proportional controller calculated in Exercise 100 reject the disturbance? In other words, is the regulation made without any static error? (ii) We would now like to accelerate the feedback system in a manner to obtain a unity gain angular frequency of* 1 *rad/s and also retain a phase lag margin of* $60°$. *Determine an appropriate controller. (iii) Suppose now that a constant disturbance is added to the* input *of* **P**. *Is the answer to question (i) still exact? (iv) Determine a PID controller for which the phase margin is* $60°$ *with a unity gain angular frequency of* 1 *rad/s.*

Chapter 6

RST Controller

6.1. Structure and implementation of an RST controller

In the previous chapter, we studied a class of controllers of basic structure. Although these controllers are quite common in the industry, it is clear from Chapter 4 that they do not enable us to find an accurate robustness/performance compromise. In order to manage this compromise in a satisfactory manner – and even, in numerous cases, just to obtain a stable closed-loop in the absence of modeling error – it becomes necessary to consider a much larger class of controllers. This is the subject of the present chapter.

Consider a SISO system, governed by the differential equation

$$A(\partial)\; y = B(\partial)\; u + d \tag{6.1}$$

where $\partial = \frac{d}{dt}$ and where $A(\partial) \in \mathbb{R}[\partial]$ and $B(\partial) \in \mathbb{R}[\partial]$ are the polynomials

$$\begin{aligned} A(\partial) &= \partial^n + a_1\,\partial^{n-1} + \cdots + a_n \\ B(\partial) &= b_1\,\partial^{n-1} + \cdots + b_n; \end{aligned}$$

d is an unmeasured disturbance, y is the variable to be regulated, and u is the control. We are also given a reference signal r and we call

$$e = y - r \tag{6.2}$$

the tracking error. Our purpose is to design a controller that makes e zero in steady state – irrespective of the presence of the disturbance d – and ensures a good transient response. It is necessary to measure y, and we denote the measured variable by z, so that

$$z = y + \nu \tag{6.3}$$

where ν is the measurement noise.

An RST controller is linearly related to the input u, the measured variable z, and the reference r, as well as to a finite number of derivatives of these variables. It is thus governed by a differential equation of the form

$$\boxed{S(\partial)\, u = -R(\partial)\, z + T(\partial)\, r} \tag{6.4}$$

where $S(\partial)$, $R(\partial)$, and $T(\partial)$ are elements of $\mathbb{R}[\partial]$.

An RST controller thus has two inputs (z and r) and one output (u). Its transfer matrix is

$$H(s) = \begin{bmatrix} -\frac{R(s)}{S(s)} & \frac{T(s)}{S(s)} \end{bmatrix}. \tag{6.5}$$

It is proper if and only if the following condition holds:

$$\boxed{\max\{d^\circ(R),\, d^\circ(T)\} \leq d^\circ(S)}. \tag{6.6}$$

The structure of the RST controller, resulting from equation (6.4), is represented in Figure 6.1. It cannot be implemented as such because the differential polynomials $R(\partial)$ and $T(\partial)$ generate derivatives of signals z and r (except in the very particular case where the polynomials are of degree zero). Let $V(\partial)$ be a polynomial such that

$$\max\{d^\circ(R),\, d^\circ(T)\} \leq d^\circ(V) \leq d^\circ(S)$$

and all its roots belong to the left half-plane. In passing into the Laplace domain with zero initial conditions, equation (6.4) yields

$$\frac{S(s)}{V(s)}\hat{u}(s) = -\frac{R(s)}{V(s)}\hat{z}(s) + \frac{T(s)}{V(s)}\hat{r}(s).$$

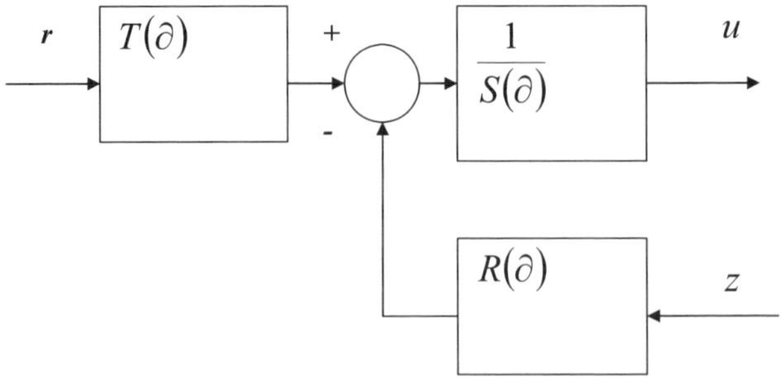

Figure 6.1. *Initial structure of the RST controller*

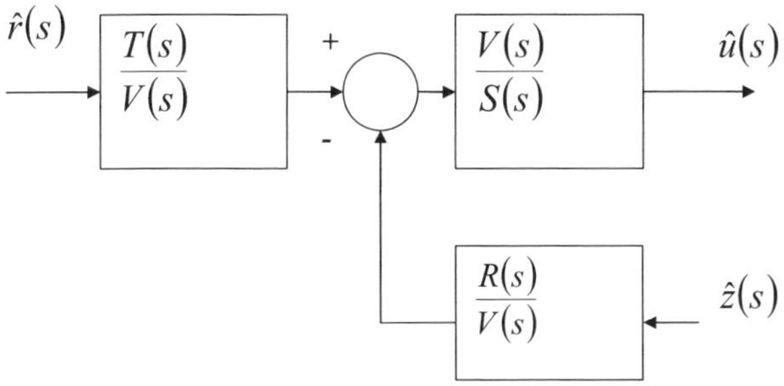

Figure 6.2. *Implementation of the RST controller*

This equation leads to the implementation diagram given in Figure 6.2, where the three transfer functions $\frac{R(s)}{V(s)}$, $\frac{T(s)}{V(s)}$, and $\frac{V(s)}{S(s)}$ are proper. Such a controller is said to be with three degrees of freedom (3-DOF) – with reference to the above three transfer functions.

If $T = R$, relation (6.4) reduces to

$$S(\partial)\, u = -R(\partial)\,(z - r)$$

and the controller is 1-DOF (see section 5.1). On the other hand, if polynomial $R(\partial)$ has all its roots in the left half-plane, and if $d^\circ(T) \leq d^\circ(R)$, we can choose $V = R$ and the controller in Figure 6.2 is reduced to two degrees of freedom (2-DOF).

6.2. Closed loop

6.2.1. *Closed-loop equations*

According to (6.1), (6.2), (6.3), and (6.4) we obtain

$$A_{cl}(\partial)\; y = S(\partial)\; d - B(\partial)\; R(\partial)\; \nu + B(\partial)\; T(\partial)\; r \tag{6.7}$$

$$A_{cl}(\partial)\; e = S(\partial)\; d + \{B(\partial)\; [T(\partial) - R(\partial)] - A(\partial)\; S(\partial)\}\; r - B(\partial)\; R(\partial)\; \nu \tag{6.8}$$

$$A_{cl}(\partial)\; u = -R(\partial)\; d + A(\partial)\; T(\partial)\; r - A(\partial)\; R(\partial)\; \nu \tag{6.9}$$

where

$$\boxed{A_{cl}(\partial) = A(\partial)\; S(\partial) + B(\partial)\; R(\partial)}\,. \tag{6.10}$$

From (6.8) and (6.9), the "characteristic polynomial of the controlled system" is $A_{cl}(\partial)$ given by (6.10), which is coherent with Theorem 76 and Definition 78 (section 4.1.3).

6.2.2. *Pole placement and stability*

Pole placement consists of fixing in advance polynomial $A_{cl}(\partial)$ or, equivalently, the roots of this polynomial, i.e. the poles of the controlled system. We are led to determine polynomials $S(\partial)$ and $R(\partial)$ such that equation (6.10) can be satisfied. This equation is a Bézout equation (see section 13.1.5). It admits a solution if and only if $A_{cl}(\partial)$ is a multiple of the greatest common divisor (gcd) of $A(\partial)$ and $B(\partial)$ (Theorem 494).

According to Definition 74 (section 4.1.3), the controlled system is stable if and only if the roots of $A_{cl}(\partial)$ are all located in the left half-plane (taking into account the fact that $d^\circ(A) < d^\circ(B)$ and condition (6.6)). As a result, it is possible to stabilize the system $\mathbf{P}$ defined by the left form

$$\boxed{A(\partial)\, y = B(\partial)\, u} \tag{6.11}$$

if and only if $\gcd(A(\partial), B(\partial))$ has all its roots in the left half-plane; in addition, it is possible to obtain arbitrarily chosen poles for the controlled system if and only if $\gcd(A(\partial), B(\partial)) = 1$. We are led to the following definition:

DEFINITION 108.– *System* (6.11) *is controllable (resp., stabilizable) if* $\gcd(A(\partial), B(\partial)) = 1$ *(resp., the roots of* $\gcd(A(\partial), B(\partial))$ *all lie in the left half-plane).*

Nevertheless, obtaining a controlled system with correctly placed poles resolves only part of the problem. It is now appropriate to consider this whole problem, and then solve it.

6.3. Usual case

The case we consider here is one where the reference signal and the disturbance have their spectrum in low frequencies, and thus are modeled as constants, and where the measurement noise has its spectrum in high frequencies. This framework has already been studied in sections 4.2.5, 4.2.6 and 5.1.1. On the other hand, system (6.11) is assumed to be controllable.

6.3.1. *Disturbance rejection*

Let us start by providing an interpretation of disturbance d : suppose that the system to be controlled is affected by two *constant* disturbances, one at input, d_u, and the other at output, d_y, as shown in Figure 5.1. Writing $P(s) = \frac{B(s)}{A(s)}$, we obtain

$$\hat{y}(s) = \hat{d}_y(s) + \frac{B(s)}{A(s)}\left(\hat{u}_c(s) + \hat{d}_u(s)\right)$$

where u_c is the control signal provided by the controller. As a result, we obtain (6.1) with $u = u_c$ and

$$d = A(\partial)\ d_y + B(\partial)\ d_u.$$

We can have two cases where $d = 0$:

(i) $d_u = 0$ and $A(0) = 0$ (integrator system: see Definition 32);

(ii) $d_y = 0$ and $B(0) = 0$ (derivator system: see Definition 32).

In these two very specific cases, there is no need for disturbance rejection, which simplifies the structure of the controller. They fall into the general context of section 6.4, but we will not be covering it in the rest of this section.

The only information at our disposal about the disturbance d is that it satisfies the equation $\partial d = 0$. According to (6.8), it is "rejected" – in other words, thanks to the control mechanism, it does not affect the error e in steady-state – if and only if $S(\partial)$ is a multiple of ∂, or equivalently if $S(\partial)$ has a zero root:

$$\boxed{S(0) = 0}. \tag{6.12}$$

The above constraint imposes the controller to have a zero pole (from expression (6.5) of the transfer matrix), or in other words to be an integrator (Definition 32).

6.3.2. *Absence of static error*

The only information we have about reference signal r is that it satisfies the equality $\partial r = 0$. Suppose condition (6.12) holds. According to (6.8), the term r does not affect the error e in steady-state if and only if $B(0)[T(0) - R(0)] = 0$. If $B(0) = 0$, from (6.9) and (6.10), $s = 0$ is a root of the characteristic polynomial $A_{cl}(s)$ and the feedback system is not stable; this possibility is to be ruled out. The static error is therefore zero (when $\nu = 0$) if and only if

$$\boxed{T(0) = R(0)}. \tag{6.13}$$

6.3.3. *Measurement noise filtering*

As mentioned in section 4.2.6, it is necessary for the controller to have limited gain at high frequencies in such a way that the measurement noise does not introduce an excessive agitation into the control variable. According to (6.6), the relative degree $\delta\left(\frac{R}{S}\right)$ is necessarily ≥ 0 and we can arrive at imposing a condition such that

$$\boxed{\delta\left(\tfrac{R}{S}\right) \geq \delta_0} \tag{6.14}$$

where $\delta_0 \geq 0$ is a non-negative integer chosen in advance. If $\delta_0 = 0$, condition (6.14) imposes nothing more than (6.6). With $\delta_0 \geq 1$, condition (6.14) imposes a *blocking zero at infinity* of order δ_0 to the transfer function $\frac{R}{S}$ of the controller (see section 2.4.7, Remark 38). According to Remark 65 (section 3.4.2), the Bode diagram of the transfer function $\frac{R}{S}$ has a slope that is at most $-20\,\delta_0$ dB/decade at high frequencies. It is thus the transfer function of a low-pass filter (for $\delta_0 \geq 1$); the larger the δ_0, the more selective the filter. The integer δ_0 is the *roll-off* of the controller (already mentioned in the preface).

Note that the transfer function that relates ν to e is $-T_o(s)$ (according to sections 4.2.6 and (6.8)) and $\delta(T_o) = \delta\left(\frac{B}{A}\right) + \delta\left(\frac{R}{S}\right)$ (see section 13.6.1). Since $\delta\left(\frac{B}{A}\right) \geq 1$, condition (6.14) implies

$$\delta(T_o) \geq \delta_0 + 1.$$

The error e is thus less sensitive to the high-frequency measurement noise than the control variable u.

6.3.4. *Problem resolution*

Stabilizability condition

Polynomial $S(\partial)$ is of the form $\partial S_I(\partial)$, where $S_I(\partial) \in \mathbb{R}[\partial]$. Put $A_I(\partial) = A(\partial)\,\partial$, i.e.

$$A_I(\partial) = \partial^{n+1} + a_1 \partial^n + \cdots + a_n \partial. \tag{6.15}$$

From (6.10), we have

$$A_I(\partial)\, S_I(\partial) + B(\partial)\, R(\partial) = A_{cl}(\partial). \tag{6.16}$$

Now let us apply the theory in section 13.1.5 with $a(s) = A_I(s)$, $b(s) = B(s)$, $x(s) = S_I(s)$, $y(s) = R(s)$, and $c(s) = A_{cl}(s)$.

PROPOSITION 109.– *There exist polynomials $S_I(\partial)$ and $R(\partial)$ such that the polynomial $A_{cl}(\partial)$ has all its roots in the left half-plane if and only if $B(0) \neq 0$, or equivalently if and only if the system (6.11) has no zero equal to 0 (i.e. is not a derivator system).*

PROOF. (i) If $B(0) \neq 0$, polynomials $A_I(\partial)$ and $B(\partial)$ are coprime. Indeed, $A(\partial)$ and $B(\partial)$ has no roots in common, for the system (6.11) is assumed to be controllable, and likewise ∂ and $B(\partial)$ do not have any root in common. As a result, according to Theorem 494 (section 13.1.5), equation (6.16) admits a solution $(S_I(s), R(\partial))$ for any polynomial $A_{cl}(\partial)$. (ii) If $B(0) = 0$, $A_{cl}(0) = 0$ because $S_I(0) = 0$. ■

Degrees of the polynomials

Assume that $B(0) \neq 0$ from now on. The polynomial $R(s)$ is of the form $y_0\, s^{\boldsymbol{v}} + y_1\, s^{\boldsymbol{v}-1} + \cdots + y_{\boldsymbol{v}}$. Since the polynomials $a(s)$ and $b(s)$ are coprime, there exists a unique solution $(x, y) = (S_I, R)$ for which $\boldsymbol{v} = \boldsymbol{\alpha} - 1$, where $\boldsymbol{\alpha} = n + 1$, thus $\boldsymbol{v} = n$. From (6.14), we have $d^\circ(S) = n + \delta_0$, therefore $\boldsymbol{\xi} \triangleq d^\circ(S_I) = n + \delta_0 - 1$.

Since $\boldsymbol{\beta} = n - 1$ and $\delta_0 \geq 0$, we have $\boldsymbol{\xi} \geq \boldsymbol{\beta}$, and therefore the simplification indicated at the end of section 13.1.5 can be made. The polynomials S_I and A_{cl} are chosen to be monic. We have $d^\circ(A\, S) = 2n + \delta_0$ and $d^\circ(B\, R) \leq 2n - 1$, thus $d^\circ(A_{cl}) = 2n + \delta_0$. Put

$$\begin{aligned} S(\partial) &= \partial^{n+\delta_0} + \sigma_1\, \partial^{n+\delta_0-1} + \cdots + \sigma_{n+\delta_0-1}\, \partial, \\ R(\partial) &= r_0\, \partial^n + r_1\, \partial^{n-1} + \cdots + r_n, \\ A_{cl}(\partial) &= \partial^{2n+\delta_0} + c_1\, \partial^{2n+\delta_0-1} + \cdots + c_{2n+\delta_0}. \end{aligned}$$

Sylvester system

Taking into account the above-mentioned simplification, the Sylvester system (13.13) (section 13.1.5) can be written as

$$\begin{bmatrix} 1 & 0 & \cdots & \cdots & 0 & 0 & \cdots & \cdots & 0 \\ a_1 & 1 & \ddots & & \vdots & \vdots & \ddots & & \vdots \\ \vdots & a_1 & \ddots & \ddots & \vdots & 0 & & \ddots & \vdots \\ a_n & & \ddots & \ddots & 0 & b_1 & \ddots & & 0 \\ 0 & \ddots & & \ddots & 1 & \vdots & \ddots & 0 & \vdots \\ \vdots & & \ddots & & a_1 & b_n & & b_1 & 0 \\ \vdots & & & \ddots & \vdots & 0 & \ddots & \vdots & b_1 \\ \vdots & & & & a_n & \vdots & \ddots & b_n & \vdots \\ 0 & \cdots & \cdots & \cdots & 0 & 0 & \cdots & 0 & b_n \end{bmatrix} \begin{bmatrix} \sigma_1 \\ \sigma_2 \\ \vdots \\ \vdots \\ \sigma_{n+\delta_0-1} \\ r_0 \\ r_1 \\ \vdots \\ r_n \end{bmatrix} = \begin{bmatrix} c_1 - a_1 \\ \vdots \\ c_n - a_n \\ c_{n+1} \\ \vdots \\ \vdots \\ \vdots \\ \vdots \\ c_{2n+\delta_0} \end{bmatrix}. \tag{6.17}$$

Let M be the matrix given above on the left. Its order is $2n + \delta_0$. The number of columns containing terms a_i (resp. b_i) is equal to $n + \delta_0 - 1$ (resp. $n + 1$). The number of zeros below the last a_n (resp. above the first b_1) is equal to 1 (resp. δ_0).

The matrix M is of the form

$$\begin{bmatrix} 1 & 0 & \dots & 0 & \dots & 0 \\ * & \ddots & \ddots & & & \vdots \\ \vdots & \ddots & \ddots & \ddots & \ddots & \vdots \\ \vdots & & \ddots & 1 & 0 & 0 \\ * & \dots & \dots & * & \Sigma(a,b) & * \\ 0 & \dots & \dots & \dots & 0 & b_n \end{bmatrix}$$

where $*$ each time stands for a non-specified submatrix. Since $\Sigma(a,b)$ is invertible and $b_n \neq 0$, this matrix is invertible. This confirms that the system (6.17) admits a unique solution.

6.3.5. *Choice of the poles*

Introduction

The integral action of the controller ensures a large gain of $L(s)$ at low frequencies. By imposing $\delta_0 > 1$, we ensure a small gain of $L(s)$ and of $K(s) = \frac{R(s)}{S(s)}$ at high frequencies. The stability of the closed-loop system is guaranteed (in the absence of modeling error) as long as the roots of $A_{cl}(\partial)$ are chosen in the left half-plane. A number of specifications presented in section 4.2 are therefore satisfied. But some among these are not yet considered, in particular: (i) How to obtain a sufficient modulus margin? (ii) How to obtain a sufficient delay margin? (iii) How to obtain a sufficient rapidity? To meet these requirements, *the closed-loop poles must be chosen in an appropriate way.*

For this analysis, we consider the closed-loop in the equivalent form shown in Figure 6.3.

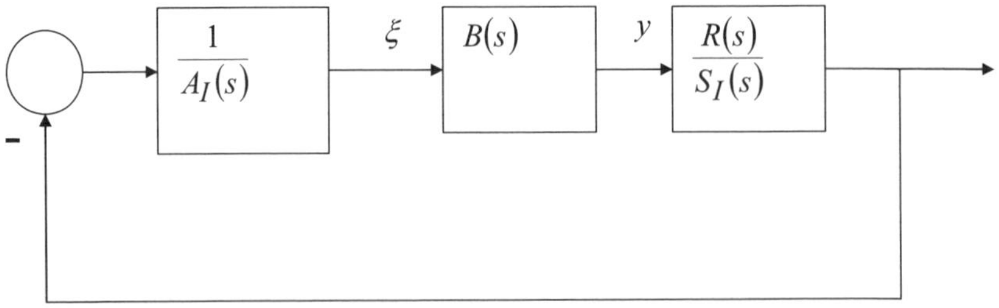

Figure 6.3. *Closed-loop (equivalent diagram)*

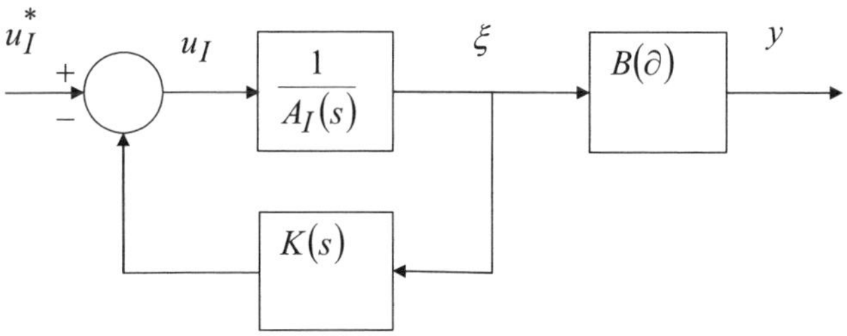

Figure 6.4. *Partial state feedback*

Partial state feedback

As a first step, assume that the partial state ξ of the minimal system with transfer function $\frac{B(s)}{A_I(s)}$ is available to provide the feedback needed (see section 2.3.5). This system (the input of which is $u_I = \partial u$) is represented by the right form

$$\begin{cases} B(\partial)\,\xi = y \\ A_I(\partial)\,\xi = u_I \end{cases}$$

where $A_I(\partial)$ is the polynomial of degree $n+1$ defined by (6.15). The control variable u_I of the above system is chosen to be of the form

$$u_I = u_I^* - K(\partial)\,\xi$$

where u_I^* is a signal provided by the command and $K(\partial)$ is a polynomial such that the characteristic polynomial of the closed-loop is a monic polynomial A_c of the same degree as A_I, which is $n+1$ (see Figure 6.4). The polynomial A_c is thus of the form

$$A_c(\partial) = \partial^{n+1} + \gamma_1\,\partial^n + \cdots + \gamma_{n+1}.$$

The closed-loop is governed by the differential equation $A_I(\partial)\,\xi = u_I^* - K(\partial)\,\xi$, from which we get

$$\begin{aligned} A_c(\partial)\,\xi &= u_I^*, \\ A_c(\partial) &= A_I(\partial) + K(\partial). \end{aligned}$$

(It is essential to note that the transfer function $B(s)$ is not part of the loop.) As a result, the polynomial $K(\partial)$ is determined by

$$\begin{aligned} K(\partial) &= A_c(\partial) - A_I(\partial) \\ &= (\gamma_1 - a_1)\,\partial^n + \cdots + (\gamma_n - a_n)\,\partial + \gamma_{n+1}. \end{aligned} \tag{6.18}$$

The open-loop transfer function corresponding to Figure 6.4 is given by $L_e(s) = \frac{K(s)}{A_I(s)}$. Therefore, the corresponding sensitivity function is given by

$$S_e(s) = \frac{1}{1 + L_e(s)} = \frac{A_I(s)}{A_c(s)}. \tag{6.19}$$

Let $p_1, ..., p_{n+1}$ be the roots of $A_I(s)$ and $\pi_1, ..., \pi_{n+1}$ be the roots of $A_c(s)$. The roots π_k $(1 \leq k \leq n+1)$ are of course assumed to have negative real parts, so that $S_e(s) \in \Re\mathcal{H}_\infty$. It is not hard to prove the following lemma:

LEMMA 110.– *A sufficient condition for the inequality* $\left|\frac{i\omega - p_k}{i\omega - \pi_k}\right| \leq 1$ *to be satisfied for any root* p_k *and at any frequency* ω *is:* $\operatorname{Im} \pi_k = \operatorname{Im} p_k$ *and* $\operatorname{Re} \pi_k \leq -|\operatorname{Re} p_k|$.

Lemma 110 provides a simple way of choosing the roots of $A_c(s)$ as a function of those of $A_I(s)$ in order to get $\|S_e\|_\infty = 1$, and thus a modulus margin equal to 1, when the partial state is used in the feedback loop. Indeed, we have

$$|S_e(i\omega)| = \prod_{1 \leq k \leq n+1} \left|\frac{i\omega - p_k}{i\omega - \pi_k}\right|.$$

The following rule thus follows, for any integer k, $1 \leq k \leq n+1$:

RULE 111.– *Choose* $\operatorname{Im} \pi_k = \operatorname{Im} p_k$ *and* $\operatorname{Re} \pi_k \leq -|\operatorname{Re} p_k|$ *(with* $\operatorname{Re} \pi_k < 0$*).*

The criterion that presides over the above rule is uniquely the modulus margin. Other considerations of course need to be taken into account, in particular the rapidity and the delay margin, as well as, possibly, the damping coefficient of each pole.

On the other hand, this rule is closely related to the "linear quadratic control" (see [20], Part I, and Theorem 247 of section 8.1.4).

Supplementary poles

Case of a minimum phase system

In reality, the partial state is in general unavailable for the control (except in the particular case where $0 \neq B(\partial) \in \mathbb{R}$). That is why it is necessary to come back to the Bézout equation (6.16). We choose a polynomial $A_{cl}(\partial)$ of the form

$$\boxed{A_{cl} = A_c\, B\, F} \tag{6.20}$$

where $A_c(\partial)$ is a monic polynomial of degree $n+1$, the roots of which are chosen according to Rule 111, and $F(\partial)$ is a polynomial of degree $m = \delta\left(\frac{B}{A}\right) + \delta_0 - 1$, with roots in the left half-plane, and such that the product $B(\partial)\, F(\partial)$ is a monic polynomial. That way, $A_{cl}(\partial)$ is indeed a monic polynomial of degree $2n + \delta_0$, all the roots of which are in the left half-plane since, from hypothesis, the roots of $B(\partial)$ satisfy this last condition.

The Bézout equation (6.16) can thus be written as

$$A_I\, S_I = B\,(A_c\, F - R)\,.$$

Since A_I and B are coprime, there exists a polynomial Q such that

$$S_I = B\,Q.$$

This polynomial Q is necessarily of degree $\delta\left(\frac{B}{A}\right) + \delta_0 - 1 = d^\circ(F)$; we get

$$R = A_c F - A_I Q. \tag{6.21}$$

The open-loop transfer function is therefore

$$L = \frac{B}{A_I}\frac{R}{S_I} = \frac{B}{A_I}\frac{A_c\, F - A_I\, Q}{B\,Q} = \frac{F}{Q}\frac{A_c}{A_I} - 1$$

from which we deduce the sensitivity function:

$$S_o = \frac{1}{1+L} = \frac{Q}{F}\frac{A_I}{A_c}.$$

Equation (6.21) is a Bézout equation with unknown (R, Q) for given A_c, A_I, and F. Previously, we have seen that this equation admits a unique solution such that $d^\circ(R) \leq n$ and $d^\circ(Q) = \delta\left(\frac{B}{A}\right) + \delta_0 - 1 = m$. This solution (R, Q) depends linearly on F and is denoted as $\Psi(F)$. The linear mapping Ψ is continuous (in the sense specified in Theorem 112).

Let now (F_k) be a sequence of polynomials having the same properties as the polynomial F above and such that as $k \to +\infty$, all roots of F_k tend to $-\infty$ while remaining on the real axis. There exists a sequence of non-zero real numbers (λ_k) such that $\lambda_k F_k \to 1$ (the sequence converges in the sense specified in Theorem 112). For any integer k, $(R_k, Q_k) = \Psi(F_k)$. According to (6.21), we have

$$\lambda_k R_k + A_I \lambda_k Q_k = A_c \lambda_k F_k\,. \tag{6.22}$$

Since $K + A_I = A_c$, we deduce that $\lambda_k\, R_k \to K$ and $\lambda_k\, Q_k \to 1$. Therefore, $\frac{Q_k}{F_k} \to 1$ and the sensitivity function $S_{o,k} = \frac{Q_k}{F_k}\frac{A_I}{A_c}$ converges to the sensitivity function S_c given by (6.19).

The following theorem details what has just been discussed:

THEOREM 112.– *Let the characteristic polynomial of the closed-loop be defined by* (6.20)*, where the roots of* A_c *are chosen according to Rule* 111*, and assume that* F *is replaced by* F_k*, where* (F_k) *is a sequence of polynomials as specified above. Then,* $\left(\|S_{o,k} - S_e\|_\infty\right) \to 0$.

PROOF. * 1) Let $\mathbf{X}$ (resp., $\mathbf{Y}$) be the subspace of the $\mathbb{R}$-vector $\mathbb{R}[\partial]$ consisting of all polynomials with degree $\leq m$ (resp. $\leq n$). Consider the linear mapping $\Psi : \mathbf{X} \to \mathbf{X} \times \mathbf{Y}$ defined above. The spaces $\mathbf{X}$, $\mathbf{Y}$ are finite-dimensional, therefore Ψ is continuous (see section 12.1.3). Let D be an infinite compact set included in the closed right half-plane. Then $\sup_{s \in D} |f(s)| = \|f\|$ ($f \in \mathbf{X}$, resp. $f \in \mathbf{Y}$) is a norm on $\mathbf{X}$ (resp. $\mathbf{Y}$), and $\|(f,g)\| = \sup(\|f\|, \|g\|)$ ($f \in \mathbf{X}, g \in \mathbf{Y}$) is a norm on $\mathbf{X} \times \mathbf{Y}$. Let $-\frac{1}{\tau_{1,k}}, \ldots, -\frac{1}{\tau_{m,k}}$ be the roots of F_k, where $\tau_{i,k} > 0$ ($\tau_{i,k} \in \mathbb{R}$). We have

$$\lambda_k F_k(s) = \prod_{1 \leq j \leq m} (1 + \tau_{j,k} s) \tag{6.23}$$

and $\lim_{k \to +\infty} \tau_{j,k} = 0$. As a result, $(\lambda_k F_k) \to 1$ uniformly on D, i.e. $(\|\lambda_k F_k - 1\|) \to 0$. We have $\Psi(1) = (K, 1)$, and thus $(\|\lambda_k R_k - K\|) \to 0$ and $(\|\lambda_k Q_k - 1\|) \to 0$ because of the continuity of Ψ. This means that $(\lambda_k R_k) \to K$ and $(\lambda_k Q_k) \to 1$ uniformly on D. We have $S_{o,k} - S_e = \left(\frac{Q_k}{F_k} - 1\right) S_e$, and thus

$$\sup_{s \in D} |S_{o,k}(s) - S_e(s)| \leq \sup_{s \in D} \left| \frac{Q_k(s)}{F_k(s)} - 1 \right| \|S_e\|_\infty,$$

and this quantity tends to 0 as $k \to +\infty$.

2) We have from (6.22)

$$\frac{\lambda_k R_k(s)}{A_c(s)} + \frac{A_I(s)}{A_c(s)} \lambda_k Q_k(s) = \lambda_k F_k(s).$$

For $k \to +\infty$, the roots of $\lambda_k R_k$ converge to those of K according to 1) and $d^\circ(\lambda_k R_k) < d^\circ(A_c)$; therefore, for $|s| \to +\infty$ (s remaining in the closed right half-plane), $\frac{\lambda_k R_k(s)}{A_c(s)} \to 0$ uniformly with respect to k, thus $\lambda_k Q_k(s) - \lambda_k F_k(s) \to 0$ uniformly with respect to k because $\frac{A_I(s)}{A_c(s)} \to 1$. Now let $s = i\omega$, $\omega \to +\infty$; we have $|1 + \tau_{j,k} i\omega| \geq 1$ for any $j \in \{1, \ldots, m\}$ and any natural integer k, therefore $|\lambda_k F_k(i\omega)| \geq 1$ according to (6.23); as a result, $\left|\frac{Q_k(i\omega)}{F_k(i\omega)} - 1\right| \to 0$ uniformly with respect to k. For any $\varepsilon > 0$, it follows that there exists a frequency $\Omega > 0$, only depending on ε, such that $|S_{o,k}(i\omega) - S_e(i\omega)| \leq \varepsilon$, for any integer k, as long as $\omega \geq \Omega$. On the other hand, according to 1), there exists an integer k_0 such that $|S_{o,k}(i\omega) - S_e(i\omega)| \leq \varepsilon$ for any $k \geq k_0$ and any $\omega \in [0, \Omega]$ (with $D = [0, \Omega]$); this integer k_0 only depends on ε. For any $k \geq k_0$, $\|S_{o,k} - S_e\| \leq \varepsilon$ (see section 13.6.2), and this proves the theorem. * ■

REMARK 113.– *(i) This theorem shows that* one can obtain a modulus margin as close to 1 as one would like *by choosing a polynomial for F which is of degree $m = \delta\left(\frac{B}{A}\right) + \delta_0 - 1$ and the roots of which are negative real with sufficiently large absolute value. According to relations* (4.9), (4.10), *for a modulus margin of* 1, *we have a guaranteed gain margin* $(-6\,\mathrm{dB}, +\infty)$ *and a guaranteed phase margin* $(-60°, 60°)$. *(ii) This theorem is closely related to the "LTR method" developed in [38]. Nevertheless, in the cited reference, only the* simple convergence *of* $(S_{o,k})$ *to* S_e *is obtained, which does not allow one to conclude anything about the behavior of the modulus margin. See section 9.1.4 for more details.*

REMARK 114.– *Let* $S_{I,n} = B\,Q_n$. *The transfer function of the controller obtained above is* $\frac{R_n(s)}{s\,S_{I,n}(s)} = \frac{\lambda_n R_n(s)}{s\,\lambda_n S_{I,n}(s)} \to \frac{K(s)}{s\,B(s)}$. *The transfer function obtained by taking the limit is thus improper (except if* $d°(B) = n - 1$*). Thus a compromise needs to be found between the modulus margin (which will be all the larger as the roots of F are "faster", i.e. will have a larger absolute value) and the sensitivity of the control to the measurement noise.*

Case of a non-minimum phase system

The procedure just described is obviously not applicable to the case of a non-minimum phase system since the roots of $B(s)$ are then unstable.

(i) If $B(s)$ has at least one root z with positive real part, we have seen in section 4.2.8 that we could replace the zero z by the stable zero $-z$. We will thus proceed this way for each zero z such that $\operatorname{Re} z > 0$.

(ii) If $B(s)$ has at least one root on the imaginary axis (or so close to the imaginary axis that it can be considered as located on it), we can proceed as indicated in Remark 91 (section 4.2.8) by replacing this zero z by $-z - \alpha$, where $\alpha > 0$ is such that condition (4.22) holds; this procedure can be carried out for each non-zero root of $B(s)$ located on the imaginary axis.

We thus construct from $B(s)$ a polynomial $B^*(s)$, all roots of which belong to the left half-plane. As already mentioned in section 4.2.8, the multiplicative error made when replacing $B(s)$ by $B^*(s)$ is small if all unstable zeros of $B(s)$ are "fast" with respect to the maximum unity gain frequency of the closed-loop system. This consideration may lead to make the closed-loop system slow, thus with a poor performance, as indicated in this same section.

Lastly, the choice made for the polynomial $A_{cl}(\partial)$ is following:

$$\boxed{A_{cl} = A_c\,B^*\,F}. \tag{6.24}$$

Of course, the Bézout equation to be solved is still (6.16).

Choice of the polynomial $T(\partial)$

The polynomial $T(\partial)$ can be chosen in such a way so that some undesirable dynamics can be cancelled in the transfer function relating the reference signal r to the output y. This transfer function is, according to (6.7), $\frac{B\,T}{A_{cl}}$. The two only constraints that $T(\partial)$ needs to obey is the degree condition (6.6), which is $d^\circ(T) \leq n + \delta_0$, and equality (6.13). It is difficult to make more precise general recommendations, as shown in the following examples.

REMARK 115.– *With $d^\circ(T) = n + \delta_0$ we obtain $\delta(B\,T/A_{cl}) = d^\circ(B) + n + \delta_0 - (2n + \delta_0) = \delta(B/A)$. Thus by choosing T a divisor of A_{cl}, the controlled system behaves (in the absence of any disturbance and modeling error) like a system of the same order as system* **P**.

6.3.6. *Examples*

PID designed by pole placement

Consider the minimal system **P** with transfer function $P(s) = 2\,\frac{1-0.1\,s}{(1+0.5\,s)(1+s)}$ (see sections 5.3.2, 5.4.1 and 5.4.2). This system is defined by the left form

$$(\partial + 1)(\partial + 2)\,y = -\frac{4}{10}\,(\partial - 10)\,u. \tag{6.25}$$

This system is non-minimum phase, but its zero can be considered "fast". Let us apply the above theory using $A_c(\partial) = (\partial + 2)^3$ and $A_{cl}(\partial) = A_c(\partial)(\partial + 10)$ (whence $\delta_0 = 0$). We obtain

$$R(\partial) = (\partial + 2)(\partial + 1.26)\,\frac{10}{1.26} \tag{6.26}$$

$$S(\partial) = \partial\,(\partial + 16.18). \tag{6.27}$$

We can put the transfer function $\frac{R(s)}{S(s)}$ in the classic form (5.17) of the transfer function of a PID controller with : $k = 1.53$; $T_i = 1.23$ s; $T_d = 0.26$ s; $N = 4.22$. With the time unit in seconds, the Black plot of the open-loop transfer function $L(s) = \frac{B(s)}{A(s)}\frac{R(s)}{S(s)}$ is shown in Figure 6.5. By choosing $T(\partial) = R(\partial)$, the controller has 1-DOF, that is it is a PID in the most traditional form. We can see the behavior of the feedback system in Figure 6.6 with regard to the following events: (i) unit step command at time $t = 0$; (ii) unit step disturbance of amplitude 0.3 added to y at time $t = 5$ s. * The measurement noise is a Gaussian white noise with standard deviation 0.1 (the computation step is 0.1 s). [1] * The time responses of the controlled

1. It is a discrete-time white noise, the computation step is considered as being the sampling period (see section 11.1).

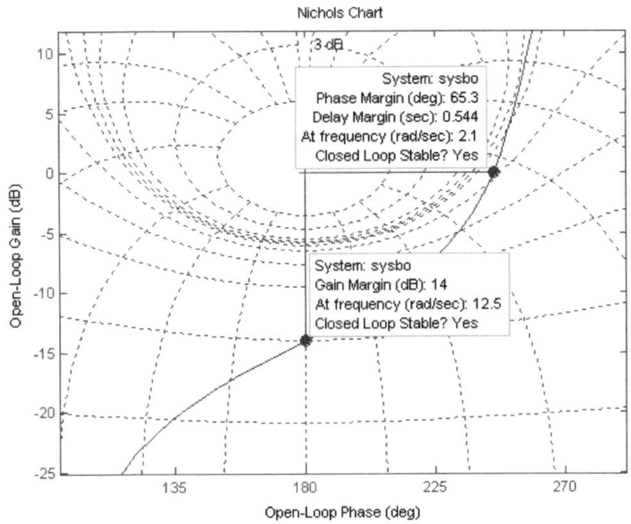

Figure 6.5. *Black plot (open-loop with PID)*

system with the PID designed by pole placement (-) and of that with the PID designed using the geometric method of sections 5.4.1 and 5.4.2 (- -) are simulated over 10 s of time. The two controlled systems have practically identical behavior. In both cases, the noise filtering is poor, this is one of the major inconveniences of the PID (due to the derivative action, even if the derivative is filtered!).

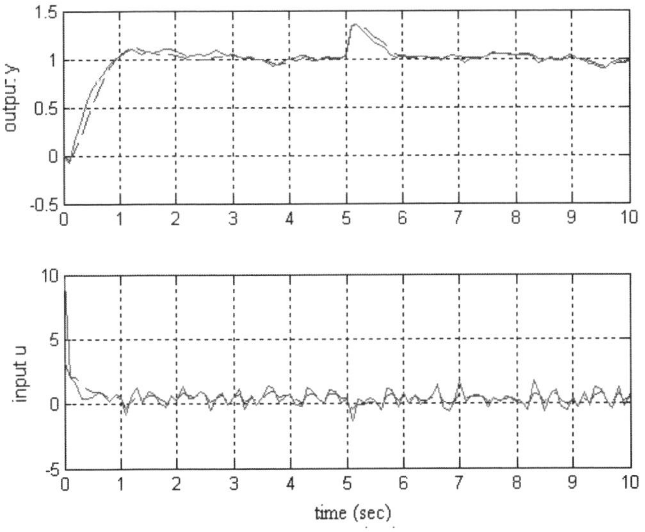

Figure 6.6. *Time domain responses with the two PIDs*

Adding blocking zeros at infinity

Considering the same system as before, i.e. defined by the left form (6.25), take $A_c(\partial) = (\partial + 2)^3$ again and consider the following two cases:

(i) $A_{cl}(\partial) = A_c(\partial)(\partial + 10)^2$ (from which $\delta_0 = 1$);

(ii) $A_{cl}(\partial) = A_c(\partial)(\partial + 10)^3$ (from which $\delta_0 = 2$).

In both cases, we choose $T = R$. In case (i), we obtain

$$R(\partial) = (\partial + 2)(\partial + 1.23)\frac{100}{1.22} \quad (6.28)$$

$$S(\partial) = \partial(\partial + \alpha)(\partial + \bar{\alpha}) \quad (6.29)$$

with $\alpha = 11.5 + 7.83\,i$. In case (ii), we obtain

$$R(\partial) = (\partial + 2)(\partial + 1.20)\frac{1000}{1.20} \quad (6.30)$$

$$S(\partial) = \partial(\partial + \beta)(\partial + \bar{\beta})(\partial + 18.46) \quad (6.31)$$

with $\beta = 7.27 + 8.35\,i$.

Call $\mathbf{K}_2$ the RST controller, the polynomials of which are defined by (6.26) and (6.27), $\mathbf{K}_3$ the one, the polynomials of which are defined by (6.28) and (6.29), and $\mathbf{K}_4$ the one, the polynomials of which are defined by (6.30) and (6.31) ($\mathbf{K}_1$ denotes the PID determined in sections 5.4.1 and 5.4.2). With $\mathbf{K}_2$, the phase lag margin is $65.3°$ and the delay margin is 0.5 s (as we can see in Figure 6.5). With $\mathbf{K}_3$ (resp. $\mathbf{K}_4$), the phase lag margin is $62.5°$ (resp. $60.4°$) and the delay margin is 0.6 s (resp. 0.67 s). Adding fast supplementary poles to the closed-loop system thus has a tendency to slightly reduce the phase lag margin and increase the delay margin (and therefore decelerating the closed-loop system). These variations, hardly significant, are due to the fact that the supplementary poles are at -10, which are still quite far from infinity... The comparison of the Bode plots of the open-loop transfer functions is shown in Figure 6.7 for the controllers $\mathbf{K}_2$ (-), $\mathbf{K}_3$ (- -) and $\mathbf{K}_4$ (-.).

For the time responses in Figure 6.8 the events simulated are the same as the previous ones (see Figure 6.6, where the conventions for the lines are the same as above). We see that the Bode plots are almost identical up to the unity gain frequency (which is of the order of 2 rad/s) and become very different at frequencies higher than 10 rad/s (corresponding to the absolute value of the "fast poles" added). The greater the roll-off δ_0, the better is the measurement noise filtered (this is particularly clear on the control signal). Other than this (even though that is important, even crucial), the three time responses are very similar.

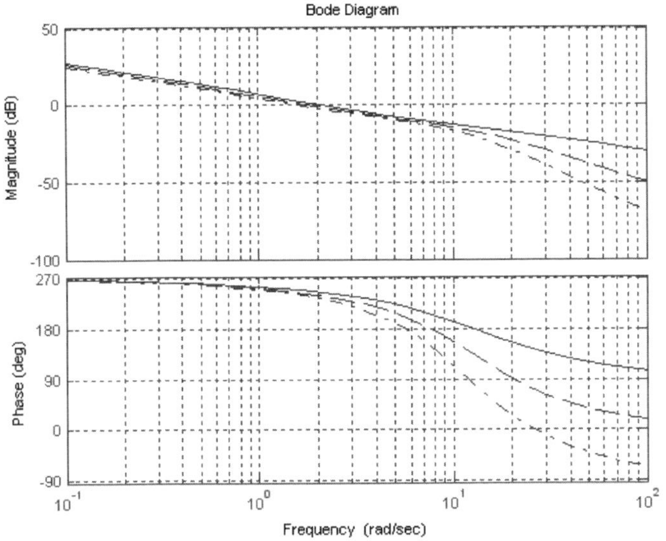

Figure 6.7. *Bode plots of the three controllers*

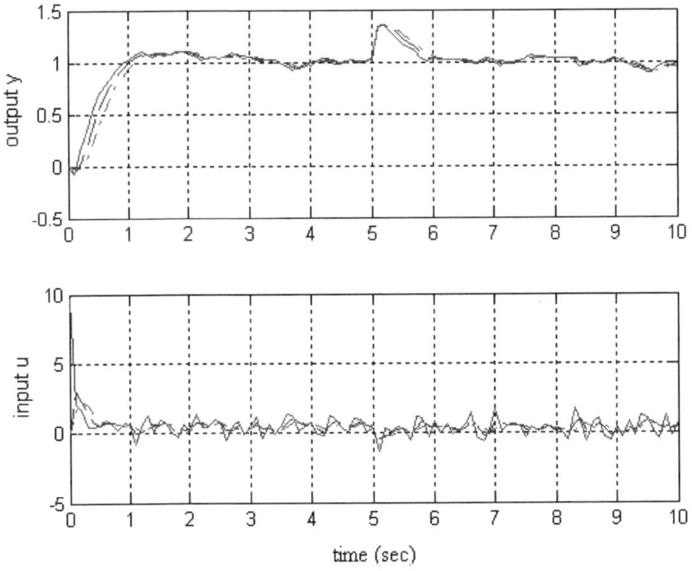

Figure 6.8. *Time domain responses with the three controllers*

A difficult case

Let a system now be defined by the left form

$$\left(\partial^2 + 1\right) y = (\partial - 5)\, u.$$

This system is purely oscillatory and non-minimum phase. The difficulty, in the present case, is that if we damp the poles $\pm i$ too much, the phase margin will degrade. After some of trial and error, the following choice turns out to be correct:

$$\begin{aligned} A_c(\partial) &= (\partial + 1)(\partial + 0.5 + i)(\partial + 0.5 - i) \\ A_{cl}(\partial) &= A_c(\partial)(\partial + 5)(\partial + 10), \end{aligned}$$

where the pole -10 is added to get $\delta_0 = 1$. The polynomial $T(\partial)$ is chosen in such a way so as to cancel both poorly damped complex conjugate poles of the closed-loop and the rapid pole. One obtains

$$\begin{aligned} R(\partial) &= -\frac{12.5}{|\gamma|^2}(\partial + \gamma)(\partial + \bar{\gamma}) \\ S(\partial) &= \partial(\partial + \lambda)(\partial + \bar{\lambda}) \\ T(\partial) &= \mu(\partial + 0.5 + i)(\partial + 0.5 - i)(\partial + 10) \end{aligned}$$

with $\gamma = 0.15 + 0.69\, i$, $\lambda = 8.50 + 5.84\, i$, and where μ is such that $T(0) = R(0) = -12.5$.

The Bode plot of the open-loop transfer function $L(s)$ is shown in Figure 6.9. We note that the phase lag margin is approximately $44°$, which is reasonable. The same conclusion is for the gain margin which is in the range $(-\infty, 10\,\mathrm{dB})$.

The time responses in Figure 6.10 show the system behavior faced with the following events: (i) unit step command at $t = 0$; (ii) step disturbance of amplitude 0.3 adding to y at time $t = 10$ s. We can see the excellent quality of the step response, without overshoot and having a "negative start" (see section 3.6, Exercise 72). The response to the disturbance has very different dynamics because, contrary to the reference signal, this disturbance excites the oscillatory dynamics of the closed-loop. Of course, these oscillatory dynamics can only be cancelled by $T(s)$ in the *absence of modeling error*. This example illustrates the advantage of a 3-DOF controller over a 1-DOF controller, whose response to a step command would be very poor in the present case

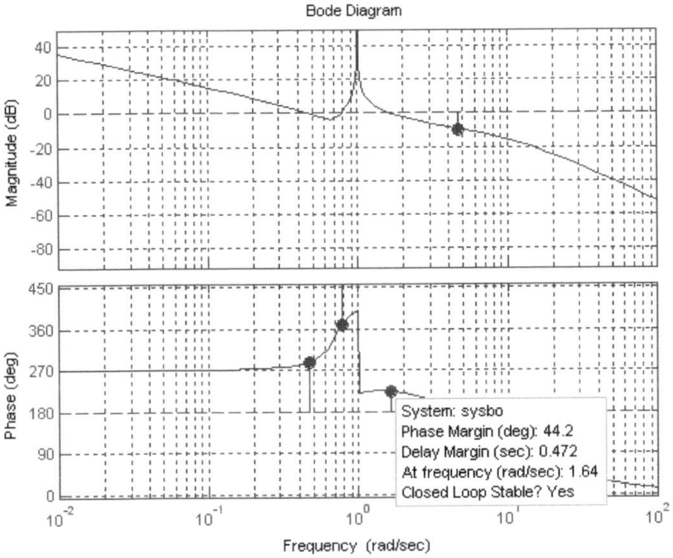

Figure 6.9. *Bode plot (case of the pure oscillatory system)*

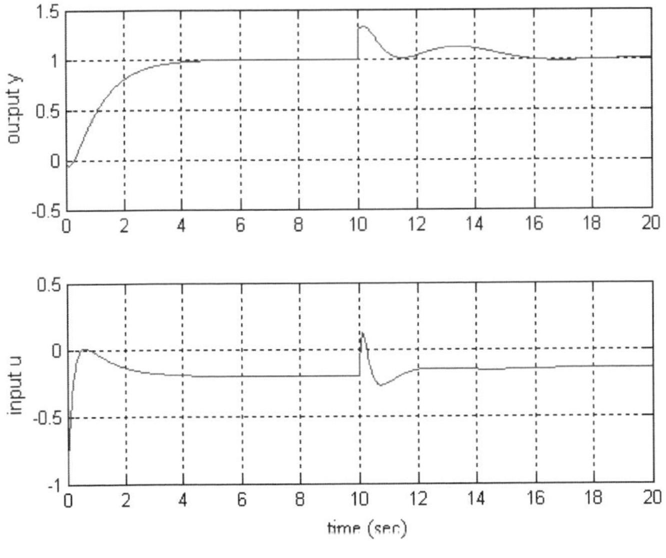

Figure 6.10. *Time domain response (system originally oscillatory)*

6.4. *General case

Suppose now that the disturbance d and the reference signal r are governed by linear differential equations with constant coefficients of the form

$$D_1(\partial)\,d = 0, \quad D_2(\partial)\,r = 0.$$

The "usual case" previously considered was a particular case of what is now discussed, with $D_1(\partial) = D_2(\partial) = \partial$. Here, the reference signal can be, for example, a ramp (in which case, $D_2(\partial) = \partial^2$) and the disturbance can be sinusoidal (by putting $D_1(\partial) = \partial^2 + \omega_0^2$). The disturbance can also be the superposition of a ramp and a sinusoid ($D_1(\partial) = \partial^2\left(\partial^2 + \omega_0^2\right)$), etc.

The measurement noise is again assumed to have its spectrum in high frequencies and system (6.11) is assumed to be controllable.

6.4.1. *Disturbance rejection*

According to (6.8), the disturbance d is rejected if and only if

$$S \in (D_1) \tag{6.32}$$

where (D_1) denotes the ideal in $\mathbb{R}[\partial]$ generated by D_1 (see section 13.1.1). This signifies that the left member of the relation (6.32) is a multiple of the polynomial D_1. This condition generalizes (6.12).

6.4.2. *Reference tracking*

In expression (6.8), the term dependent on r becomes zero if and only if

$$B\,(T - R) - A\,S \in (D_2)\,. \tag{6.33}$$

If we wish to obtain a good *robustness of performance*, condition (6.33) needs to be satisfied even when the polynomials $A(\partial)$ and $B(\partial)$ are uncertain. As a result, (6.33) leads to the double condition

$$S \in (D_2)\,, \tag{6.34}$$

$$T - R \in (D_2)\,. \tag{6.35}$$

Condition (6.35) generalizes (6.13).

6.4.3. *Internal model principle*

Conditions (6.32) and (6.35) are both satisfied if and only if $S \in (D_1) \cap (D_2) = (D)$ where $D = \operatorname{lcm}(D_1, D_2)$ (see section 13.1.3). This means there exists a polynomial S_I such that

$$S = S_I\, D. \tag{6.36}$$

Condition (6.36) is the "Internal Model Principle" [119].[2]

We make the following hypothesis:

ASSUMPTION 116.– *D has all its roots on the imaginary axis.*

The reason for Assumption 116 is as follows:

(i) If D has roots in the left half-plane, one can write this polynomial in the form $D = D_s\, D_i$, where D_s is a polynomial which has *all* its roots in the left half-plane and where D_i is a polynomial prime with D_s. Thus every signal w (disturbance or reference signal) satisfying the differential equation $D(\partial)\, w = 0$ can be decomposed as $w = w_s + w_i$, where $D_s(\partial)\, w_s = 0$ and $D_i(\partial)\, w_i = 0$ (see section 2.3.8, Theorem 449). Since $w_s(t) \to 0$ as $t \to +\infty$, there is no need to account for this signal which is zero in steady state. This case is thus to be excluded.

(ii) If D has roots in the right half-plane, there exists a signal w (disturbance or reference signal) satisfying the differential equation $D(\partial)\, w = 0$ and which diverges exponentially. A control signal thus can only cancel out the error e asymptotically if it itself also diverges exponentially. This is why this case is to be excluded as well.

It is also important to note the following: for a *bounded* control signal to be able to asymptotically cancel out the error e while the system is excited by a signal w (disturbance or reference signal), this signal itself must be bounded too. This signal is characterized by the fact that it is a solution to the differential equation $D(\partial)\, w = 0$; it is therefore necessary that the roots of $D(s)$ be *simple* and located on the imaginary axis (section 12.5.2, Theorem 449). Nevertheless, in some control problems, the requirements call for the tracking of a reference ramp signal for a *finite duration* of time. This is why, contrary to some authors, we keep the possibility that $D(s)$ may have *multiple roots on the imaginary axis* open (this does not however change anything to the theory).

---

2. In the quoted reference, the Internal Model Principle is stated in the framework of state-space systems. See section 8.3.

6.4.4. *Filtering of measurement noise*

The presence of measurement noise makes it necessary to impose condition (6.32); the argument developed in section 6.3.3 remains unchanged.

6.4.5. *Problem resolution*

Stability condition

Let $p = d^\circ(D)$ and $p_2 = d^\circ(D_2)$. Write

$$A_I(\partial) = A(\partial)\, D(\partial),$$

and let

$$A_I(\partial) = \partial^{n+p} + \alpha_1 \partial^{n+p-1} + \cdots + \alpha_{n+p}. \tag{6.37}$$

From (6.10) and (6.36), we are led to resolve the Bézout equation (6.16), with unknowns S_I and R, and also to apply the theory in section 13.1.5, with $a(s) = A_I(s)$, $b(s) = B(s)$, $x(s) = S_I(s)$, $y(s) = R(s)$, and $c(s) = A_{cl}(s)$.

The following is a generalization of Proposition 109.

PROPOSITION 117.– *There exist polynomials $S_I(\partial)$ and $R(\partial)$ such that the polynomial $A_{cl}(\partial)$ has all its roots in the left half-plane if and only if the polynomials $D(\partial)$ and $B(\partial)$ are coprime.*

PROOF. Let $A_f = \gcd(A_I, B)$. The Bézout equation (6.16) admits a solution (S_I, R) if and only if A_{cl} is of the form $A_{cl}(\partial) = A_f(\partial)\, A_l(\partial)$ (from Theorem 494) where $A_l(\partial)$ is an arbitrary polynomial. Now, $\gcd(A, B) = 1$; thus $A_f = \gcd(D, B)$, and according to Assumption 116, a necessary and sufficient condition for A_f to have all its roots in the left half-plane is $A_f = 1$. ■

Degrees of the polynomials

Let $\gcd(D, B) = 1$ from now on. Through the same rationale as in section 6.3.4, we arrive at the following conclusions:

$$\begin{aligned} d^\circ(R) &\leq n + p - 1, \\ d^\circ(S) &= n + p + \delta_0 - 1, \\ d^\circ(S_I) &= n + \delta_0 - 1, \\ d^\circ(A_{cl}) &= 2n + p + \delta_0 - 1. \end{aligned}$$

We have $\boldsymbol{\xi} = d^\circ\left(S_I\right) = n + \delta_0 - 1$, $\boldsymbol{\beta} = n - 1$ and $\delta_0 \geq 0$, thus

$$\boldsymbol{\xi} \geq \boldsymbol{\beta},$$

and the simplification indicated at the end of section 13.1.5 can be made. The polynomials S_I and A_{cl} are therefore chosen to be monic.

Now let

$$\begin{aligned} S_I\left(\partial\right) &= \partial^{n+\delta_0-1} + x_1\partial^{n+\delta_0-2} + \cdots + x_{n+\delta_0-1}, \\ R\left(\partial\right) &= r_0\partial^{n+p-1} + r_1\,\partial^{n+p-2} + \cdots + r_{n+p-1}, \\ A_{cl}\left(\partial\right) &= \partial^{2n+p+\delta_0-1} + c_1\partial^{2n+p+\delta_0-2} + \cdots + c_{2n+p+\delta_0-1}. \end{aligned}$$

Sylvester system

With the above-mentioned simplification, the Sylvester system (13.13) can be written as

$$\begin{bmatrix} 1 & 0 & \cdots & 0 & \cdots & \cdots & \cdots & \cdots & 0 \\ \alpha_1 & \ddots & \ddots & & \ddots & \vdots & \vdots & & \vdots \\ \alpha_2 & \ddots & 1 & \ddots & \vdots & 0 & & & \vdots \\ \vdots & \ddots & \alpha_1 & \ddots & 0 & b_1 & 0 & & \vdots \\ \alpha_{n+p} & \vdots & \alpha_2 & \ddots & 1 & \vdots & b_1 & \ddots & \vdots \\ 0 & \ddots & \vdots & \ddots & \alpha_1 & b_n & \vdots & \ddots & 0 \\ \vdots & \ddots & \alpha_{n+p} & \ddots & \alpha_2 & 0 & b_n & \vdots & b_1 \\ \vdots & \vdots & 0 & \ddots & \vdots & \vdots & 0 & \ddots & \vdots \\ 0 & \cdots & \cdots & 0 & \alpha_{n+p} & 0 & \cdots & 0 & b_n \end{bmatrix} \begin{bmatrix} x_1 \\ \vdots \\ \vdots \\ \vdots \\ x_{n+\delta_0-1} \\ r_0 \\ \vdots \\ \vdots \\ r_{n+p-1} \end{bmatrix}$$

$$= \begin{bmatrix} c_1 - \alpha_1 \\ \vdots \\ \vdots \\ \vdots \\ \vdots \\ c_{n+p} - \alpha_{n+p} \\ c_{n+p+1} \\ \vdots \\ c_{2n+p+\delta_0-1} \end{bmatrix}$$

The above-given matrix (left) is of order $2n+p+\delta_0-1$; it consists of $n+\delta_0-1$ columns containing coefficients α_i and $n+p$ columns containing coefficients b_i.

6.4.6. *Choice of poles*

Choice of $A_{cl}(\partial)$

The approach in section 6.3.5 applies in broad outline for the calculation of $A_{cl}(\partial)$. This polynomial takes on the form (6.24), where $A_c(\partial) = \prod_{1\leq k\leq n+p} (\partial - \pi_k)$ has its roots π_k chosen according to Rule 111 as a function of the roots p_k of $A_I(\partial)$, where $B^*(\partial)$ is constructed as indicated in section 6.3.5, and where $F(\partial)$ is a polynomial of degree $\delta\left(\frac{B}{A}\right)+\delta_0-1$, and the poles of which are "fast". Following the rationale in section 6.3.5, we obtain the following:

THEOREM 118.– *If system* (6.11) *is* minimum phase, *the statement of Theorem 112 remains valid.*

On the other hand, in the case where a system is in *non-minimum phase*, the remarks in section 6.3.5 remain valid as well.

Choice of $T(\partial)$

The constraints imposed on the polynomial T are (6.35) and (6.6). Condition (6.35) means that there exists a polynomial $Q(\partial) \in \mathbb{R}[\partial]$ such that

$$T - R = -Q\,D_2. \tag{6.38}$$

As in section 6.3.5, one would like to choose polynomial $T(\partial)$ in such a way that it will cancel certain undesirable dynamics in the transfer function $\frac{B\,T}{A_{cl}}$ relating the reference r to the output y. Suppose $A_s(\partial)$ is the factor of $A_{cl}(\partial)$ that we would like to cancel. Then T has to be of the form

$$T = T_I\,A_s \tag{6.39}$$

where $T_I \in \mathbb{R}[\partial]$. According to (6.38) and (6.39), the polynomials T_I and Q have to satisfy the Bézout equation

$$\boxed{A_s\,T_I + D_2\,Q = R}. \tag{6.40}$$

This equation admits multiple solutions, since $\gcd(A_s, D_2) = 1$ (from Assumption 116).

The case $p_2 = 0$ is trivial, so let us assume that $p_2 \geq 1$. Among the solutions to (6.40), a unique one exists such that $d^\circ(T_I) \leq p_2 - 1$ (section 13.1.5, Theorem

494). We can be sure that there exists a polynomial T satisfying (6.38), (6.39) and the constraint (6.6) only if $d^\circ(A_s) + p_2 - 1 \leq n + p + \delta_0 - 1$, i.e.

$$\boxed{d^\circ(A_s) \leq n + p - p_2 + \delta_0}. \tag{6.41}$$

REMARK 119.– *With* $d^\circ(A_s) = n + p - p_2 + \delta_0$, *we have* $d^\circ(T) = n + p + \delta_0 - 1$. *As a result,*

$$\begin{aligned}
\delta\left(\frac{BT}{A_{cl}}\right) &= d^\circ(B) + d^\circ(T) - (2n + p + \delta_0 - 1) \\
&= d^\circ(B) - n + (n + p + \delta_0 - 1) - (n + p + \delta_0 - 1) \\
&= \delta\left(\frac{B}{A}\right).
\end{aligned}$$

This generalizes Remark 115. Note that T *can be chosen to be a divisor of* A_{cl} *only if* T_I *is a unit, thus of degree zero; this happens when* $p_2 = 1$.

6.4.7. *Examples*

1. Constant reference, sinusoidal disturbance

Consider again the minimal system **P** defined by the left form (6.25) and suppose we add at its output a sinusoidal disturbance d_y, with frequency 1 rad/s. The objective here is to track a constant reference signal. The relative degree δ_0 is imposed to be 1.

Theoretically, $D_1(\partial) = \partial^2 + 1$. In a way to broaden the "resonance peak" of the transfer function $\frac{1}{D(s)}$, it is preferable to use a very small but non-zero damping coefficient ς. With $\varsigma = 0.5\%$, we obtain

$$D_1(\partial) = \partial^2 + 0.01\,\partial + 1 = (\partial + \eta)(\partial + \bar{\eta})$$

with $\eta \simeq 0.005 + i$. On the other hand, $D_2(\partial) = \partial$, thus

$$D(\partial) = \partial\left(\partial^2 + 0.01\,\partial + 1\right).$$

The theory is applied with

$$\begin{aligned}
A_s(\partial) &= (\partial + 1 + i)(\partial + 1 - i)(\partial + 10)^2, \\
A_{cl}(\partial) &= A_s(\partial)(\partial + 2)(\partial + 1)^2.
\end{aligned}$$

Condition (6.41) is satisfied and we obtain

$$\begin{aligned} R(\partial) &= (\partial + 1)(\partial + 2)(\partial + \alpha)(\partial + \bar{\alpha})\frac{80}{|\alpha|^2} \\ S(\partial) &= D(\partial)(\partial + \beta)(\partial + \bar{\beta}) \\ T(\partial) &= A_s(\partial)\frac{R(0)}{A_s(0)} \end{aligned}$$

with $\alpha = 0.338 + 0.639\,i$ and $\beta = 11.50 + 8.30\,i$.

The Bode plot of the open-loop transfer function is shown in Figure 6.11. We note the large gain in low frequencies (in order to track the reference without static error) and in the neighborhood of 1 rad/s (for the rejection of sinusoidal disturbance); the roll-off accentuates from 10 rad/s (this frequency is the absolute value of the root of the polynomial $F(\partial) = \partial + 10$, used for obtaining $\delta_0 = 1$). The gain augmentation margin (about 10 dB) and the phase lag margin (about 45°) are correct, along with the delay margin (0.3 s).

The simulation of the controlled system shown in Figure 6.12 corresponds to the following events: (i) step command at $t = 0$; (ii) sinusoidal disturbance d_y of frequency 1 rad/s and amplitude 0.5 adding to y from $t = 10$ s; * (iii) all along the simulation, a Gaussian white noise with standard deviation 0.03 is added to the measurement (the computation step is of 0.1 s). * The step response is excellent; the disturbance is rejected, but with more oscillatory dynamics (because the dynamics

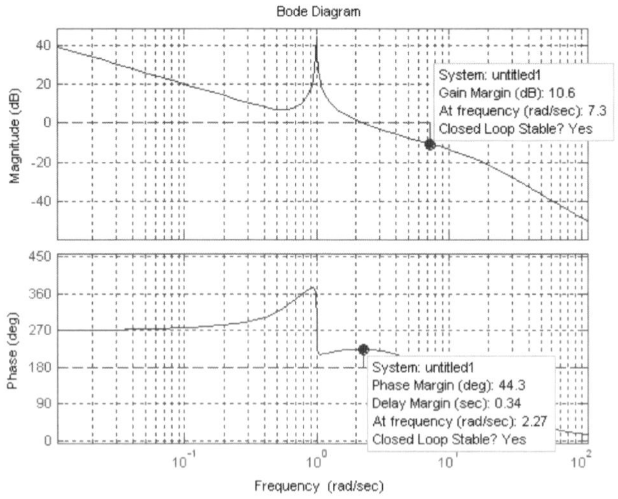

Figure 6.11. *Bode plot of L(s) (Example 1)*

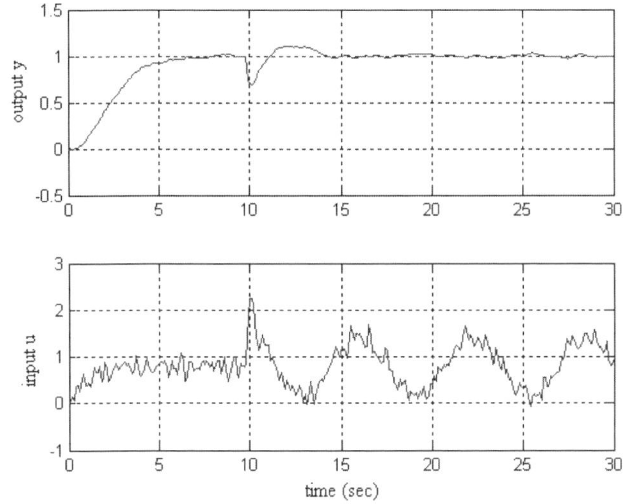

Figure 6.12. *Time domain response (Example 1)*

corresponding to the roots of $A_s(\partial)$ are excited by this disturbance). We note that from $t = 10$ s, the control becomes sinusoidal (if we leave aside the effect of the measurement noise) counteracting the disturbance d_y.

2. Ramp reference, sinusoidal disturbance

Consider again the minimal system **P** defined by the left form (6.25) and suppose that a sinusoidal disturbance d_y, with frequency 1 rad/s, is added to the output of **P**. Different from what was presented previously, the objective now is to track a reference signal which is a *ramp*. The relative degree δ_0 is fixed to be 1.

We again choose $D_1(\partial) = \partial^2 + 0.01\,\partial + 1$, but $D_2(\partial) = \partial^2$. The theory is applied using

$$\begin{aligned} A_s(\partial) &= (\partial + 0.4 + i)(\partial + 0.4 - i)(\partial + 10)^2, \\ A_{cl}(\partial) &= A_s(\partial)(\partial + 2)(\partial + 1)^3. \end{aligned}$$

Condition (6.41) is satisfied and we obtain

$$\begin{aligned} R(\partial) &= (\partial + 2)(\partial + 1)(\partial + 0.40)(\partial + \alpha)(\partial + \bar{\alpha})\frac{29}{0.40 \times |\alpha|^2} \\ S(\partial) &= D(\partial)(\partial + \beta)(\partial + \bar{\beta}) \\ T(\partial) &= \mu A_s(\partial)(\partial + 0.28) \end{aligned}$$

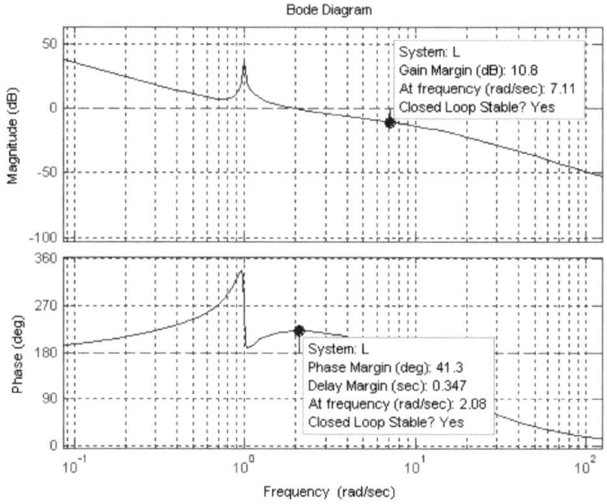

Figure 6.13. *Bode plot of L(s) (Example 2)*

where $\alpha = 0.19 + 0.88\,i, \beta = 11.40 + 8.05\,i$, and where μ is such that $T(0) = R(0) = 58$ (we also have $T'(0) = R'(0)$, where $(.)'$ denotes the derivative of the polynomial in parentheses).

The Bode plot of the open-loop transfer function is shown in Figure 6.13. The slope is -40 dB/decade in the low frequencies and the gain increases again in the neighborhood of 1 rad/s (for the rejection of the sinusoidal disturbance). The

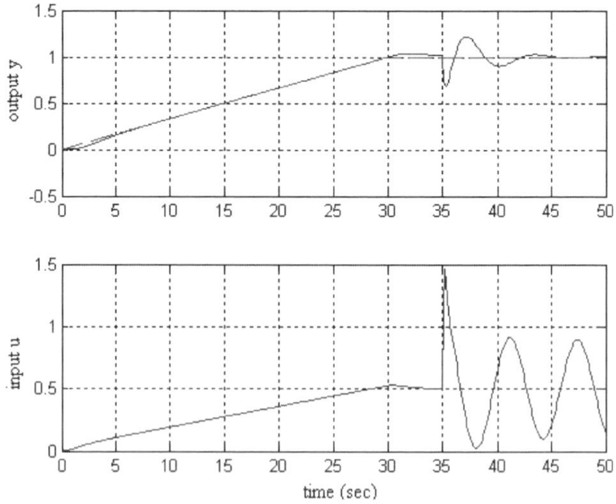

Figure 6.14. *Time domain response (Example 2)*

gain augmentation margin (about 10 dB) and the phase lag margin (about $41°$) are sufficient, along with the delay margin (0.35 s).

The simulation of the controlled system shown in Figure 6.14 corresponds to the following events: (i) reference ramp signal starting at $t = 0$ until it reaches the value of 1, and then maintaining this signal at this value; (ii) sinusoidal disturbance d_y with frequency 1 rad/s and amplitude 0.5 comes on at $t = 35$ s. The signal y and the reference r are shown in the upper part of Figure 6.14. Note that the ramp is followed *without tracking error*. The dynamics of the disturbance rejection are very different from those of the reference tracking. It was necessary to only weakly damp the closed-loop in order to obtain correct phase and gain margins (for the open-loop is purely oscillatory with a double pole at the origin and is non-minimum phase).

6.5. Exercises

EXERCISE 120.– *Detail the proof of equalities* (6.8), (6.9) *and* (6.10).

EXERCISE 121.– *Extend the theory presented in section 6.3.4 to the case where*

$$B(\partial) = b_0\,\partial^{n-1} + b_1\,\partial^n + \cdots + b_n$$

to simplify the rationale, first assume that $\delta_0 > 0$ *when* $b_0 \neq 0$ *and then consider the case* $\delta_0 = 0$ *and* $b_0 \neq 0$.

EXERCISE 122.– *Justify the choice made for polynomial* $A_{cl}(\partial)$ *in the four examples given in section 6.3.6.*

EXERCISE 123. *Let* $\mathbf{P}$ *be a minimal system with transfer function* $G(s) = \frac{s+2}{s\,(s+1)}$. *1) Determine an RST controller for this system that has the following properties: (i) it rejects the constant disturbances* d_u *and* d_y *that are added to the input and output of* $\mathbf{P}$*, respectively; (ii) it ensures the tracking of a constant reference signal without static error; (iii)* $A_{cl}(s)$ *has roots* -1 *(triple root),* -2*, and* -10*; (iv) the only pole of the transfer function between reference* r *and output* y *is* -1 *(double pole). 2) How is such a choice of* A_{cl} *justified? 3) What is the relative degree of the transfer function* $\frac{R}{S}$*?*

EXERCISE 124.– *Let* $\mathbf{P}$ *be a minimal system with transfer function* $P(s) = \frac{s+2}{(s+1)^2}$, *for which we design an RST controller having Properties (i) and (ii) of Exercise 123. We choose* -1 *(triple pole) and* -2 *to be the poles of the closed-loop. (i) Determine* R *and* S*, and then* T *of the form* $\lambda(\partial + 1)$*, where* λ *is such that the static error is zero. (ii) Explicitly determine the sensitivity function* $S_o(s)$*, and then the modulus margin. (iii) Calculate the transfer function between the reference signal* r *and the output* y*, and then explicitly determine the step response. What can we conclude?*

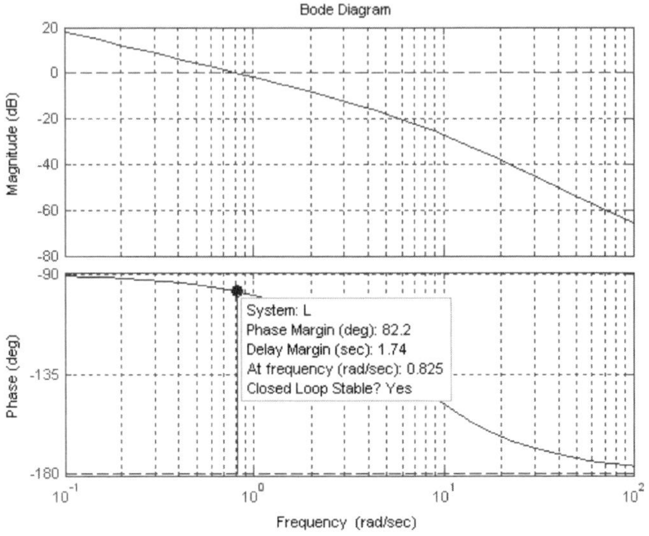

Figure 6.15. *Bode plot of L(s) (Exercise 126)*

EXERCISE 125.– *This exercise is a continuation of the previous one. This time, we choose a closed-loop having, compared to that of Exercise 124, a supplementary pole* $-\alpha$*. (i) What seems to be a reasonable value of* α*? (Defend your viewpoint!). (ii) Choose a value of* α *that seems pertinent, then determine the RST controller such that the step response is the same as in Exercise 124. (iii) What are the advantages and inconveniences of the two RST controllers?*

EXERCISE 126.– *This exercise is a continuation of the previous one. Choosing, from Exercise 125,* $\alpha = 5$*, we obtain the open-loop transfer function* $L(s)$*, the Bode plot of which is shown in Figure 6.15. (i) Is this result coherent? (Explain!) (ii) Suppose now that* $\mathbf{P}$ *is a simplified model of a system of fourth order whose transfer function is* $\tilde{P}(s) = P(s) E(s)$ *with* $E(s) = \frac{\omega_0^2}{s^2 + 2\varsigma\,\omega_0\, s + \omega_0^2}$ *where* $\varsigma = 5.10^{-3}$ *and* $\omega_0 \geq 30$ *rad/s. Show the shape of the Bode plot of* $E(s)$ *and in particular calculate the maximum of* $|E(i\omega)|$ *and at which frequency the maximum is attained. (iii) Supposing we control the complete system using the RST controller determined in Exercise 125 (with* $\alpha = 5$*), is the closed-loop system stable? What will happen if we use the controller determined in Exercise 124? (Defend your argument.)*

EXERCISE 127.– *Consider a controllable and observable second-order system with transfer function* $G(s) = 1/\left(s^2+s-1\right)$. *(i) Design an RST controller for this system, placing all the poles at* -1*, such that the transfer of the controlled system between reference and output is of the order of* 2*, and such that* $\delta_0 = 0$. *(ii) Same question with* $\delta_0 = 1$. *(iii) In the above two cases, is the choice of the poles pertinent?*

EXERCISE 128.– *Extend the theory presented in section 6.4.5 to the case where*

$$B(\partial) = b_0\,\partial^{n-1} + b_1\,\partial^{n} + \cdots + b_n$$

(see Exercise 121)

Chapter 7

Systems Theory (II)

The theory of linear time-invariant systems turned out to be essentially algebraic, which is the approach this chapter will be following. Given its general nature and significance, such an approach will not be possible unless we use the "language of modules" as presented in section 13.4. From the point of view of *applications*, it is sufficient, in most cases, to consider "state-space systems" (section 2.3.6). Fundamentally, however, this framework is too narrow.

Take an example: one of the most important notions of systems theory is *controllability*. We have already come across this notion in section 6.2.2 (see Definition 108) concerning SISO systems represented by left forms. However, there does not exist two types of controllability, one for these left forms and another one for state-space systems. It is therefore necessary to define controllability for a class of systems that encompasses these two types of representations. There are also systems that are described by a Rosenbrock representation, which is completely general as shown in section 2.3.5. Therefore, it appears that it is sufficient to properly define the controllability of a Rosenbrock representation. But is it a particular representation of the system which is controllable or is it the system itself? The works of Willems and Fliess ([117], [42]), dating back to the beginning of the 1990s, have clearly demonstrated that *this is a property of the system*, and not of one of its representations. Controllability is therefore a concrete and objective property. Henceforth, this is the viewpoint that deserves to be expounded. Paradoxically, to achieve this, an often abstract language is necessary.

7.1. Structure of a linear system

7.1.1. **The notion of a linear system*

Let us begin by recalling what has been established in the remarks of Chapter 2. In what follows, the term "system" signifies "linear time-invariant system", unless otherwise stated.

(i) A system Σ can be identified with a finitely presented $\mathbf{R}$-module[1] M, where $\mathbf{R} = \mathbb{R}[\partial]$ (Remark 8, section 2.2.2). This module M is defined by an equation such that (2.6)

$$E(\partial)\, w = 0.$$

The variables $w_1, ..., w_k$ generate M, and we write $M = [w]_{\mathbf{R}}$, where $w = [w_1 \ ... \ w_k]^T$. The matrix $E(\partial)$ is a *matrix of definition* of M (or of the system $\mathbf{\Sigma}$). We can assume that $E(\partial)$ is left-regular (i.e. full row rank) because $\mathbf{R}$ is a principal ideal domain (see sections 13.1.4 and 13.4.2), and then $E(\partial) \in \mathbf{R}^{r \times k}$, where $r = \operatorname{rk} E(\partial)$. The system Σ (identified with the module M) becomes a *control system* once the input variables $u_1, ..., u_m$ and the output variables $y_1, ..., y_p$ are chosen. A finite sequence of variables $u = [u_1 \ ... \ u_m]^T$ is a possible input for $\mathbf{\Sigma}$ if (i) u is *independent*, i.e. $[u]_{\mathbf{R}}$ is a free $\mathbf{R}$-module of rank m, and (ii) the module $M/[u]_{\mathbf{R}}$ is torsion (Remark 12, section 2.3.1). The only *a priori* constraint on the output variables is that they belong to M (Remark 15, section 2.3.5). One can always represent a system in a Rosenbrock form like the one defined by relation (2.1.1) of section 2.3.5 (Remark 15, section 2.3.5). The poles of the control system Σ are the Smith zeros of the torsion module $M/[u]_{\mathbf{R}}$ (Remark 19, section 2.3.7).

7.1.2. *State-space representation*

We owe to Fliess [42] the following result:

THEOREM 129.– *Every control system admits a state-space representation (of the form* (2.20)*, section 2.3.6).*

PROOF. * 1) Since the module $T = M/[u]_{\mathbf{R}}$ is torsion, it is a finite-dimensional $\mathbb{R}$-vector space (according to Theorem 565 of section 13.4.3). Let $\bar{\eta} = (\bar{\eta}_i)_{1 \leq i \leq n}$ be a basis of this vector space. Since $\partial \bar{\eta}_i$ is an $\mathbb{R}$-linear combination of the $\bar{\eta}_j$s, $1 \leq j \leq n$,

1. Or, what comes to the same thing (because $\mathbf{R}$ is Noetherian), a finitely generated $\mathbf{R}$-module. We can nonetheless consider a linear system defined over a non-Noetherian ring (see section 13.2.3, Remark 500(ii)); such a system can be defined as (or be identified with) a finitely presented module.

there exists a matrix $A \in \mathbb{R}^{n \times n}$ such that $\partial\bar{\eta} = A\,\bar{\eta}$. There also exist n elements $\eta_i \in M$ such that $\bar{\eta}_i$ is the canonical image of η_i in T $(1 \leq i \leq n)$. As a result, there exist matrices $B_j \in \mathbb{R}^{n \times m}$ $(0 \leq j \leq s$, s finite) such that $B_s \neq 0$ and

$$\partial\eta = A\,\eta + \sum_{0 \leq j \leq s} B_j\,\partial^j u. \tag{7.1}$$

If $s \geq 1$, then let $\eta^* = \eta - B_s\,\partial^{s-1} u$. This yields

$$\partial\eta^* = A\,\eta^* + \left(\sum_{0 \leq j \leq s-2} B_j\,\partial^j + B'_{s-1} \right) u$$

where $B'_{s-1} = A\,B_s + B_{s-1}$. Iterating this procedure, we obtain the form

$$\partial x = A\,x + B\,u. \tag{7.2}$$

2) It follows from this expression that the module T is defined by the equation

$$\partial\bar{x} = A\,\bar{x}$$

where $\bar{x} = (\bar{x}_i)_{1 \leq i \leq n}$ and $\bar{x}_i$ is the canonical image of x_i in T $(1 \leq i \leq n)$. Therefore, $T = [\bar{x}]_{\mathbf{R}}$, and as the quantities $\bar{x}_i$ are $\mathbb{R}$-linearly independent, $\bar{x}$ is a basis of the vector $\mathbb{R}$-space T. Let $\bar{y}_i$ be the canonical image of $y_i \in M$ in T. There exists a matrix $W(\partial) \in \mathbf{R}^{p \times m}$ such that

$$y = C\,x + W(\partial)\,u, \tag{7.3}$$

and expression (2.20) of section 2.3.6 is obtained.* ■

REMARK 130.– *As mentioned in section 2.5.3, we are only interested in* proper *control systems in practice. From Theorem 48, a proper system Σ admits a state-space representation of the form*

$$\left\{ \begin{array}{l} \partial x = A\,x + B\,u \\ y = C\,x + D\,u \end{array} \right. \tag{7.4}$$

where $A \in \mathbb{R}^{n \times n}$, $B \in \mathbb{R}^{n \times m}$, $C \in \mathbb{R}^{p \times n}$, and $D \in \mathbb{R}^{p \times m}$; in addition, $D = 0$ if and only if, Σ is strictly proper. Such a state-space representation is denoted by $\{A, B, C, D\}$ in what follows, and as $\{A, B, C\}$ if $D = 0$. As to the possibility of obtaining such a representation systematically, see Remark 49 (section 2.5.3).

DEFINITION 131 – *The matrices A, B, C and D above are called, respectively, the* state matrix, *the* input matrix, *the* output matrix *and the* direct term matrix *(or, more succinctly, the* direct term*). The first of the equations of equation (7.4) is the* state equation, *the second is the* output equation.

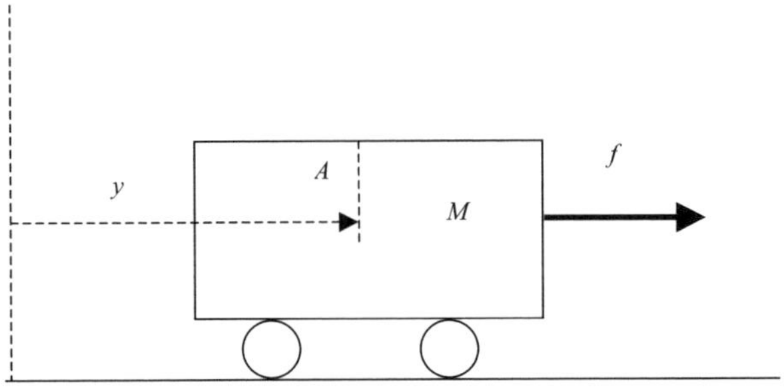

Figure 7.1. *Carriage*

REMARK 132.– *Suppose we change the basis in $\mathbb{R}^n$ according to $x = P\eta$, where the matrix $P \in \mathbb{R}^{n \times n}$ is invertible. In the new basis, the representation is*

$$\left\{ \begin{array}{r} \partial\eta = P^{-1}\,A\,P\,\eta + P^{-1}\,B\,u, \\ y = C\,P\,\eta + D\,u. \end{array} \right. \tag{7.5}$$

EXAMPLE 133.– *Consider a carriage of mass M rolling frictionless on a horizontal plane and subjected to a force f (see Figure 7.1).*

Assuming that the wheels have negligible mass and moment of inertia, we have according to Newton's law (equation (1.12)*, section 1.2.1)*

$$f = M\,\dot{v}$$

where $v = \dot{y}$ is the velocity of the carriage (y is the position with respect to a fixed point). The control system with input $u = f$ and output y is thus described by the state-space representation

$$\begin{aligned} \dot{x} &= \begin{bmatrix} 0 & 0 \\ 1 & 0 \end{bmatrix} x + \begin{bmatrix} \frac{1}{M} \\ 0 \end{bmatrix} u, \\ y &= \begin{bmatrix} 0 & 1 \end{bmatrix} x, \end{aligned}$$

where $x \triangleq \begin{bmatrix} v & y \end{bmatrix}^T$. In this case, all the state components have a physical significance. This is not always so.

7.1.3. *Controllability*

A system is *controllable* if and only if, none of the variables of this system satisfies a non-trivial homogenous differential equation; in other words, *if and only if, all its variables are free* (see sections 13.4.1 and 13.4.3). Indeed, suppose there exists a non-invertible polynomial $a(\partial)$ and a variable $v \in M$ such that $a(\partial)\,v = 0$. The evolution of a variable $v \in \mathcal{E}(\mathbb{R})$ satisfying the same differential equation only depends on initial conditions, and hence no possible action can cause any inflexion on the evolution of this variable. Conversely, the evolution of a free variable (not determined by initial conditions) can be governed by an appropriate action.

*To translate what we have just said into mathematical words, we begin by the following remark:

REMARK 134.– *Let M be the module associated with a system Σ. According to Corollary 555 (section 13.4.2), the module M is free if and only if, it is torsion-free.*

We are now led to the following definition, due to Fliess [42]:

DEFINITION 135.– *A system Σ is controllable if the associated module M is free.**

REMARK 136.– *There exists a "behavioral" definition of controllability, due to Willems [117] (see also [96], section 5.2). The study of the equivalence between the controllability "à la Willems" and that of Definition 135 was carried out by Fliess [43]. According to these two definitions, the controllability of a system is a concept neither related to the type of representation chosen for the system nor related to a particular choice of the control variables.*

THEOREM 137.– * *A control system Σ is controllable if and only if, it can be represented by a "right form" (see section 2.3.5).**

PROOF. * A control system $\mathbf{\Sigma}$ is representable by the equalities (2.19) of section 2.3.5 if and only if, the associated module is free (with basis ξ).* ∎

THEOREM 138.– *Let $\mathbf{\Sigma}$ be a system defined by a Rosenbrock representation $\{D, N, Q, W\}$ (Definition 14, section 2.3.5); $\mathbf{\Sigma}$ is controllable if and only if, the matrices D and N are left-coprime.*

PROOF. This is obvious according to Theorem 505 (section 13.2.6) and Remark 134. ∎

COROLLARY 139.– "Popov–Belevitch–Hautus (PBH) test for controllability". *A system $\mathbf{\Sigma}$ defined by a Rosenbrock representation $\{D, N, Q, W\}$ is controllable if and only if, $\mathrm{rk}_{\mathbb{C}} \begin{bmatrix} D(s) & N(s) \end{bmatrix} = r$ for every $s \in \mathbb{C}$ (where $r \triangleq \mathrm{rk}_{\mathbf{R}} D(\partial)$). In particular, a state-space system $\{A, B, C, D\}$ is controllable if and only if, $\mathrm{rk}_{\mathbb{C}} \begin{bmatrix} sI_n - A & B \end{bmatrix} = n$ for every $s \in \mathbb{C}$.*

PROOF. The matrices $\{D, N\}$ are left-coprime if and only if the Smith form of $\begin{bmatrix} D & N \end{bmatrix}$ is $\begin{bmatrix} I_r & 0 \end{bmatrix}$ (according to Theorem 505 of section 13.2.6), i.e. when the stated condition holds. ∎

REMARK 140.– *The necessary and sufficient condition expressed by Theorem 138 generalizes Definition 108 of section 6.2.2, valid only in the case of an SISO system defined by a left form.*

THEOREM 141.– "Kalman criterion for controllability" *[65]. A state-space system* $\{A, B, C, D\}$ *is controllable if and only if,* $\operatorname{rk}\Gamma = n$ *, where* n *is the dimension of the state vector (section 2.3.6, Remark 16) and where* Γ *is the "controllability matrix", defined by*

$$\boxed{\Gamma = \begin{bmatrix} B & AB & \dots & A^{n-1}B \end{bmatrix}}. \tag{7.6}$$

PROOF. * 1) As we have indicated in section 2.3.6, a state-space representation $\{A, B, C, *\}$ is a Rosenbrock representation $\{D, N, Q, *\}$ of a particular type, where $D(\partial) = \partial I_n - A$, $N(\partial) = B$ and $Q(\partial) = C$. According to Theorem 138, the system defined by this representation is controllable if and only if, the matrices $\partial I_n - A$ and B are left-coprime, that is to say if $\begin{bmatrix} \partial I_n - A & B \end{bmatrix}$ is right-invertible according to Definition 506 (section 13.2.6). This amounts to saying that if $v^T = \begin{bmatrix} v_1 & \dots & v_n \end{bmatrix}$ is a row of elements of a right **R**-module, the equality $v^T \begin{bmatrix} \partial I_n - A & B \end{bmatrix} = 0$ implies $v^T = 0$. The first of these equalities is equivalent to (a) $\partial v^T = v^T A$ and (b) $v^T B = 0$. Multiplying (b) by ∂ on the right, we obtain $v^T \partial B = 0$, and hence $v^T A B = 0$ according to (a). This last equality, multiplied on the right by ∂, implies that $v^T \partial A B = 0$, and hence $v^T A^2 B = 0$ according to (a), etc. As a result, the equality $v^T \begin{bmatrix} \partial I_n - A & B \end{bmatrix} = 0$ implies $v^T \Gamma = 0$. If $\operatorname{rk}\Gamma = n$, this implies $v^T = 0$. Therefore, the condition $\operatorname{rk}\Gamma = n$ is *sufficient* for the system to be controllable. 2) Let us show by contradiction that this is also a necessary condition. Let $\operatorname{rk}\Gamma = \rho < n$. Let $P = \begin{bmatrix} P_1 & P_2 \end{bmatrix} \in \mathbb{R}^{n \times n}$, where the submatrix P_1 consists of ρ linearly independent columns of Γ and where P_2 is a submatrix chosen such that P will be invertible. Such a matrix exists according to the "theorem of the incomplete basis" (Corollary 517, section 13.3.1). Indeed, the columns of P_1 are identified with linearly independent vectors x_i $(\rho + 1 \leq i \leq n)$ in the canonical basis of $\mathbb{R}^n$. We can determine the vectors x_i $(\rho + 1 \leq i \leq n)$ in such a way that $(x_i)_{1 \leq i \leq n}$ is a basis of $\mathbb{R}^n$. The vectors x_i $(\rho + 1 \leq i \leq n)$ are identified with column vectors in the canonical basis of $\mathbb{R}^n$ and form the matrix P_2 which is being sought after. Let $\mathbf{u}$ and $\mathbf{b}$ be the homomorphisms represented by the matrices A and B, respectively, when the bases chosen in $\mathbb{R}^n$ and $\mathbb{R}^m$ are the canonical bases. The subspace of $\mathbb{R}^n$ spanned by the vectors x_i $(1 \leq i \leq \rho)$ is $E_1 = \sum_{0 \leq i \leq n-1} \mathbf{u}^i \operatorname{im} \mathbf{b}$ (where $\mathbf{u}^0 = I_n$). According to the Cayley–Hamilton theorem (Theorem 537, section 13.3.4), $\mathbf{u}^n$ is an $\mathbb{R}$-linear combination of the endomorphisms $\mathbf{u}^i$, $0 \leq i \leq n - 1$, and, as a result, E_1 is $\mathbf{u}$-invariant. As shown in section 13.3.2 (see equation (13.24)), the matrix $\tilde{A}$ representing $\mathbf{u}$ in the basis $(x_i)_{1 \leq i \leq n}$, i.e. $\tilde{A} = P^{-1} A P$, is therefore of the form

$$\tilde{A} = \begin{bmatrix} A_c & * \\ 0 & A_{\bar{c}} \end{bmatrix}$$

where the square matrices $\tilde{A}_c$ and $\tilde{A}_{\bar{c}}$ are of order ρ and $n-\rho$, respectively. On the other hand, $\operatorname{im}\mathbf{b} \subset E_1$, and thus the matrix $\tilde{B}$ representing $\mathbf{b}$ in the canonical basis of $\mathbb{R}^m$ and the basis $(x_i)_{1\leq i\leq n}$ of $\mathbb{R}^n$, i.e. $\tilde{B} = P^{-1}B$ (see equation (7.5), section 7.1.2), is of the form

$$\tilde{B} = \begin{bmatrix} B_c \\ 0 \end{bmatrix}$$

where $B_c \in \mathbb{R}^{\rho\times m}$. Putting $x = P\,\eta$, where $\eta = \begin{bmatrix} x_c \\ x_{\bar{c}} \end{bmatrix}$, we obtain

$$\boxed{\left\{ \begin{array}{r} \partial x_c = A_c\,x_c + *\,x_{\bar{c}} + B_c\,u \\ \partial x_{\bar{c}} = A_{\bar{c}}\,x_{\bar{c}} \end{array} \right.} \tag{7.7}$$

where $*$ is an unspecified matrix. Consequently, the submodule $[\eta_{\bar{c}}]_{\mathbf{R}}$ of M generated by the components of $x_{\bar{c}}$ is torsion and not reduced to 0. The module M is therefore not free and the associated system is therefore non-controllable.* ■

REMARK 142.– *(i) Since the controllability matrix* Γ *defined by equation* (7.6) *depends only on* A *and* B*, from now on we will call it the "controllability matrix of* (A,B)*" and denote it as* $\Gamma(A,B)$*. We will express that it is of rank* n *by saying that "*(A,B) *is controllable". (ii) An elementary calculation shows that for any matrix* $P \in GL_n(\mathbb{R})$,[2]

$$\Gamma\left(P^{-1}A\,P, P^{-1}B\right) = P^{-1}\,\Gamma(A,B)\,. \tag{7.8}$$

As a result, the controllability of (A,B) *only involves the homomorphisms* $\mathbf{u}$ *and* $\mathbf{b}$ *defined by these matrices in the chosen bases [119]. (iii) Originally, controllability has been defined* for state-space systems *by Kalman [65] in the following manner: "A state-space system* $\{A,B,C,D\}$ *is controllable if for any initial condition* (t_0,x_0) *and any point* x^* *in the state space, there exists an instant* $t_1 > t_0$ *and a control* u *defined on* $[t_0,t_1]$ *for which the state* x *passes between the instant* t_0 *and the instant* t_1 *from the value* x_0 *to the value* x^**". Kalman showed in the cited reference that a necessary and sufficient condition for the system* $\{A,B,C,D\}$ *to be controllable, in the sense as specified just now, is the one given in Theorem 141 (in the case of discrete-time systems, see Definition 315 and Theorem 317 (section 10.4.2)).*

REMARK 143.– *It is easy to show using Theorem 141 that the carriage of Example 133 is a controllable system. It is equally easy to show that from any initial instant*

2. $GL_n(\mathbb{R})$ is the general linear group of the square matrices of order n over $\mathbb{R}$, i.e. the multiplicative group consisting of all invertible matrices of order n with real entries: see section 13.1.4.

t_0 and any initial state $x_0 = \begin{bmatrix} v_0 & y_0 \end{bmatrix}^T$ ($x_0 = x(t_0)$), one can bring the carriage to an arbitrarily chosen state $x^* = \begin{bmatrix} v^* & y^* \end{bmatrix}^T$ at any time $t_1 > t_0$ ($x(t) = x^*$) by modulating the force f on the interval $[t_0, t_1]$ in an appropriate manner, which is Kalman's definition of controllability (Remark 142(iii)).

*Using the notation in the proof of Theorem 141, we obtain the following result:

PROPOSITION 144.– *The torsion submodule of M is $\mathcal{T}(M) = [x_{\bar{c}}]_{\mathbf{R}}$.*

PROOF. 1) We know that $[x_{\bar{c}}]_{\mathbf{R}} \subset \mathcal{T}(M)$. 2) To prove the converse, consider the quotient module $M/[x_{\bar{c}}]_{\mathbf{R}} = [\bar{x}_c, \bar{u}]_{\mathbf{R}}$, where the components of $\bar{x}_c$ and of $\bar{u}$ are the canonical images of those of x_c and of u, respectively. We obtain

$$\partial \bar{x}_c = A_c\, \bar{x}_c + B_c\, \bar{u}. \tag{7.9}$$

We have $\operatorname{rk} \Gamma\left(\tilde{A}, \tilde{B}\right) = \rho$ and

$$\Gamma\left(\tilde{A}, \tilde{B}\right) = \begin{bmatrix} B_c & A_c\, B_c & \cdots & A_c^{n-1}\, B_c \\ 0 & 0 & \cdots & 0 \end{bmatrix},$$

and hence $\rho = \operatorname{rk} \begin{bmatrix} B_c & A_c\, B_c & \ldots & A_c^{n-1}\, B_c \end{bmatrix} = \operatorname{rk} \Gamma(A_c, B_c)$; the last equality is a consequence of the Cayley–Hamilton theorem (see above the proof of Theorem 141). Therefore, the module $M/[x_{\bar{c}}]_{\mathbf{R}}$ is free, and thus $\mathcal{T}(M) \subset [x_{\bar{c}}]_{\mathbf{R}}$ * ■

DEFINITION 145.– *We call* the controllable quotient system *the system with equation* (7.9) (**associated with the quotient module $M/\mathcal{T}(M)$**)[3].

The above proves that (A_c, B_c) is controllable. Furthermore, one can represent the decomposition (7.7), called "Kalman controllability decomposition", by the diagram in Figure 7.2:

7.1.4. *Observability*

Intuitively, a control system is observable if, when both its input and output are identically zero, all its variables are necessarily zero (in other terms, every evolution of the system variables can be detected from the input and output).

3. *Since a system is associated with a finitely presented module, a *quotient system* is associated with a *quotient* of this module [22].*

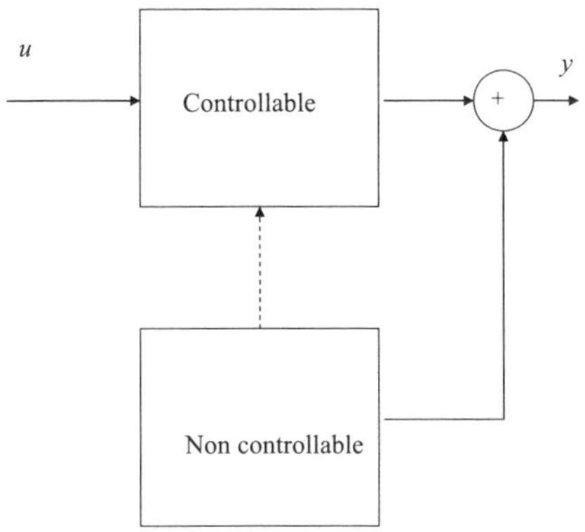

Figure 7.2. *Kalman controllability decomposition*

*To express what has been just said in mathematical terms, the language of modules is, once again, the best suited one: the control system considered is observable if and only if,

$$\frac{M}{[y,u]_{\mathbf{R}}} = 0, \tag{7.10}$$

and we are thus led to the following definition [42] :

DEFINITION 146.– *A control system Σ, with input u and output y, is observable, if the associated module M satisfies the equality $M = [y,u]_{\mathbf{R}}$.**

THEOREM 147.– *A control system Σ is observable if and only if, it can be represented by a "left form" (see section 2.3.5).*

PROOF. * A control system Σ is representable by the equality (2.17) of section 2.3.5 if and only if, the associated module is $M = [y,u]_{\mathbf{R}}$.* ■

THEOREM 148.– *Let Σ be a control system defined by a Rosenbrock representation $\{D, N, Q, W\}$; Σ is observable if and only if, the matrices D and Q are right-coprime.*

PROOF. * The system Σ is observable if and only if, the equality (7.10) is satisfied. Let ξ be the partial state of Σ and $\overline{\overline{\xi}}$ be the column matrix; the entries are the canonical images of the components of ξ in $M/[y,u]_{\mathbf{R}}$. We get

$$\begin{bmatrix} D(\partial) \\ Q(\partial) \end{bmatrix} \overline{\overline{\xi}} = 0. \tag{7.11}$$

Σ is observable if and only if, this equality implies $\bar{\bar{\xi}} = 0$, which is the case if and only if, the matrix $\begin{bmatrix} D(\partial) \\ Q(\partial) \end{bmatrix}$ is left-invertible. This means that the matrices $\{D, Q\}$ are right coprime (see section 13.2.6).* ■

Following a similar line of reasoning as in the proof of Corollary 139 and of Theorem 141, we obtain the following results (with the same notation):

COROLLARY 149.– "Popov–Belevitch–Hautus (PBH) test for observability". *A system Σ defined by a Rosenbrock representation $\{D, N, Q, W\}$ is observable if and only if,* $\mathrm{rk}_{\mathbb{C}} \begin{bmatrix} D^T(s) & Q^T(s) \end{bmatrix} = r$ *for every* $s \in \mathbb{C}$. *In particular, if Σ is a state-space system $\{A, B, C, D\}$, Σ is controllable if and only if,* $\mathrm{rk}_{\mathbb{C}} \begin{bmatrix} s I_n - A^T & C^T \end{bmatrix} = n$ *for every* $s \in \mathbb{C}$.

THEOREM 150.– "Kalman criterion for observability" *[65]. A state-space system $\{A, B, C, D\}$ is observable if and only if* $\mathrm{rk}\, \Omega = n$, *where n is the dimension of the state vector and where Ω is the "observability matrix" defined by*

$$\boxed{\Omega = \begin{bmatrix} C^T & A^T C^T & \ldots & \left(A^T\right)^{n-1} C^T \end{bmatrix}^T}. \tag{7.12}$$

REMARK 151.– *Since the observability matrix Ω only depends on A and C, it can be denoted by $\Omega(A, C)$, and we can express that it is of rank n by saying that "(C, A) is observable". (One can also refer more intrinsically to the homomorphisms defined by the matrices C and A in the chosen bases, in the spirit of Remark 142 of section 7.1.3.)*

REMARK 152.– *It is easy to show, using Theorem 150, that the carriage of Example 133 (with the specified input and output) is observable. We can interpret this property in the following manner: suppose that the system evolves over a time interval $[t_0, t]$, $t > t_0$, and we collect measurements of the input and the output over this interval. From these data, it is possible to reconstruct the initial state $x_0 = x(t_0)$. (If the output is velocity v instead of position y, such a reconstruction is impossible, since the input and output remain unchanged if one adds an arbitrary constant to the position y; the control system is therefore non-observable.) Such an approach of observability is due to Kalman [65] (in case of discrete-time state-space systems, see Definition 327 and Theorem 328 (section 10.4.3)). The interpretation of observability due to Willems is very similar but not related to the state-space representation ([96], section 5.3).*

The following is deduced from Theorem 150:

COROLLARY 153.– Controllability $\leftrightarrow$ observability duality. *(C, A) is observable if and only if, $\left(A^T, C^T\right)$ is controllable.*

PROOF. It is clear that $\Omega^T(A,C) = \Gamma(A^T, C^T)$, and hence $\operatorname{rk}\Omega(A,C) = \operatorname{rk}\Gamma(A^T,C^T)$ according to Proposition 534 (section 13.3.4). ■

The above *duality* has important consequences; in particular, the following: let $\boldsymbol{\Sigma}$ be a state-space system $\{A,B,C,D\}$ of order n and which is non-observable. Let $\omega < n$ be the rank of its observability matrix Ω. Let $P = \begin{bmatrix} P_1 & P_2 \end{bmatrix} \in \mathbb{R}^{n\times n}$, where P_1 is formed by ω linearly independent columns of Ω^T and P_2 is a sub-matrix chosen in such a way that P is invertible (see Part 2 of the proof of Theorem 141). The matrices $\breve{A}^T = P^{-1}A^T P$ and $\breve{C}^T = P^{-1}C^T$ are of the form

$$\breve{A}^T = \begin{bmatrix} A_o^T & * \\ 0 & A_{\bar{o}}^T \end{bmatrix}, \quad \breve{C}^T = \begin{bmatrix} C_o^T \\ 0 \end{bmatrix}$$

where $(\breve{A}_o^T, \breve{C}_o^T)$ is controllable. As a result, $\breve{A} = P^T A P^{-T}$ and $\breve{C} = C P^{-T}$ (where P^{-T} means $(P^{-1})^T$) satisfy

$$\breve{A} = \begin{bmatrix} A_o & 0 \\ * & A_{\bar{o}} \end{bmatrix}, \quad \breve{C} = \begin{bmatrix} C_o & 0 \end{bmatrix}$$

where (C_o, A_o) is observable. In the new bases, the state equations take the form

$$\boxed{\begin{cases} \partial \begin{bmatrix} x_o \\ x_{\bar{o}} \end{bmatrix} = \begin{bmatrix} A_o & 0 \\ * & A_{\bar{o}} \end{bmatrix} \begin{bmatrix} x_o \\ x_{\bar{o}} \end{bmatrix} + \begin{bmatrix} * \\ * \end{bmatrix} u \\ y = \begin{bmatrix} C_o & 0 \end{bmatrix} \begin{bmatrix} x_o \\ x_{\bar{o}} \end{bmatrix} + [*]\, u. \end{cases}} \tag{7.13}$$

*We deduce the following result:

PROPOSITION 154.– *The "non-observable quotient system" is associated with the quotient module $M/[y,u]_{\mathbf{R}}$ (see Footnote 3 of section 7.1.3) defined by the equation*

$$\partial \bar{\bar{x}}_{\bar{o}} = A_{\bar{o}}\, \bar{\bar{x}}_{\bar{o}} \tag{7.14}$$

where the components of $\bar{\bar{x}}$ are the canonical images of the components of $x_{\bar{o}}$ in $M/[y,u]_{\mathbf{R}}$.

PROOF. We have

$$\begin{bmatrix} \partial I_\omega - A_o \\ C_o \end{bmatrix} \bar{\bar{x}}_o = 0$$

and thus $\bar{\bar{x}}_o = 0$ because the matrix $\begin{bmatrix} \partial I_\omega - A_o \\ C_o \end{bmatrix}$ is left-invertible (from the fact that (C_o, A_o) is observable and according to Theorem 148). Thus, $M/[y,u]_{\mathbf{R}} = [\bar{\bar{x}}_{\bar{o}}]_{\mathbf{R}}$, and this last module is defined by equation (7.14).* ■

We can represent the decomposition (7.13) by the diagram in Figure 7.3:

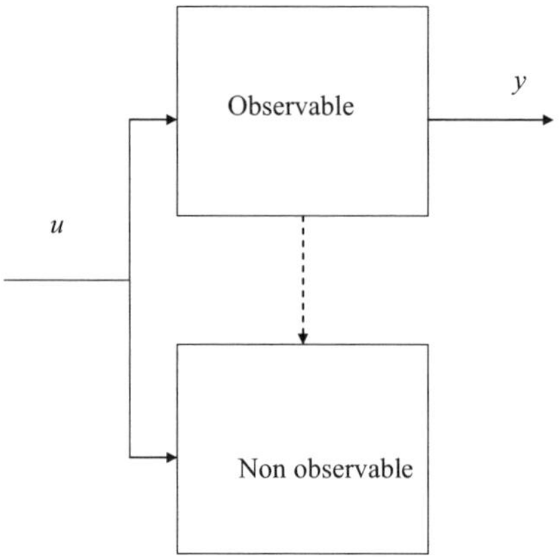

Figure 7.3. *Kalman observability decompostion*

7.1.5. *Canonical structure of a system*

Consider a control system $\boldsymbol{\Sigma}$ in the state-space form $\{A, B, C\}$,[4] which can be non-controllable and/or non-observable. By decomposing this system according to controllability, one obtains the representation (7.7), where (A_c, B_c) is controllable and where $A_{\bar{c}}$ is the empty matrix if $\boldsymbol{\Sigma}$ is controllable. The output y can be expressed as

$$y = \begin{bmatrix} C_c & C_{\bar{c}} \end{bmatrix} \begin{bmatrix} x_c \\ x_{\bar{c}} \end{bmatrix}.$$

Now let us consider the controllable quotient system (7.9) and its "output equation"

$$\tilde{y} = C_c \, \tilde{x}_c$$

where the components of $\tilde{y}$ are the canonical images of those of y in the quotient $M/\left[x_{\bar{c}}\right]_{\mathbf{R}}$. We can decompose this system according to observability (decomposition including empty matrices if this system is observable). The quotient system $M/\left[x_c, u\right]_{\mathbf{R}}$ can be decomposed in the same way. We finally obtain the following result:

THEOREM 155.– "Canonical structure of a state-space system" *[65]. (i) There exists a change of basis matrix $P \in \mathbb{R}^{n \times n}$ such that the matrices $\tilde{A} = P^{-1} A P$,*

4. The direct term matrix D does not play any role in the next theorem and can be assumed to be zero without loss of generality.

$\tilde{B} = P^{-1} B$, *and* $\tilde{C} = C P$ *take the form*

$$\boxed{\begin{cases} \tilde{A} = \begin{bmatrix} A_{c\bar{o}} & * & * & * \\ 0 & A_{co} & 0 & * \\ 0 & 0 & A_{\bar{c}\bar{o}} & * \\ 0 & 0 & 0 & A_{\bar{c}o} \end{bmatrix}, & \tilde{B} = \begin{bmatrix} B_{c\bar{o}} \\ B_{co} \\ 0 \\ 0 \end{bmatrix} \\ \tilde{C} = \begin{bmatrix} 0 & C_{co} & 0 & C_{\bar{c}o} \end{bmatrix} & \end{cases}} \tag{7.15}$$

where

$$\left(\begin{bmatrix} A_{c\bar{o}} & * \\ 0 & A_{co} \end{bmatrix}, \begin{bmatrix} B_{c\bar{o}} \\ B_{co} \end{bmatrix} \right)$$

is controllable and

$$\left(\begin{bmatrix} C_{co} & C_{\bar{c}o} \end{bmatrix}, \begin{bmatrix} A_{co} & * \\ 0 & A_{\bar{c}o} \end{bmatrix} \right)$$

is observable. (ii) The state-space system $\{A_{co}, B_{co}, C_{co}\}$ *is both controllable and observable.*

PROOF. (i) is obtained by applying the above indicated method (for more details, see [65], ([63], section 3.4.3) or ([96], section 5.4)). (ii) Let $n_{c\bar{o}}$ (resp., n_{co}) be the order of the matrix $A_{c\bar{o}}$ (resp., A_{co}). According to the Popov–Belevitch–Hautus test (Corollary 139, section 7.1.3), we get

$$\mathrm{rk}_{\mathbb{C}} \begin{bmatrix} s\, I_{n_{c\bar{o}}} - A_{c\bar{o}} & * & B_{c\bar{o}} \\ 0 & s\, I_{n_{co}} - A_{co} & B_{co} \end{bmatrix} = n_{c\bar{o}} + n_{co}$$

for any $s \in \mathbb{C}$. This implies

$$\mathrm{rk}_{\mathbb{C}} \begin{bmatrix} s\, I_{n_{co}} - A_{co} & B_{co} \end{bmatrix} = n_{co}$$

for any $s \in \mathbb{C}$, and thus (A_{co}, B_{co}) is controllable. The observability of (C_{co}, A_{co}) can be shown in a similar manner. ■

The decomposition expressed in the above theorem can be represented by the diagram in Figure 7.4.

REMARK 156.– *In general, it is not true that* $(A_{c\bar{o}}, B_{c\bar{o}})$ *will be controllable nor* $(C_{\bar{c}o}, A_{\bar{c}o})$ *will it be observable (see Exercise 213, section 7.6).*

PROPOSITION 157.– *The transfer matrix of the system* $\mathbf{\Sigma}$ *is*

$$\boxed{G(s) = C\,(sI - A)^{-1}\, B = C_{co}\,(sI - A_{co})^{-1}\, B_{co}}. \tag{7.16}$$

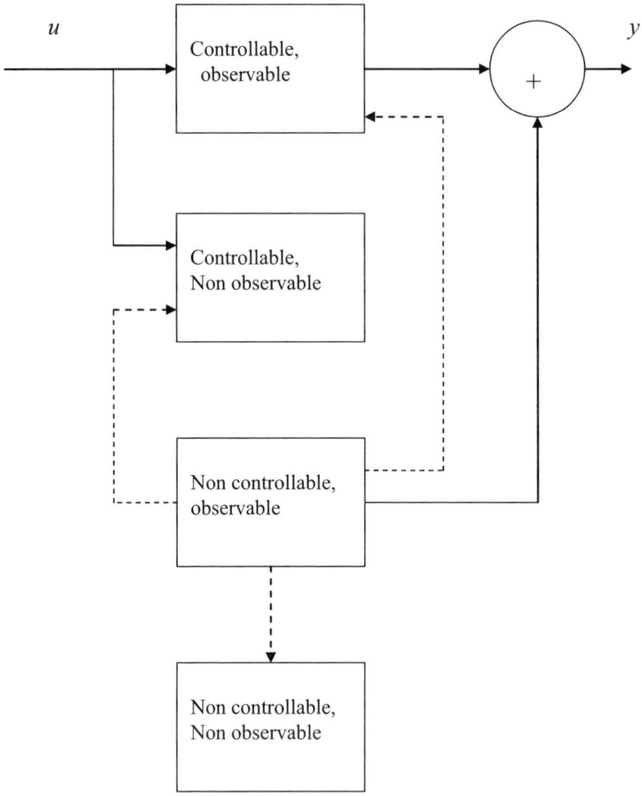

Figure 7.4. *Kalman general decomposition*

PROOF. The first equality is an immediate consequence of equality (2.48) (section 2.5.3) since $W(s) = 0$. The second equality is a consequence of the structure of the matrices in equation (7.15). ■

THEOREM 158.– *A control system Σ is minimal (see section 2.4.6) if, and only if it is both controllable and observable.*

PROOF. * 1) According to equation (7.16) and equality (2.37) of section 2.4.6, the transmission order of Σ cannot exceed the order of the matrix A_{co}. Thus, Σ is not minimal if it is non-observable or non-controllable. 2) Conversely, suppose Σ is observable. It is representable by a left form $D(\partial)\, y = N(\partial)\, u$ according to Theorem 147 (section 7.1.4). If Σ is also controllable, then the matrices $\{D, N\}$ are left-coprime according to Theorem 138 (section 7.1.3). Therefore, according to Remark 31 (section 2.4.5), the transmission poles of Σ are the Smith zeros of $D(s)$ and they coincide with the poles of Σ according to Remark 19 of section 2.3.7. As a result, the control system Σ is minimal.* ■

7.2. Zeros of a system

We have seen in section 2.4.7 that we can define *finite* transmission poles and zeros and also poles and zeros at *infinity*. This is also true for all other types of poles and zeros. The reader can find a complete exposition of the different types of poles and zeros at infinity in [21] and [22]. In what follows, we will be interested only in *finite* poles and zeros.

7.2.1. *Invariant zeros and transmission zeros*

Intuitively, the *invariant zeros* $(i.z.)$ of a control system $\mathbf{\Sigma}$, with input u and output y, characterize the dynamics of this system when its output is maintained at 0; the transmission zeros have been defined in sections 2.4.4 and 2.4.5 from the transfer matrix (see Definition 28).

*Now let us express the definitions in the language of modules [19]. Consider the finitely presented $\mathbf{R}$-module M associated with $\mathbf{\Sigma}$ and its decomposition (13.42) (section 13.4.2), i.e.

$$M = \mathcal{T}(M) \oplus \Phi \tag{7.17}$$

where Φ is a free submodule. In what follows, the notation and terminology are those used in sections 13.4.1 and 13.4.2. According to Proposition 561, one can choose $\Phi = \Phi_{[y,u]_{\mathbf{R}}}$, where $\Phi_{[y,u]_{\mathbf{R}}}$ is such that

$$[y,u]_{\mathbf{R}} = \mathcal{T}([y,u]_{\mathbf{R}}) \oplus \left([y,u]_{\mathbf{R}} \cap \Phi_{[y,u]_{\mathbf{R}}}\right).$$

It is this choice that we will use below.

DEFINITION 159.– *(i) The module of invariant zeros is $\mathcal{T}(M/[y]_{\mathbf{R}})$. (ii) The module of transmission zeros is $\mathcal{T}(\Phi \cap [y,u]_{\mathbf{R}}) / \mathcal{T}(\Phi \cap [y]_{\mathbf{R}})$. (iii) The invariant zeros (resp., the transmission zeros) are the Smith zeros of the first (resp., of the second) module.* *

Let system $\mathbf{\Sigma}$ be described by a Rosenbrock representation $\{D, N, Q, W\}$ (section 2.3.5). Let $\bar{\bar{\xi}} = \left(\bar{\bar{\xi}}_i\right)_{1 \le i \le r}$ and $\bar{\bar{u}} = \left(\bar{\bar{u}}_i\right)_{1 \le i \le m}$, where $\bar{\bar{\xi}}_i$ and $\bar{\bar{u}}_i$ are the canonical images of ξ_i and u_i, respectively, in $M/[y]_{\mathbf{R}}$ ($\xi = (\xi_i)_{1 \le i \le r}$ is the partial state and $u = (u_i)_{1 \le i \le m}$ is the input). One obtains

$$\begin{bmatrix} D(\partial) & -N(\partial) \\ Q(\partial) & W(\partial) \end{bmatrix} \begin{bmatrix} \bar{\bar{\xi}} \\ \bar{\bar{u}} \end{bmatrix} = 0. \tag{7.18}$$

DEFINITION 160.– *The matrix*

$$\boxed{R(\partial) = \begin{bmatrix} D(\partial) & -N(\partial) \\ Q(\partial) & W(\partial) \end{bmatrix}} \tag{7.19}$$

is called the Rosenbrock matrix *(or* system matrix*) of the control system* $\mathbf{\Sigma}$.

From the above two definitions, equation (7.18), and section 13.2.5, we get the following result:

THEOREM 161.– *The invariant zeros of* $\mathbf{\Sigma}$ *are the Smith zeros of its Rosenbrock matrix.*

REMARK 162.– *Suppose* $\mathbf{\Sigma}$ *is a state-space system* $\{A, B, C, D\}$ *(taking care not to confuse the matrix* D *of this state-space representation with the matrix* $D(\partial)$ *of the Rosenbrock representation). Then, the Rosenbrock matrix of the system is*

$$R(s) = \begin{bmatrix} s\,I_n - A & -B \\ C & D \end{bmatrix}. \tag{7.20}$$

PROPOSITION 163.– *(i) Let* $\rho(.)$ *be the rank of the matrix in parentheses over the field of rational functions* $\mathbf{F} = \mathbb{R}(s)$*. The rank of the Rosenbrock matrix* (7.19) *over* $\mathbf{F}$ *satisfies the equality*

$$\rho(R) = \rho(D) + \rho(G) \tag{7.21}$$

where G *is the transfer matrix of* $\mathbf{\Sigma}$*. (ii) In particular, if* R *is given by* (7.20) *and if* G *is semi-regular (section 13.1.4), we obtain*

$$\rho(R) = n + \min\{p, m\}. \tag{7.22}$$

PROOF. Notice that (with $r = \rho(D)$):

$$\begin{bmatrix} I_r & 0 \\ -Q\,D^{-1} & I_p \end{bmatrix} R(\partial) = \begin{bmatrix} D & -N \\ 0 & G \end{bmatrix}$$

according to equation (2.27) (section 2.4.2), from which we obtain the equality (7.21) because the matrix on the left is invertible. ∎

7.2.2. *Input-decoupling zeros*

Input-decoupling zeros are also called "non-controllable poles". They are defined independently of the choice of the input of the system.

*Let us first provide an abstract definition by considering the $\mathbf{R}$-module M associated with system $\mathbf{\Sigma}$ [19]:

DEFINITION 164.– *The module of input-decoupling zeros ($i.d.z.$) is $\mathcal{T}(M)$. The $i.d.z.s$ are the Smith zeros of this module.**

The result below is an immediate consequence of Definitions 135 (section 7.1.3) and 164:

THEOREM 165.– *A system is controllable if and only if, it does not have $i.d.z.s$.*

Suppose $\mathbf{\Sigma}$ is a control system defined by a Rosenbrock representation $\{D, N, Q, W\}$. We then have the following:

PROPOSITION 166.– *The $i.d.z.s$ are the Smith zeros of the matrix $\begin{bmatrix} D & N \end{bmatrix}$.*

PROOF. * This is obvious since $M = [\xi, u]_{\mathbf{R}}$, where

$$\begin{bmatrix} D(\partial) & N(\partial) \end{bmatrix} \begin{bmatrix} -\xi \\ u \end{bmatrix} = 0.*$$

■

Suppose now that $\mathbf{\Sigma}$ is a state-space system $\{A, B, C, D\}$, and consider its controllability decomposition (7.7) (section 7.1.3).

PROPOSITION 167.– *The $i.d.z.s$ are the eigenvalues of the matrix $A_{\bar{c}}$.*

PROOF. * This is an immediate consequence of the second equality of (7.7) and of Proposition 144 (section 7.1.3).* ■

7.2.3. *Output-decoupling zeros*

*Let us begin by an abstract definition [19], considering the $\mathbf{R}$-module M associated with the control system $\mathbf{\Sigma}$ with input u and output y.

DEFINITION 168.– *The module of output-decoupling zeros ($o.d.z.$) is $M/[y, u]_{\mathbf{R}}$. The $o.d.z.s$ are the Smith zeros of this module.**

Output-decoupling zeros are also called *non-observable poles*. The following theorem can be derived immediately from the definitions :

THEOREM 169.– *A control system is observable if and only if, it has no $o.d.z.$*

Suppose $\mathbf{\Sigma}$ is defined by a Rosenbrock representation $\{D, N, Q, W\}$. One has the following:

PROPOSITION 170.– *The o.d.z.s are the Smith zeros of the matrix* $\begin{bmatrix} D \\ Q \end{bmatrix}$.

PROOF. * It suffices to remember that $M/\left[y,u\right]_{\mathbf{R}}$ is defined by equation (7.11) (section 7.1.4).* ■

Last, suppose Σ is a state-space system $\{A, B, C, D\}$ and let us consider its observability decomposition (7.13) (section 7.1.4).

PROPOSITION 171.– *The o.d.z.s are the eigenvalues of the matrix* $A_{\bar{o}}$.

PROOF. * This is obvious according to Proposition 154.* ■

7.2.4. *Input–ouput decoupling zeros*

*Let us begin by an abstract definition [19]:

DEFINITION 172.– *The module of input–output decoupling zeros (i.o.d.z.) is*

$$\frac{\mathcal{T}(M)}{\mathcal{T}\left(\left[y,u\right]_{\mathbf{R}}\right)}.$$

*The i.o.d.z.s are the Smith zeros of this module.**

Let $\Sigma = \{A, B, C, D\}$ be a state-space system and consider its canonical structure given by Theorem 155 (section 7.1.5). We obtain the following result:

PROPOSITION 173.– *The i.o.d.z.s are the eigenvalues of* $A_{\bar{c}\bar{o}}$*, and* $\{i.o.d.z.\} \subset \{i.d.z.\} \cap \{o.d.z.\}$.

PROOF. * The module $\mathcal{T}(M)/\mathcal{T}\left(\left[y,u\right]_{\mathbf{R}}\right)$ is described by the equation $\partial\breve{x}_{\bar{c}\bar{o}} = A_{\bar{c}\bar{o}}\,\breve{x}_{\bar{c}\bar{o}}$, where the components of $\breve{x}_{\bar{c}\bar{o}}$ are the canonical images of the components of $x_{\bar{c}\bar{o}}$ (which belong to $\mathcal{T}(M)$) in $\mathcal{T}(M)/\mathcal{T}\left(\left[y,u\right]_{\mathbf{R}}\right)$* ■

7.2.5. *Hidden modes*

*Let us begin by giving an abstract but "intrinsic" definition of hidden modes ([19], Definition 16):

DEFINITION 174.– *The module of the hidden modes (h.m.) is* $M/\left(\Phi \cap \left[y,u\right]_{\mathbf{R}}\right)$. *The hidden modes are the Smith zeros of this module.**

The hidden modes are directly derived from the $i.d.z.$s, the $o.d.z.$s, and the $i.o.d.z.$s. We have indeed the following result, where $\varepsilon(.)$ denotes the set of elementary divisors (taking into account multiplicities) of the module or of the matrix in parentheses (Lemma 559, section 13.4.2).

THEOREM 175.– *The following equality holds:*

$$\boxed{\varepsilon(h.m.) = \varepsilon(i.d.z.)\,\dot{\cup}\,\varepsilon(o.d.z.) \setminus \varepsilon(i.o.d.z.)}$$

where $\dot{\cup}$ is the "disjoint union" (section 13.4.2, Lemma 559) and $A \setminus B$ is the complement of B in A (when $B \subset A$).

PROOF. * We have according to Theorem 538 (section 13.4.1)

$$\frac{M}{\Phi \cap [y,u]_{\mathbf{R}}} = \frac{\mathcal{T}(M) \oplus \Phi}{\Phi \cap [y,u]_{\mathbf{R}}} \cong \mathcal{T}(M) \oplus \frac{\Phi}{\Phi \cap [y,u]_{\mathbf{R}}},$$

$$\begin{aligned}\frac{M}{[y,u]_{\mathbf{R}}} &= \frac{\mathcal{T}(M) \oplus \Phi}{(\mathcal{T}(M) \cap [y,u]_{\mathbf{R}}) \oplus (\Phi \cap [y,u]_{\mathbf{R}})} \\ &\cong \frac{\mathcal{T}(M)}{\mathcal{T}([y,u]_{\mathbf{R}})} \oplus \frac{\Phi}{\Phi \cap [y,u]_{\mathbf{R}}}.\end{aligned} \tag{7.23}$$

According to Lemma 559(ii) (section 13.4.2), we obtain from the first equality

$$\varepsilon(h.m.) = \varepsilon(i.d.z.)\,\dot{\cup}\,\varepsilon(\Phi/\Phi \cap [y,u]_{\mathbf{R}})$$

and from the second equality,

$$\varepsilon(o.d.z.) = \varepsilon(i.o.d.z.)\,\dot{\cup}\,\varepsilon(\Phi/\Phi \cap [y,u]_{\mathbf{R}})$$

from which we get the desired result.* ■

We immediately deduce from the above the following classic result:

COROLLARY 176.– *The following equality holds:*

$$\{h.m.\} = \{i.d.z.\}\,\dot{\cup}\,\{o.d.z.\} \setminus \{i.o.d.z.\},$$

where $\{.\}$ denotes the set of elements in brackets (multiplicities taken into account).

Let $\mathbf{\Sigma} = \{A, B, C, D\}$ be a state-space system and consider its canonical structure given by Theorem 155 (section 7.1.5). The following result is a direct consequence of Theorem 175:

PROPOSITION 177.– *The hidden modes are the eigenvalues of the matrix $A_{c\bar{o}} \oplus A_{\bar{c}\bar{o}} \oplus A_{\bar{c}o}$ (counting multiplicities).*[5]

5. $A_{c\bar{o}} \oplus A_{\bar{c}\bar{o}} \oplus A_{\bar{c}o}$ is a "diagonal sum" of matrices (section 13.1.4).

7.2.6. *Relationships between poles and zeros*

Denote the system poles by $s.p.$, the transmission poles by $t.p.$, and the transmission zeros by $t.z.$. The following definition is due to Rosenbrock [101]:

DEFINITION 178.– *The* system zeros *($s.z.$) are defined by the equality*

$$\{s.z.\} = \{t.z.\} \dot{\cup} \{h.m.\}.$$

The following theorem describes the various relations among the different kinds of poles and zeros; most of those are owed to Rosenbrock [100], [101]; the first inclusion of equation (7.25) was established in [19].

THEOREM 179.– *(i)*

$$\{s.p.\} = \{t.p.\} \dot{\cup} \{h.m.\}, \tag{7.24}$$

$$\{t.z.\} \dot{\cup} \{i.o.d.z.\} \subset \{i.z.\} \subset \{s.z.\}. \tag{7.25}$$

(ii) If the transfer matrix is left-regular,

$$\{t.z.\} \dot{\cup} \{i.d.z.\} \subset \{i.z.\}.$$

(iii) If the transfer matrix is right-regular,

$$\{t.z.\} \dot{\cup} \{o.d.z.\} \subset \{i.z.\}.$$

(iv) If the transfer matrix is square and regular,

$$\{s.z.\} = \{i.z.\}.$$

PROOF. * We are going to prove (i) (for the other points, see Exercise 214, or [19]). Equality (7.24) is derived from Theorem 155 (since, according to Theorem 129 of section 7.1.2, the problem can always come down to the case where the system is given by a state–state representation). For (7.25), we observe that

$$\frac{M}{[y]_{\mathbf{R}}} \cong \frac{\mathcal{T}(M)}{\mathcal{T}([y]_{\mathbf{R}})} \oplus \frac{\Phi}{\Phi \cap [y]_{\mathbf{R}}}, \qquad \frac{\Phi}{\Phi \cap [y]_{\mathbf{R}}} \supset \frac{\Phi \cap [y,u]_{\mathbf{R}}}{\Phi \cap [y]_{\mathbf{R}}},$$

and according to Lemma 559(i) (section 13.4.2),

$$\mathcal{Z}\left(\frac{\mathcal{T}(M)}{\mathcal{T}([y]_{\mathbf{R}})}\right) \supset \mathcal{Z}\left(\frac{\mathcal{T}(M)}{\mathcal{T}([y,u]_{\mathbf{R}})}\right)$$

from which we deduce the first inclusion. According to equation (7.23), one obtains $M/[y,u]_{\mathbf{R}} \cong T_1$, where

$$T_1 \supset \frac{\mathcal{T}(M)}{\mathcal{T}([y,u]_{\mathbf{R}})} \oplus T_2, \quad T_2 = \mathcal{T}\left(\frac{\Phi}{\Phi \cap [y,u]_{\mathbf{R}}}\right),$$

and as a result,

$$\{o.d.z.\} \supset \mathcal{Z}\left(T_2\right) \dot{\cup} \{i.o.d.z.\}. \tag{7.26}$$

On the other hand, according to Theorem 538(iii) (section 13.4.1) and Lemma 559(i) (section 13.4.2),

$$T_2 \cong \mathcal{T}\left(\frac{\Phi}{\Phi \cap [y]_{\mathbf{R}}}\right) / \mathcal{T}\left(\frac{\Phi \cap [y,u]_{\mathbf{R}}}{\Phi \cap [y]_{\mathbf{R}}}\right), \quad \mathcal{Z}\left(T_2\right) = \mathcal{Z}\left(\frac{\Phi}{\Phi \cap [y]_{\mathbf{R}}}\right) \setminus \{t.z.\}.$$

Last,

$$\mathcal{Z}\left(\frac{\Phi}{\Phi \cap [y]_{\mathbf{R}}}\right) = \{z.i\} \backslash \mathcal{Z}\left(\frac{\mathcal{T}(M)}{\mathcal{T}([y]_{\mathbf{R}})}\right), \quad \mathcal{Z}\left(\mathcal{T}(M) / \mathcal{T}([y]_{\mathbf{R}})\right) \subset \{i.d.z.\}$$

from which we get

$$\mathcal{Z}\left(T_2\right) \dot{\cup} \{t.z.\} \dot{\cup} \{i.d.z.\} \supset \{z.i\}. \tag{7.27}$$

Here, the second inclusion of (7.25) is a consequence of (7.26), (7.27) and of Corollary 176.* ■

7.3. Stability, stabilizability and detectability

The notion of stability, essential to control theory, so far has been defined only for minimal systems (Definition 61, section 3.1.1). In the general case, one uses the following definition:

DEFINITION 180.– *A control system Σ (assumed to be linear time-invariant) is* stable *if all its poles are located in the left half-plane.*

REMARK 181.– *According to section 2.3.8, a control system is stable if and only if, all variables of its free behavior tend to 0 as $t \to +\infty$. (In a "behavioral approach" [96], this property is taken as* definition *of stability of a control system; with this definition, the following is a* theorem: *a system is stable if and only if, all its poles belong to the left half-plane.*[6]*)*

Suppose Σ is a state-space system $\{A, B, C, D\}$. Then the poles of Σ are the eigenvalues of A (Theorem 18, section 2.3.7). We will then use the following definition:

6. This remark appears to be made based on very subtle distinctions. The approach according to which a linear system is characterized by a finitely presented module and that in which such a system is characterized by a "behavior" are not identical. The equivalence between these two approaches was deeply studied by Oberst [94] (for more recent results, see [23] and [22]).

DEFINITION 182.– *A matrix* $A \in \mathbb{R}^{n\times n}$ *is called a* stability matrix *if all its eigenvalues belong to the left half-plane.*

We can complete Definition 180 by the following:

DEFINITION 183.– *A control system* Σ *(assumed to be linear time-invariant) is* marginally stable *if all its poles lie in the closed left half-plane, those belonging to the imaginary axis (if any) having all their structural indices equal to 1.*

REMARK 184.– *According to Theorem 23 (section 3.1.1), a control system is marginally stable if and only if, all variables of its free behavior are bounded. (In a "behavioral approach", this property is used as the* definition *of a marginally stable control system). With this definition, the following is a* theorem*: a control system is marginally stable if and only if, all its poles belong to the closed left half-plane, those that belong to the imaginary axis (if any) having all their structural indices equal to 1.)*

The notions that follow will show all their significance when we get to the subject of control by state feedback and the theory of observers, in Chapters 8 and 9, respectively.

DEFINITION 185.– *A control system is stabilizable (resp., detectable) if none of its* $i.d.z.s$ *(resp., its* $o.d.z.s$*) belongs to the closed right half-plane* $\bar{\mathbb{C}}_+$.

We deduce immediately from Propositions 166 (section 7.2.2) and 170 (section 7.2.4) the following result:

PROPOSITION 186.– "Popov–Belevitch–Hautus *test for stabilizability and detectability". Let* Σ *be a system defined by a Rosenbrock representation* $\{D, N, Q, W\}$*; this system is stabilizable (resp., detectable) if and only if,* $\mathrm{rk}_{\mathbb{C}} \begin{bmatrix} D(s) & N(s) \end{bmatrix} = r$ *(resp.,* $\mathrm{rk}_{\mathbb{C}} \begin{bmatrix} D^T(s) & Q^T(s) \end{bmatrix} = r$*) for any* $s \in \bar{\mathbb{C}}_+$ *(where* $r \triangleq \mathrm{rk}_{\mathbf{R}} D(\partial)$*). In particular, a state-space system* $\{A, B, C, D\}$ *is stabilizable (resp., detectable) if and only if,* $\mathrm{rk}_{\mathbb{C}} \begin{bmatrix} sI_n - A & B \end{bmatrix} = n$ *(resp.,* $\mathrm{rk}_{\mathbb{C}} \begin{bmatrix} sI_n - A^T & C^T \end{bmatrix} = n$*) for all* $s \in \bar{\mathbb{C}}_+$.

The following is an immediate consequence of the above proposition and is related to Corollary 153 (section 7.1.4).

COROLLARY 187.– Stabilizability $\leftrightarrow$ detectability duality. (C, A) *is detectable if and only if,* $\left(A^T, C^T\right)$ *is stabilizable.*

7.4. Realization

7.4.1. *Introduction*

"Realization" means putting a control system into a *state-space form*. This term is a little ambiguous: Theorem 129 (section 7.1.2) shows how to obtain a realization of a general control system. Very often, a state-space system comes naturally from physical equations (see the exercises). In this section, we are going to examine how to obtain a realization of a left form as well as that of a right form. Then, it becomes easy to determine a "minimal realization"[7] from a transfer matrix because such a matrix admits a left-coprime factorization (corresponding to a controllable left form) and a right-coprime factorization (corresponding to an observable right form): see Remark 31 (section 2.4.5).

Before doing so, notice the following: let $\{A, B, C, D\}$ be a state-space representation of a proper control system. For any invertible matrix P, $\{P^{-1} A P, P^{-1} B, C P, D\}$ is also a state-space representation of the same control system (Remark 7.5, section 7.1.2). This basis transformation matrix P can be chosen such that $P^{-1} A P$ is, for example, a rational canonical form of A (section 13.4.3), and therefore there exist some forms of realization that are more remarkable than others. In addition, if they have the property of *uniqueness*, they are called *canonical forms*. The latter have the particularity of having the simplest possible structure. On the other hand, when the system is of a high order, these canonical forms are to be avoided because their state matrix is in general ill-conditioned (see section 13.5.7).

7.4.2. *SISO systems*

All systems considered in this paragraph are SISO.

Case of an observable system

Consider a control system which is observable, and which can thus be described by the left form (2.17) (section 2.3.5) according to Theorem 147 (section 7.1.4). Assuming that this system is proper, we obtain

$$D(\partial)\, y = N(\partial)\, u, \tag{7.28}$$

$$D(\partial) = \partial^n + a_1\, \partial^{n-1} + \ldots + a_n, \tag{7.29}$$

$$N(\partial) = b_0\, \partial^n + b_1\, \partial^{n-1} + \ldots + b_n. \tag{7.30}$$

Performing the Euclidean division of $N(\partial)$ by $D(\partial)$ yields

$$N(\partial) = b_0\, D(\partial) + N'(\partial)$$

7. That is a realization in the form of a minimal state-state system, in the sense of Definition 33 (section 2.4.6).

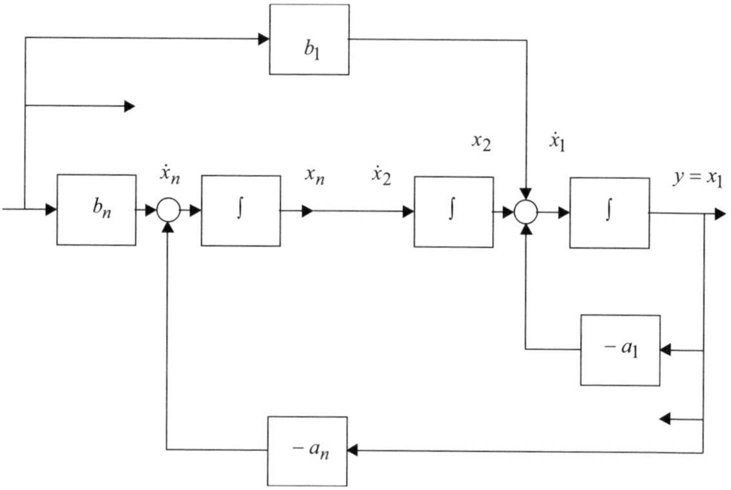

Figure 7.5. *Diagram of an observable canonical form*

where $d^\circ(N') < d^\circ(D)$. Let $\xi = y - b_0\, u$; then

$$\begin{aligned} D(\partial)\,\xi &= N'(\partial)\,u \\ y &= \xi + b_0\, u. \end{aligned} \tag{7.31}$$

It now suffices to determine a realization of a strictly proper left form (7.31). In order to simplify the calculations, we assume in what follows, without loss of generality, that the system considered is strictly proper.

One can proceed as in section 12.5.1: the left form (7.28), (7.29), (7.30) (with $b_0 = 0$) is identical to the linear differential equation with constant coefficients (12.97). Defining the states $x_1, ..., x_n$ by the relations (12.98), (12.99), one obtains the system (12.100), (12.101), i.e.

$$\dot{x} = \begin{bmatrix} -a_1 & 1 & 0 & \cdots & 0 \\ -a_2 & 0 & \vdots & \ddots & \vdots \\ \vdots & 0 & \vdots & \ddots & 0 \\ \vdots & \vdots & \vdots & 0 & 1 \\ -a_n & 0 & 0 & \cdots & 0 \end{bmatrix} x + \begin{bmatrix} b_1 \\ \vdots \\ \vdots \\ \vdots \\ b_n \end{bmatrix} u, \tag{7.32}$$

$$y = \begin{bmatrix} 1 & 0 & \cdots & \cdots & 0 \end{bmatrix} x. \tag{7.33}$$

These state equations can be represented by the diagram in Figure 7.5, the states being the outputs of the integrators.

Note that the state-space form (7.32), (7.33) is only another way of writing the left form (7.28). As this is observable (Theorem 147, section 7.1.4), so is the state-space system (7.32), (7.33) (as can be shown using the Kalman criterion (Theorem 150, 7.1.4). Another remarkable property of the state-space form (7.32), (7.33) is its simplicity: this realization $\{A_o, B_o, C_o\}$ is such that A_o is a companion matrix of the polynomial $D(\partial)$, B_o has coefficients same as those of the polynomial $N(\partial)$, and the matrix C_o expresses the relation $y = x_1$. It is a "canonical form", in the sense specified in section 7.4.1.

DEFINITION 188.– *The realization* (7.32), (7.33) *is called the* observable canonical form *associated with the left form* (7.28).

According to the remarks preceding Definition 188 and Theorem 138 (section 7.1.3), we have the following:

PROPOSITION 189.– *The observable canonical form* (7.32), (7.33) *is always observable; it is controllable if and only if, the polynomials* $\{D, N\}$ *are coprime.*

Case of a controllable system

Consider a control system which is controllable, and thus can be described by the right form (2.19) (section 2.3.5) according to Theorem 137 (section 7.1.3), and suppose this system is strictly proper. We thus have the equations

$$\left\{ \begin{array}{l} y = N(\partial)\,\xi, \\ u = D(\partial)\,\xi, \end{array} \right. \tag{7.34}$$

$$D(\partial) = \partial^n + a_1\,\partial^{n-1} + \ldots + a_n, \tag{7.35}$$

$$N(\partial) = b_1\,\partial^{n-1} + \ldots + b_n. \tag{7.36}$$

We can write the first line of (7.34) explicitly as:

$$\partial^n \xi = -\left(a_1\,\partial^{n-1} + \ldots + a_n\right)\xi + u.$$

Let $x = \left[\begin{array}{ccc} \partial^{n-1}\xi & \ldots & \xi \end{array}\right]^T$; then, it follows that

$$\dot{x} = \left[\begin{array}{ccccc} -a_1 & -a_2 & \cdots & \cdots & -a_n \\ 1 & 0 & \cdots & 0 & 0 \\ 0 & 1 & \ddots & \vdots & \vdots \\ \vdots & \ddots & \ddots & 0 & \vdots \\ 0 & 0 & 0 & 1 & 0 \end{array}\right] x + \left[\begin{array}{c} 1 \\ 0 \\ \vdots \\ \vdots \\ 0 \end{array}\right] u, \tag{7.37}$$

$$y = \left[\begin{array}{ccccc} b_1 & b_2 & \cdots & \cdots & b_n \end{array}\right] x. \tag{7.38}$$

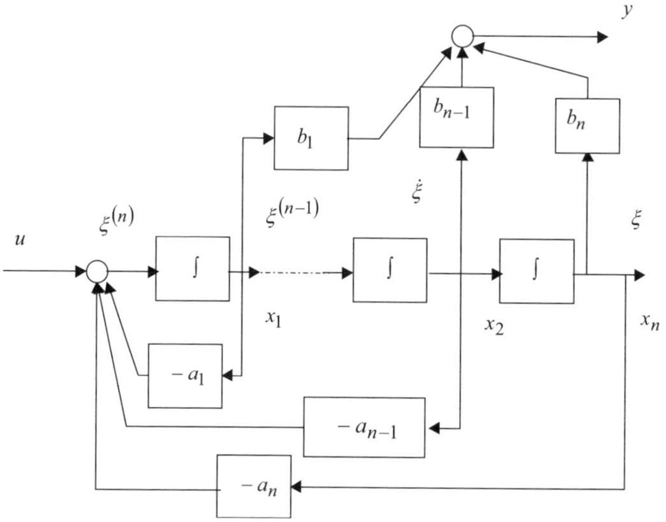

Figure 7.6. *Diagram of a controllable canonical form*

These equations, which are those of a realization $\{A_c, B_c, C_c\}$, can be represented by the diagram in Figure 7.6 below; the realization $\{A_c, B_c, C_c\}$ is very special, like the observable canonical form described above: A_c is a companion matrix of the polynomial $D(\partial)$, the coefficients of C_c are those of the polynomial $N(\partial)$, and B_c has a structure that cannot be further simplified. Since every right form is controllable and equations (7.37) and (7.38) represent only a rewriting (in a different formalism) of a right form, the state-space system $\{A_c, B_c, C_c\}$ is necessarily controllable.

DEFINITION 190.– *The realization* $\{A_c, B_c, C_c\}$ *is called the* controllable canonical form *associated with the right form* (7.34).

The following proposition is obtained by a similar reasoning that leads to Proposition 189.

PROPOSITION 191.– *The controllable canonical form* $\{A_c, B_c, C_c\}$ *is always controllable; it is observable, if, and only if, the polynomials* $\{D, N\}$ *are coprime.*

We can verify the controllability of (A_c, B_c) using the Kalman criterion (Theorem 141, section 7.1.3). The controllability matrix $\Gamma(A_c, B_c)$ (see Remark 142) is formed by iteration in the following manner: its first column is B_c and its $(i+1)$th column $(1 \leq i \leq n-1)$ is equal to its ith column left-multiplied by A_c.

REMARK 192.– *One should always follow the above procedure to calculate the controllability matrix. This avoids needlessly calculating the powers of the state matrix, which is very cumbersome.*

From the above method, we obtain a controllability matrix of the form

$$\Gamma_c = \begin{bmatrix} 1 & * & \dots & * \\ 0 & 1 & \ddots & \vdots \\ \vdots & \ddots & \ddots & * \\ 0 & \dots & 0 & 1 \end{bmatrix}$$

which is obviously of rank n. More precisely, a rather simple calculation shows that $\Gamma_c \mathcal{A}_+ = I_n$, where $\mathcal{A}_+$ is the upper triangular *Toeplitz matrix*

$$\mathcal{A}_+ = \begin{bmatrix} 1 & a_1 & \cdots & a_{n-1} \\ 0 & \ddots & & \\ \vdots & \ddots & \ddots & a_1 \\ 0 & \cdots & 0 & 1 \end{bmatrix}.$$

As a result,

$$\Gamma_c = \mathcal{A}_+^{-1}. \tag{7.39}$$

Other realizations of a controllable system

The realizations of the right form (7.34) are all of the form $\{A, B, C\} = \{P A_c P^{-1}, P B_c, C_c P^{-1}\}$, according to Remark 132 (section 7.1.3), by using a change of basis matrix P^{-1}. According to Remark 142(ii), we have $\Gamma(A, B) = P\Gamma_c$, and therefore, according to equation (7.39),

$$P = \Gamma(A, B) \mathcal{A}_+. \tag{7.40}$$

It follows from the above that there exists a unique realization $\{A_{cc}, B_{cc}, C_{cc}\}$ of equation (7.34) such that $\Gamma(A_{cc}, B_{cc}) = I_n$. Indeed,

$$\{A_{cc}, B_{cc}, C_{cc}\} = \{\mathcal{A}_+ A_c \mathcal{A}_+^{-1},\ \mathcal{A}_+ B_c,\ C_c \mathcal{A}_+^{-1}\}. \tag{7.41}$$

Let

$$A_{cc} = \begin{bmatrix} 0 & 0 & \cdots & 0 & -a_n \\ 1 & 0 & \cdots & 0 & 0 \\ 0 & 1 & \ddots & \vdots & \vdots \\ \vdots & \ddots & \ddots & 0 & -a_2 \\ 0 & 0 & 0 & 1 & -a_1 \end{bmatrix}, \quad B_{cc} = \begin{bmatrix} 1 \\ 0 \\ \vdots \\ \vdots \\ 0 \end{bmatrix}, \tag{7.42}$$

$$C_{cc} = \begin{bmatrix} b_1 & b_2 & \cdots & \cdots & b_n \end{bmatrix} \mathcal{A}_+^{-1}. \tag{7.43}$$

One can easily verify that $\mathcal{A}_+ A_c = A_{cc} \mathcal{A}_+$, and thus relation (7.41) is satisfied.

DEFINITION 193.– *The realization* $\{A_{cc}, B_{cc}, C_{cc}\}$ *is called the* canonical form of controllability.

Gathering the results obtained, we have the following theorem:

THEOREM 194.– *(i) Let* $\{A, B, C\}$ *be any realization (with state* x*) of the right form* (7.34)*. The state* x_c *of the controllable canonical form* $\{A_c, B_c, C_c\}$ *is determined by the relation* $x = P\,x_c$*, where the change of basis matrix* P *satisfies equation* (7.40)*. (ii) The canonical form of controllability* $\{A_{cc}, B_{cc}, C_{cc}\}$ *is the unique realization of the right form* (7.34)*, the controllability matrix of which is* I_n*. It relates to the controllable canonical form through the relation* (7.41) *and is given by equations* (7.42) *and* (7.43)*;* A_{cc} *is a companion matrix of the polynomial* $D(\partial)$ *(section 13.4.3).*

Duality between canonical forms

We have the following "duality relations" between the observable and controllable canonical forms (these relations are linked to Corollary 153, section 7.1.4):

$$A_c = A_o^T, \quad B_c = C_o^T, \quad C_c = B_o^T. \tag{7.44}$$

We would like to emphasize the fact that the observability $\leftrightarrow$ controllability duality, and thus the observable canonical form $\leftrightarrow$ controllable canonical form duality, is strictly formal, in the following sense: let a physical system be described by an observable canonical form (this can be, for example, the RLC circuit in Figure 1.1, section 1.1.1, described by the left form (2.12) of section 2.3.3, because such a description is *equivalent* to an observable canonical form); the "dualized" system according to the formulas (7.44) does not, in general, correspond to *any physical reality*.

7.4.3. **MIMO systems*

We will now deal with the subject of MIMO systems, where the canonical forms are numerous and more complicated than in the SISO case [64].

Realization of a controllable system

In the following, only the case of a controllable system is considered because the realization of an observable system can be derived by duality (Corollary 153, section 7.1.4). We can assume the system is described by the right form (2.19) (section 2.3.5):

$$\begin{cases} u = D(\partial)\,\xi, \\ y = Q(\partial)\,\xi, \end{cases} \tag{7.45}$$

where $D(\partial) \in \mathbf{R}^{r\times r}$ is a matrix of rank r. There exist matrices $U(\partial)$ and $V(\partial)$, invertible over $\mathbf{R}$, such that $\Delta = U^{-1}\,D\,V$ is a diagonal matrix $\operatorname{diag}\{\beta_1, ...\beta_r\}$

according to Theorem 497 (section 13.2.3), where $0 \neq \beta_i \in \mathbf{R}$ is a monic polynomial for every index i. The matrix Δ can be the Smith form of D, but as shown by the proof of the aforementioned theorem, obtaining the diagonal form, without divisibility condition on the elements β_i, requires less calculation.

Let $\epsilon = V^{-1}(\partial)\,\xi$, $v = U^{-1}(\partial)\,u$, and $\tilde{Q} = Q\,V$. Relations (7.45) can be put in the form

$$\begin{cases} v = \Delta(\partial)\,\epsilon, \\ y = \tilde{Q}(\partial)\,\epsilon. \end{cases} \tag{7.46}$$

Here, as seen above, ϵ is a column vector with r entries ϵ_i $(1 \leq i \leq r)$.

Suppose that in the list $(\beta_i)_{1 \leq i \leq r}$, the first j elements, and only those, are equal to 1. For $j+1 \leq i \leq r$, we write the polynomial β_i in the form

$$\beta_i(\partial) = \partial^{n_i} + \beta_{i,1}\,\partial^{n_i-1} + \ldots + \beta_{i,n_i}.$$

On the other hand, putting

$$X_i = \begin{bmatrix} \partial^{n_i-1} \\ \vdots \\ \partial \\ 1 \end{bmatrix} \epsilon_{j+i}, \quad 1 \leq i \leq r-j,$$

we obtain, according to the first row of equation (7.46),

$$\partial X_i = C(\beta_i)\; X_i + \mathbf{1}_{n_i}\, v_{j+i}, \quad 1 \leq i \leq r-j,$$

where $C(\beta_i)$ is the companion matrix of the polynomial β_i having the structure of the state matrix of a canonical controllable form (see equation (7.37), section 7.4.2) and where $\mathbf{1}_{n_i}$ is a column matrix with n_i rows, whose only non-zero entry is the first one which is equal to 1. Setting

$$X = \begin{bmatrix} X_1 \\ \vdots \\ X_{r-j} \end{bmatrix}, \quad \tilde{v} = \begin{bmatrix} v_{j+1} \\ \vdots \\ v_r \end{bmatrix},$$

one obtains the equation

$$\partial\, X = A\, X + \operatorname{diag}\{\mathbf{1}_{n_i}\}_{1 \leq i \leq r-j}\; \tilde{v}$$

with

$$A = \operatorname{diag}\{C(\beta_i)\}_{1 \leq i \leq r-j}.$$

Writing $v = U^{-1}(\partial)\, u$, we obtain an expression of the form (7.1) (section 7.1.2) with $X = \eta$. There only remains to implement the procedure already used in the proof of Theorem 129 (section 7.1.2) in order to get the state equation (7.2) (section 7.1.2). Finally, the output $y = \tilde{Q}(\partial)\, \epsilon$ is written in the form of equation (7.3) (section7.1.2), taking into account the equalities $\epsilon_i = v_i,\, 1 \leq i \leq j$.

EXAMPLE 195.– *Consider the transfer matrix*

$$G(s) = \begin{bmatrix} \frac{s}{(s+1)^2} & \frac{-s^2}{(s+1)^2(s-1)} \\ \frac{1}{(s+1)^2} & \frac{s^2+s+1}{(s+1)^2(s-1)} \end{bmatrix}.$$

Proceeding as in Remark 31(ii) (section 2.4.5), we obtain the right-coprime factorization $(Q(s), D(s))$ *of* $G(s)$ *with*

$$Q(s) = \begin{bmatrix} s & 0 \\ 1 & 1 \end{bmatrix}, \quad D(s) = \begin{bmatrix} (s+1)^2 & s \\ 0 & s-1 \end{bmatrix}.$$

The Smith form of $D(s)$ *is* $\Sigma = U^{-1}\, D\, V = \operatorname{diag}\left\{1, (s+1)^2\,(s-1)\right\}$ *with*

$$U^{-1}(s) = \begin{bmatrix} 1 & -1 \\ 1-s & s \end{bmatrix}, \quad V = \begin{bmatrix} 0 & -1 \\ 1 & (s+1)^2 \end{bmatrix},$$

which we can obtain by using the method of "marking out elementary operations" (see section 13.2.3). The steps of obtaining a minimal realization *of* $G(s)$ *are now as follows: (i) Write the equality*

$$\begin{bmatrix} v_1 \\ v_2 \end{bmatrix} = U^{-1}(\partial) \begin{bmatrix} u_1 \\ u_2 \end{bmatrix} = \begin{bmatrix} u_1 - u_2 \\ u_1 - \partial\,(u_1 - u_2) \end{bmatrix}. \tag{7.47}$$

(ii) According to the first line of equation (7.46), *one gets* $v_1 = \epsilon_1$ *and* $(\partial + 1)^2\,(\partial - 1)\,\epsilon_2 = v_2$. *This last equality leads us to put* $X = \begin{bmatrix} \partial^2 \epsilon_2 & \partial \epsilon_2 & \epsilon_2 \end{bmatrix}^T$, *and we get* $\partial X = A\,X + \mathbf{1}_3 v_2$ *with*

$$A = \begin{bmatrix} -1 & 1 & 1 \\ 1 & 0 & 0 \\ 0 & 1 & 0 \end{bmatrix}. \tag{7.48}$$

According to equation (7.47), $\partial X = A\,X + B_0\, u + B_1\, \partial u$, *where*

$$B_0 = \begin{bmatrix} 1 & 0 \\ 0 & 0 \\ 0 & 0 \end{bmatrix}, \quad B_1 = \begin{bmatrix} -1 & 1 \\ 0 & 0 \\ 0 & 0 \end{bmatrix}.$$

(iii) Following the method used in the proof of Theorem 129, let us put $x = X - B_1\,u$. *We get* $\partial x = A\,x + B\,u$ *with* $B = A\,B_1 + B_0$*, i.e.*

$$B = \begin{bmatrix} 2 & -1 \\ -1 & 1 \\ 0 & 0 \end{bmatrix}. \tag{7.49}$$

(iv) At the end, $y = \tilde{Q}\,(\partial)\,\epsilon$ *with*

$$\tilde{Q}\,(\partial) = Q\,(\partial)\,V = \begin{bmatrix} \partial & 0 \\ 1 & 1 \end{bmatrix} \begin{bmatrix} 0 & -1 \\ 1 & (\partial+1)^2 \end{bmatrix} = \begin{bmatrix} 0 & -\partial \\ 1 & \partial^2 + 2\partial \end{bmatrix}.$$

Therefore, $y_1 = -x_2$ *and* $y_2 = \epsilon_1 + \partial^2\epsilon_2 + 2\partial\epsilon_2 = u_1 - u_2 + (x_1 - u_1 + u_2) + 2x_2$*, from which we have* $y = C\,x$ *with*

$$C = \begin{bmatrix} 0 & -1 & 0 \\ 1 & 2 & 0 \end{bmatrix}. \tag{7.50}$$

Controllability indices

Thanks to the above procedure, we obtain a realization whose state matrix A has quite a particular form (and is actually in a rational canonical form, like in Example 195), but the B matrix does not have a privileged structure (see equation (7.49)). The form obtained is therefore not canonical. We now are going to examine how we can go from any realization to a "canonical form of controllability".

Let $\{A, B, C\}$ be a state-space system of order n and b_i $(1 \leq i \leq m)$ be the ith column of $B \in \mathbb{R}^{n\times m}$. Traverse the controllability matrix $\Gamma\,(A, B)$ from left to right and eliminate, as one goes along, the columns that are linearly dependent on the preceding ones. By re-arranging the remaining columns, we obtain a matrix of the form

$$\begin{bmatrix} b_1 & \ldots & A^{k_1-1}b_1 & b_2 & \ldots & A^{k_2-1}b_2 & \ldots & b_m & \ldots & A^{k_m-1}b_m \end{bmatrix}$$

using the convention that $A^j b_i$ is the empty column if $j < 0$.

DEFINITION 196.– *The natural integers* k_i, $1 \leq i \leq m$, *are the* controllability indices *of the state-space system* $\{A, B, C\}$.

REMARK 197.– *There exists a zero controllability index* k_i *if, and only if,* $\Gamma\,(A, B)$ *does not contain any term that is dependent on* b_i*. As a result, if the matrix* B *is right-regular (i.e.* $\operatorname{rk} B = m$*), all the* k_i*s are positive.*

DEFINITION 198.– *[44] The state-space system* $\{A, B, C\}$*, where* $A \in \mathbb{R}^{n\times n}$, $B \in \mathbb{R}^{m\times n}$*, and* $C \in \mathbb{R}^{p\times n}$*, is said to be* well-formed *if* $\operatorname{rk} B = m$ *and* $\operatorname{rk} C = p$.[8]

8. In the cited reference, only the matrix B is considered for this definition.

In the following, the state-space system is assumed to be well-formed.

REMARK 199.– *The above hypothesis does not generate any loss of generality in reality. Indeed, suppose that* $\operatorname{rk} B < m$ *and* $\operatorname{rk} C = p$*, and to clarify ideas let* $m = 3$ *and* $\operatorname{rk} B = 2$*. One of the columns of* B*,* b_3 *for example, is thus an* $\mathbb{R}$*-linear combination of the other columns, and so there exist real numbers* λ_1 *and* λ_2 *such that* $b_3 = \lambda_1 b_1 + \lambda_2 b_2$*. Consequently,* $B\,u = b_1\,(u_1 + \lambda_1 u_3) + b_2\,(u_2 + \lambda_2 u_3)$*. Substituting* $u_1' = u_1 + \lambda_1 u_3$ *and* $u_2' = u_2 + \lambda_2 u_3$*, one is led to the case of a well-formed state-space system with two inputs* u_1' *and* u_2'*. A similar reasoning can justify that the matrix* C *can always be assumed to be of rank* p*. See, in [44], complementary interpretations of the hypothesis that* $\operatorname{rk} B = m$*.*

REMARK 200.– *Assuming that a state-space system is well-formed, there exists a permutation* σ *of* $\{1, ..., m\}$ *such that* $\mu_1 \triangleq k_{\sigma(1)} \geq \mu_2 \triangleq k_{\sigma(2)} \geq ... \geq \mu_m \triangleq k_{\sigma(m)} \geq 1$*;* σ *corresponds to a right-multiplication of* B *by a permutation matrix* $Q_{cc} \in GL_m(\mathbb{R})$*. The integers* $\{\mu_1, ..., \mu_m\}$ *are again the controllability indices, but arranged in a "canonical order" with which the matrix*

$$P_{cc} = \begin{bmatrix} b_1' & \dots & A^{\mu_1 - 1} b_1' & \dots & b_m' & \dots & A^{\mu_m - 1} b_m' \end{bmatrix} \tag{7.51}$$

is associated, where $b_i' = b_{\sigma(i)}$*. In what follows, we will call the finite sequence* $(\mu_i)_{1 \leq i \leq m}$ the list of controllability indices. *This list is invariant under change of basis in the state space and permutation of system inputs.*

The proof of the theorem below is deduced in an obvious manner from the construction of the list of controllability indices $(\mu_i)_{1 \leq i \leq m}$ in Remark 200 and from Theorem 141 (section 7.1.3):

THEOREM 201.– *(i) The largest controllability index* μ_1 *is the smallest integer* μ *such that* $\rho(\mu) \triangleq \operatorname{rk} \begin{bmatrix} B & AB & \dots & A^{\mu-1} B \end{bmatrix}$ *is stationary for* $\mu \geq \mu_1$*. (ii) We have* $\rho(\mu_1) = \sum_{1 \leq i \leq m} \mu_i$ *and the system is controllable if and only if,* $\rho(\mu_1) = n$*.*

The reader can find a "geometrical interpretation" of the controllability indices in ([119], section 5.7).

Canonical form of controllability

DEFINITION 202.– *The canonical form of controllability (with input* $u' = Q_{cc}^{-1} u$*) of the well-formed controllable state-space system* $\{A, B, C\}$ *is* $\{A_{cc}, B_{cc}, C_{cc}\}$ *where*

$$A_{cc} = P_{cc}^{-1} A\, P_{cc}, \quad B_{cc} = P_{cc}^{-1} B\, Q_{cc}, \quad C_{cc} = C\, P_{cc}. \tag{7.52}$$

This form is the following when $m = 3$ and $\{\mu_1, \mu_2, \mu_3\} = \{4, 3, 2\}$ ([64], section 6.4):

$$A_{cc} = \begin{bmatrix} 0 & & & * & & & * & & * \\ 1 & & & * & & & * & & * \\ & 1 & & * & & & * & & * \\ & & 1 & * & & & * & & * \\ & & & * & 0 & & * & & * \\ & & & * & 1 & & * & & * \\ & & & * & & 1 & * & & * \\ & & & * & & & * & 0 & * \\ & & & * & & & * & 1 & * \end{bmatrix}, \quad B_{cc} = \begin{bmatrix} 1 & & \\ 0 & & \\ 0 & & \\ 0 & & \\ & 1 & \\ & 0 & \\ & 0 & \\ & & 1 \\ & & 0 \end{bmatrix},$$

and the matrix C_{cc} does not have *a priori* any particular form. The unmarked entries contain a zero and the stars designate possibly non-zero elements. It is easy to show that $\Gamma(A_{cc}, B_{cc}) = I_n$; this form is thus a generalization to the MIMO case of the canonical form of controllability (7.42) (section 7.4.2).

EXAMPLE 203.– *Consider again Example 195; we have* $P_{cc} = \begin{bmatrix} b_1 & A b_1 & b_2 \end{bmatrix}$, *thus the controllability indices are* $\{\mu_1, \mu_2\} = \{2, 1\}$. *Explicitly,*

$$P_{cc} = \begin{bmatrix} 2 & -3 & -1 \\ -1 & 2 & 1 \\ 0 & -1 & 0 \end{bmatrix}$$

and by applying (7.52) *we deduce the following:*

$$\begin{aligned} A_{cc} &= \begin{bmatrix} 0 & -1 & 0 \\ 1 & -2 & -1 \\ 0 & 0 & 1 \end{bmatrix}, \quad B_{cc} = \begin{bmatrix} 1 & 0 \\ 0 & 0 \\ 0 & 1 \end{bmatrix}, \\ C_{cc} &= \begin{bmatrix} 1 & -2 & -1 \\ 0 & 1 & 1 \end{bmatrix}. \end{aligned}$$

Controllable canonical form

In what follows, the controllable canonical form becomes more useful than the canonical form of controllability, but it is a little more difficult to obtain in the MIMO case (contrary to the SISO case). The following construction is due to Popov [97] (see also [26], Chapter 7). Consider again a well-formed state-space system $\{A, B, C\}$.

Consider the invertible matrix P_{cc} given by (7.51). Let $M = P_{cc}^{-1}$ and we write (assuming clarity of ideas that $m = 3$)

$$M = \begin{bmatrix} e_{1,1} \\ \vdots \\ e_{1,\mu_1} \\ e_{2,1} \\ \vdots \\ e_{2,\mu_2} \\ e_{3,1} \\ \vdots \\ e_{3,\mu_3} \end{bmatrix}.$$

Let P_c be the change of basis matrix defined by

$$P_c^{-1} = \begin{bmatrix} e_{1,\mu_1}\, A^{\mu_1 - 1} \\ \vdots \\ e_{1,\mu_1} \\ e_{2,\mu_2}\, A^{\mu_2 - 1} \\ \vdots \\ e_{2,\mu_2} \\ e_{3,\mu_3}\, A^{\mu_3 - 1} \\ \vdots \\ e_{3,\mu_3} \end{bmatrix}. \tag{7.53}$$

Let $\{P_c^{-1} A\, P_c, P_c^{-1} B\, Q_{cc}, C\, P_c\} = \{A_c, B_c', C_c\}$ (where Q_{cc} is the permutation matrix defined in Remark 200). The matrices A_c and B_c' have the following structures (for $m = 3$ and $\{\mu_1, \mu_2, \mu_3\} = \{4, 3, 2\}$):

$$A_c = \begin{bmatrix} * & * & * & * & * & * & * & * & * \\ 1 & & & & & & & & \\ & 1 & & & & & & & \\ & & 1 & 0 & & & & & 0 \\ * & * & * & * & * & * & * & * & * \\ & & & & 1 & & & & \\ & & & & & 1 & 0 & & \\ * & * & * & * & * & * & * & * & * \\ & & & & & & & 1 & \end{bmatrix}, \quad B_c' = \begin{bmatrix} 1 & b_1 & b_2 \\ 0 & 0 & 0 \\ 0 & 0 & 0 \\ 0 & 0 & 0 \\ 0 & 1 & b_3 \\ 0 & 0 & 0 \\ 0 & 0 & 0 \\ 0 & 0 & 1 \\ 0 & 0 & 0 \end{bmatrix}, \tag{7.54}$$

where the unmarked entries are zero and the matrix C_c does not have *a priori* any particular form.

Now let

$$\begin{aligned} \breve{B} &= \begin{bmatrix} 1 & b_1 & b_2 \\ 0 & 1 & b_3 \\ 0 & 0 & 1 \end{bmatrix}, \\ v &= \breve{B}\, u' = \breve{B}\, Q_{cc}\, u, \end{aligned} \tag{7.55}$$

where Q_{cc} is the permutation matrix defined in Remark 200 and u' is as in Definition 202. We obtain $\left\{P_c^{-1}\, A\, P_c, P_c^{-1}\, B\, Q_{cc}\, \breve{B}^{-1}, C\, P_c\right\} = \{A_c, B_c, C_c\}$, where A_c is given by equation (7.54) and B_c has the same structure as B_c' but with all the entries b_i $(1 \leq i \leq 3)$ equal to zero and where C_c does not have *a priori* any particular structure.

DEFINITION 204.– *$\{A_c, B_c, C_c\}$ is the controllable canonical form (with input $v = \breve{B}\, Q_{cc}^{-1}\, u$) of the well-formed state-space system $\{A, B, C\}$.*

EXAMPLE 205.– *Let us continue with Example 195. The change of basis matrix given by equation (7.53) is*

$$P_c = \begin{bmatrix} 2 & 1 & 1 \\ -1 & 0 & 0 \\ 0 & -1 & 0 \end{bmatrix} \tag{7.56}$$

and we obtain with $Q_{cc} = I_3$

$$A_c = \begin{bmatrix} -2 & -1 & -1 \\ 1 & 0 & 0 \\ 0 & 0 & 1 \end{bmatrix}, \quad B_c' = \begin{bmatrix} 1 & -1 \\ 0 & 0 \\ 0 & 1 \end{bmatrix}. \tag{7.57}$$

$$C_c = \begin{bmatrix} 1 & 0 & 0 \\ 0 & 1 & 1 \end{bmatrix}. \tag{7.58}$$

As a result,

$$\breve{B} = \begin{bmatrix} 1 & -1 \\ 0 & 1 \end{bmatrix}, \quad B_c = \begin{bmatrix} 1 & 0 \\ 0 & 0 \\ 0 & 1 \end{bmatrix}. \tag{7.59}$$

7.5. Flatness

The notion of a flat system was conceived by Fliess *et al.* [48] in the framework of nonlinear systems; it solves the problem of *motion planning*, which is the determination of an open-loop control that, when applied to a system in the absence of any disturbance and modeling error, has the effect of having this system follow an appropriate trajectory.

7.5.1. *Flatness of nonlinear systems

Let $\mathbf{\Sigma}$ be a nonlinear system, defined by equation (2.4), section 2.2.2, for example, where each component of F is a vector-valued rational function with coefficients in a differential field $\mathbf{K}$.

DEFINITION 206.– *A variable $\sigma = (\sigma_1, ..., \sigma_m)$ is called a* flat output *of $\mathbf{\Sigma}$ if: (i) the components σ_i of σ can be expressed rationally over $\mathbf{K}$ as a function of the variables w_j of $\mathbf{\Sigma}$ $(1 \leq j \leq k)$ and a finite number of their derivatives; (ii) conversely, all the variables w_j of $\mathbf{\Sigma}$ can be expressed rationally over $\mathbf{K}$ as a function of the components σ_i of σ and a finite number of their derivatives; (iii) σ is an independent variable (section 2.3.1). A* flat system *is a system that has a flat output.*

7.5.2. Flatness of linear systems

Let $\mathbf{\Sigma}$ be a *linear* system, associated with a finitely presented $\mathbf{R}$-module M, where $\mathbf{R} = \mathbf{K}[\partial]$ (with $\mathbf{K} = \mathbb{R}$ if, in addition, $\mathbf{\Sigma}$ is *time-invariant*). Then, we can replace the adverb "*rationally*" everywhere in Definition 206 by "*linearly*"; the theorem below follows from Remark 134 (section 7.1.3):

THEOREM 207.– *A flat output of a linear system $\mathbf{\Sigma}$ is a basis of its associated module M. A linear system is flat if and only if, it is controllable.*

Systematic determination of a flat output

Let $\mathbf{\Sigma}$ be the linear controllable system defined by equation (2.6) (section 2.2.5), where $E(\partial) \in \mathbf{R}^{r \times k}$ is left-regular (i.e. of rank r). Since the module M is free, the Smith form of $E(\partial)$ is $\begin{bmatrix} I_r & 0_{m \times r} \end{bmatrix}$ (see section 13.4.2), where $r = k - m$ is the rank of the module M. As a result, there exist matrices $U(\partial)$ and $V(\partial)$, invertible over $\mathbf{R}$, such that

$$U^{-1}(\partial) E(\partial) V(\partial) = \begin{bmatrix} I_r & 0_{m \times r} \end{bmatrix}.$$

Let $V^{-1}(\partial) w = \begin{bmatrix} \varsigma \\ \sigma \end{bmatrix}$, where ς and σ have, respectively, r and m components. Then, equation (2.6) is equivalent to $\varsigma = 0$, and σ is a basis of M; therefore, σ is a flat output of $\mathbf{\Sigma}$.

"Natural" flat outputs

A controllable system often has a flat output whose physical significance is clear. Consider, for example, the linearized inverted pendulum (2.7) (section 2.2.7). The abscissa $y_1 = y + l\theta$ of the mass m is a flat output. Indeed, from the second equality of equation (2.7), we get $\theta = \ddot{y}_1/g$, and therefore $y = y_1 - \ddot{y}_1\, l/g$. From the first equality of (2.7) we obtain

$$f = \ddot{y}_1 (M + m) - y_1^{(4)}\, M\, l/g. \tag{7.60}$$

The variable y_1 is a basis of the module M since (i) it is $\mathbf{R}$-linearly independent (which is trivial in the present case because it has only one component that is not a torsion element), (ii) all the elements of M are $\mathbf{R}$-linear combinations of y_1. From a physical point of view, y_1 is the quantity that provides the most information about the system (a juggler who would like to balance the inverted pendulum through small movements of the carriage would have his/her eyes fixed on mass m throughout his/her number). Note that the *nonlinear* inverted pendulum is not flat.

Application to motion planning

Suppose now we would like to move the inverted pendulum from one equilibrium point to another between the instants t_0 and $t_1 > t_0$, while y changes from value $y(t_0) = 0$ to $y(t_1) > 0$; suppose also that we would like to realize this movement by imposing that the abscissa y_1 be strictly increasing (and thus without any oscillation of the mass m). The equilibrium conditions, expressed as a function of the variable y_1, can be written for $i \in \{0, 1\}$ as:

$$y_1(t_i) = y(t_i), \qquad y_1^{(\beta)}(t_i) = 0 \quad (1 \leq \beta \leq 4).$$

These can be satisfied if we choose a polynomial $P(t)$ of degree 9 and appropriate coefficients for the trajectory $t \mapsto y_1(t)$ (following Hermite's interpolation method: see Exercise 223, section 7.6). The open-loop control to be applied to the system is then obtained by equation (7.60). (In the present case, the system being unstable, it is also necessary to stabilize it by feedback around the "nominal trajectory" as defined here.) The trajectory of the flat output with $t_0 = 0$, $t_1 = 1$, $y(0) = 0$, and $y(1) = 1$ is represented in Figure 7.7.

7.6. Exercises

EXERCISE 208.– *Consider the DC motor defined by equation* (2.14) *(section 2.3.3). (i) Put this system in state-space form using input* $u = V$, *state* $x = \begin{bmatrix} \theta & \omega & i \end{bmatrix}^T$, *and output of* $y = \theta$. *(ii) Show that this system is controllable and observable.*

EXERCISE 209.– *(i) From the equations of the linearized inverted pendulum* ((2.7), *section 2.2.7), determine a state-space realization of that system by choosing state* $x = \begin{bmatrix} \dot{y} & l\theta & l\dot{\theta} & y \end{bmatrix}^T$ *and input* $u = f/M$. *In the remaining part of this exercise, we put* $\sigma = \sqrt{\frac{g}{l}}$ *and* $\varepsilon = \frac{m}{M}$ *and we assume that* $0 < \varepsilon < 1$. *(ii) Determine the poles of this system; is this system stable? (iii) Show that this system is controllable.*

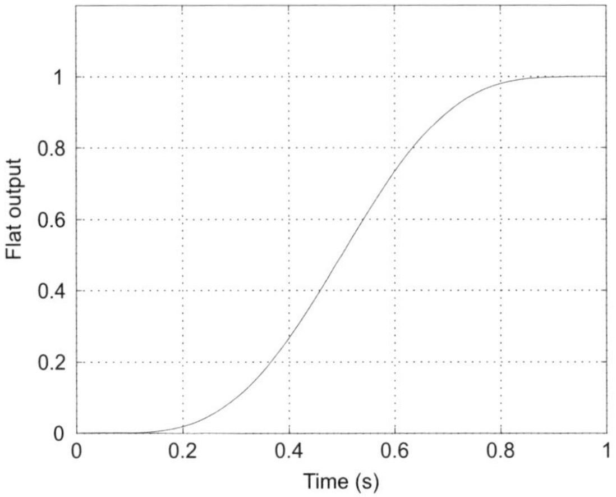

Figure 7.7. *Trajectory of the flat output*

(iv) Supposing that the output is an $\mathbb{R}$-linear combination of x_1, x_2, and x_3, is the system observable? Is it detectable? (v) What if the output is x_4? (vi) If x_4 is taken as output, determine the transmission zeros of the system. What happens when $\varepsilon \ll 1$?

EXERCISE 210.– *Consider the train defined by equations* (1.19) *(section 1.11). (i) Put this system in a state-space form with $x = \begin{bmatrix} \dot{z}_1 & \dot{z}_2 & z_1 & z_2 \end{bmatrix}^T$, $\lambda = \frac{k}{m_1}$, $\rho = \frac{m_1}{m_2}$, and control $u = \frac{f}{m_1}$. (ii) Using this state-space representation, determine the poles of the system. Is there coherence with the expression* (2.13) *(section 2.3.3)? (iii) Is the system controllable? (iv) If one chooses x_2 as output, is the system observable? How would you interpret this? (v) Determine the hidden modes when x_2 is the output. (vi) What happens if we choose x_4 as the output? (vii) With the last choice, does the control system have transmission zeros*[9]*?*

EXERCISE 211.– *Consider the RLC circuit described by the right form* (2.18) *(section 2.3.5). (i) Put this system in controllable canonical form. (ii) Calculate its observability matrix. What happens if the capacitor has a capacitance that tends to $+\infty$? Interpretation?*

EXERCISE 212.– *Consider the linearized inverted double pendulum with equations determined in Exercise 56 (Section 2.7). (i) Show that this system admits a state-space representation with state $x = \begin{bmatrix} \theta_1 & \theta_2 & \dot{\theta}_1 & \dot{\theta}_2 \end{bmatrix}^T$ and input $u = -\frac{f}{M}$. (ii) What are the necessary and sufficient conditions on the lengths l_1 and l_2 for the system to be controllable? (iii) Choosing x_1 as output, will the system be observable?*

9. As mentioned in the beginning of section 7.2, we are only concerned with finite transmission zeros.

EXERCISE 213.– *Let* $\{A, B, C\}$ *be a state-space system with*

$$A = \begin{bmatrix} 0 & 0 \\ 1 & 0 \end{bmatrix}, \quad B = \begin{bmatrix} 1 \\ 0 \end{bmatrix}, \quad C = \begin{bmatrix} 1 & 0 \end{bmatrix}.$$

Noticing that this system already has the canonical structure in Theorem 155 (section 7.1.5), show that $(A_{c\bar{o}}, B_{c\bar{o}})$ *is not controllable.*

EXERCISE 214.– * *We call* Laplace functor *[45] the functor* $\mathbf{K} \otimes_{\mathbf{R}} -$, *where* $\mathbf{K}$ *is the field of fractions of* $\mathbf{R}$ *(see section 13.6.5). Let* y *be a column vector whose entries* y_i *belong to an* $\mathbf{R}$*-module* M*; we see that* $\hat{y}_i$ *is the image of* y_i *under the canonical homomorphism (i.e.* $\hat{y}_i = \frac{1}{1}\, y_i$*) and that* $\hat{y}$ *is the column vector constituted of the* $\hat{y}_i$*s. (a) Consider a linear control system associated with a finitely presented* $\mathbf{R}$*-module* M*, with input* u *and output* y *having* m *and* p *components, respectively, with transfer matrix* G*. Show that* $\hat{y} = G\,\hat{u}$*. (b) If* $\operatorname{rk}_{\mathbf{K}} G = p$*, using results from the last part of section 13.6.5, show that* $[y]_{\mathbf{R}}$ *is a free module and deduce Part (ii) of Theorem 179 (by adaptation of the proof of Part (i)). (c) If* $\operatorname{rk}_{\mathbf{K}}\, G = m$*, show that the module* $M / [y]_{\mathbf{R}}$ *is torsion and, by a similar approach as here above, prove Part (iii) of Theorem 179. (d) Finally, prove Part (iv) of this theorem.*

EXERCISE 215.– * *Let* $\mathbf{R}$ *be an integral domain and let* (M, u, y) *be a control system whose ring of operators is* $\mathbf{R}$*. (i) Using the Laplace functor (see Exercise 214), define the transfer matrix of this system (using* (13.66)*). (ii) Assuming that* $\mathbf{R}$ *is a commutative elementary divisor ring (see section 13.2.3) define the Smith–MacMillan form of a transfer matrix, thus generalizing Definition 29.*

EXERCISE 216.– *(i) Consider the state space system* $\dot{x}_1 = u_1$, $\dot{x}_2 = x_2 + u_2$, $y = x_1$*. Calculate its* o.d.z.s *and its* i.z.s*. (ii) Consider the state-space system* $\dot{x}_1 = u$, $\dot{x}_2 = x_2$, $y_1 = x_1$, $y_2 = x_2$ *. Calculate its* i.d.z.s *and its* i.z.s*. (iii) Is there coherence with Theorem 179?*

EXERCISE 217.– *Let there be the transfer matrix*

$$G(s) = \begin{bmatrix} \frac{1}{s+1} & \frac{1}{s+2} \\ \frac{1}{s+3} & \frac{1}{s+4} \end{bmatrix}.$$

(i) Show that $(Q(s), D(s))$ *is a right-coprime factorization of* $G(s)$ *with*

$$Q(s) = \begin{bmatrix} s+3 & s+4 \\ s+1 & s+2 \end{bmatrix}, \quad D(s) = \begin{bmatrix} s^2+4s+3 & 0 \\ 0 & s^2+6s+8 \end{bmatrix}.$$

(ii) Deduce a minimal realization of $G(s)$ *in controllable canonical form. Is it unique? If not, why?*

EXERCISE 218.– *(i) Analogous to controllability indices, define the observability indices of an MIMO state system* $\{A, B, C\}$*. (ii) Assuming that this system is well-formed, describe its canonical form of observability* $\{A_{oo}, B_{oo}, C_{oo}\}$ *and the observable canonical form* $\{A_o, B_o, C_o\}$*. (iii) How does one go from* $\{A, B, C\}$ *to* $\{A_{oo}, B_{oo}, C_{oo}\}$ *and then to* $\{A_o, B_o, C_o\}$*?*

EXERCISE 219.– *Let there be the transfer matrix*

$$G(s) = \begin{bmatrix} \frac{s}{(s+1)^2 (s+2)^2} & \frac{-s}{(s+2)^2} \\ \frac{s}{(s+2)^2} & \frac{-s}{(s+2)^2} \end{bmatrix}.$$

(i) Show that $(D(s), N(s))$ *is a left-coprime factorization of* $G(s)$ *with*

$$D(s) = \begin{bmatrix} 0 & (s+2)^2 \\ -(s+1)^2 (s+2) & s+2 \end{bmatrix}, \quad N(s) = \begin{bmatrix} s & -s \\ 0 & s^2 \end{bmatrix}.$$

(ii) Deduce a minimal realization of $G(s)$ *in observable canonical form (see Exercise 218). Is it unique?*

EXERCISE 220.– *Let* $\mathbf{\Sigma} = \{A, B, C\}$ *be the state-space system with*

$$A = \begin{bmatrix} -1 & 0 & 0 \\ 1 & -3 & 0 \\ 1 & -4 & 1 \end{bmatrix}, \quad B = \begin{bmatrix} 0 \\ 1 \\ 0 \end{bmatrix},$$

$$C = \begin{bmatrix} 1 & 0 & 0 \\ -1 & 1 & 0 \end{bmatrix}.$$

(i) Study the controllability and the observability of Σ*. (ii) Determine the poles of this system. (iii) Calculate (if any) its* i.d.z.s*, its* o.d.z.s*, its* i.o.d.z.s*, and its hidden modes. Deduce its transmission poles. (iv) Is* Σ *stabilizable? detectable? (v) Calculate its invariant zeros.*

EXERCISE 221.– *Same questions as in Exercise 220 when*

$$A = \begin{bmatrix} -1 & 0 & 0 \\ 1 & 1 & 0 \\ 0 & 0 & 0 \end{bmatrix}, \quad B = \begin{bmatrix} 0 \\ 0 \\ 1 \end{bmatrix},$$

$$C = \begin{bmatrix} -1 & 0 & 1 \end{bmatrix}.$$

(before answering Question (iii), calculate the transfer function of $\{A, B, C\}$*).*

EXERCISE 222.– *(i) Show that a state-space system* $\mathbf{\Sigma} = \{A, B, C, D\}$ *is controllable if and only if, there is no left eigenvector of* A *that will annihilate* B*, i.e. a row vector* $v^T \neq 0$ *such that* $v^T A = \lambda A$ *and* $v^T B = 0$*. (ii) State a criterion of the same type, for observability.*

EXERCISE 223.– *Let there be two distinct real numbers* y_0 *and* y_1 *and two instants* t_0 *and* $t_1 = t_0 + \Delta$, $\Delta > 0$. *Show that there exists a unique polynomial* $P(t)$ *of degree* $2n+1$ *that satisfies the following conditions:* $P(t_i) = y_i$, $P^{(\beta)}(t_i) = 0$, $1 \leq \beta \leq n$, $i \in \{0,1\}$. *Find the system of linear equations that will determine the coefficients of this polynomial. ("Hermite's interpolation".)*

EXERCISE 224.– *(i) Show that* z_2 *is "natural flat output" of the train defined by equations* (1.19) *(section 1.11). (ii) We wish to transfer this train from one equilibrium position to another between instants* t_0 *and* $t_1 > t_0$, *while* z_2 *moves from value* $z_2(t_0) = z_{20}$ *to* $z_2(t_1) = z_{21} \neq z_{20}$. *Write down the equilibrium conditions as a function of the flat output* z_2.

Chapter 8

State Feedback

The state-space formalism is very useful in providing both a simple and complete system representation. This type of representation is indeed simpler than a "Rosenbrock representation": see section 2.3.6. On the other hand, within this formalism, a complete description of the system is possible (if the latter's "structure at infinity" [22] is left aside). In addition, hidden modes are not overlooked unlike the representation by a transfer matrix (Proposition 157, section 7.1.5). But the state-space formalism is especially interesting in that it is particularly well adapted to solving control problems. This is the subject of this and the following chapters.

First, we will study the control by an "elementary" state feedback. This essential part sheds new light on the notions of controllability and stabilizability. But this is only a preliminary step. We will then explain how state feedback can be used *in practice* to solve control problems.

8.1. Elementary state feedback

8.1.1. *General principle*

Let there be a state-space system

$$\dot{x} = Ax + Bu, \quad x(t) \in \mathbb{R}^n, \quad u(t) \in \mathbb{R}^m, \tag{8.1}$$

$$y = Cx, \quad y(t) \in \mathbb{R}^p. \tag{8.2}$$

In order to control this system by state feedback, we will assume in the rest of this chapter that *all the state components are measured.* This hypothesis is rather strict (and we will see in the next chapter how to do without it), but there are in practice quite a number of cases where this is satisfied.

Let us consider the simplest example that we have encountered so far: that of the carriage (Example 133, section 7.1.2). The state of this system consists of its position and velocity; in order to control this system using state feedback, we will assume that it is equipped with a position sensor (which is necessary anyway if we want to control this position) and a speed sensor. We may, however, dispense with a speed sensor by estimating the speed v using the position y. We have $v = \dot{y}$, and hence, in passing into the Laplace domain, $\hat{v}(s) = s\,\hat{y}(s)$ if $y(0^-) = 0$. The derivative operator has transfer function s, which is improper, but one can replace it with a filtered derivative operator with transfer function $\frac{s}{1+\tau s}$, $\tau > 0$. The "estimated speed" $v_e = \mathcal{L}^{-1}\left(\frac{s}{1+\tau\, s}\,\hat{y}(s)\right)$ will be as close to the real speed v as the time constant τ is small.

A second example is the DC motor studied in section 1.3, which can be put in state-space form with $x = \begin{bmatrix} \theta & \omega & i \end{bmatrix}^T$ (see Exercise 208, section 7.6). If we put in a sensor for an angular position θ and a sensor for coil current i, one can consider that the state is measured because the angular speed ω is derived from θ, within a good approximation, thanks to the procedure described above.

A state feedback for system (8.1) is of the form

$$u = v - K\,x \tag{8.3}$$

where $K \in \mathbb{R}^{m \times n}$ is the *gain matrix* (or more succinctly, the *gain*) of the state feedback and v is an external signal, calculated from the reference signal. The diagram of the feedback system is shown in Figure 8.1.

Note that the state feedback (in the elementary version presented in this section) does not depend – at least not directly – on the output y. The equation of the feedback

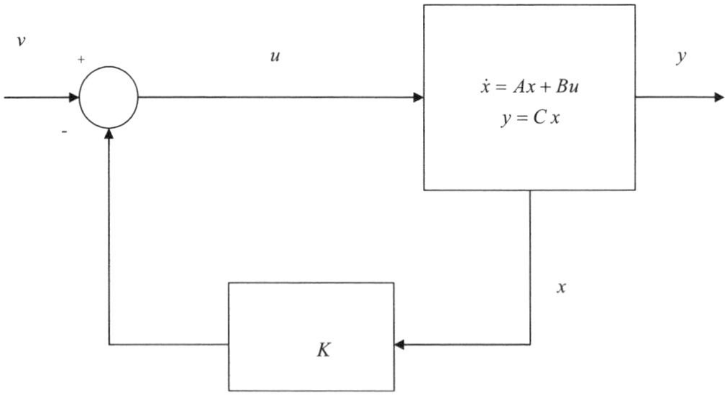

Figure 8.1. *Control by state feedback*

system is obtained by replacing the control u in equation (8.1) by its expression (8.3), and one obtains

$$\dot{x} = (A - B\,K)\;x + B\,v. \tag{8.4}$$

This expression is of the form of equation (8.1) with A replaced by $A - BK$ and u replaced by v.

8.1.2. *Pole placement by state feedback*

LEMMA 225.– *The* $i.d.z.$s *of a system are invariant under state feedback.*

PROOF. Changing the basis in the state-space if necessary, we can assume that system (8.1) can be decomposed according to controllability (see section 7.1.3, equation (7.7) and Figure 7.2). In other terms, we can suppose that matrices A and B are of the form

$$A = \begin{bmatrix} A_c & A_{12} \\ 0 & A_{\bar{c}} \end{bmatrix}, \quad B = \begin{bmatrix} B_c \\ 0 \end{bmatrix}$$

where (A_c, B_c) is controllable and the submatrix $A_{\bar{c}}$ is empty if and only if, (A, B) is controllable. Let $K = \begin{bmatrix} K_c & K_{\bar{c}} \end{bmatrix}$ be the gain matrix of the state feedback (decomposed in coherence with the structure of the matrices A and B above). One obtains

$$A - B\,K = \begin{bmatrix} A_c - B_c\,K_c & A_{12} - B_c\,K_{\bar{c}} \\ 0 & A_{\bar{c}} \end{bmatrix}. \tag{8.5}$$

The $i.d.z.$s are the eigenvalues of $A_{\bar{c}}$ (section 7.2.2, Proposition 167), and hence they are invariant under state feedback. ■

The poles of a linear time-invariant system with *real* coefficients always have the following symmetry property: if $p \in \mathbb{C}$ is a pole of the system, then the conjugate $\bar{p}$ of p is also a pole of the system. This leads us to adopt the following definition, in accordance with ([119], section 0.8):

DEFINITION 226.– *A subset* $\mathcal{P}$ *of* $\mathbb{C}$ *is said to be* symmetric *(about the real axis) if: for every* $p \in P$, *we have* $\bar{p} \in \mathcal{P}$.

LEMMA 227. *Consider the system* Σ *defined by the state equation* (8.1) *with* $m = 1$. *Let* $\mathcal{P} \subset \mathbb{C}$ *be a symmetric set of* n *elements. If* Σ *is controllable, then there exists a* unique *state feedback such that the poles of the feedback system are the elements of* $\mathcal{P}$.

PROOF. Since Σ is controllable, the matrix P defined by (7.40) (section 7.4.2) is invertible, and $(A_c, B_c) = (P^{-1} A P, P^{-1} B)$ is in a controllable canonical form; in other words, the matrices A_c and B_c are, respectively, the state matrix and control matrix of system (7.37) (section 7.4.2). Let x_c be the state of this controllable canonical form. We have $x = P x_c$, and hence the control (8.3) is written as $u = v - K P x_c$. Let $K_c = K P = \begin{bmatrix} k_{c_1} & k_{c_2} & \dots & k_{c_n} \end{bmatrix}$. The feedback system, in the new basis, is expressed as $\dot{x}_c = (A_c - B_c K_c) x_c + B_c u$ with

$$A_c - B_c K_c = \begin{bmatrix} -a_1 - k_{c_1} & -a_2 - k_{c_2} & \cdots & \cdots & -a_n - k_{c_n} \\ 1 & 0 & \cdots & 0 & 0 \\ 0 & 1 & \ddots & \vdots & \vdots \\ \vdots & \ddots & \ddots & 0 & \vdots \\ 0 & 0 & 0 & 1 & 0 \end{bmatrix}.$$

The matrix $A_c - B_c K_c$ is a companion of the polynomial

$$f(s) = s^n + f_1 s^{n-1} + \dots + f_n, \tag{8.6}$$

$$f_i = a_i + k_{c_i},\ 1 \leq i \leq n; \tag{8.7}$$

$f(s)$ is thus the characteristic polynomial of $A_c - B_c K_c$ (section 13.4.3, Proposition 562(i)), and therefore of $A - B K$. Let

$$f(s) = \prod_{p \in \mathcal{P}} (s - p)$$

and we write $f(s)$ in the form of equation (8.6). According to equation (8.6) and the expression (7.40) (section 7.4.2), $\mathcal{P}$ is the set of poles of the feedback system if, and only if,

$$\boxed{K = \begin{bmatrix} f_1 - a_1 & f_2 - a_2 & \dots & f_n - a_n \end{bmatrix} \left[\Gamma(A, B) \mathcal{A}_+\right]^{-1}}. \tag{8.8}$$

■

REMARK 228.– *Expression* (8.8) *of the gain matrix of the state feedback is called the* Bass–Gura formula. *It is, in general, quite cumbersome to use in practice, but it has the interesting point of showing K can be expressed* linearly *as a function of the coefficients $f_i - a_i$, $1 \leq i \leq n$.*

LEMMA 229.– *Consider the system Σ defined by the state equation* (8.1) *with $m > 1$. Let $\mathcal{P} \subset \mathbb{C}$ be a symmetric set of n elements. If Σ is controllable, then there exists a* non-unique *state feedback control such that poles of the feedback system are the elements of $\mathcal{P}$.*

PROOF. Since Σ is controllable, we can put the system in controllable canonical form by change of basis in the state-space and in the control space, i.e. in $\mathbb{R}^n$ and $\mathbb{R}^m$, respectively (see section 7.4.3). To clarify ideas, suppose this canonical form is (7.54); denote by x_c its state and by υ its control (see Definition 204, section 7.4.3). Let the state feedback control be equation (8.3), which can be put in the form $\upsilon = \nu - K_c\, x_c$ by change of variable (7.55) (section 7.4.3). (1) One first method consists of choosing real numbers α_i $(1 \leq i \leq 4)$, β_i $(1 \leq i \leq 3)$, and γ_i $(1 \leq i \leq 2)$ and then determine K_c in such a way that

$$A_c - B_c\, K_c = \begin{bmatrix} -\alpha_1 & -\alpha_2 & -\alpha_3 & -\alpha_4 & 0 & 0 & 0 & 0 & 0 \\ 1 & & & & & & & & \\ & 1 & & & & & & & \\ & & 1 & 0 & & & & & 0 \\ 0 & 0 & 0 & 0 & -\beta_1 & -\beta_2 & -\beta_3 & 0 & 0 \\ & & & & 1 & & & & \\ & & & & & 1 & 0 & & \\ 0 & 0 & 0 & 0 & 0 & 0 & 0 & -\gamma_1 & -\gamma_2 \\ & & & & & & & 1 & \end{bmatrix}$$

(this determination is not difficult due to the form of B_c). Therefore, $\{A_c - B_c\, K_c, B_c\}$ has the same controllability indices as $\{A, B\}$ and

$$\begin{aligned} \det\left(s\, I_9 - A_c + B_c\, K_c\right) &= \alpha(s)\, \beta(s)\, \gamma(s), \\ \alpha(s) &= s^2 + \alpha_1\, s + \alpha_2, \\ \beta(s) &= s^3 + \beta_1\, s^2 + \beta_2\, s + \beta_3, \\ \gamma(s) &= s^4 + \gamma_1\, s^3 + \ldots + \gamma_4. \end{aligned}$$

Polynomials $\alpha(s)$, $\beta(s)$, and $\gamma(s)$ can be arbitrarily chosen, and their roots, which are the poles of the feedback system, can be arbitrarily chosen too. (2) The same poles can be obtained with different methods, giving different gain matrices, and we are going to show one case here below: it consists of choosing the real numbers α_i $(1 \leq i \leq 9)$ and determining K_c in such a way that

$$A_c - B_c\, K_c = \begin{bmatrix} -\alpha_1 & -\alpha_2 & -\alpha_3 & -\alpha_4 & -\alpha_5 & -\alpha_6 & -\alpha_7 & -\alpha_8 & -\alpha_9 \\ 1 & & & & & & & & \\ & 1 & & & & & & & \\ & & 1 & 0 & & & & & 0 \\ 0 & 0 & 0 & 1 & 0 & 0 & 0 & 0 & 0 \\ & & & & 1 & & & & \\ & & & & & 1 & 0 & & \\ 0 & 0 & 0 & 0 & 0 & 0 & 1 & 0 & 0 \\ & & & & & & & 1 & \end{bmatrix};$$

the determination of K_c is not difficult, for the same reason as given before. This time, the matrix $A_c - B_c K_c$ is cyclic since it is a companion of the polynomial $\alpha(s) = s^9 + \sum_{1 \leq i \leq 9} \alpha_i s^{9-i}$. The poles of the feedback system are the roots of the polynomial $\alpha(s)$ and thus can arbitrarily be chosen. ■

The following theorem is one of the most important in the theory of state-space systems:

THEOREM 230.– *Consider the system* Σ *defined by equation* (8.1). *(i) The following two conditions (a) and (b) are equivalent: (a) for any symmetric set* $\mathcal{P} \subset \mathbb{C}$ *of* n *elements, there exists a state feedback* (8.3) *for which the poles of the closed-loop system are the elements of* $\mathcal{P}$*; (b) the system* Σ *is controllable. (ii) There exists a state feedback for which the closed-loop system is stable if, and only if,* Σ *is stabilizable.*

PROOF. 1) (a) $\Rightarrow$ (b): Let us prove this implication by contradiction. If system (8.1) is not controllable, the set of its $i.d.z.$s is non-empty (Theorem 165, section 7.2.2). These $i.d.z.$s are poles of the system and are invariant by feedback according to Lemma 8.1, and thus Condition (a) does not hold. 2) (b) $\Rightarrow$ (a) according to Lemmas 227 and 229. 3): Assume, without loss of generality, that the system Σ is decomposed according to controllability. Since the feedback system is given by equation (8.5), the set of its poles is $\mathrm{Sp}(A_c - B_c K_c) \dot{\cup} \mathrm{Sp}(A_{\bar{c}})$ (where Sp and $\dot{\cup}$ denote the spectrum and the disjoint union, respectively: see sections 13.3.3 and 7.2.5). The system Σ is stabilizable if and only if all its $i.d.z.$s (which are the eigenvalues of $A_{\bar{c}}$ as shown in the proof of Lemma 225) belong to the left half-plane (Definition 185, section 7.3). On the other hand, (A_c, B_c) is controllable according to Proposition 144 (section 7.1.3), and hence the eigenvalues of $A_c - B_c K_c$ can be arbitrarily assigned (provided that they form a symmetric subset of $\mathbb{C}$ whose cardinal is n) by choice of K_c according to (i). ■

REMARK 231.– Brunovski canonical form. *We use the first method in the proof of Lemma 229 and we choose the coefficients* α_i, β_i, *and* γ_i *to be zero. The feedback system thus is reduced to three chains of integrators, which are:*

$$\begin{aligned} \dot{x}_1 &= v_1, \quad \dot{x}_{i+1} = x_i, \ 1 \leq i \leq 3 \\ \dot{x}_5 &= v_2, \quad \dot{x}_{i+1} = x_i, \ 5 \leq i \leq 6 \\ \dot{x}_8 &= v_3, \quad \dot{x}_9 = x_8. \end{aligned}$$

This state representation is called the Brunovski canonical form *of the well-formed system considered. We encounter this, in particular, in the context of nonlinear system, when we use the technique of "linearization by feedback and diffeomorphism" [62]; a robust version of this technique was developed in [49]. For further details on the Brunovski canonical form, see [44].*

8.1.3. *Choice of the pole placement in the SISO case*

Let $\mathbf{\Sigma}$ be a controllable SISO state-space system of order n. According to Lemma 227, if one chooses a symmetric set $\mathcal{P} \subset \mathbb{C}$ of n elements, there exists a unique state feedback control for which the poles of the closed-loop system are the elements of $\mathcal{P}$. The choice of the pole placement is thus reduced to the choice of the *poles* of the closed-loop system. By change of basis in the state-space – if necessary – the problem comes down to the case where $\mathbf{\Sigma}$ is in controllable canonical form. The state of this canonical form is $x_c = \begin{bmatrix} \partial^{n-1}\xi & \dots & \xi \end{bmatrix}^T$ where ξ is the partial state (see section 7.4.2), and hence the state feedback control

$$u = v - K_c\, x_c = v - \begin{bmatrix} k_{c_1} & \dots & k_{c_n} \end{bmatrix} x_c$$

can be written in the form

$$\begin{aligned} u &= v - \mathbf{K}(\partial)\,\xi, \\ \mathbf{K}(\partial) &= \sum_{1 \le i \le n} k_{c_i}\,\partial^{n-i}. \end{aligned} \tag{8.9}$$

Expression (8.9) is that of a "partial state feedback control" (see section 6.3.5). To obtain a modulus margin Mm_i equal to 1, it is thus sufficient to choose the poles π_k of the feedback system as a function of the n poles p_k $(1 \le k \le n)$ of $\mathbf{\Sigma}$ according to Rule 111 (section 6.3.5).

EXAMPLE 232.– *The example presented here, even though it is academic, allows us to highlight several important points. Let* $\mathbf{\Sigma} = \{A, B\}$ *be a state-space system with*

$$A = \begin{bmatrix} -1 & 2 & 0 \\ 1 & 0 & 0 \\ 1 & 1 & 0 \end{bmatrix}, \quad B = \begin{bmatrix} 1 \\ 0 \\ 0 \end{bmatrix}.$$

The controllability matrix is

$$\Gamma(A, B) = \begin{bmatrix} 1 & -1 & 3 \\ 0 & 1 & -1 \\ 0 & 1 & 0 \end{bmatrix};$$

its determinant is 1, *and hence* $\mathbf{\Sigma}$ *is controllable. The poles of* $\mathbf{\Sigma}$ *are* $\{-2, 0, 1\}$. *Rule 111 (section 6.3.5) is thus respected if we choose the poles of the closed-loop system to be* $\{-2, -1, -1\}$. *The characteristic polynomial of the feedback system is then* $P_c(s) = (s+2)(s+1)^2 = s^3 + 4s^2 + 5s + 2$. *Now let the state feedback be equation* (8.3) *(section 8.1.1) with* $K = \begin{bmatrix} k_1 & k_2 & k_3 \end{bmatrix}$. *According to* (8.4), *the state matrix of the feedback system is*

$$A - BK = \begin{bmatrix} -1-k_1 & 2-k_2 & -k_3 \\ 1 & 0 & 0 \\ 1 & 1 & 0 \end{bmatrix}$$

and its characteristic polynomial is

$$f(s) = s^3 + s^2(k_1+1) + s(k_2+k_3-2) + k_3.$$

By identifying $P_c(s)$ and $f(s)$ term by term, we obtain $K = \begin{bmatrix} 3 & 5 & 2 \end{bmatrix}$. The method used here to calculate K requires significantly fewer computations (when they are "handcalculated") than the "Bass-Gura formula" (see Remark 228). The sensitivity function S_i is given by the relation $S_i = (I+L_i)^{-1}$ with $L_i = K(sI_n - A)^{-1}B$. Since the feedback system is stable, we have $S_i \in \Re\mathcal{H}_\infty$. The Bode plot of S_i is shown in Figure 8.2.[1] *Since $|S_i(i\omega)| \le 1$ and $\lim_{\omega\to+\infty} |S_i(i\omega)| = 1$, we have $\|S_i\|_\infty = 1$, and hence $Mm_i = 1$ according to the expression (4.23) of section 4.2.9. According to Theorem 93 (section 4.2.9), this input modulus margin guarantees an input gain margin of $(-6\ dB, +\infty)$ and an input phase margin of $(-60°, 60°)$ (Definition 92). We can verify this point on the Bode plot of L_i, shown in Figure 8.3. The* actual *input gain margin and* actual *input phase margin are $(-9.5\ dB, +\infty)$ and $(-64°, 64°)$, respectively. They are thus, in this particular case, quite close to the gain margin and phase margin guaranteed by the modulus margin (which is related to the fact that $\mathbf{\Sigma}$ is unstable and thus the Nyquist plot of $L_i(s)$ encircles the critical point -1 in the anti-clockwise sense: see Figure 8.4).*

8.1.4. **Choice of the pole placement in the MIMO case*

Let $\mathbf{\Sigma}$ be a state-space system of order n, well-formed, and with m control variables.

The difficulties

As shown by Lemma 229, the choice of the poles of the feedback system *does not uniquely determine* the state feedback if $m > 1$.

EXAMPLE 233.– *Consider the state system $\mathbf{\Sigma} = \{A, B, C\}$ defined in Example 195 (section 7.4.3) and, more precisely, by the equalities (7.48), (7.49), and (7.50). According to Example 205 (section 7.4.3), by putting $x = P_c\, x$ and $v = \breve{B}\, u$, where P_c and $\breve{B}$ are given by (7.56) and (7.59), respectively, we obtain the controllable canonical form $\{A_c, B_c, C_c\}$, with state x_c and control v, given by (7.57), (7.58), and (7.59). The poles of $\mathbf{\Sigma}$ are $\{-1,-1,1\}$. Therefore, Rule 111 of section 6.3.5 is being complied with if we choose the poles of the feedback system to be $\{-3,-3,-5\}$. (1) We will follow this pole placement according to Method (1) of the proof of Lemma 229. We obtain*

$$A_c - B_c K_c = \begin{bmatrix} -6 & -9 & 0 \\ 1 & 0 & 0 \\ 0 & 0 & -5 \end{bmatrix}, \quad K_c = \begin{bmatrix} 4 & 8 & -1 \\ 0 & 0 & 6 \end{bmatrix}.$$

1. The curves related to section 8.1 are placed at the end of this section.

The gain matrix of the state feedback (8.3) *is*

$$K = \breve{B}^{-1} K_c P_c^{-1} = \begin{bmatrix} 5 & 6 & -3 \\ 6 & 12 & 6 \end{bmatrix}.$$

The singular values of the sensitivity function S_i *are plotted in Figure 8.5 as a function of the frequency* ω. *We note that* $\|S_i\|_\infty \simeq 0.008$ dB. *This quantity is larger than 1 (by just a little), and hence the property that holds in the SISO case (i.e.* $\|S_i\|_\infty = 1$*) is lost. (2) We will now use the same pole placement according to Method (2) of the proof of Lemma 229. We get*

$$(s+3)^2 (s+5) = s^3 + 11s^2 + 39s + 45.$$

We obtain

$$A_c - B_c K_c = \begin{bmatrix} -11 & -39 & -45 \\ 1 & 0 & 0 \\ 0 & 1 & 0 \end{bmatrix}, \quad K_c = \begin{bmatrix} 9 & 38 & 44 \\ 0 & -1 & 1 \end{bmatrix}.$$

The gain matrix of the state feedback (8.3) *is*

$$K = \breve{B}^{-1} K_c P_c^{-1} = \begin{bmatrix} 45 & 81 & 8 \\ 1 & 2 & 2 \end{bmatrix}.$$

The singular values of the sensitivity function S_i *are plotted in Figure 8.6 as a function of* ω. *This time,* $\|S_i\|_\infty \simeq 11.9$ *dB, or* $\|S_i\|_\infty \simeq 4$. *The input modulus margin* Mm_i *is only* 0.25, *and Theorem 93 (section 4.2.9) gives a guaranteed gain margin of* $(-1.9 \text{ dB}, 2.5 \text{ dB})$ *and a guaranteed phase margin of* $(-14.4°, 14.4°)$. *These are very poor values.*

REMARK 234.– *(i) The above example shows that for identical poles of the closed-loop system, the state feedback control can have robustness properties very dependent on the* method used *for designing the pole placement. (ii) We can conjecture that Method (1) provides a better modulus margin than Method (2). The reader, however, should not believe that Method (1) will always provide a modulus margin close to* 1 *as long as Rule 111 of section 6.3.5 is being abided by, with respect to the choice of the closed-loop poles. For example, for the choice* $\{-1, -1, -1\}$, *which complies with this rule, we obtain*

$$K = \begin{bmatrix} 1 & 2 & 1 \\ 2 & 4 & 2 \end{bmatrix}, \quad Mm_i \simeq 0.62.$$

(α) *method*

We will provide in what follows a solution to the difficulty shown above. We will call this solution the "(α) method".

Let $P \in \mathbb{R}^{n \times n}$ be a symmetric real matrix and $A_\alpha = A + \alpha I_n$, $\alpha \in \mathbb{R}$. Consider the state feedback (8.3) where

$$K = B^T P. \tag{8.10}$$

Let $\{\lambda_1, ..., \lambda_n\}$ be the eigenvalues of $\tilde{A}_\alpha \triangleq A_\alpha - BK$ (repeating each eigenvalue as many times as its multiplicity). Assume that $\tilde{A}_\alpha$ is diagonalizable, and let $\{\xi_1, ..., \xi_n\}$ be a basis of associated eigenvectors. Let $\eta_i = P\xi_i$ $(1 \le i \le n)$, which is equivalent to $\eta = P\xi$, and, since ξ is invertible, to

$$P = \eta\,\xi^{-1} \tag{8.11}$$

where

$$\eta = \begin{bmatrix} \eta_1 & \cdots & \eta_n \end{bmatrix}, \quad \xi = \begin{bmatrix} \xi_1 & \cdots & \xi_n \end{bmatrix}; \tag{8.12}$$

and then let

$$\nu_i \triangleq \begin{bmatrix} \eta_i \\ \xi_i \end{bmatrix}, \quad 1 \le i \le n. \tag{8.13}$$

REMARK 235.– *The hypothesis that $\tilde{A}_\alpha$ is diagonalizable is only made to simplify the discussion. The reader can verify that the rationale which follows is along the main lines and is still valid if this hypothesis is removed and we denote a basis of* generalized *eigenvalues of $\tilde{A}_\alpha$ by $\{\xi_1, ..., \xi_n\}$ (see Remark 531, section 13.3.4). *From a numerical point of view, it is preferable to use the* Schur form *than the Jordan form ([2], Annex E).**

LEMMA 236.– *(i) Matrix P is a solution of the equation*

$$P A_\alpha + A_\alpha^T P - P B B^T P = 0, \tag{8.14}$$

(called an algebraic Riccati equation*) if and only if, for any $i \in \{1, ..., n\}$, the following equality holds:*

$$\begin{bmatrix} A_\alpha & -B B^T \\ 0 & -A_\alpha^T \end{bmatrix} \nu_i = \lambda_i \nu_i. \tag{8.15}$$

The matrix appearing on the left of this equality (denoted by H in what follows) is called the Hamiltonian matrix, *and equation (8.15) expresses the fact that ν_i is an eigenvector of H associated with the eigenvalue λ_i. (ii) Let σ be a permutation of $\{1, ..., n\}$, $\eta_\sigma = \begin{bmatrix} \eta_{\sigma(1)} & \cdots & \eta_{\sigma(n)} \end{bmatrix}$, $\xi_\sigma = \begin{bmatrix} \xi_{\sigma(1)} & \cdots & \xi_{\sigma(n)} \end{bmatrix}$, and $P_\sigma = \eta_\sigma \xi_\sigma^{-1}$; and we have $P = P_\sigma$.*

PROOF. (i) For any $i \in \{1, ..., n\}$, we have $\lambda_i \xi_i = (A_\alpha - B B^T P)\,\xi_i = A_\alpha \xi_i - B B^T \eta_i$. (1) If P is a solution of the algebraic equation (8.14), then for any

$i \in \{1, ..., n\}$ we have $\lambda_i \, \eta_i = P \left(A_\alpha - B \, B^T \, P \right) \xi_i = -A_\alpha^T \, P \, \xi_i = -A_\alpha^T \, \eta_i$, and hence equation (8.15) is satisfied. (2) Conversely, if for all $i \in \{1, ..., n\}$, the equality (8.15) is satisfied, we have $A_\alpha \, \xi_i - B \, B^T \, P \, \xi_i = \lambda_i \, \xi_i$ and $-A_\alpha^T \, \eta_i = \lambda_i \, \eta_i$ with $\eta_i = P \, \xi_i$, from which we get $\left(P \, A_\alpha + A_\alpha^T \, P - P \, B \, B^T \, P \right) \xi_i = 0$. As a result, P is a solution of (8.14). (ii) The permutation σ corresponds to a right-multiplication of η and ξ by a permutation matrix Π, i.e. $\eta_\sigma = \eta \, \Pi$ and $\xi_\sigma = \xi \, \Pi$. Therefore, $P_\sigma = (\eta \, \Pi) \, (\xi \, \Pi)^{-1} = P$. ■

The following result is classic and is proved, for example, in ([64], A.41) and ([119], section 12.4).

LEMMA 237.– *Let $Z \in \mathbb{R}^{n \times n}$ be a symmetric non-negative definite matrix (i.e. $Z \geq 0$)*[2] *and let $F \in \mathbb{R}^{n \times n}$ be a stability matrix (Definition 182, section 7.3). Then, the* Lyapunov equation

$$F^T \, P + P \, F = -Z$$

has a unique solution $P \in \mathbb{R}^{n \times n}$, and $P \geq 0$. If, in addition, $\left(\sqrt{Z}, F \right)$ is observable[3]*, then $P > 0$.*

THEOREM 238.– *(i) If λ is an eigenvalue of H, so is $-\lambda$. (ii) Let $\{\nu_i, 1 \leq i \leq n\}$ be a set of linearly independent eigenvectors of H, let η_i and ξ_i $(1 \leq i \leq n)$ be the vectors defined by equation* (8.13)*, and let η and ξ be the matrices defined by equation* (8.12)*; if ξ is invertible, P defined by* (8.11) *is a solution of the algebraic Riccati equation* (8.14)*. (iii) Of the following conditions, ((a)&(b))⇔(c): (a) (A_α, B) is stabilizable; (b) H has no imaginary eigenvalues; (c) among the matrices P defined by equation* (8.11)*, there exists a matrix $\check{P}$ for which $A_\alpha - B \, K$ (with K given by equation* (8.10)*) is a stability matrix. (iv) When Condition (c) holds, the matrix $\check{P}$ is unique and is symmetric real non-negative definite; it is the matrix P obtained when the λ_is $(1 \leq i \leq n)$ are the eigenvalues $\check{\lambda}_i$ of H belonging to the left half-plane. (v) Let $\{\pi_1, ..., \pi_n\}$ be the poles of $\mathbf{\Sigma}$. The poles of the closed-loop system with the state feedback control* (8.3) *(where $K = -B^T \, \check{P}$) are $\left\{ \check{\lambda}_1, ..., \check{\lambda}_n \right\}$ such that*

$$\check{\lambda}_i = \begin{cases} \pi_i \text{ if } \operatorname{Re}(\pi_i) < -\alpha, \\ -\pi_i - 2\alpha \text{ if } \operatorname{Re}(\pi_i) > -\alpha. \end{cases} \tag{8.16}$$

PROOF. (i) Is obvious since the eigenvalues of H are those of A_α plus those of $-A_\alpha$. (ii) Is a reformulation of Lemma 236. (iii) and (iv) According to (i), if H has

2. For a square matrix Z with real entries, the symbol $Z \geq 0$ (resp., $Z > 0$) signifies that Z is symmetric non negative definite (resp., symmetric positive definite). see Definition 575 (section 13.5.6). The same holds when Z has complex entries, changing *symmetric* to *Hermitian*.

3. $\sqrt{Z}$ is the Hermitian (real symmetric in the present case) square root of Z (see section 13.5.6). But any square root of Z can be used.

no imaginary eigenvalues, this matrix has n eigenvalues belonging to the left half-plane and n eigenvalues belonging to the right half-plane. If, in addition, (A_α, B) is stabilizable, there exists a unique matrix $P = \check{P}$ of the form (8.11) for which $A_\alpha - B\,K$ is a stability matrix: see ([69], Theorem 1) where the converse is also proved. According to equation (8.14), we have

$$\check{P}\,(A_\alpha - B\,K) + (A_\alpha - B\,K)^T\,\check{P} = -Z$$

with $Z = \check{P}\,B\,B^T\,\check{P} \geq 0$. As a result, according to Lemma 237, $\check{P} \geq 0$. (v) The eigenvalues of H are $\pi_i + \alpha$ and $-\pi_i - \alpha$, and hence the eigenvalues of $A_\alpha - B\,K$ are $\pi_i + \alpha$ if $\mathrm{Re}\,(\pi_i) < -\alpha$ and $-\pi_i - \alpha$ if $\mathrm{Re}\,(\pi_i) > -\alpha$. It remains now to subtract α in order to obtain the eigenvalues of $A - B\,K$. ■

EXAMPLE 239.– *Consider the same system as in Example 233. By taking $\alpha = 2$, the poles of the feedback system placed according to the above method are $\{-3, -3, -5\}$, which are exactly the poles chosen in the example considered. We can also choose $\alpha = 1$; although, in this case, the Hamiltonian matrix H has eigenvalues at the origin, the state feedback with gain matrix* (8.10) *and P given by equation* (8.11) *are well-defined, and the poles are then placed at $\{-1, -1, -3\}$. These are values that satisfy* (8.16).

The interest in all previous development can be found in the theorem below.

THEOREM 240.– *Let there be the state feedback in Theorem 238(iv). If $\alpha \geq 0$, the input modulus margin is $Mm_i = 1$.*

PROOF. Let $L\,(s) = K\,(s\,I_n - A)\,B$, $L_\alpha\,(s) = L\,(s - \alpha) = K\,(s\,I_n - A_\alpha)\,B$, and $\Phi_\alpha\,(s) = (sI_n - A_\alpha)^{-1}$. Use the following notation:

$$(.)^\sim\,(s) = (.)^T\,(-s)$$

where the term in parentheses is any transfer matrix. And finally, let $P = \check{P}$. By multiplying equation (8.14) by -1, then subtracting $s\,P$ to $-A_\alpha\,P$ in the obtained equation, then adding onto the term $-A_\alpha^T P$ the same quantity, and last multiplying the newly obtained equation on the left by $B^T\,\Phi_\alpha^\sim\,(s)$ and by $\Phi_\alpha\,(s)\,B$ on the right, we obtain:

$$(I_m + L_\alpha)^\sim\,(I_m + L_\alpha) = I_m; \tag{8.17}$$

as a result,

$$\sup_{\mathrm{Re}(s) \geq 0} \bar{\sigma}\left(\left(I_m + L_\alpha\,(s)\right)^{-1}\right) \leq 1$$

since the poles of the transfer matrix $(I_m + L_\alpha(s))^{-1}$ are $\{\check{\lambda}_1, ..., \check{\lambda}_n\}$, which belong to the left half-plane. We have, on the other hand, since $S_i = (I_m + L)^{-1}$,

$$\|S_i\|_\infty = \sup_{\mathrm{Re}(s)\geq 0} \bar{\sigma}\left((I_m + L(s))^{-1}\right) = \sup_{\mathrm{Re}(s)\geq -\alpha} \bar{\sigma}\left((I_m + L_\alpha(s))^{-1}\right),$$

and hence if $\alpha \geq 0$,

$$\|S_i\|_\infty \leq \sup_{\mathrm{Re}(s)\geq 0} \bar{\sigma}\left((I_m + L_\alpha(s))^{-1}\right) \leq 1.$$

Since $\lim_{|s|\to+\infty} S_i(s) = I_m$, we have $\|S_i\|_\infty = 1$, and therefore $Mm_i = 1$. ■

DEFINITION 241.– *The* (α) *method consists of determining the gain matrix* K *of the state feedback according to equation* (8.10) *with* $P = \check{P}$, *for a real number* $\alpha \geq 0$ *such that (i)* (A_α, B) *is stabilizable, (ii)* A_α *(or, in an equivalent manner,* H*) has no imaginary eigenvalues if* $\alpha = 0$, *(iii)* ξ *is invertible (this last condition is always satisfied if* H *has no imaginary eigenvalues).*

EXAMPLE 242.– *Continuing from Example 239, we obtain*

$$\check{P} = \begin{bmatrix} 9.9 & 16.1 & 3.9 \\ & 28.5 & 11.2 \\ & & 13.8 \end{bmatrix}, \quad K = \begin{bmatrix} 3.7 & 3.6 & -3.3 \\ 6.2 & 12.5 & 7.2 \end{bmatrix},$$

with $\check{P} = \check{P}^T$. *The singular values of the sensitivity function* S_i *are plotted in Figure 8.7 as a function of the angular frequency* ω. *This figure illustrates well the fact that* $Mm_i = 1$.

REMARK 243.– *(i) The* (α) *method is closely related to the "linear quadratic optimal control" (see Remark 245(ii) below). This is nevertheless, in the formulation proposed here, a method of pole placement. (ii) In the case of an SISO system, Theorem 240 does not contradict equality* (4.17) *of section 4.2.8 because the relative degree of the open-loop transfer function* $L(s)$ *is equal to* 1. *On the other hand, according to equality* (4.18) *of the above-mentioned paragraph, all zeros of* $L(s)$ *belong to the closed left half-plane. This remark remains valid in the case of an MIMO system (see section 4.2.9).*

Extension of the (α) *method: the LQR method*

The "(α) method" is a particular case of the "LQR method",[4] the algebraic characteristics of which we will now describe, without insisting on the "optimal

4. LQR stands for "Linear Quadratic Regulator"; see Remark 245 below. This regulator corresponds to the *LQ control* (already mentioned in the preface).

control" aspect (in accordance with what is said in the preface). Suppose that (A_α, B) is stabilizable and let there be a state feedback with gain matrix K given by

$$K = R^{-1} B^T P \tag{8.18}$$

where $R \in \mathbb{R}^{m \times m}$, $R > 0$ and where P is a real symmetric matrix; the expression (8.18) is a generalization of equation (8.10). Consider also the *algebraic Riccati equation*

$$\boxed{P A_\alpha + A_\alpha^T P - P B R^{-1} B^T P + Q = 0} \tag{8.19}$$

$(Q \in \mathbb{R}^{n \times n}, Q \geq 0)$, which is a generalization of equation (8.14). And finally, consider the *Hamiltonian matrix*

$$\boxed{H = \begin{bmatrix} A_\alpha & -B R^{-1} B^T \\ -Q & -A_\alpha^T \end{bmatrix}}. \tag{8.20}$$

Let E be a square root of Q (section 13.5.6), i.e. a matrix such that $Q = E^T E$.

THEOREM 244.– *(i) Parts (i)-(iv) of the statement of Theorem 238 remain valid (the ν_is, $1 \leq i \leq n$, are the* generalized *eigenvalues of H if this matrix is not diagonalizable: see Remarks 235 (section 8.1.4) and 531 (section 13.3.4)) when the expressions* (8.10) *and* (8.14) *are replaced, respectively, by equations* (8.18) *and* (8.19) *and H is defined by equation* (8.20). *(ii) Suppose from now on that Condition (a) of Theorem 238 holds. In order that Condition (b) of this same theorem be satisfied (for Condition (c) to hold), it suffices that the following condition (d) be satisfied: (d)* (E, A_α) *is detectable. (iii) In this case, there exists a unique solution to the algebraic Riccati equation* (8.19) *for which $A_\alpha - B K$ is a stability matrix and it is the above matrix $\check{P}$; $\check{P}$ is also the unique non-negative definite solution to equation* (8.19). *(iv) If in addition* (E, A_α) *is observable, then* $\check{P} > 0$.

PROOF. (i) is an easy generalization of the corresponding parts of the discussion of Theorem 238 (see Exercise 268). For (ii), see ([119], section 12.3) and ([72], section 3.4.3). (iii) is a consequence of Lemma 237. ■

REMARK 245.– *(i) Condition (d) is not necessary for the existence of $\check{P}$, and that is why the (α) method applies to unstable systems (see Example 242). Condition (d) is only necessary for the uniqueness in Theorem 244(iv) ([69], Theorem 2). (ii) Conditions (a) and (d) are necessary and sufficient for the control $u = -K x$ (with K satisfying equation* (8.18) *and $P = \check{P}$) to minimize the quadratic criterion*

$$J_\alpha = \int_0^{+\infty} e^{2\alpha t} \left(x^T(t)\, Q\, x(t) + u^T(t)\, R\, u(t) \right) dt$$

([72], [2]). The matrices Q and R can be interpreted as weighting matrices, *allowing weighting at various levels of certain $\mathbb{R}$-linear combinations of the state variables and of the control variables (the more such a linear combination is weighted, the smaller its variations will be over time).*

Suppose from now on that $\alpha \geq 0$ and Conditions (a) and (b) of Theorem 238 hold. Let there be the state feedback (8.18) with $P = \check{P}$. Using the same notation as in the proof of Theorem 240, we obtain (by an identical approach) the generalization below of equality (8.17), a generalization which is called the *Kalman equality*:

$$\boxed{[I_m + L_\alpha(s)]^\sim \, R\,[I_m + L_\alpha(s)] = R + B^T\,\Phi_\alpha^\sim(s)\,Q\,\Phi_\alpha(s)\;B}\,. \qquad (8.21)$$

Let $s = i\omega$; we have $B^T\,\Phi_\alpha^\sim(i\omega)\;Q\,\Phi_\alpha(i\omega)\;B \geq 0$, and as a result

$$\bar{\sigma}\left(\sqrt{R}\,(I_m + L_\alpha(i\omega))^{-1}\sqrt{R^{-1}}\right) \leq 1, \quad \forall\omega.$$

We deduce the following generalization of Theorem 240 (which corresponds to the case $Q = 0$ and $R = I_m$):

THEOREM 246.– *With the state feedback defined above and, for any $\alpha \geq 0$, we have $\left\|\sqrt{R}\,S_i\,\sqrt{R^{-1}}\right\|_\infty = 1$. In particular, if $R = I_m$, or more generally, if R is a* scalar matrix,[5] *we have $Mm_i = 1$.*

The question that remains to be examined is the nature of the pole placement carried out using the LQR method. In order to do so, the matrix R is replaced in what follows by ρR, where $\rho > 0$ is a real number. Let

$$G(s) = E\Phi(s)\,B, \quad \Phi(s) = (s\,I_n - A)^{-1};$$

where $G(s)$ is the transfer matrix of the state-space system $\{A, B, E\}$. Also, let there be the following hypothesis (H):

(H): $G(s) \in \mathbb{R}(s)^{m\times m}$, $\mathrm{rk}_{\mathbb{R}(s)}\,G(s) = m$ and the state-space system $\{A, B, E\}$ is minimal.

THEOREM 247.– *(i) Let $\{\pi_1, ..., \pi_n\}$ be the poles of $\mathbf{\Sigma}$ (counting multiplicities). As $\rho \to +\infty$, the poles $\check{\lambda}_k$ $(1 \leq k \leq n)$ of the closed-loop system tend to*

$$\begin{cases} \pi_k \text{ if } \mathrm{Re}(\pi_k) \leq -\alpha, \\ -\pi_k - 2\alpha \text{ if } \mathrm{Re}(\pi_k) > -\alpha. \end{cases}$$

5. A "scalar matrix" $R \in \mathbb{R}^{m\times m}$ is of the form rI_m, where $r \in \mathbb{R}$. Such a matrix is obviously > 0 if and only if, $r > 0$. Note that if R is a scalar matrix, we come back to the case $R = I_m$ by replacing P by P/r and Q by Q/r in equation (8.19).

(ii) Suppose that Hypothesis (H) is in force. Let $\{\nu_1, ..., \nu_q\}$ be the MacMillan zeros of $G(s)$ (counting multiplicities). As $\rho \to 0^+$, q poles $\check{\lambda}_k$ $(1 \le k \le q)$ of the closed-loop system tend to

$$\begin{cases} \nu_k \text{ if } \operatorname{Re}(\nu_k) \le -\alpha, \\ -\nu_k - 2\alpha \text{ if } \operatorname{Re}(\nu_k) > -\alpha. \end{cases}$$

The remaining $n - q$ poles $\check{\lambda}_{k+q}$ $(1 \le k \le n - q)$ are unbounded and when $m = 1$, $\check{\lambda}_{k+q} \sim s_k$, where

$$s_k = e^{i\left(\frac{1}{2} + \frac{-1+2k}{2(n-q)}\right)\pi} \left(\frac{\beta^2}{\rho}\right)^{\frac{1}{2(n-q)}}, \quad \beta \ne 0. \tag{8.22}$$

PROOF. Let $a(s) = |s I_n - A|$ and $f(s) = |s I_n - A + B K|$ (the roots of $a(s)$ and of $f(s)$ are the poles of the open-loop system and of the closed-loop system, respectively). We obtain, by considering the determinant of the two members of equation (8.21) and by putting $\aleph_\alpha(s) = \aleph(s - \alpha)$ for any matrix $\aleph$ of rational functions

$$f_\alpha^\sim f_\alpha = a_\alpha^\sim a_\alpha \det\left(I_m + \frac{1}{\rho}\sqrt{R^{-1}}\, G_\alpha^\sim G_\alpha \sqrt{R^{-1}}\right). \tag{8.23}$$

Let $M = \sqrt{R^{-1}}\, G_\alpha^\sim G_\alpha \sqrt{R^{-1}}$ and $r = \operatorname{rk}_{\mathbb{R}(s)} G(s) \le m$. We have from Lemma 519 (section 13.3.3)

$$\det(\rho I_m + M(s)) = \rho^m + \sum_{k=1}^{r} \Delta_k(s)\, \rho^{m-k}$$

where $\Delta_k(s)$ is the sum of the principal minors of order k of $M(s)$. Therefore,

$$f_\alpha^\sim(s)\, f_\alpha(s) = a_\alpha^\sim(s)\, a_\alpha(s) \left(1 + \sum_{k=1}^{r} \Delta_k(s)\, \rho^{-k}\right). \tag{8.24}$$

(i) $\rho \to +\infty$. Then, we get $f_\alpha^\sim f_\alpha \to a_\alpha^\sim a_\alpha$, $f(s)$ being a polynomial, all roots of which belong to the closed left half-plane. We now only need to apply the same rationale as in the proof of Theorem 238(iii). (ii) $\rho \to 0^+$. (a) Let $\varsigma(s)$ be the unique monic polynomial, the roots of which are the transmission zeros of $\{A, B, E\}$, and let $q = \deg \varsigma(s) < n$. Considering the Smith–MacMillan form of $G(s)$ (section 2.4.5), we get $G(s) = U^{-1}(s)\, \boldsymbol{\epsilon}(s)\, \boldsymbol{\Psi}^{-1}(s)\, V(s)$, where the polynomial matrices $U(s)$ and $V(s)$ are invertible over $\mathbb{R}[s]$. Supposing Hypothesis (H) is in force, we have $r = m$ and

$$\begin{aligned} \boldsymbol{\epsilon}(s) &= \operatorname{diag}(\varepsilon_1(s), ..., \varepsilon_m(s)), \quad \boldsymbol{\Psi}(s) = \operatorname{diag}(\psi_1(s), ..., \psi_m(s)), \\ \varsigma(s) &= \det \boldsymbol{\epsilon}(s) = \prod\nolimits_{i=1}^{m} \varepsilon_i(s), \quad a(s) = \det \boldsymbol{\Psi}(s) = \prod\nolimits_{i=1}^{m} \psi_i(s), \\ \Delta_m(s) &= \det M = \frac{\beta^2}{\det R} \frac{\varsigma_\alpha^\sim \varsigma_\alpha}{a_\alpha^\sim a_\alpha}, \end{aligned}$$

where $\beta = \det U^{-1}(s) \det V(s)$. As a result, from equation (8.24),

$$\begin{aligned} f_\alpha^\sim(s)\, f_\alpha(s) &= \frac{1}{\rho^m}\left[\frac{\beta^2}{\det R}\varsigma_\alpha^\sim(s)\,\varsigma_\alpha(s) + \ldots\right. \\ &\left. \ldots + a_\alpha^\sim(s)\, a_\alpha(s)\left(\rho^m + \sum_{k=1}^{m-1}\Delta_k(s)\,\rho^{m-k}\right)\right]. \end{aligned} \quad (8.25)$$

(b) Let $\mathbf{K}$ be a compact subset of the complex plane, which does not contain any MacMillan pole of $G_\alpha^\sim G_\alpha$, and the interior of which contains all roots of $\varsigma_\alpha^\sim(s)\,\varsigma_\alpha(s)$. The $\Delta_k(s)$ are bounded on $\mathbf{K}$, and hence the roots of $f_\alpha^\sim(s)\, f_\alpha(s)$ remaining in $\mathbf{K}$ converge to the roots of $\varsigma_\alpha^\sim(s)\,\varsigma_\alpha(s)$ as $\rho \to 0^+$. (c) $n-q$ roots of $f(s)$ are unbounded as $\rho \to 0^+$, according to the above. In the case $m=1$, we can assume without loss of generality that $R=1$ and the expression (8.25) becomes

$$f_\alpha^\sim(s)\, f_\alpha(s) = \frac{1}{\rho}\left(\beta^2\varsigma_\alpha^\sim(s)\,\varsigma_\alpha(s) + \rho\, a_\alpha^\sim(s)\, a_\alpha(s)\right). \quad (8.26)$$

Since $a(s)$ is a monic polynomial of degree n, we have

$$a_\alpha^\sim(s)\, a_\alpha(s) = (-1)^n\, s^{2n} + s^{2n}\omega_1(1/s) \quad (8.27)$$

where $\omega_1(1/s)$ tends to 0 and so does $1/s$. The same rationale is valid for $\varsigma(s)$, and hence

$$\varsigma_\alpha^\sim(s)\,\varsigma_\alpha(s) = (-1)^q\, s^{2q} + s^{2q}\omega_2(1/s) \quad (8.28)$$

where $\omega_2(1/s)$ tends to 0 and so does $1/s$. According to equations (8.26), (8.27), and (8.28), the roots we are looking for asymptotically satisfy

$$(-1)^n\, s^{2n} + \frac{\beta^2}{\rho}(-1)^q\, s^{2q} = 0.$$

The roots of this equation which belong to the closed left half-plane are the s_ks defined by equation (8.22). ■

REMARK 248.– *(i) The configuration of the roots s_k, well-known in the theory of filters, is called the* Butterworth configuration *of order $n-q$: the roots s_k $(1 \le k \le n-q)$ are equally distributed in the complex plane on a circle centered at the origin, and have argument $\left(\frac{1}{2} + \frac{-1+2k}{2(n-q)}\right)\pi$. (ii) For $m > 1$ and under Hypothesis (H), the number of poles of the feedback system having absolute value tending to $+\infty$ is still equal to $n-q$, i.e. to the degree of zero at infinity of $G(s)$ since the* defect *of the transfer matrix $G(s)$ is zero (see Remark 40, section 2.4.7). These $n-q$ poles have generally a more complex behavior than in the case $m=1$; their asymptotic directions can be distributed according to a combination of several Butterworth configurations of different orders: see ([72], section 3.8, Example 3.19).*

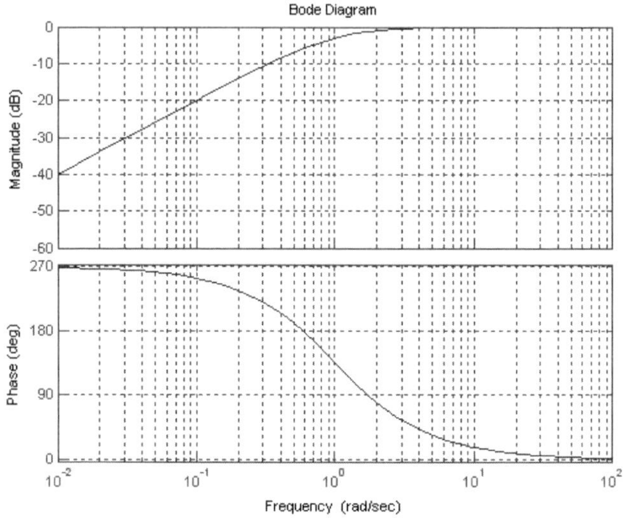

Figure 8.2. *Bode plot of the sensitivity function*

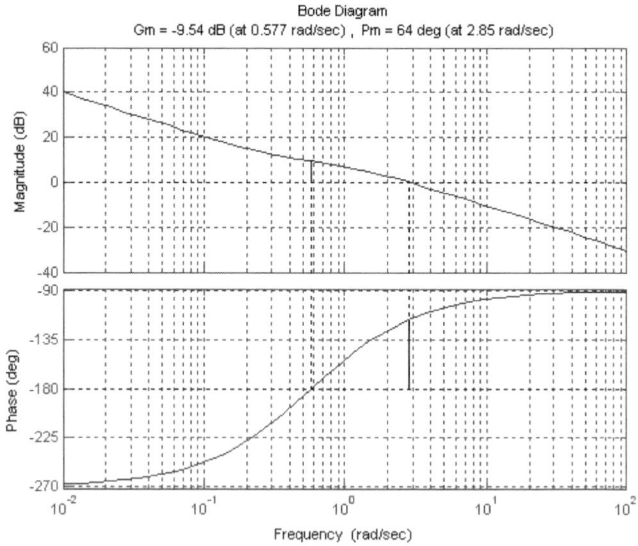

Figure 8.3. *Bode plot of Li(s)*

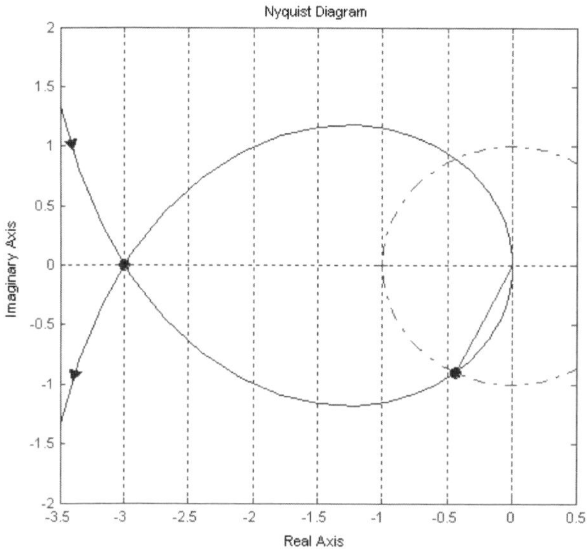

Figure 8.4. *Nyquist plot of Li(s)*

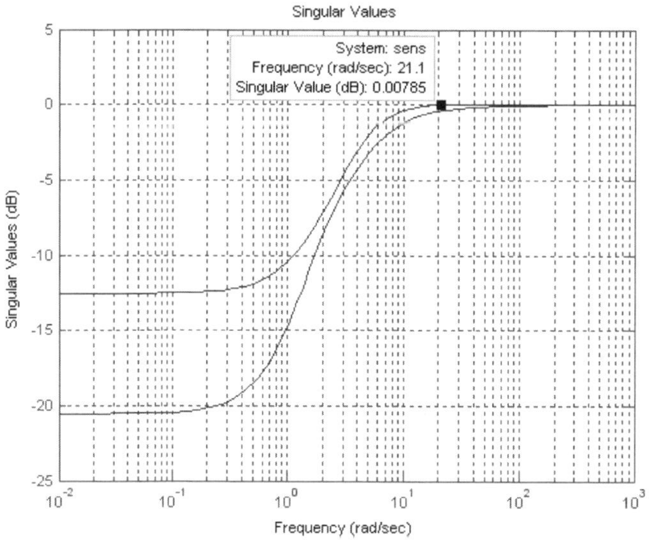

Figure 8.5. *Singular values of Si(s) – method 1*

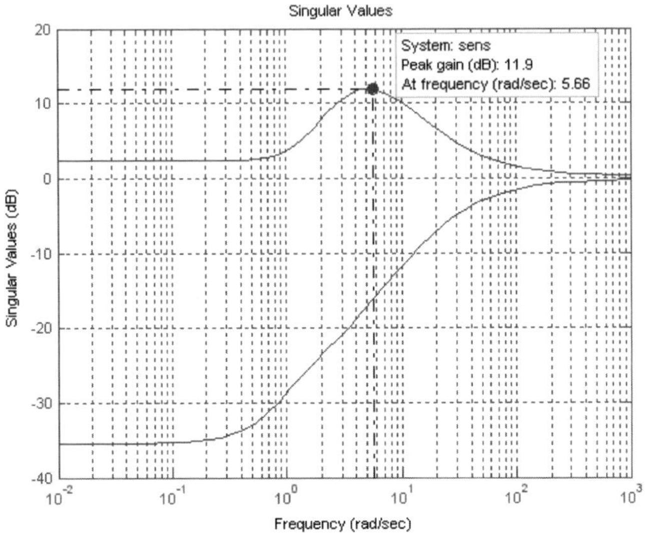

Figure 8.6. *Singular values of Si(s) – method 2*

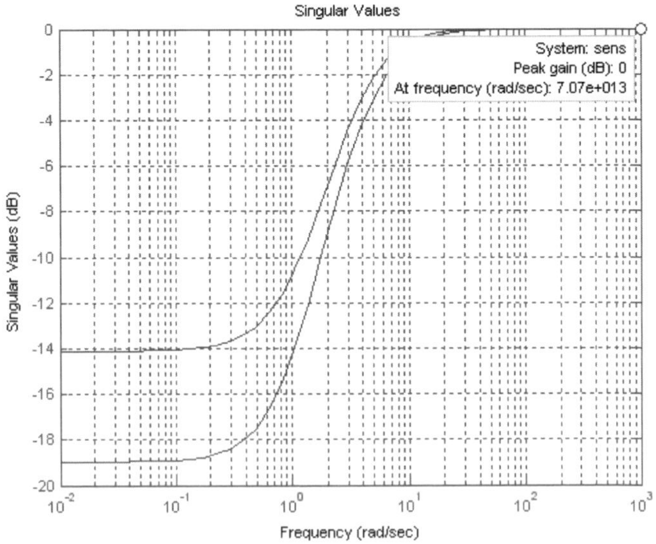

Figure 8.7. *Singular values of Si*

REMARK 249.– *(i) Suppose Hypothesis (H) is replaced by the following hypothesis (H'), which is less restrictive: $G(s) \in \mathbb{R}^{r \times m}$, $\mathrm{rk}_{\mathbb{R}(s)}\, G(s) = r$ $(r \leq m)$, and the state-space system $\{A, B, E\}$ is minimal. Then, according to equation (8.24), the roots of $f_{\alpha}^{\sim}(s)\, f_{\alpha}(s)$ that remain bounded converge (as $\rho \to 0^+$) to the roots of the polynomial $a_{\alpha}^{\sim}(s)\, a_{\alpha}(s)\, \Delta_r(s)$.*

(ii) Let there be the state-space system $\{A, B, E\}$, with transfer matrix $G(s)$, and suppose that the above hypothesis (H) is satisfied as well as the following condition: $\{A, B, E\}$ does not have transmission zeros in the closed right half-plane (in other words, $\{A, B, E\}$ is minimum phase: see Remark 68). Let also P_ρ be the unique ≥ 0 solution of the algebraic Riccati equation (8.19) for $\alpha = 0$ and R replaced by ρR, $\rho > 0$. It is shown in ([72], section 3.8.3) that $P_\rho \to 0$ as $\rho \to 0^+$.

8.2. State feedback with integral action

8.2.1. *Insufficiency of state feedback*

Consider the elementary example of a 2nd-order system having no zeros and described by the left form

$$\left(\partial^2 + a_1\, \partial + a_2\right) y = b_2\, u \quad (b_2 \neq 0).$$

We can put this system in state-space form by setting $x = \begin{bmatrix} \dot{y} & y \end{bmatrix}^T$. A state-feedback control $u = v - K\, x$, where $K = \begin{bmatrix} k_1 & k_2 \end{bmatrix}$, can be written as

$$u = v - k_1 \dot{y} - k_2 y.$$

It is a control of the proportional and derivative kind, in which an integral action is missing and therefore a zero static error cannot be guaranteed. What follows aims remedying such a flaw.

8.2.2. *Feedback control in the presence of disturbances*

Let there be the system

$$\left\{ \begin{aligned} \dot{x} &= A\, x + B\, u + d_1, \\ y &= C\, x + d_2, \end{aligned} \right. \tag{8.29}$$

where $x(t) \in \mathbb{R}^n$, $u(t) \in \mathbb{R}^m$, $y(t) \in \mathbb{R}^p$, $d_1 \in \mathbb{R}^n$, and $d_2 \in \mathbb{R}^p$, where d_1 and d_2 are constant disturbances (an algebraic manner of expressing that these are signals, the whole spectrum of which lies in the low frequencies). These disturbances are supposed to be *unmeasured and unknown*.

On the other hand, the constant reference signal $r \in \mathbb{R}^p$ is assumed to be *known*. The state x and the output y are supposed to be measured at each instant.

Let e be the regulation error defined by

$$e = y - r. \tag{8.30}$$

The objective of what follows is to design a control law for which Property (P) below is satisfied:

– $\lim_{t \to +\infty} e(t) = 0$,

– $\lim_{t \to +\infty} x(t) = const.$,

– $\lim_{t \to +\infty} u(t) = const.$,

$\forall\, d_1 \in \mathbb{R}^n$, $d_2 \in \mathbb{R}^p$, both of them unknown, $\forall r \in \mathbb{R}^p$ known, and with any initial conditions.

The hypotheses imposed on the state-space system $\{A, B, C\}$ are the following:

• (H1) (A, B) is controllable;

• (H2) the transfer matrix $G(s)$ of the system $\{A, B, C\}$ is semiregular over $\mathbb{R}(s)$ (see section 13.1.4).

8.2.3. *Resolution of the static problem*

The "static problem" can be stated in the following manner: Does there exist (for any constant disturbances d_1 and d_2 and any constant reference signal r) an equilibrium state x_e and an equilibrium control u_e for which the regulation error is zero? This translates into the equations

$$\begin{cases} 0 = A\,x_e + B\,u_e + d_1 \\ r = C\,x_e + d_2 \end{cases}$$

which are equivalent to

$$\begin{bmatrix} -A & -B \\ C & 0 \end{bmatrix} \begin{bmatrix} x_e \\ u_e \end{bmatrix} = \begin{bmatrix} d_1 \\ r - d_2 \end{bmatrix}. \tag{8.31}$$

The right-hand side of this equation is any vector in $\mathbb{R}^{n+p}$; as a result, the static problem admits a solution if and only if, the matrix on the left of the above equality is of rank $n + p$. This condition can be written in the form

$$\operatorname{rk} \begin{bmatrix} s\,I_n - A & -B \\ C & 0 \end{bmatrix}_{s=0} = n + p$$

in a way that reveals the Rosenbrock matrix $R(s)$ of $\{A, B, C\}$. From Hypothesis (H2) and Proposition 163 (section 7.2.1), the rank $\rho(R)$ of $R(s)$ over $\mathbb{R}(s)$ is $n + \min(p, m)$. Finally, according to Theorem 161 (section 7.2.1), we obtain the following result:

PROPOSITION 250.– *The static problem admits a solution if and only if, the following two conditions hold: (i) $m \geq p$; (ii) $s = 0$ is not an invariant zero of $\{A, B, C\}$.*

REMARK 251.– *Taking into account Hypothesis (H2), Condition (i) of Proposition 250 expresses the fact that there are at least as many linearly independent controls as linearly independent regulated outputs. This is thus a matter of* number of degrees of freedom. *Condition (ii) expresses the fact that one cannot force the output of a derivator system to be an arbitrary constant. It is similar to the necessary and sufficient conditions expressed by Proposition 109 (section 6.3.4).*

8.2.4. *Resolution of the dynamic problem*

By differentiating equations (8.29) and (8.30), and by putting

$$\chi = \begin{bmatrix} \partial x \\ e \end{bmatrix}, \quad \upsilon = \partial u \tag{8.32}$$

we obtain the state-space system

$$\begin{aligned} \dot{\chi} &= F\chi + G\upsilon, \\ F &= \begin{bmatrix} A & 0 \\ C & 0 \end{bmatrix}, \quad G = \begin{bmatrix} B \\ 0 \end{bmatrix}. \end{aligned} \tag{8.33}$$

According to Theorem 230(ii), there exists a state feedback

$$\upsilon = -K\chi \tag{8.34}$$

for which the closed-loop system is stable if and only if, (F, G) is stabilizable.

LEMMA 252.– *Let there be the control* (8.34). *Property (P) in section 8.2.2 is satisfied if and only if, $F - GK$ is a stability matrix.*

PROOF. Let $\chi_0 = \chi(0)$. We have for every $t \geq 0$

$$\chi(t) = e^{(F - GK)t}\chi_0.$$

Property (P) is thus satisfied if and only if, $F - GK$ is a stability matrix. ■

PROPOSITION 253.– *(F, G) is stabilizable if and only if, Conditions (i) and (ii) in Proposition 250 both hold. In that case, (F, G) is controllable.*

PROOF. According to Proposition 186 (section 7.3), (F, G) is stabilizable if and only if,

$$\mathrm{rk}\begin{bmatrix} s\,I_{n+p} - F & G \end{bmatrix} = n + p, \quad \forall s \in \bar{\mathbb{C}}_+.$$

Notice that

$$\begin{bmatrix} s\,I_{n+p} - F & G \end{bmatrix} = \begin{bmatrix} s\,I_n - A & 0 & B \\ C & s\,I_p & 0 \end{bmatrix} \sim \begin{bmatrix} s\,I_n - A & -B & 0 \\ C & 0 & s\,I_p \end{bmatrix}.$$

(i) For $s \neq 0$, the rank of the matrix on the right is equal to $p + \mathrm{rk}\begin{bmatrix} s\,I_n - A & B \end{bmatrix} = p + n$ since (A, B) is controllable. (ii) For $s = 0$, the rank of this same matrix is equal to the rank, for this same value of s, of the Rosenbrock matrix of $\{A, B, C\}$, a rank which is the subject of Proposition 250. If Conditions (i) and (ii) of this proposition both hold, we have $\mathrm{rk}\begin{bmatrix} s\,I_{n+p} - F & G \end{bmatrix} = n + p$, $\forall s \in \mathbb{C}$, and (F, G) is therefore controllable. ■

Therefore, we can assume that Conditions (i) and (ii) of Proposition 250 are satisfied. Theorem 230 shows that one can choose the control (8.34) in such a manner as to arbitrarily assign the eigenvalues of $F - G\,K$ in the left half-plane (obeying the symmetry property with respect to the real axis in the case of imaginary eigenvalues). We put

$$K = \begin{bmatrix} K_p & K_i \end{bmatrix}$$

where $K_p \in \mathbb{R}^{m \times n}$ and $K_i \in \mathbb{R}^{m \times p}$. Taking into account equation (8.32), equation (8.34) can be written as

$$\partial u = -K_p\,\partial x - K_i\,e,$$

and by integrating this expression, we obtain

$$u(t) = -K_p x(t) - K_i \int e(t)\,dt + const. \tag{8.35}$$

The diagram of this control law, called *state feedback with integral action*, is shown in Figure 8.8.

REMARK 254.– *(i) Property (P) is still satisfied even in the presence of modeling errors and, more precisely, if the matrices A, B, and C of the system are replaced by $A + \Delta A$, $B + \Delta B$, and $C + \Delta C$, respectively, where ΔA, ΔB, and ΔC are errors such that $F + \Delta F - (G + \Delta G)\,K$ (where ΔF and ΔG are the additive errors on F and G, respectively, resulting from the additive errors ΔA, ΔB, and ΔC on A,*

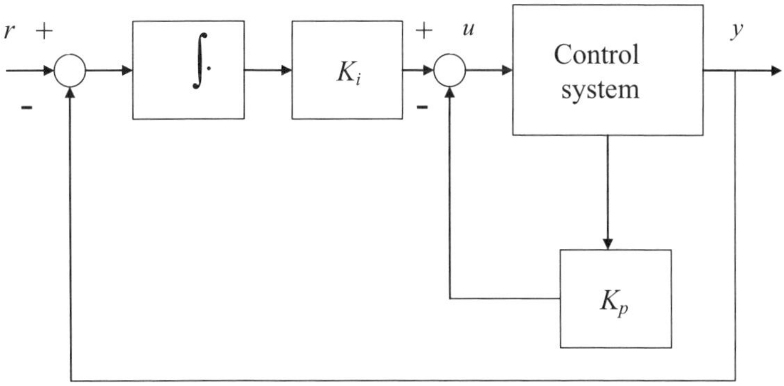

Figure 8.8. *State feedback with integral action*

B, and C) remains a stability matrix. (ii) Suppose system $\{A, B, C\}$ is well-formed. If $m > p$, there are several solutions to the static problem (8.31)*. As a result, even with zero disturbances d_1 and d_2 and a fixed reference r, the equilibrium state x_e and control u_e are not continuous functions of A, B, and C. Very small errors ΔA, ΔB, and ΔC can induce a very important change to these equilibrium values and, consequently, can lead to such a change in the behavior of the feedback system. This is to be avoided and it is thus preferable, even indispensable, in practice, to impose that $\{A, B, C\}$ be well-formed and $m = p$ [92].*

EXAMPLE 255.– *For the second order system of section 8.2.1, the control* (8.35) *can be put in the form*

$$u(t) = -K_{p_1}\,\dot{y}(t) - K_{p_2}\,y(t) - K_i \int e(t)\,dt + const.$$

and we obtain the traditional PID controller.

EXAMPLE 256.– *Let there be the state-space system $\mathbf{\Sigma} = \{A, B, C\}$ defined in Example 195 (section 7.4.3) and again considered in Examples 233, 239, and 242 of section 8.1.4. This system has a transmission zero at $s = 0$ (since $\operatorname{rk} G(0) = 1$), and hence the above control is not applicable.*

EXAMPLE 257.– *Consider the state-space system $\mathbf{\Sigma} = \{A, B, C\}$ with*

$$A = \begin{bmatrix} -4 & -3 & 0 & 0 \\ 1 & 0 & 0 & 0 \\ 0 & 0 & -6 & -8 \\ 0 & 0 & 1 & 0 \end{bmatrix}, \quad B = \begin{bmatrix} 1 & 0 \\ 0 & 0 \\ 0 & 1 \\ 0 & 0 \end{bmatrix},$$

$$C = \begin{bmatrix} 1 & 3 & 1 & 4 \\ 1 & 1 & 1 & 2 \end{bmatrix}$$

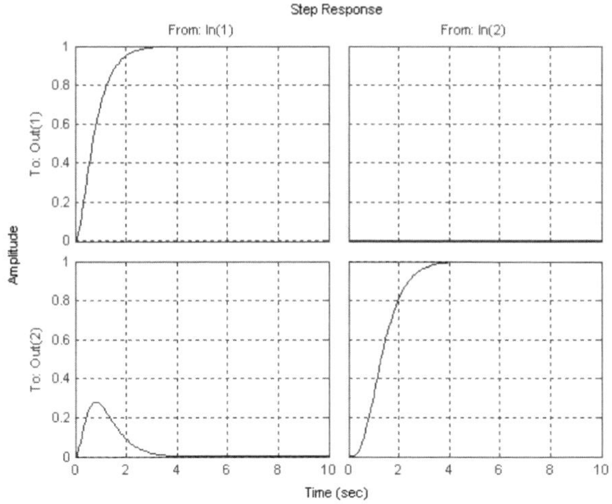

Figure 8.9. *Step responses of the feedback system*

which is a minimal realization in controllable canonical form of the transfer matrix of Exercise 217 (Section 7.6). Conditions (i) and (ii) of Proposition 250 are both satisfied. The eigenvalues of A are $\{-4,-3,-2,-1\}$. Apply the (α) method of section 8.1.4 to the system (8.33). With $\alpha = 3/2$, we obtain

$$K_p = \begin{bmatrix} 4 & 36 & 0 & 18 \\ 0 & -24 & 3 & -6 \end{bmatrix}, \quad K_i = \begin{bmatrix} 18 & -36 \\ -12 & 36 \end{bmatrix}$$

and the eigenvalues of $F - G\,K$ are $\{-4,-3,-3,-3,-2,-2\}$. The step responses of the closed-loop system and the corresponding controls are represented in Figures 8.9 and 8.10 – see below. The curves on the left (resp., on the right) are the variations of y_1 and y_2 (for Figure 8.9) and of u_1 and u_2 (for Figure 8.10) when r_1 is a unit step and $r_2 = 0$ (resp., when $r_1 = 0$ and r_2 is a unit step). We observe not only that the static error is zero, which is in line with the theory, but also an absence of overshoot and a good decoupling between the two outputs (even though $\mathbf{\Sigma}$ is heavily coupled).

Now it remains to verify that when the gain matrix K is determined as in Example 257, the input modulus margin of $\mathbf{\Sigma}$ is good (and, if possible, equal to 1). The input sensitivity function $\tilde{S}_i$ of the system (8.33) fed back by the control (8.34) is such that $\left\|\tilde{S}_i\right\|_\infty = 1$, according to Theorem 240, since $\alpha \geq 0$.

THEOREM 258.– *Let $\mathbf{\Sigma}$ be fed by the control (8.35). Assume that the gain matrix $K = \begin{bmatrix} K_p & K_i \end{bmatrix}$ is chosen according to the (α) method (or, more generally, the LQR method) in section 8.1.4 applied to system (8.33) with $\alpha \geq 0$. Then, the input sensitivity function S_i is such that $\|S_i\|_\infty = 1$.*

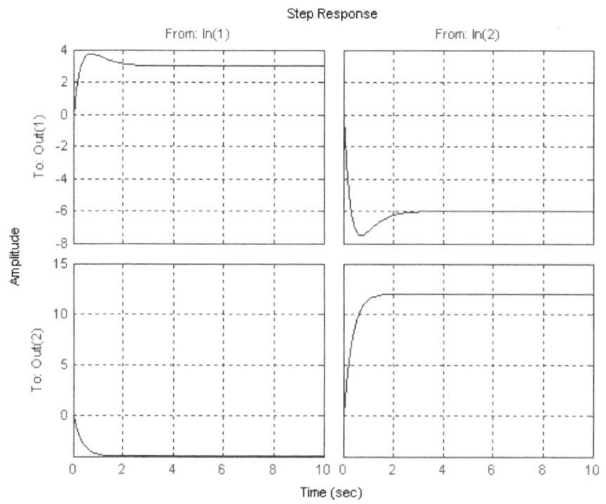

Figure 8.10. *Corresponding controls*

PROOF. We have $S_i = (I_m + L_i)^{-1}$ with $L_i(s) = \left(K_p + \frac{1}{s} K_i C\right)(s I_n - A)^{-1} B$. On the other hand, $\tilde{S}_i = \left(I_m + \tilde{L}_i\right)^{-1}$ with

$$L_i(s) = \begin{bmatrix} K_p & K_i \end{bmatrix} \begin{bmatrix} s I_n - A & 0 \\ -C & s I_p \end{bmatrix}^{-1} \begin{bmatrix} B \\ 0 \end{bmatrix}.$$

According to equation (13.9) (section 13.1.4), we have

$$\begin{bmatrix} s I_n - A & 0 \\ -C & s I_p \end{bmatrix}^{-1} = \begin{bmatrix} (s I_n - A)^{-1} & 0 \\ (1/s)\, C\, (s I_n - A)^{-1} & (1/s)\, I_p \end{bmatrix},$$

and hence $L_i = \tilde{L}_i$. Thus $S_i = \tilde{S}_i$. ■

EXAMPLE 259.– *(Example 257 continued). The singular values of the sensitivity function S_i and those of L_i are plotted in Figures 8.11 and 8.12, respectively, as a function of angular frequency ω. We note that $\|S_i\|_\infty = 1$ in accordance with what was confirmed in Theorem 258, and that the loopshaping of the singular values of $L_i(i\omega)$ is similar to that in Figure 4.15 of section 4.2.9. The balancing of the singular values in the neighborhood of 0 dB is correct.*

8.3. *Internal model principle

8.3.1. *Problem setting*

We consider again system Σ defined by equation (8.29) and the error e defined by equation (8.30), but in a more general context. We also consider the disturbances d_i

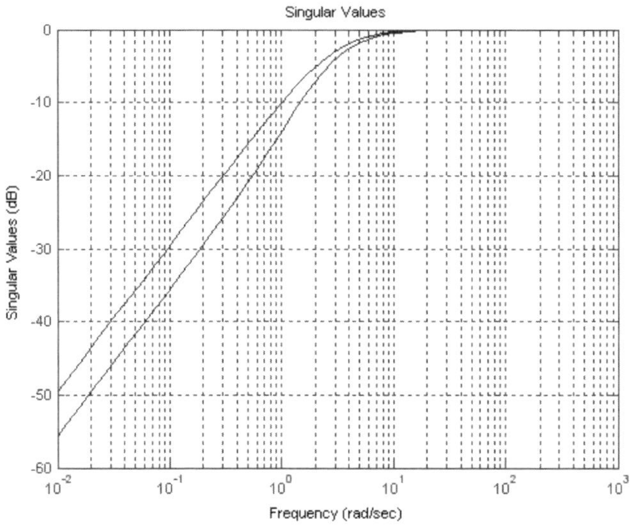

Figure 8.11. *Singular values of Si*

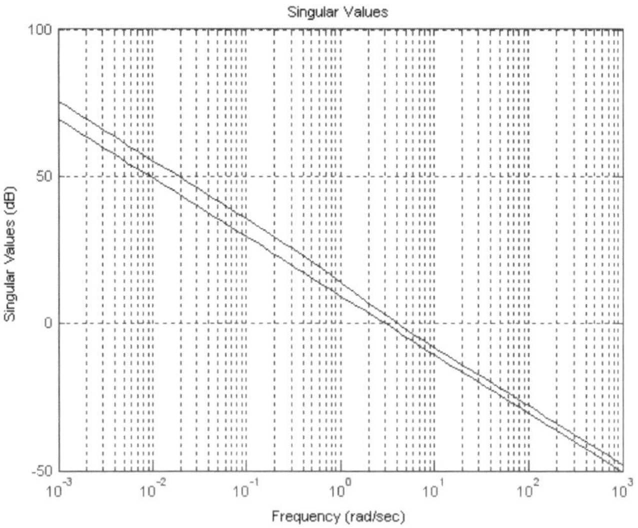

Figure 8.12. *Singular values of Li*

$(i = 1, 2)$ as well as the reference signal r which are torsion elements over the ring $\mathbb{R}[\partial]$ (see section 13.4.1). The set of all annihilating differential polynomials of these disturbances and of this reference signal is an ideal $\mathfrak{I}$ of $\mathbb{R}[\partial]$; the ideal $\mathfrak{I}$ is principal and generated by a unique monic polynomial denoted by $D(\partial)$.[6]

The situation considered in section 8.2 is a special case of this, obtained with $D(\partial) = \partial$. On the other hand, this polynomial $D(\partial)$ corresponds to what is considered in section 6.4. It is assumed that all its roots lie on the imaginary axis (see section 6.4.3, Hypothesis 116), and that these roots are simple (the latter condition makes it possible to simplify the problem setting: see the last paragraph of section 6.4.3).

We have thus by hypothesis

$$D(\partial)\, d_i = 0 \quad (i = 1, 2), \quad D(\partial)\, r = 0. \tag{8.36}$$

Also, Hypotheses (H1) and (H2) of section 8.2.2 are in force.

The objective of what follows is to design a control law for which the following property (P') is satisfied:

- $\lim_{t\to+\infty} e(t) = 0$,
- x is bounded,
- u is bounded,

for any disturbances d_i $(i = 1, 2)$ and any reference signal r satisfying equation (8.36).

8.3.2. *Solution*

Left-multiplying (8.29) by $D(\partial)$, we obtain, putting $\eta = D(\partial)\, x$, $\varsigma = D(\partial)\, e$, and $\upsilon = D(\partial)\, u$,

$$\begin{cases} \dot{\eta} = A\,\eta + B\,\upsilon, \\ \varsigma = C\,\eta. \end{cases}$$

We have $e = D(\partial)^{-1} I_p \varsigma$. Let $\{A_1, B_1, C_1\}$ be a minimal realization of $D(s)^{-1}$. Then, a minimal realization of $D(s)^{-1} I_p$ is $\{A_\varepsilon, B_\varepsilon, C_\varepsilon\}$ with

$$A_\varepsilon = \bigoplus_{1\le i\le p} A_i, \quad B_\varepsilon = \bigoplus_{1\le i\le p} B_i \quad C_\varepsilon = \bigoplus_{1\le i\le p} C_i$$

6. This polynomial $D(\partial)$ should not be confused with the direct term matrix D of the state-space system.

where $A_i = A_1$, $B_i = B_1$, and $C_i = C_1$ for any $i \in \{1, ..., p\}$ and where $\oplus$ denotes the diagonal sum (see section 13.1.4). The tracking error e is the output of system $\{A_\varepsilon, B_\varepsilon, C_\varepsilon\}$ with input ς. Denoting by η_ε the state of that system, we have

$$\begin{cases} \dot{\eta}_\varepsilon = A_\varepsilon\, \eta_\varepsilon + B_\varepsilon\, \varsigma, \\ \quad e = C_\varepsilon\, \eta_\varepsilon. \end{cases}$$

Now let

$$\chi = \begin{bmatrix} \eta \\ \eta_\varepsilon \end{bmatrix}.$$

Then, equation (8.33) is satisfied, as well as the output equation

$$e = H\,\chi$$

with

$$\begin{aligned} F &= \begin{bmatrix} A & 0 \\ B_\varepsilon\, C & A_\varepsilon \end{bmatrix}, \quad G = \begin{bmatrix} B \\ 0 \end{bmatrix}, \\ H &= \begin{bmatrix} 0 & C_\varepsilon \end{bmatrix}. \end{aligned}$$

The following result can be easily proved:

LEMMA 260.– *The statement of Lemma 252 is still valid in the present context, replacing (P) by* (P').

Proposition 253 (section 8.2.4) can be generalized as follows:

PROPOSITION 261.– *The pair* (F, G) *is stabilizable if, and only if, the following two conditions hold: (i)* $m \geq p$*; (ii)* $\{A, B, C\}$ *has no invariant zeros that are roots of* $D(s)$*. Then,* (F, G) *is controllable.*

PROOF. Let $q = d^\circ(D)$. According to Proposition 186 (section 7.3), (F, G) is stabilizable if, and only if, for any $s \in \bar{\mathbb{C}}_+$,

$$\mathrm{rk}_{\mathbb{C}} \begin{bmatrix} s\, I_{n+pq} - F & G \end{bmatrix} = n + pq. \tag{8.37}$$

Let $V(s) = \begin{bmatrix} s\, I_{n+pq} - F & G \end{bmatrix}$. (1) If s is not an eigenvalue of A_d (i.e. not a root of $D(s)$), we immediately see that Condition (8.37) holds, due to the fact that (A, B) is controllable. (2) Suppose now that s is an eigenvalue of A_d. We have $V(s) = V_1(s)\, V_2(s)$ with

$$\begin{aligned} V_1(s) &= \begin{bmatrix} I_n & 0 & 0 \\ 0 & -B_\varepsilon & s\, I_{pq} - A_\varepsilon \end{bmatrix}, \\ V_2(s) &= \begin{bmatrix} s\, I_n - A & 0 & B \\ C & 0 & 0 \\ 0 & I_{pq} & 0 \end{bmatrix}. \end{aligned}$$

In addition, $\mathrm{rk}_{\mathbb{C}}\, V_1\,(s) = n + pq$ because $(A_\varepsilon, B_\varepsilon)$ is controllable. On the other hand,

$$\mathrm{rk}_{\mathbb{C}}\, V_2\,(s) = pq + \mathrm{rk}_{\mathbb{C}} \begin{bmatrix} s\, I_n - A & -B \\ C & 0 \end{bmatrix}.$$

According to Hypothesis (H2) and Proposition 163 (section 7.2.1),

$$\mathrm{rk}_{\mathbb{R}(s)} \begin{bmatrix} s\, I_n - A & -B \\ C & 0 \end{bmatrix} = n + \min\{m, p\},$$

and hence $\mathrm{rk}_{\mathbb{C}}\, V_2\,(s) \leq pq + n + \min\{m, p\}$, with equality if and only if, s is not an invariant zero of $\{A, B, C\}$. According to the Sylvester inequality (13.7) (section 13.1.4), we have

$$\mathrm{rk}_{\mathbb{C}}\, V\,(s) \geq \mathrm{rk}_{\mathbb{C}}\, V_1\,(s) + \mathrm{rk}_{\mathbb{C}}\, V_2\,(s) - (n + p + pq).$$

Therefore, the quantity on the right-hand side of this inequality is equal to $n + pq$ if and only if, Conditions (i) and (ii) stated are satisfied. In this case, we have $\mathrm{rk}_{\mathbb{C}} \begin{bmatrix} s\, I_{n+pq} - F & G \end{bmatrix} = n + pq$ for all $s \in \mathbb{C}$. ■

Assuming that Conditions (i) and (ii) of Proposition 261 hold, we can thus stabilize (8.33) by a feedback of the form (8.34), i.e.

$$v = -\begin{bmatrix} K_p & K_\varepsilon \end{bmatrix} \begin{bmatrix} \eta \\ \eta_\varepsilon \end{bmatrix}. \tag{8.38}$$

This expression can be further developed in the following manner:

$$D\,(\partial)\; u = -K_p\, D\,(\partial)\, x - K_\varepsilon\, \eta_\varepsilon.$$

Let

$$x_\varepsilon = D\,(\partial)^{-1}\, I_m \eta_\varepsilon.$$

According to the previous discussion, we obtain the expression of the "state feedback with internal model" controller

$$\dot{x}_\varepsilon = A_\varepsilon\, x_\varepsilon + B_\varepsilon\, e, \tag{8.39}$$

$$u = -K_p\, x - K_\varepsilon\, x_\varepsilon. \tag{8.40}$$

The "internal model" is the system (8.39): it is a model of the disturbances and of the reference signal, duplicated p times, and having the tracking error e as input.

REMARK 262.– *(i) The state x_d is only defined up to a term annihilated by $D\,(\partial)$ and the same is true for the control u. (ii) The control (8.40) is the exact generalization of equation (8.35) (where the only term just as mentioned above is a constant of integration). (iii) The diagram obtained for the control law is identical to that in Figure 8.8, with the only difference that the integrator needs to be replaced by the internal model. (iv) Remark 254 remains valid. (v) The statement of Theorem 258 remains valid in the more general context that is considered here.*

8.4. Exercises

EXERCISE 263.– *Show that the invariant zeros of a state-space system* $\{A, B, C, D\}$ *are invariant under state feedback.*

EXERCISE 264.– *Let there be a state-space system* $\Sigma = \{A, B, C\}$ *with*

$$A = \begin{bmatrix} 0 & 2 \\ 1 & -1 \end{bmatrix}, \quad B = \begin{bmatrix} 1 \\ -2 \end{bmatrix}, \quad C = \begin{bmatrix} 2 & 1 \end{bmatrix}.$$

(i) Study controllability and observability of Σ. *(ii) Calculate the transfer function* $G(s)$ *of* Σ, *its poles, and its invariant zeros. (iii) Without making any additional calculation, indicate whether it is possible to maintain the system output to a constant reference by way of a state feedback with integral action. (iv) Determine a state feedback with integral action for which the poles of the closed-loop system are* $\{-1, -1, -2\}$. *(v) Trace the shape of the Nyquist plot of the open-loop transfer function* $L_i(s)$.

EXERCISE 265.– *Answer the same questions as those of Exercise 264 when* Σ *is the linearized inverted pendulum whose state-space representation has been determined in Exercise 209 (Section 7.6), assuming that the controlled variable is the position* y *of the carriage, and given the following:* $M = 10\,kg$, $m = 1\,kg$, $l = 0.981\,m$. *For question (iv), choose the poles of the closed-loop system to be* $\{-2, -3, -4, -5, -6\}$.

EXERCISE 266.– *Consider the system below, consisting of two identical tanks, the first one draining into the second one in Figure 8.13. The system input is the discharge of water going into the 1st tank, the output is the water level* x_2 *in the 2nd tank.*

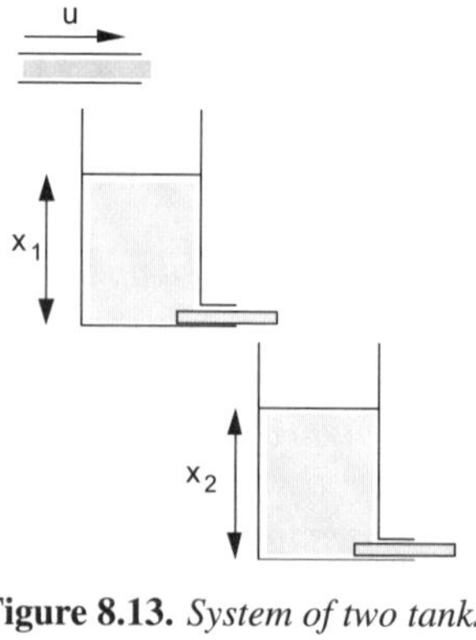

Figure 8.13. *System of two tanks*

(i) Justify the fact that we have a state-space system $\{A, B, C\}$, *where* A *and* B *are of the form*

$$A = \begin{bmatrix} -\alpha_2 & 0 \\ \alpha_2 & -\alpha_3 \end{bmatrix}, \quad B = \begin{bmatrix} \alpha_1 \\ 0 \end{bmatrix}$$

and determine the matrix C. *Assume from now on that* $\alpha_1 = \alpha_3 = 0.01$ *and* $\alpha_2 = 0.02$. *(ii) Is this system controllable? observable? (iii) What are the poles of this system? Is the system stable? (iv) Does this system have zeros (if yes, what are they)? What is its static gain? (v) Suppose all the components of the state are measured, and determine a state feedback with integral action for which all the poles of the closed-loop are placed at* -0.05. *How can such a choice be justified?*

EXERCISE 267.– *Consider the hot air balloon in Figure 8.14. Let:*

– θ *be the variation of temperature in the balloon with respect to a reference temperature;*

– u *be the variation of the quantity of heat provided by the balloon (which is the control variable);*

– w *be the rising speed of the wind (which is a disturbance);*

– h *be the variation of the altitude of the balloon with respect to a reference altitude;*

– v *be the vertical velocity of the balloon.*

A rough model leads to the following equations:

$$\begin{cases} \tau_1\,\dot{\theta} = -\theta + \tau_1\,u \\ \tau_2\,\dot{v} = -v + \sigma\,\tau_2\,\theta + w \\ \dot{h} - v \end{cases}$$

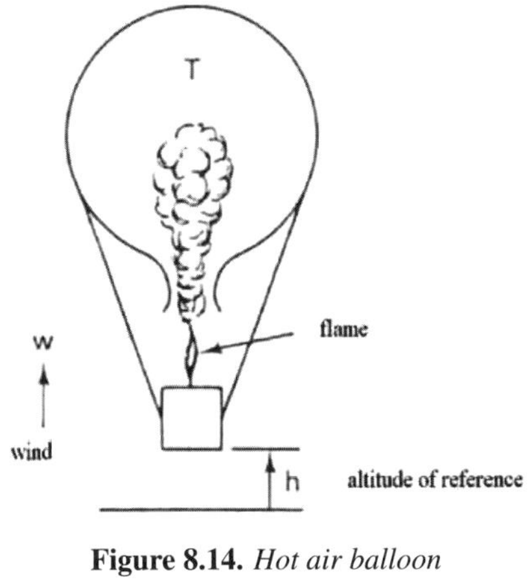

Figure 8.14. *Hot air balloon*

where τ_1 *and* τ_2 *are time constants with values of* 10 s *and* 1 s*, respectively;* σ *is a coupling factor of value* 1*, taking into account the chosen units. (i) Write this system in state-space form with input* $\begin{bmatrix} u \\ w \end{bmatrix}$ *and output* h*. (ii) Determine the poles of this system. Is it stable? (iii) We at first assume that* $u = 0$*. (a) Write the system thus obtained in state-space form with input* w *and output* h*. (b) Is this system observable? (c) Interpret the results physically. (iv) We now assume that* $w = 0$*. (a) Repeat the questions of (iii) by replacing* w *by* u*. (b) Determine the state feedback that assigns all closed-loop poles at* -1*.*

EXERCISE 268.– * *(i) Let* P *be a solution of the algebraic Riccati equation* (8.19)*, let* $T = \begin{bmatrix} I & 0 \\ P & I \end{bmatrix}$ *and let* $F = A - B\,K$*. (a) Calculate* $T^{-1}\,H\,T$*, where* H *is the Hamiltonian matrix* (8.20)*. (b) Show that*

$$H\begin{bmatrix} I \\ P \end{bmatrix} = \begin{bmatrix} I \\ P \end{bmatrix} F.$$

*(ii) Using (i), prove Parts (i)-(iii) of Theorem 244. (The reader can find a precise discussion of the solution of this exercise in ([64], section 3.4, Exercises 3.4-9 and 3.4-10.)**

EXERCISE 269.– *Extend the theories developed in sections 8.2 and 8.3 to the case where the system to be controlled has a direct term, i.e. the 2nd equation of equation* (8.29) *is* $y = C\,x + D\,u + d_2$.

Chapter 9

Observers

In the previous chapter, we have seen how to use the system state to control the latter when all state components are measured, thus known at each instant. In this chapter, we consider the situation where the state $x(t)$ is entirely or partly unknown at each instant t.

9.1. Full-order observers

9.1.1. *General principle*

Consider the state-space system $\{A, B, C, D\}$:

$$\begin{cases} \dot{x} = A\,x + B\,u \\ y = C\,x + D\,u \end{cases} \tag{9.1}$$

where $x(t) \in \mathbb{R}^n$, $u(t) \in \mathbb{R}^m$, and $y(t) \in \mathbb{R}^p$. The state x is unknown, while the output y and the input u are assumed to be known. We suggest to determine from these data a variable $t \mapsto \hat{x}(t)$ such that the error

$$\boxed{\tilde{x}(t) = x(t) - \hat{x}(t)} \tag{9.2}$$

tends to 0 when $t \to +\infty$. The determination of $\hat{x}$ is often called the "reconstruction of the system state" and the terms "state observer" and "state reconstructor" are synonymous. The quantity $\tilde{x}$ defined by (9.2) is called the *reconstruction error* or the *observation error*.

The form of a full-order observer is

$$\partial\,\hat{x} = (A\,\hat{x} + B\,u) + \tilde{K}\,(y - C\,\hat{x} - D\,u) \tag{9.3}$$

$$= \left(A - \tilde{K}\,C\right)\,\hat{x} + \tilde{K}\,y + \left(B - \tilde{K}\,D\right)\,u \tag{9.4}$$

where $\tilde{K} \in \mathbb{R}^{n\times p}$ is the "gain matrix of the observer" (sometimes called, more succinctly, the "observer gain"). For $\tilde{K} = 0$, (9.3) is a simple simulation of the state-space system $\{A, B\}$. The term $y - C\,\hat{x} - D\,u$ is the error we see when the simulation is poor. Left-multiplied by $\tilde{K}$, this error provides a "correction term".

By subtracting (9.3) from the first line of (9.1) and taking into account of the other equations, we obtain the differential equation of the observation error $\tilde{x}$:

$$\boxed{\partial\,\tilde{x} = \left(A - \tilde{K}\,C\right)\,\tilde{x}}. \tag{9.5}$$

REMARK 270.– *Equation* (9.5) *is obtained in the "ideal case" where the system, the state of which we seek to reconstruct, is perfectly linear time-invariant, with matrices* A*,* B*,* C*, and* D *perfectly known. If we only know approximations* $\hat{A}$*,* $\hat{B}$*,* $\hat{C}$*, and* $\hat{D}$ *of these matrices, equation* (9.3) *becomes*

$$\partial\,\hat{x} = \left(\hat{A}\,\hat{x} + \hat{B}\,u\right) + \tilde{K}\,\left(y - \hat{C}\,\hat{x} - \hat{D}\,u\right),$$

which complicates a lot the differential equation of the observation error. We can already conclude that we will certainly be faced with the problem of robustness of the observer. *This point will be dealt with later.*

DEFINITION 271.– *(i) The state observer* (9.3) *is said to be* stable *if* $\tilde{x}\,(t) \to 0$ *as* $t \to +\infty$*, for any initial error* $\tilde{x}\,(0)$*. (ii) The eigenvalues of* $A - \tilde{K}\,C$ *are called the* observer poles.

THEOREM 272.– *The state observer* (9.3) *is stable if and only if, all its poles lie in the left half-plane.*

PROOF. This is an immediate consequence of Remark 181 (section 7.3). ■

THEOREM 273.– *Let there be the state observer* (9.3)*. (i) The following conditions* (a) *and* (b) *are equivalent: (a) for any symmetric set* $P \subset \mathbb{C}$ *of* n *elements, there exists a gain matrix* $\tilde{K}$ *for which the observer poles are the elements of* P*; (b) the pair* (C, A) *is observable. (ii) There exists a gain matrix* $\tilde{K}$ *with which the observer is stable if and only if,* (C, A) *is detectable.*

PROOF. The eigenvalues of $A - \tilde{K}\,C$ are identical to those of $\left(A - \tilde{K}\,C\right)^T = A^T - C^T\,\tilde{K}^T$ (section 13.3.4, Proposition 534). The theorem is thus an immediate consequence of Corollary 153 (section 7.1.4), Corollary 187 (section 7.3), and Theorem 230 (section 8.1.2). ■

REMARK 274.– State feedback $\leftrightarrow$ observer duality. *As a result of the proof of Theorem 273, the determination of the gain matrix of an observer and that of a state feedback are two "dual problems". More precisely, the gain matrix $\tilde{K}$ of an observer can be determined based on a fictitious state-space system $\left\{A^T, C^T\right\}$. For this system, we determine a fictitious state feedback with gain matrix $\tilde{K}^T$, by using one of the methods studied in Chapter 8. Then, we set $\tilde{K} = \left(\tilde{K}^T\right)^T$.*

REMARK 275.– *Expression (9.4) shows that an observer is a feedback system: the "open-loop system" $\partial\hat{x} = A\,\hat{x} + e$ is fed back according to $e = -\tilde{K}\,C\,\hat{x} + z$, where $z = \tilde{K}\,y + \left(B - \tilde{K}\,D\right)\,u$. The reader is asked to draw a diagram of this loop.*

9.1.2. *State feedback/observer synthesis*

Let there be a state-space system $\{A, B, C, D\}$ of the form (9.1), where (A, B) is stabilizable and (C, A) is detectable.

1) Since (A, B) is stabilizable, we can design, if the state $x\,(t)$ is known at each instant t, a stabilizing state feedback control

$$u = v - K\,x.$$

If (A, B) is controllable, we can moreover choose the gain matrix K in a way that arbitrarily assigns the eigenvalues of $A - B\,K$ in the left half-plane (with the usual symmetry condition with respect to the real axis): see Theorem 230 (section 8.1.2).

2) Assume now that only the output $y\,(t)$ and the control $u\,(t)$ are known at each instant t. The hypothesis of detectability of (C, A) allows us to affirm that, according to Theorem 273 (section 9.1.1), we can reconstruct the state using a *stable* full-order observer of the form (9.3). If (C, A) is observable, we can moreover choose the gain matrix $\tilde{K}$ in a manner as to arbitrarily assign the observer poles in the left half-plane (with the symmetry condition with respect to the real axis as recalled above).

3) Consider the control

$$\boxed{u = v - K\,\hat{x}} \tag{9.6}$$

where $\hat{x}$ is the estimated state provided by the observer. We obtain the following theorem, sometimes called the "separation principle".[1]

THEOREM 276.– *The poles of the closed-loop system with control* (9.6) *are the elements of*

$$\mathrm{Sp}\,(A - B\,K)\,\dot{\cup}\,\mathrm{Sp}\left(A - \tilde{K}\,C\right)$$

(counting multiplicities), where Sp *denotes the spectrum – see section 13.3.3 – and* $\dot{\cup}$ *denotes the disjoint union – see section 13.4.2, Lemma 559.*

PROOF. According to (9.2) and (9.6), we have

$$u = v - K\,(x - \tilde{x})\,.$$

We therefore have, from (9.1),

$$\dot{x} = A\,x + B\,(v - K\,(x - \tilde{x})) = (A - B\,K)\,x + B\,K\,\tilde{x} + B\,v.$$

Finally, according to (9.5), the state of a closed-loop system is $\left[\begin{array}{c} x \\ \tilde{x} \end{array}\right]$ and this system is governed by

$$\begin{aligned} \partial\left[\begin{array}{c} x \\ \tilde{x} \end{array}\right] &= \underbrace{\left[\begin{array}{cc} A - B\,K & B\,K \\ 0 & A - \tilde{K}\,C \end{array}\right]}_{A_a}\left[\begin{array}{c} x \\ \tilde{x} \end{array}\right] + \underbrace{\left[\begin{array}{c} B \\ 0 \end{array}\right]}_{B_a} v, \\ y &= \underbrace{\left[\begin{array}{cc} C & 0 \end{array}\right]}_{C_a}\left[\begin{array}{c} x \\ \tilde{x} \end{array}\right]. \end{aligned}$$

We have

$$\det\left(\partial\,I_{2n} - A_a\right) = \det\left(\partial\,I_n - A + B\,K\right)\,\det\left(\partial\,I_n - A + \tilde{K}\,C\right)$$

(see section 13.1.4), which proves the desired result. ■

REMARK 277.– *It is clear according to* (9.5) *that the dynamics of the observer are not controllable (in the absence of any modeling error and any disturbance).*

1. *Some authors reserve this terminology for the "linear quadratic gaussian control", which is shown to be a "linear quadratic control"/"Kalman filter" synthesis in the context of optimal control. The term "certainty equivalence principle" is also used in that case.*

We will now determine the transfer matrix $H(s)$ of a controller designed by state feedback/observer synthesis. It is a transfer function $y \rightarrow -u$ when $v = 0$ and when the control system is suppressed from the loop (the $-$ sign is due to the fact that by convention, the feedback is always negative; see Figure 4.1 in section 4.1.1). According to equations (9.6) and (9.4)

$$\partial \hat{x} = \left(A - \tilde{K} C - \left(B - \tilde{K} D K\right)\right) \hat{x} + \tilde{K} y,$$

and hence the transfer matrix $H(s)$ is given by

$$\boxed{H(s) = K \left(s I_n - A + B K + \tilde{K} C - \tilde{K} D K\right)^{-1} \tilde{K}}. \tag{9.7}$$

9.1.3. *State feedback/observer synthesis and RST controller*

It is worthwhile seeking a relationship between the state feedback/observer synthesis and the RST controller as studied in Chapter 6 *in the case of an SISO control system*, assumed to be strictly proper in order to simplify the notation. To avoid confusion, the control system is denoted by $\{F, G, H\}$ in what follows. This system is therefore governed by the equations

$$\begin{cases} \dot{x} = F x + G u \\ y = H x \end{cases} \tag{9.8}$$

whereas the observer is given by

$$\partial \hat{x} = (F \hat{x} + G u) + \tilde{K} (y - H \hat{x}) \tag{9.9}$$

and the control satisfies equation (9.6).

The transfer function $\Theta(s)$ of the system $\{F, G, H\}$ is given by

$$\Theta(s) = H (s I_n - F)^{-1} G$$

(see Proposition 157, section 7.1.5); therefore, by equation (13.8) (section 13.1.4), we have

$$\begin{aligned} \Theta(s) &= \frac{B(s)}{A(s)}, \\ A(s) &= \det(s I_n - F), \quad B(s) = H \operatorname{adj}(\partial I_n - F) G \end{aligned} \tag{9.10}$$

where adj (.) is the "classical adjoint matrix" of (.).

Suppose $\{F, G, H\}$ is *minimal*. In this case, its order and its transmission order coincide (section 2.4.6), and thus the rational function (9.10) is irreducible, i.e. the polynomials $B(s)$ and $A(s)$ are coprime.

On the other hand, let

$$A_c(\partial) = \det(\partial I_n - F + G K), \quad A_o(\partial) = \det\left(\partial I_n - F + \tilde{K} H\right).$$

According to (9.9), we have

$$\left(\partial I_n - F + \tilde{K} H\right) \hat{x} = G u + \tilde{K} y \tag{9.11}$$

and we know that

$$\left(\partial I_n - F + \tilde{K} H\right)^{-1} = \frac{\operatorname{adj}\left(\partial I_n - F + \tilde{K} H\right)}{A_o(\partial)}. \tag{9.12}$$

Let

$$Q(\partial) = K \operatorname{adj}\left(\partial I_n - F + \tilde{K} H\right) G, \quad R(\partial) = K \operatorname{adj}\left(\partial I_n - F + \tilde{K} H\right) \tilde{K}.$$

We get from equations (9.6), (9.11) and (9.12)

$$S(\partial) u = -R(\partial)\, y + T(\partial)\, v$$

with $S(\partial) = A_o(\partial) + Q(\partial)$ and $T(\partial) = A_o(\partial)$.

A state feedback/observer synthesis is therefore identical to a particular RST controller.

According to section 6.2.1, the characteristic polynomial of the closed-loop system is

$$A_{cl}(\partial) = A(\partial) S(\partial) + B(\partial) R(\partial)$$

and by Theorem 276 (section 9.1.1), we have (with the new notation)

$$A_{cl}(\partial) = A_c(\partial) A_o(\partial). \tag{9.13}$$

This last expression is to be compared with equalities (6.20) and (6.24) of section 6.3.5.

The transfer function $\Theta_{bf}(s)$ of the closed-loop system (with input v and output y) is given by

$$\Theta_{bf}(s) = \frac{B(s)\, T(s)}{A(s)\, S(s) + B(s)\, R(s)} = \frac{B(s)}{A_c(s)}.$$

The poles of the observer are thus hidden modes of the closed-loop system, which is consistent with Remark 277 (section 9.1.2). (As a complement to this analysis, see Exercise 301.)

9.1.4. *LTR method*

Introduction

We have shown in sections 8.1.3 and 8.1.4 that it is possible to design a state feedback control for an SISO or an MIMO state-space system Σ, such that it will stabilize the closed-loop system with a certain robustness, and, more specifically, an input modulus margin equal to 1. Two conditions are required for this: the controllability (or by default the stabilizability) of Σ, and the possibility to know all the state components at each instant. A state observer has the role of relaxing this last constraint. But there is the question of knowing under what condition (if it exists) the presence of an observer will not destroy the robustness of the closed-loop.

This is a problem that has already been touched upon in the framework of RST controllers. Indeed, suppose Σ is a *minimal* SISO state-space system. A state feedback control can be written in the form of a "partial state feedback control" (see section 8.1.3). As a result, according to (9.13), the choice, in section 6.3.5, of (6.20) as a characteristic polynomial of the closed-loop, corresponds to that of *a state observer, the poles of which are the transmission zeros of* Σ, *plus fast poles*. It is this choice which allows us to conserve in an approximate manner the modulus margin obtained by state feedback (Theorem 112, section 6.3.5). A necessary and sufficient condition for the synthesis of such an observer to be possible is that Σ should *be minimum phase*. According to Proposition 90 (section 4.2.8), this limitation is in the nature of things and is not related to the method used.

Case of an SISO system

For an *SISO system* Σ (assumed to be minimal), it is possible to design by pole placement a full-order observer like the one above. This observer is indeed entirely determined by its characteristic polynomial. Let

$$A_o(\partial) = B^*(\partial) F(\partial) \tag{9.14}$$

be this characteristic polynomial (with the notation used in equation (6.24) (section 9.1.3). Let $L(s)$ (resp., $L_{sf}(s)$) be the transfer function of the open loop when the control used is the state feedback/observer synthesis (resp., the state feedback).

(i) If Σ has no transmission zeros in the closed right half-plane, then $B^*(\partial) = B(\partial)$, and hence the hypotheses of Theorem 112 are satisfied, and $\left\| \frac{1}{1+L} - \frac{1}{1+L_{sf}} \right\|_\infty \to 0$ when the roots of the polynomial $F(\partial)$ tend to $-\infty$ on the real axis.

(ii) If the only transmission zeros Σ possibly has in the closed right half-plane have an absolute value that is much larger than the maximum cutoff frequency ω_c of the closed-loop system, then, when the roots of $F(\partial)$ have the above behavior, $L(i\omega) \to L_{sf}(i\omega)$ *with a good approximation* at frequencies ω for which $B^*(i\omega) \simeq B(i\omega)$, i.e. such that $\omega \leq \omega_c$.

Case of an MIMO system

For a *system* Σ *with several outputs*, we encounter the same difficulty as in section 8.1.4, i.e. a full-order observer is not determined by its poles (or by its characteristic polynomial). We are thus led to consider a *dual* of the LQR solution of section 8.1.4. Let us find out the exact nature of this duality. In the table below, the data that are necessary for the synthesis of a state feedback are shown in the left column, those that are necessary for the synthesis of an observer are shown in the right column (the matrix denoted by $\check{P}$ in section 8.1.4 is denoted by P in what follows).

A	A^T
B	C^T
K	$\tilde{K}^T$
Q	$\tilde{Q}$
R	$\tilde{R}$
P	$\tilde{P}$

The first three rows express the duality already mentioned in Remark 274 (section 9.1.1). The next three rows are interpreted as follows: consider the expression (8.18), where K, R, B, and P are replaced by $\tilde{K}^T$, $\tilde{R}$, C^T, and $\tilde{P}$ respectively. We have $\tilde{K}^T = \tilde{R}^{-1} C \tilde{P}$, from which we get

$$\boxed{\tilde{K} = \tilde{P} C^T \tilde{R}^{-1}}. \tag{9.15}$$

Let's work the same way as in expression (8.19), with $\alpha = 0$ and replacing A by A^T. We obtain

$$\boxed{\tilde{P} A^T + A \tilde{P} - \tilde{P} C^T \tilde{R}^{-1} C \tilde{P} + \tilde{Q} = 0}. \tag{9.16}$$

REMARK 278.– * *(i) Expressions* (9.3), (9.15) *and* (9.16) *are those of the "Kalman filter", dual of the "linear quadratic control" [2]*. (ii) By "dualizing" Theorem 244 of section 8.1.4 (with* $\alpha = 0$*), we obtain the theorem below.*

Let there be the Hamiltonian matrix

$$\boxed{\tilde{H} = \begin{bmatrix} A^T & -C^T \tilde{R}^{-1} C \\ -\tilde{Q} & -A \end{bmatrix}}$$

and let $\tilde{E}$ be a square root of $\tilde{Q}$. Also let,

$$\left\{ \tilde{\nu}_i \triangleq \begin{bmatrix} \tilde{\xi}_i \\ \tilde{\eta}_i \end{bmatrix}, 1 \leq i \leq n \right\}$$

be a set of generalized linearly independent eigenvectors of $\tilde{H}$, where $\tilde{\nu}_i$ is associated with the eigenvalue $\tilde{\lambda}_i$ of $\tilde{H}$. Last, let

$$\tilde{\eta} = \begin{bmatrix} \tilde{\eta}_1 & \dots & \tilde{\eta}_n \end{bmatrix}, \quad \tilde{\xi} = \begin{bmatrix} \tilde{\xi}_1 & \dots & \tilde{\xi}_n \end{bmatrix}$$

and, supposing that $\tilde{\xi}$ is invertible, let

$$\tilde{P} = \tilde{\eta}\,\tilde{\xi}^{-1}. \tag{9.17}$$

THEOREM 279.– *(i) The matrix $\tilde{P}$ defined by* (9.17) *is a solution of the algebraic Riccati equation* (9.16). *(ii) The following conditions (a) and (b) are equivalent: (a) there exists a matrix $\tilde{P}$ of the form* (9.17) *such that $A - \tilde{K}\,C$ is a stability matrix (this matrix $\tilde{P}$, which is then unique and satisfies $\tilde{P} \geq 0$, is obtained by choosing the n eigenvalues $\tilde{\lambda}_i$ in the left half-plane); (b) (C, A) is detectable and $\tilde{H}$ has no imaginary eigenvalues. (iii) Suppose* from here onward *that (C, A) is detectable. For Condition (a) or (b) to be satisfied, it is sufficient that the following condition holds: (c) $\left(A, \tilde{E}\right)$ is stabilizable. (iv) Suppose Condition (c) holds. Then, there exists a unique solution $\tilde{P}$ to the algebraic Riccati equation* (9.16) *for which $A - \tilde{K}\,C$ is a stability matrix; $\tilde{P}$ is also the unique non-negative definite solution to* (9.16). *(v) If, in addition, $\left(A, \tilde{E}\right)$ is controllable, then $\tilde{P} > 0$.*

The so-called "LTR" method[2] consists of choosing the matrix $\tilde{Q}$ as a function of a real number $q \geq 0$ (it is therefore preferable to denote this matrix by $\tilde{Q}_q$) according to the relation

$$\boxed{\tilde{Q}_q = \tilde{Q}_0 + q^2\, B\, V\, B^T} \tag{9.18}$$

where $V \in \mathbb{R}^{m \times m}$, $V > 0$, and of letting q tend to $+\infty$. We have first the result below, which is a direct consequence of ([119], section 3.10, Theorem 3.6):

LEMMA 280.– *(i) Let $\tilde{E}_0$ be a square root of $\tilde{Q}_0$. If $\left(A, \tilde{E}_0\right)$ is stabilizable (resp., controllable), then so is $\left(A, \sqrt{\tilde{Q}_q}\right)$ for any $q \geq 0$. (ii) If (A, B) is stabilizable (resp., controllable), then so is $\left(A, B\,\sqrt{V}\right)$.*

Now consider the following hypothesis $(\mathbf{H_{LTRu}})$,[3] where $\tilde{E}_0$ is a square root of $\tilde{Q}_0$ and $G(s) = C\,(s\,I_n - A)^{-1}\,B$:

$(\mathbf{H_{LTRu}})$: (a) The state-space system $\mathbf{\Sigma} = \{A, B, C\}$ is minimal; (b) $A - B\,K$ is a stability matrix; (c) $\left(A, \tilde{E}_0\right)$ is controllable; (d) $G(s) \in \mathbb{R}(s)^{m \times m}$;

2. LTR : acronym of "Loop Transfer Recovery" (see Remark 283 below).

3. Hypothesis $(\mathbf{H_{LTRu}})$ is stronger than Hypothesis $(\mathbf{H_{LTR}})$ usually considered (see Remark 283 below); it guarantees uniform – instead of simple – convergence on the compact subsets of $\bar{\mathbb{C}}_+$.

(e) $\text{rk}_{\mathbb{R}(s)}\, G(s) = m$; (f) $G(s)$ has no MacMillan zeros in the closed right half-plane $\bar{\mathbb{C}}_+$.

Assuming that Conditions (a) and (c) of Hypothesis $(\mathbf{H_{LTRu}})$ hold, the algebraic Riccati equation (9.16) (where $\tilde{Q} = \tilde{Q}_q$) admits, for any $q \geq 0$, a unique non-negative definite solution (this solution is denoted by $\tilde{P}_q$ in what follows) and $\tilde{P}_q > 0$ (according to Theorem 279 and Lemma 280). Let $\tilde{K}_q = \tilde{P}_q\, C^T\, \tilde{R}^{-1}$ be the corresponding observer gain.

The transfer matrix of the open-loop at the input of $\mathbf{\Sigma}$, usually denoted by $L_i(s)$, will be denoted by $L_q(s)$ in what follows just to clearly indicate its dependence with respect to the parameter q; it is given by

$$L_q(s) = H_q(s)\, G(s)$$

where $H_q(s)$ is the transfer matrix $H(s)$ satisfying equation (9.7) (see section 4.1.2 where $H(s)$ and $G(s)$ are denoted by $K(s)$ and $P(s)$, respectively). The *input* sensitivity function (or matrix) is thus given by

$$S_q(s) = (I_m + L_q(s))^{-1}.$$

The transfer matrix $L_q(s)$ is to be distinguished from what we would obtain with a state feedback control, which is

$$L_{sf}(s) = K\Phi(s)\, B$$

where $\Phi(s) = (s\, I_n - A)^{-1}$, and to which corresponds the input sensitivity function (or matrix)

$$S_{sf}(s) = (I_m + L_{sf}(s))^{-1}.$$

We have the following theorem, which describes the "uniform LTR effect":

THEOREM 281.– *Let Hypothesis* $(\mathbf{H_{LTRu}})$ *be in force. Then, for any compact set* $\mathbf{K} \subset \bar{\mathbb{C}}_+$, $S_q(s) - S_{sf}(s) \to 0$ *uniformly with respect to* $s \in \mathbf{K}$ *as* $q \to +\infty$.

PROOF. 1) For any $q > 0$, the algebraic Riccati equation (9.16) can be written in the form

$$\frac{\tilde{P}_q}{q^2}\, A^T + A\, \frac{\tilde{P}_q}{q^2} - \frac{\tilde{P}_q}{q^2}\, C^T \left(\frac{\tilde{R}}{q^2}\right)^{-1} C\, \frac{\tilde{P}_q}{q^2} + \frac{\tilde{Q}_0}{q^2} + \left(\sqrt{V} B^T\right)^T \left(\sqrt{V} B^T\right) = 0. \tag{9.19}$$

Condition (a) of Hypothesis $(\mathbf{H_{LTRu}})$ implies that $\left(A^T, C^T\right)$ is controllable and $\left(\sqrt{V}\, B^T, A^T\right)$ is observable (according to Corollary 153 of section 7.1.4 and Lemma

280(ii) above). According to Remark 249(ii) of section 8.1.4 (with $\rho = \frac{1}{q^2}$ and $P_\rho = \frac{\tilde{P}_q}{q^2}$),

$$\frac{\tilde{P}_q}{q^2} \to 0 \text{ as } q \to +\infty.$$

Therefore, as $q \to +\infty$,

$$\left(\frac{\tilde{P}_q}{q^2}\right) C^T \left(\frac{\tilde{R}}{q^2}\right)^{-1} C \left(\frac{\tilde{P}_q}{q^2}\right) \to B\, V\, B^T.$$

According to (9.15), $\tilde{K}_q$ is therefore such that

$$\frac{1}{q^2}\, \tilde{K}_q\, \tilde{R}\, \tilde{K}_q^T \to B\, V\, B^T$$

and $\operatorname{rk} \tilde{K}_q = m$ because $\tilde{P}_q > 0$. From Corollary 586 (section 13.5.7), there exists for any $q > 0$ an orthogonal matrix $U_q \in \mathbb{R}^{m \times m}$ and a matrix $\epsilon_1(1/q) \in \mathbb{R}^{n \times m}$ such that

$$\sqrt{\tilde{R}}\, \tilde{K}_q^T = q \left(U_q \sqrt{V}\, B^T + \sqrt{\tilde{R}}\, \epsilon_1(1/q)^T\right)$$

and $\epsilon_1(1/q) \to 0$ as $q \to +\infty$. Putting $W_q = \sqrt{V}\, U_q^T \sqrt{\tilde{R}^{-1}}$, it follows that

$$\tilde{K}_q = q\, (B\, W_q + \epsilon_1(1/q)). \tag{9.20}$$

2) Let $\Psi(s) = (s\, I_n - A + B\, K)^{-1}$. We have

$$H_q(s) = K\, \Psi(s) \left(I_m + \tilde{K}_q\, C\, \Psi(s)\right)^{-1} \tilde{K}_q$$

and according to Lemma 520 (section 13.3.3)

$$H_q(s) = K\, \Psi(s)\, \tilde{K}_q \left(I_m + C\, \Psi(s)\, \tilde{K}_q\right)^{-1} = K\, \Psi(s)\, J(q, s) \tag{9.21}$$

where

$$J(q, s) = \frac{\tilde{K}_q}{q} \left(\frac{I_m}{q} + C\, \Psi(s)\, \frac{\tilde{K}_q}{q}\right)^{-1}. \tag{9.22}$$

From (9.20),

$$J(q, s) = (B\, W_q + \epsilon_1(1/q)) \left(\frac{I_m}{q} + C\, \Psi(s)\, (B\, W_q + \epsilon_1(1/q))\right)^{-1},$$

and hence, with $G_c(s) = C\,\Psi(s)\,B$,

$$\begin{aligned} J(q,s) &= B\,G_c^{-1}(s)\left(I_m + (G_c(s)\,W_q)^{-1}\left(C\,\Psi(s)\,\epsilon_1(1/q) + \frac{I_m}{q}\right)\right) \\ &\quad +\epsilon_1(1/q)\,G_c^{-1}(s)\,. \\ &\quad .\left(I_m + (G_c(s)\,W_q)^{-1}\left(C\,\Psi(s)\,\epsilon_1(1/q) + \frac{I_m}{q}\right)\right). \end{aligned} \tag{9.23}$$

Let $\mathbf{K} \subset \bar{\mathbb{C}}_+$ be a compact set. According to Conditions (d), (e) and (f) of Hypothesis $(\mathbf{H_{LTRu}})$, $G_c(s)$ is invertible in $\mathbf{K}$ (see Chapter 8, Exercise 263) and is bounded in this set. From Condition (b) of Hypothesis $(\mathbf{H_{LTRu}})$, the transfer matrix $\Psi(s)$ is also bounded in $\mathbf{K}$. As a result, according to (9.23), $J(q,s) = B\,G_c^{-1}(s)\,(I_m + \epsilon_2(1/q,s))$, where $\epsilon_2(1/q,s) \rightarrow 0$ uniformly with respect to $s \in \mathbf{K}$ as $q \rightarrow +\infty$. Using this expression in (9.21), we obtain (by an identical rationale)

$$H_q(s) = (I_m + \epsilon_3(1/q,s))\,K\,\Psi(s)\,B\,G_c^{-1}(s)$$

where $\epsilon_3(1/q,s) \rightarrow 0$ uniformly with respect to $s \in \mathbf{K}$ as $q \rightarrow +\infty$. On the other hand,

$$\Phi(s)\,B = \Psi(s)\left(I_m - B\,K\,(s\,I_n - A + B\,K)^{-1}\right)^{-1},$$

and hence from Lemma 520 (section 13.3.3),

$$\Phi(s)\,B = \Psi(s)\,B\,(I_m - K\,\Psi(s)\,B)^{-1}.$$

Therefore,

$$L_q(s) = (I_m + \epsilon_3(1/q,s))\,K\,\Psi(s)\,B\,(I_m - K\,\Psi(s)\,B)^{-1}$$

and hence

$$S_q(s) = (I_m - K\,\Psi(s)\,B)\,(I_m + \epsilon_3(1/q,s)\,K\,\Psi(s)\,B)^{-1}.$$

In addition,

$$\begin{aligned} S_{sf}(s) &= (I_m + K\,\Phi(s)\,B)^{-1} \\ &= \left(I_m + K\,\Psi(s)\,B\,(I_m - K\,\Psi(s)\,B)^{-1}\right)^{-1} \\ &= I_m - K\,\Psi(s)\,B. \end{aligned}$$

Since the transfer matrix $K\,\Psi(s)\,B$ is bounded in $\mathbf{K}$, $S_q(s) \rightarrow S_{sf}(s)$ uniformly with respect to $s \in \mathbf{K}$ as $q \rightarrow +\infty$. ■

REMARK 282.– *It has not been proven in the above theorem that* $S_q(s) \to I_m$ *as* q *and* $|s|$ *both tend to* $+\infty$ *(with* $s \in \bar{\mathbb{C}}_+$*). We thus cannot assert that* $\|S_q - S_{sf}\|_\infty \to 0$ *as* $q \to +\infty$ *(the author does not, however, know of any counterexample to this property of which we may conjecture).*

REMARK 283.– *The "LTR" method, popularized by Doyle [38] from a theory due to Kwakernaak and Sivan [71], can be described as follows (see [93] for a rigorous proof): let there be Hypothesis* $(\mathbf{H_{LTR}})$ *below: (a) The state-space system* $\{A, B, C\}$ *is stabilizable and detectable; (b)* $A - BK$ *is a stability matrix; (c)* $\left(A, \tilde{E}_0\right)$ *is stabilizable; (d)* $G(s) \in \mathbb{R}(s)^{p \times m}$ *; (e)* $\mathrm{rk}_{\mathbb{R}(s)} G(s) = m$*; (f)* $G(s)$ *has no MacMillan zeros in the closed right half-plane* $\bar{\mathbb{C}}_+$*. Under this hypothesis,* $L_q(s) \to L_{sf}(s)$ *as* $q \to +\infty$ *for any* $s \in \mathbb{C}$ *which is not a pole of system* $\{A, B, C\}$*.*

EXAMPLE 284.– *Consider the minimal system with transfer function* $G(s) = \frac{1}{s^4}$ *and its controllable canonical realization* $\{A, B, C\}$ *(see section 7.4.2). (1) Let* x *be a measured state. The gain matrix of the state feedback control for which the poles of the closed-loop system are placed at* $\{-1, -1, -1, -1\}$ *is:* $K = \begin{bmatrix} 4 & 6 & 4 & 1 \end{bmatrix}$*. This choice of the poles of the closed-loop system complies with Rule 111 (section 6.3.5). (2) Now suppose we have the state observer specified above with* $\tilde{Q}_0 = I_4$ *and* $V = 1$*. The Bode plots of the open-loop transfer functions are represented in Figure 9.2 in the following order (going from low to high in the high frequencies)*[4]*: with the state feedback/observer synthesis and* $q = 0$, $q = 10^2$, $q = 10^4$, $q = 10^6$*, and then with the complete state feedback. The corresponding Bode plots of the sensitivity functions are shown in Figure 9.3 with a reverse order in the low frequencies. One can see that* $\|S_q - S_{sf}\|_\infty \to 0$ *as* $q \to +\infty$ *(see Remark 282). The Bode plots of the transfer functions* $H_q(s)$ *of the controller, for the above-specified values of* q*, are shown in Figure 9.4, in the same order as in Figure 9.2. Let*

$$\omega_q = \arg\max_{\omega} |H_q(i\omega)|.$$

As $q \to +\infty$*, we see that* $|H_q(i\omega_q)| \to +\infty$ *and* $\omega_q \to +\infty$*. This phenomenon is one of the things that makes the analysis of the behavior of* $|H_q(i\omega) G(i\omega)|$ *difficult as both* q *and* ω *tend toward* $+\infty$*. Remember that a controller having a large gain in the high frequencies renders the control very sensitive to measurement noise (see section 4.2.6), which is to be avoided. It is therefore recommended to increase parameter* q *until a sufficient modulus margin is obtained, and then not to go beyond that value.*

4. The curves related to sections 9.1 and 9.2 are assembled at the end of section 9.2.

9.2. State feedback/observer synthesis with integral action

9.2.1. *Problem setting*

In this section, the control method by state feedback with integral action studied in section 8.2 is combined with a full-order observer. Consider the following system $\boldsymbol{\Sigma}$:

$$\left\{ \begin{aligned} \partial x &= A\,x + B\,u + d_1 \\ y &= C\,x + d_2 \\ z &= E\,y \end{aligned} \right. \tag{9.24}$$

where $x(t) \in \mathbb{R}^n$ is the state, $u(t) \in \mathbb{R}^m$ is the control, $y(t) \in \mathbb{R}^k$ is the measured vector, and $z(t) \in \mathbb{R}^p$ is the regulated variable; $d_1 \in \mathbb{R}^n$ and $d_2 \in \mathbb{R}^k$ are constant disturbances, unmeasured and unknown (see section 8.2.2). The matrices A, B and C are uncertain in practice, but matrix E is assumed to be perfectly known. The third line of (9.24) can, for example, signify that z is constituted of the first p rows of y and in that case $E = \left[\begin{array}{cc} I_p & 0 \end{array}\right]$.

Let $r \in \mathbb{R}^p$ be the reference signal, assumed to be *known*, and let the regulation error be

$$e = z - r.$$

Our aim in what follows is to design a control law for which Property (P) of section 8.2.2 is satisfied.

Let us write

$$z = E\,(C\,x + d_2) = H\,x + d_3$$

with

$$H = E\,C, \quad d_3 = E\,d_2.$$

The problem considered here would reduce to that of section 8.2.2 if the state were measured. Therefore, the same hypotheses as for this last problem are in force:

(H1) (A, B) is controllable;

(H2) the transfer matrix $G(s)$ of System $\{A, B, H\}$ is semiregular over $\mathbb{R}(s)$.

A supplementary step, in the present case, consists in reconstructing the state using the measured vector y, which leads us to add the hypothesis below:

(H3) (C, A) is observable.

9.2.2. *Algebraic solution*

We define the variables χ and υ as in equation (8.32) of section 8.2.4. We get state equation (8.33) with

$$F = \begin{bmatrix} A & 0 \\ H & 0 \end{bmatrix}, \quad G = \begin{bmatrix} B \\ 0 \end{bmatrix}.$$

Based on Hypotheses (H1) and (H2) and Proposition 253, we have the following:

PROPOSITION 285.– *(F, G) is stabilizable if and only if, the following two conditions hold: (i) $m \geq p$; (ii) $s = 0$ is not an invariant zero of system $\{A, B, H\}$. In that case, (F, G) is controllable.*

Assuming that Conditions (i) and (ii) of Proposition 285 hold, there exists a gain matrix $K \in \mathbb{R}^{m \times (n+p)}$ for which $F - G\,K$ is a stability matrix. More precisely, by choosing K in an appropriate manner, the eigenvalues of $F - G\,K$ can be placed in an arbitrary symmetric subset with $n+p$ elements of the left half-plane (see section 8.1.2, Theorem 230).

Let

$$\eta = \partial\, x$$

and consider the observer below, with aims at reconstructing η:

$$\partial\,\hat{\eta} = A\,\hat{\eta} + B\,\upsilon + \tilde{K}\,(\partial\,y - C\,\hat{\eta})\,. \tag{9.25}$$

In differentiating the first two equations of (9.24), we obtain

$$\begin{cases} \partial\,\eta = A\,\eta + B\,\upsilon, \\ \partial\,y = C\,\eta; \end{cases} \tag{9.26}$$

as a result, putting $\tilde{\eta} = \eta - \hat{\eta}$, we get from equations (9.25) and (9.26)

$$\partial\,\tilde{\eta} = \left(A - \tilde{K}\,C\right)\,\tilde{\eta}. \tag{9.27}$$

According to Hypothesis (H3), there exists a gain matrix $\tilde{K} \in \mathbb{R}^{n \times k}$ for which $A - \tilde{K}\,C$ is a stability matrix.

Now write the gain matrix K in the form

$$K = \begin{bmatrix} K_p & K_i \end{bmatrix},$$

$K_p \in \mathbb{R}^{m \times n}$, $K_i \in \mathbb{R}^{m \times p}$, and consider the control law

$$\upsilon = -K_p\,\hat{\eta} - K_i\,e. \tag{9.28}$$

Let

$$\xi = \begin{bmatrix} \eta \\ e \\ \tilde{\eta} \end{bmatrix}.$$

We obtain from equations (9.26), (9.11) and (9.28)

$$\partial \xi = \begin{bmatrix} A - B\,K_p & -B\,K_i & B\,K_p \\ H & 0 & 0 \\ 0 & 0 & A - \tilde{K}\,C \end{bmatrix} \xi \triangleq M\xi.$$

Since the matrix M is of the form

$$M = \begin{bmatrix} F - G\,K & * \\ 0 & A - \tilde{K}\,C \end{bmatrix},$$

we have

$$\mathrm{Sp}\,(M) = \mathrm{Sp}\,(F - G\,K)\,\dot{\cup}\,\mathrm{Sp}\left(A - \tilde{K}\,C\right), \tag{9.29}$$

which proves that M is a stability matrix.

Let

$$\hat{x}\,(t) = \int \hat{\eta}\,(t)\;dt, \quad \tilde{x} = x - \hat{x}.$$

By the same rationale as in the proof of Lemma 252 (section 8.2.4), we obtain the following result:

THEOREM 286.– *Using the control law* (9.28), *we have:*

$$\begin{aligned} \lim_{t\to+\infty} e\,(t) &= 0, \\ \lim_{t\to+\infty} x\,(t) &= const., \\ \lim_{t\to+\infty} \tilde{x}\,(t) &= const. \\ \lim_{t\to+\infty} u\,(t) &= const. \end{aligned}$$

The problem which remains to be solved is the implementation of the control law (9.28). Since the gain matrices K_p, K_i, and $\tilde{K}$ are constant,[5] we obtain by integrating

5. The case where the gains are scheduled is quite different [80]. Then, equations (9.28) and (9.25) must be kept and the control u must be calculated by integrating v, using an integrator placed just at the input of the control system.

equations (9.28) and (9.25)

$$u(t) = -K_p \hat{x}(t) - K_i \int e(t)\, dt + c, \quad (9.30)$$

$$\partial \hat{x} = A\hat{x} + B\,u + \tilde{K}\,(y - C\,\hat{x}) + c' \quad (9.31)$$

where c and c' are integration constants.

REMARK 287.– *(i) Taking* $c' = 0$*, equation* (9.31) *is the classic expression of an observer of state* x*.*

(ii) The control law (9.30) *is nothing but the "control by state feedback and integral action"* (8.35) *of section 8.2.4, where* x *has been replaced by* $\hat{x}$*. Theorem 286 and equality* (9.29) *can be considered as an extension of the "separation principle" (Theorem 276, section 9.1.2).*

The diagram of the closed-loop system is shown in Figure 9.1. Note that the regulation error $e = z - r$ is not reconstructed (neither is the controlled variable z).

9.2.3. *Extension of the LTR method*

Suppose we choose the gain matrix K using the (α) method or the LQR method of section 8.1.4. Then, according to Theorem 258 (section 8.2.4), we obtain using the control law (8.35) (assuming that the state is known) a closed-loop system whose input sensitivity function S_i satisfies $\|S_i\|_\infty = 1$. Like in section 9.1.4, the question on hand now is to know how to design the state observer (9.25) (or, in an equivalent manner, the state observer (9.31)) in a way as not to degrade (too much) this robustness property. We are going to show that the LTR method, detailed in section 9.1.4, extends to the situation now considered. The observer gain now depends on parameter $q \geq 0$ and is denoted by $\tilde{K}_q$.

Let us first determine the transfer matrix $\begin{bmatrix} H_{q,y}(s) & H_{q,e}(s) \end{bmatrix}$ of the controller with input $\begin{bmatrix} y \\ e \end{bmatrix}$ and output u, assuming that the external signals (i.e. the disturbances d_1 and d_2 and the reference signal r) are zero. We have

$$\left(\partial I_n - A + \tilde{K}_q C + B K_p\right) \hat{\eta} = \tilde{K}_q \partial y - B K_i e.$$

As a result,

$$H_{q,y}(s) = K_p \left(s I_n - A + B K_p + \tilde{K}_q C\right)^{-1} \tilde{K}_q,$$

$$H_{q,e}(s) = \left[I_m - K_p \left(s I_n - A + B K_p + \tilde{K}_q C\right)^{-1} B\right] \frac{K_i}{s}.$$

We can also consider that only the measured vector y goes into the controller (taking into account the relation $z = E\,y$); this controller thus has as a transfer matrix

$$\boxed{H_q(s) = H_{q,y}(s) + H_{q,e}(s)\ E}.$$

Suppose Hypothesis $(\mathbf{H_{LTR}})$ of Remark 282 (section 9.1.4) is in force and let $\mathbf{P}$ be the locus of the controller poles (i.e. $\mathrm{Sp}\left\{A - B\,K_p - \tilde{K}_q\,C\right\} \dot{\cup} \{0\}$) when q varies from 0 to $+\infty$.

THEOREM 288.– *Let $q \to +\infty$. (i) For any $s \in \bar{\mathbb{C}}_+ \setminus \left(\mathbf{P} \cap \bar{\mathbb{C}}_+\right)$, $s\,H_{q,e}(s) \to K_i$. (ii) For any $s \in \mathbb{C}$ which is not an eigenvalue of A,*

$$H_{q,y}(s)\,C\,(s\,I_n - A)^{-1}\ B \to K_p\,(s\,I_n - A)^{-1}\,B.$$

PROOF. For the sake of simplicity, assume that Hypothesis $(\mathbf{H_{LTRu}})$ is in force (if it is the replaced Hypothesis $(\mathbf{H_{LTR}})$, which is less restrictive, we must make use of the approach developed in [93]). (i): We have from equation (9.20)

$$\frac{\tilde{K}_q}{q}\,W_q^{-1} \sim B.$$

Let $\Psi(s) = (s\,I_n - A + B\,K_p)^{-1}$ and $\Delta_q(s) = \left(s\,I_n - A + B\,K_p + \tilde{K}_q\,C\right)^{-1}\ B$. For any $s \in \bar{\mathbb{C}}_+ \setminus \left(\mathbf{P} \cap \bar{\mathbb{C}}_+\right)$, s is not a pole of $\Psi(s)$ and $\left(I_n + \tilde{K}_q\,C\,\Psi(s)\right)^{-1}$ is bounded. Therefore, we have the following equivalents as $q \to +\infty$:

$$\begin{aligned}
\Delta_q(s) &= \Psi(s)\left(I_n + \tilde{K}_q C\,\Psi(s)\right)^{-1}\ B \\
&\sim \Psi(s)\left(I_m + \tilde{K}_q C\Psi(s)\right)^{-1}\frac{\tilde{K}_q}{q}W_q^{-1} \\
&\sim \Psi(s)\,\tilde{K}_q\left(I_m + C\Psi(s)\,\tilde{K}_q\right)^{-1}\frac{W_q^{-1}}{q}.
\end{aligned}$$

Using the same rationale as in the proof of Theorem 281, we thus obtain

$$\Delta_q(s) \sim \frac{1}{q}\,\Psi(s)\ B\,(C\Psi(s)\,B)^{-1}\ W_q^{-1}$$

where $W_q^{-1} = \sqrt{\tilde{R}}\,U_q\,\sqrt{V^{-1}}$. The last quantity is bounded, and hence $\Delta_q(s) \to 0$ as $q \to +\infty$, and $s\,H_{q,e}(s) \to K_i$. The proof of (ii) can clearly be derived from that of Theorem 281. ■

EXAMPLE 289.– *Consider again the system in Example 257 (section 8.2.4). The gain matrices K_p and K_i have already been calculated, but we now assume that the state x is not measured: the measured vector is $y = C\,x$, and $z = y$. Let $G(s) = C\,\Phi(s)\,B$, where $\Phi(s) = (s\,I_4 - A)^{-1}$. Using the state feedback/observer synthesis with integral action, the open-loop transfer matrix at* input *of* Σ *is*

$$\boxed{L_q(s) = H_q(s)\,G(s)}$$

while using the control law (8.35) *it is equal to*

$$\boxed{L_{sf}(s) = K_p \Phi(s)\,B + \frac{K_i}{s} G(s)}.$$

The matrix $\tilde{Q}_0$ is chosen to be the identity, as well as $\tilde{R}$. The two singular values of $L_q(i\omega)$ are plotted as a function of ω in Figure 9.5 for $q = 100$ (– -)[6]*, $q = 10^4$ (..)*[7]*, and $q = 10^6$ (- -). The two singular values are plotted in the same Figure (–). The two singular values of $S_q(i\omega) = (I_2 + L_q(i\omega))^{-1}$ and those of $S_{sf}(i\omega) = (I_2 + L_{sf}(i\omega))^{-1}$ are plotted in Figure 9.6 with the same conventions. We observe that as $q \to +\infty$, $S_q(i\omega) \to S_{sf}(i\omega)$ for any ω; in addition, in the case considered, $\|S_q - S_{sf}\|_\infty \to 0$ (see Remark 282 in section 9.1.4). With $q = 10^2$, the modulus margin is already sufficient (about -3 dB: see section 4.2.9), thus this value of q is retained. It is now relevant to verify the time-domain qualities of the feedback system. The following events are simulated: (i) a step command of amplitude 1 for the first variable and of amplitude -1 for the second, at instant $t = 1$; (ii) a step disturbance of amplitude 0.2 on the first output at $t = 5$ and of amplitude -0.3 on the second output at $t = 7$. The first output (–) and the second one (- -) as a function of time are shown in Figure 9.9. The corresponding controls are shown in Figure 9.10, with the same conventions. The responses to step commands are identical to those obtained with a control by state feedback and integral action (see Remark 277, section 9.1.4). It is not the same for the responses to disturbances, which nevertheless are satisfactory with the designed control. If we chose a much larger value of q (to increase the modulus margin), it would result in a controller having larger gains (the two singular values of $H_q(i\omega)$ are plotted as a function of ω in Figure 9.7 for $q = 100$, $q = 10^4$, and $q = 10^6$ with the already adopted conventions), from which, as a response to disturbances, we have on the one hand a control having too large amplitudes, and on the other hand a strong coupling between the two outputs. Another question that still remains to be answered is whether the* output *modulus margin is suitable for the value retained for parameter q (i.e. $q = 100$). Indeed, the* output *sensitivity function is $(I + G\,H_q)^{-1}$; it is different from the* input *sensitivity function $(I + L_q)^{-1} = (I + H_q\,G)^{-1}$ considered*

6. Alternate long and short dashes.

7. Dotted.

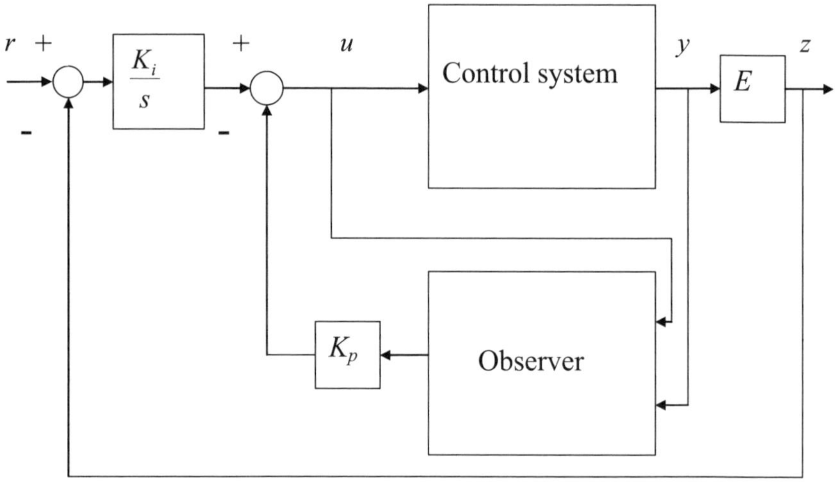

Figure 9.1. *Closed-loopsystem*

up till now. The singular values of these two sensitivity functions are shown in Figure 9.8: that at the input of Σ *(..) and that at its output (- -). The output modulus margin of* Σ*, about* -8 *dB, is clearly not as good as that at the input modulus margin, but for an MIMO system this value is still reasonable.*

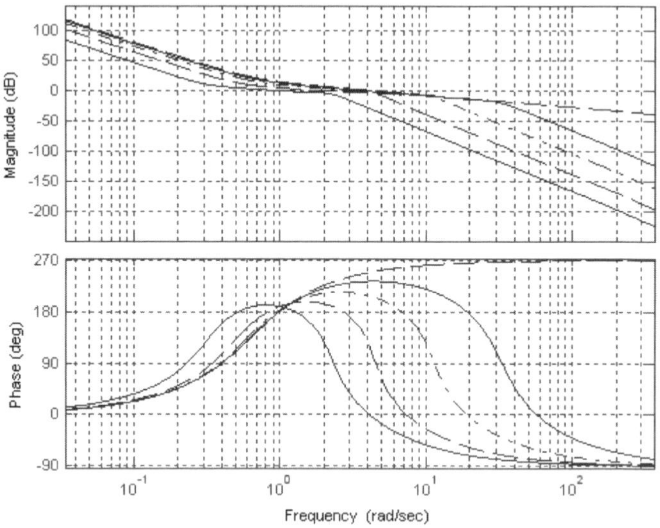

Figure 9.2. *LTR effect (1)*

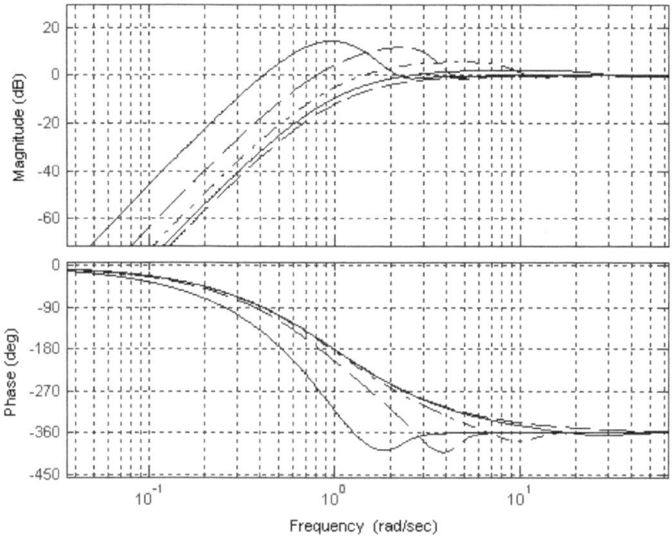

Figure 9.3. *LTR effect (2)*

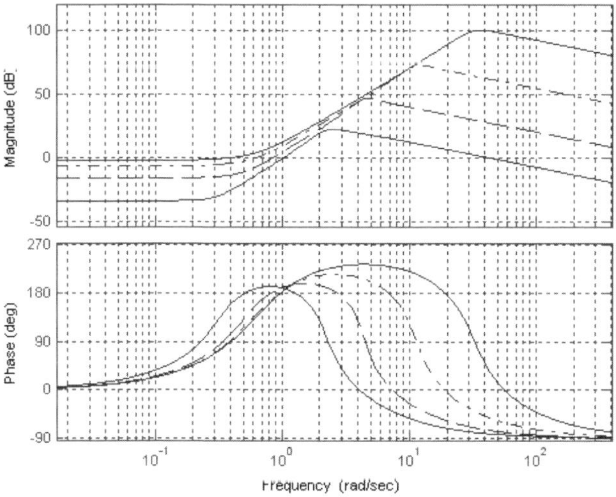

Figure 9.4. *LTR effect (3)*

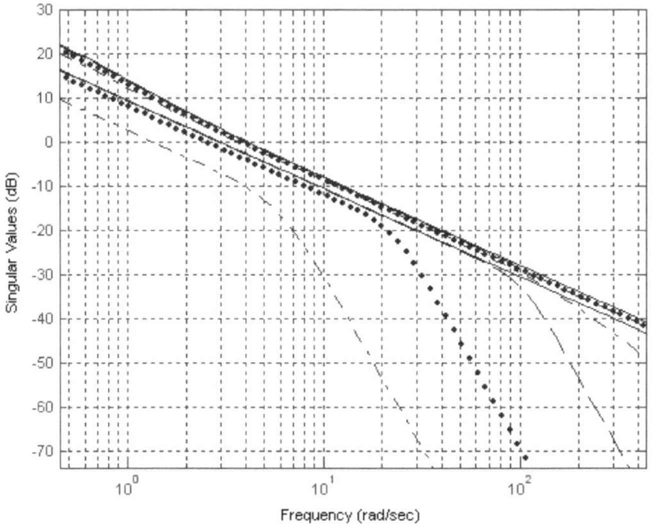

Figure 9.5. *Singular values of Li*

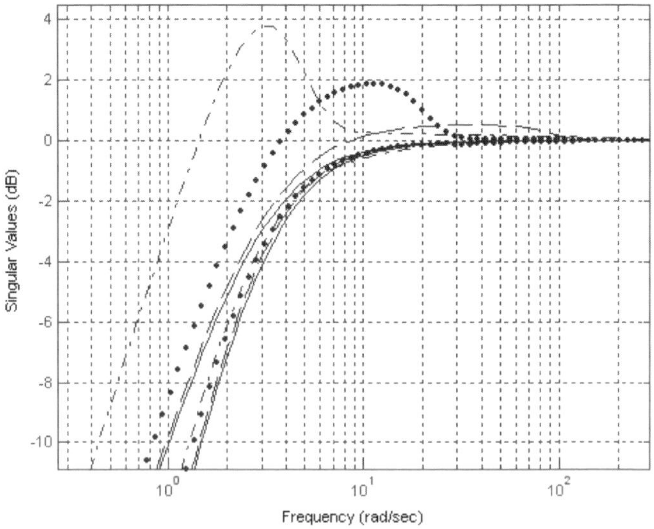

Figure 9.6. *Singular values (input sensitivity function)*

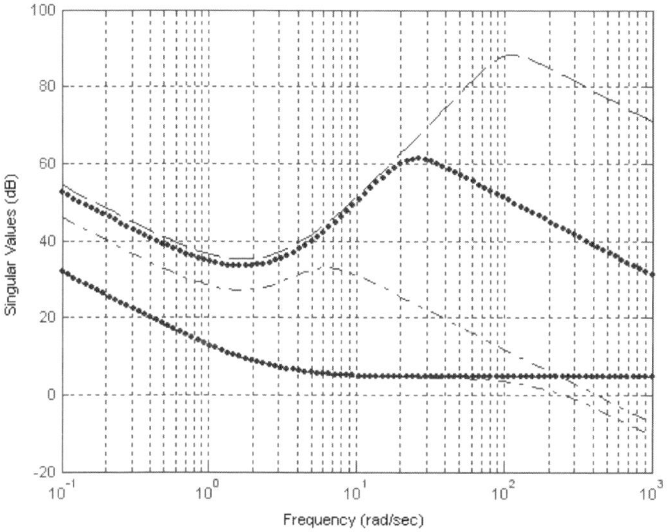

Figure 9.7. *Singular values (controller)*

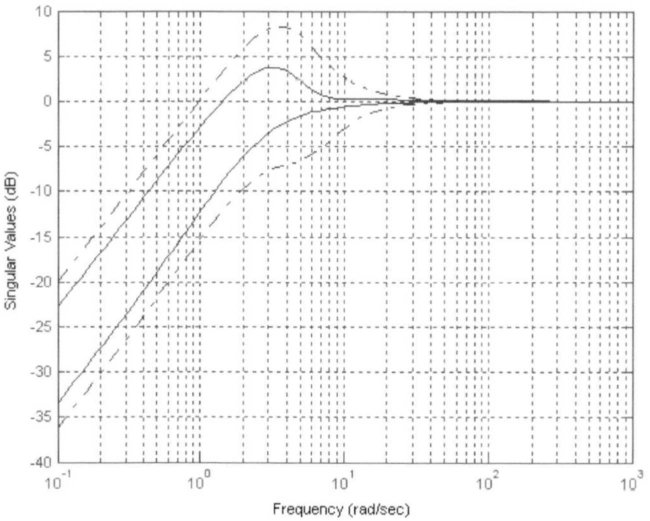

Figure 9.8. *Input and output sensitivity functions*

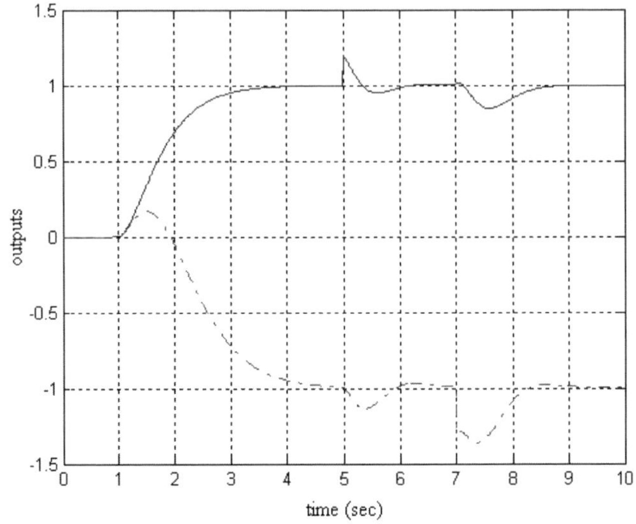

Figure 9.9. *Time responses (Example 289)*

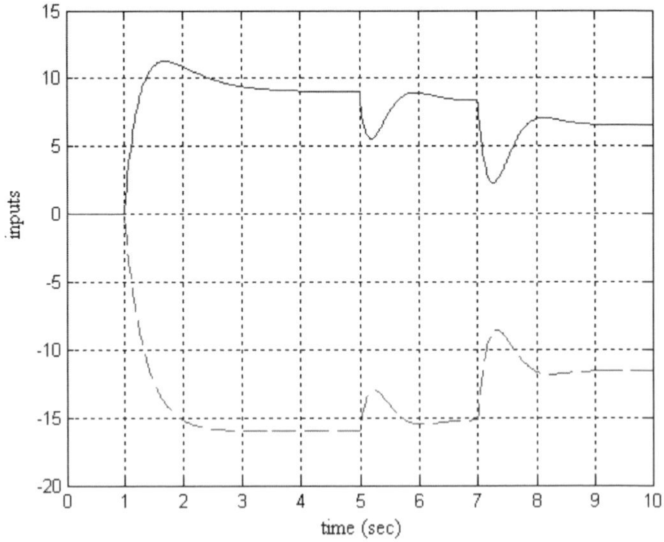

Figure 9.10. *Time responses (Example 289)*

9.3. *General theory of observers

The general theory of observers is essentially due to Luenberger [87], [88].

9.3.1. *Reduced-order observer*

Consider the state-space system $\{A, B, C\}$:

$$\begin{cases} \dot{x} = A\,x + B\,u \\ y = C\,x \end{cases} \tag{9.32}$$

where $x(t) \in \mathbb{R}^n$, $u(t) \in \mathbb{R}^m$, and $y(t) \in \mathbb{R}^p$. The matrix C is assumed to be left-regular.

REMARK 290.– *What follows easily extends to the case where the above state-space system has a direct term, i.e. where the second equation of* (9.32) *is* $y = C\,x + D\,u$. *The only thing to do is to replace* y *by* $\bar{y} = y - D\,u$ *everywhere.*

If the signal y is very noisy, it is useful for the controller to have a small gain in high frequencies (see section 4.2.6). This is what we get using a full-order observer, since then the transfer matrix $H(s)$ given by equation (9.7) (with $D = 0$) is such that $H(i\omega) \to 0$ as $\omega \to +\infty$.

Suppose now that the signal y is less noisy and that the components of y are part of the components of x. We can always come back to this case. Indeed, let $J \in \mathbb{R}^{n \times (n-p)}$ be a matrix such that

$$T = \begin{bmatrix} J \\ C \end{bmatrix}$$

is invertible. Putting $X = T\,x$, we have obviously $y = \begin{bmatrix} 0_{n \times (n-p)} & I_p \end{bmatrix} X$.

It is thus assumed in what follows that the above algebraic operation has already been made, and therefore that the state x is of the form

$$x = \begin{bmatrix} x_r \\ y \end{bmatrix} \tag{9.33}$$

where $x_r(t) \in \mathbb{R}^{n-p}$.

REMARK 291.– *Nevertheless, two cases are to be distinguished: the case where the state* x *is "by nature" of the form of equation* (9.33) *and the case where this structure is obtained through the above algebraic operation. The latter does not produce an exact result if the matrix* C *is uncertain; the "reduced-order observer" that will be discussed can then lead to a not very robust control law.*

The following decomposition of system (9.32) corresponds to decomposition (9.33) of the state:

$$\begin{bmatrix} \dot{x}_r \\ \dot{y} \end{bmatrix} = \begin{bmatrix} A_r & B_r \\ C_r & D_r \end{bmatrix} \begin{bmatrix} x_r \\ y \end{bmatrix} + \begin{bmatrix} G_r \\ H_r \end{bmatrix} u. \tag{9.34}$$

Let

$$y_r = \dot{y} - D_r\, y - H_r\, u. \tag{9.35}$$

We have

$$\left\{ \begin{array}{rl} \dot{x}_r = & A_r\, x_r + B_r\, y + G_r\, u \\ y_r = & C_r\, x_r\,. \end{array} \right. \tag{9.36}$$

LEMMA 292.– *The pair (C_r, A_r) is observable (resp., detectable) if, and only if, the pair (C, A) has the same property.*

PROOF. We have

$$\begin{bmatrix} s\,I_n - A \\ C \end{bmatrix} = \begin{bmatrix} s\,I_{n-p} - A_r & -B_r \\ -C_r & s\,I_p - D_r \\ 0 & I_p \end{bmatrix} \sim \begin{bmatrix} s\,I_{n-p} - A_r & 0 \\ C_r & 0 \\ 0 & I_p \end{bmatrix},$$

from which we deduce the stated result according to Proposition 170 and Theorem 169 (section 7.2.3), as well as Definition 185 (section 7.3). ■

In the rest of this paragraph, (C, A) is assumed to be observable.

(i) First, let us assume that the signal y_r is known (which, in reality, is not the case since its definition (9.35) involves the *derivative* of y: see section 2.5.3). In this case, a full-order observer for the system (9.36) is of the form (see equation (9.3))

$$\partial\, \hat{x}_r = A_r\, \hat{x}_r + B_r\, y + \tilde{K}_r\, \left(y_r - C_r\, \hat{x}_r\right) + G_r\, u \tag{9.37}$$

where $\tilde{K}_r \in \mathbb{R}^{(n-p)\times p}$ is such that $A_r - \tilde{K}_r\, C_r$ will be a stability matrix.

(ii) What is left to do now is to replace $\hat{x}_r$ by a variable z that is governed by a differential equation from which the variable y_r is eliminated so as to avoid the differentiation of y. Therefore, let $z = \hat{x}_r - \tilde{K}_r\, y$. We obtain

$$\begin{aligned} \dot{z} \;=\;& \left(A_r - \tilde{K}_r\, C_r\right) z + \left(B_r - \tilde{K}_r\, D_r + A_r\, \tilde{K}_r - \tilde{K}_r\, C_r\, \tilde{K}_r\right) y \\ & + \left(G_r - \tilde{K}_r\, H_r\right) u \end{aligned} \tag{9.38}$$

and the reconstructed state $\hat{x}$ furnished by the "reduced-order observer" is

$$\begin{aligned}\hat{x} &= \begin{bmatrix} \hat{x}_r \\ y \end{bmatrix} = \begin{bmatrix} z + \tilde{K}_r\, y \\ y \end{bmatrix} \\ &= \begin{bmatrix} I_{n-p} \\ 0 \end{bmatrix} z + \begin{bmatrix} \tilde{K}_r \\ I_p \end{bmatrix} y. \end{aligned} \tag{9.39}$$

Let $\tilde{x} = x - \tilde{x}$ be the observation error.

PROPOSITION 293.– *We have $\tilde{x}(t) \to 0$ as $t \to +\infty$.*

PROOF. We have $\tilde{x} = \begin{bmatrix} \tilde{x}_r \\ 0 \end{bmatrix}$ with $\tilde{x}_r = x_r - \hat{x}_r$, and according to (9.36) and (9.37), $\tilde{x}_r(t) \to 0$ as $t \to +\infty$. ■

REMARK 294.– *The observer* (9.38), (9.39) *is said to be "of reduced-order" because it is of order $n - p$, while the "full-order observer"* (9.3) *is of order n. When $n = 1$, one can show that the observer* (9.38), (9.39) *is "minimal", in the sense that there does not exist an observer of order less than $n-1$, the poles of which can be arbitrarily chosen (see [119], section 3.8).*

9.3.2. *General formalism*

We now propose to reconstruct a vector $\mu - M\,x$ (where $M \subset \mathbb{R}^{q\times n}$ is a lcft-regular matrix). The control of system (9.32) is supposed to be of the form

$$u = -L\,\hat{\mu}$$

and we write

$$K = L\,M.$$

Remark 290 remains valid.

The most general form possible for the observer is

$$\begin{cases} \dot{z} = F\,z + G\,y + H\,u \\ \hat{\mu} = E\,z + J\,y \end{cases}$$

where z is a vector with r components.

Let $T \in \mathbb{R}^{r \times n}$ be a left-regular matrix and $\xi = z - T\,x$. After some elementary calculations, we obtain the following differential equation for the closed-loop system:

$$\begin{bmatrix} \dot{x} \\ \dot{\xi} \end{bmatrix} = \begin{bmatrix} A - B\,K & -B\,L\,E \\ 0 & F \end{bmatrix} \begin{bmatrix} x \\ \xi \end{bmatrix}, \tag{9.40}$$

$$H = T\,B, \tag{9.41}$$

$$K = L\,E\,T + J\,C, \tag{9.42}$$

$$G\,C = T\,A - F\,T. \tag{9.43}$$

Equality (9.41) poses no constraint on system (9.32) ; the following is clear:

PROPOSITION 295.– *The closed-loop system is stable if and only if $A - B\,K$ and F are stability matrices, for its poles are the elements of* $\mathrm{Sp}\,(A - B\,K) \dot{\cup} \mathrm{Sp}\,(F)$.

Let us examine now under what condition the following property $(\mathbf{P})$ will be true:

$(\mathbf{P})$: The symmetric set $\mathrm{Sp}\,(A - B\,K) \dot{\cup} \mathrm{Sp}\,(F)$ can be arbitrarily chosen in the left half-plane, and equalities (9.42) and (9.43) are satisfied.

The two cases considered below are the full-order observer and the reduced-order observer.

Case of a full-order observer

In this case, $M = E = T = I_n$ and $J = 0$. Writing $G = \tilde{K}$, equation (9.43) can be written as $A - \tilde{K}\,C = F$. As a result, Property $(\mathbf{P})$ is satisfied if and only if, system $\{A, B, C\}$ is minimal.

Case of a reduced-order observer

Assuming that system (9.32) is put in the form (9.34) with $C = \begin{bmatrix} 0_{n \times (n-p)} & I_p \end{bmatrix}$, we have $M = I_n$, $T = \begin{bmatrix} I_{n-p} & -\tilde{K}_r \end{bmatrix}$, $F = A_r - \tilde{K}_r\,C_r$ and

$$E = \begin{bmatrix} K \\ 0 \end{bmatrix}, \quad J = K \begin{bmatrix} \tilde{K}_r \\ I_p \end{bmatrix}.$$

Equalities (9.42) and (9.43) are thus satisfied. According to Lemma 292 (section 9.3.1), Property $(\mathbf{P})$ is satisfied if and only if, $\{A, B, C\}$ is minimal.

REMARK 296.– *(i) According to Proposition 295, the "separation principle" (Theorem 276, section 9.1.2) is still valid when a reduced-order observer is used. (ii) Consider the RST controller of section 6.4.5 with $p = \delta_0 = 0$ (using the notation of the cited paragraph). The characteristic polynomial $A_{cl}\,(\partial)$ is then of degree*

$2n-1$. *The obtained controller is just another state feedback/observer synthesis (see section 9.1.3) when the observer used is of reduced-order (and is, in fact,* minimal *according to Remark 294 of section 9.3.1). (iii) The "LTR method" can be extended to the case where the observer used is of reduced-order [28].*

9.4. Exercises

EXERCISE 297.– *Let $\{A, B, C\}$ be defined as in Exercise 264 (section 8.4). (i) Determine a full-order observer for this system, having a double pole at -5. (ii) How can you justify this choice?*

EXERCISE 298.– *We consider the system of two tanks as defined in Exercise 266 (section 8.4). (i) Determine a full-order observer (assuming that the measured variable is the discharge of the water leaving the second tank) that has a double pole at -0.5. (ii) How can you justify this choice?*

EXERCISE 299.– *Let there be the state-space system $\mathbf{\Sigma} = \{A, B, C\}$ where*

$$A = \begin{bmatrix} -1 & 2 \\ 1 & 0 \end{bmatrix}, \quad B = \begin{bmatrix} 1 \\ 0 \end{bmatrix}, \quad C = \begin{bmatrix} 1 & \beta \end{bmatrix}.$$

(i) Determine the poles of $\mathbf{\Sigma}$ and study its stability. (ii) Calculate the transfer function of $\mathbf{\Sigma}$, as well as its transmission poles and zeros, in function of β. (iii) Is $\mathbf{\Sigma}$ controllable, and for which value(s) of β is it non-observable? (iv) Determine the invariant zeros of $\mathbf{\Sigma}$. (v) Determine a state feedback control $u = -K\,x + k_0\,r$ (where $K = \begin{bmatrix} k_1 & k_2 \end{bmatrix}$ and where r is the reference signal, assumed to be constant) having the following properties: (a) it places the poles at $\{-2, -1\}$; (b) in the absence of disturbances acting onto $\mathbf{\Sigma}$, the regulation error $e = y - r$ is zero at steady state (zero "static error"). (vi) In what follows, *$\beta = 1$. Suppose a constant but unknown disturbance d is adding to the output of $\mathbf{\Sigma}$, in such a way that the expression of y becomes $y = \begin{bmatrix} 1 & 1 \end{bmatrix} x + d$. (a) With the control calculated at Question (v), is the static error still zero? (b) Is it possible to design a state feedback with integral action for $\mathbf{\Sigma}$? (c) If yes, determine such a control so that the closed-loop poles are placed at $\{-2, -1, -1\}$. (d) How can you justify this choice? (vii) The state is non-measured, thus we decide to design a full-order observer and combine it with the control law obtained in Question (vi). We also decide to place the observer poles at $\{-1, -10\}$. (a) How can you justify this choice? (b) Calculate the gain matrix $\tilde{K}$ of this observer.*

EXERCISE 300 – *Let there be the state-space system $\mathbf{\Sigma} = \{A, B, C\}$, where*

$$A = \begin{bmatrix} 1/2 & -1/2 \\ -3/2 & -1/2 \end{bmatrix}, \quad B = \begin{bmatrix} 1 \\ -1 \end{bmatrix}, \quad C = \begin{bmatrix} 1 & 0 \end{bmatrix}.$$

(i) Study the controllability, observability, and stability of this system. (ii) Study its stabilizability. (iii) Show that, using a state feedback control $u = -K\,x$, it is possible to obtain a closed-loop system whose poles are $\{-1, \lambda\}$, where λ is arbitrarily chosen on the real axis. For a fixed λ, is there uniqueness of the solution? Give a parameterization of all solutions when $\lambda = -1$ (this value will be maintained in what follows). (iv) We assume from now on that we will not measure the state x. Determine the full-order observer whose poles are $\{-1, -10\}$. (v) Assuming that only the second component of the state is to be reconstructed, determine a minimal observer by judiciously choosing its pole. (vi) For pole placement, is an observer (even a minimal one) necessary?

EXERCISE 301.– *Consider a minimal system Σ with transfer function $G(s) = 1/\left(s^2 + s - 1\right)$. (i) Determine for Σ the RST controller with integral action (section 6.3.1) placing all closed-loop poles at $s = -1$ and such that the transfer function between reference r and output y is of order 3, in the following two cases: (a) $\delta_0 = 0$; (b) $\delta_0 = 1$. (ii) Let $\{F, G, H, J\}$ be the observable canonical form of Σ. (a) For this state-space system, determine the control law by state feedback and integral action that places all closed-loop poles at $s = -1$. (b) Determine the minimal observer (resp., the full-order observer) which has all its poles at $s = -1$. (c) The two state feedback/observer syntheses with integral action obtained in the responses to Questions (ii)(a) and (ii)(b) can be written in the form of two RST controllers. What are they? (Compare with those obtained in Question (i).) (iii) What is the possible flaw in the above controllers? How can you correct it?*

EXERCISE 302.– ** Further develop the theory discussed in section 8.3 by considering the case where the state x is not measured. Suppose that Σ is defined by (9.24), where y is the measured vector, and denote by $\varphi(\partial)$ the polynomial with minimal degree which annihilates the disturbances d_1 and d_2 as well as the reference r. Consider the case where Σ has a direct term, i.e. the second equation of (9.24) is $y = C\,x + D\,u + d_2$.**

Chapter 10

Discrete-Time Control

10.1. Introduction

The principles presented in Chapters 5, 6, 8 and 9 for the design of control laws are general, though they lead to "continuous-time" (also called "analog") controls. For a long time, in numerous branches of industry, implementation of control laws has been done through computers. The resulting controls are called "discrete-time" or "digital" because the set of instants at which the signals can be delivered to computers, and at which those computers can deliver the control, is discrete. These instants are of the form kT, where $k \in \mathbb{Z}$ and T is a real positive number; these instants are called the *sampling instants*, while T is the *sampling period* ($f_s = 1/T$ and $\omega_s = 2\pi f_s$ are then the *sampling frequency* and the *angular sampling frequency*, respectively). Digital signals have another particularity: their value at each instant is coded in a finite number of bits. This coding operation is called *quantization.*

Entire books are devoted to discrete-time control and systems, e.g. [4].

10.2. Discrete-time signals

10.2.1. *Discretization of a signal*

A continuous-time signal with values in $\mathbb{R}^n$ is a function of the real variable $x : \mathbb{R} \to \mathbb{R}^n : t \mapsto x(t)$. The discretized signal at sampling period T is the sequence x_d of elements in $\mathbb{R}^n$ defined by $x_d(k) = x(kT), k \in \mathbb{Z}$ (sequences are denoted, in this chapter, as functions defined in $\mathbb{Z}$).[1] In reality, as mentioned in section 10.1, the components of x_d are quantized, but to remain as simple as possible we will not deal with the problem of quantization – besides, its importance diminishes as the performance of computers increases; see [4] on this subject. Signal x_d is a *discrete-time signal*. It is assumed in the rest of this section that $n = 1$, except when otherwise stated. It is easy to extend the theory to the case where n can be any positive integer by doing the same rationale for each component.

10.2.2. *z-transform*

The (two-sided) z-transform of a discrete-time signal x_d is the z-transform of the sequence x_d (see sections 12.3.5 and 12.4.4) and is denoted by X in what follows. We have

$$\boxed{X(z) = \sum_{k=-\infty}^{+\infty} x_d(k)\, z^{-k}} \tag{10.1}$$

for $z \in C_c$, where C_c is the annulus of convergence of X.

10.2.3. *Sampled signal*

Let $x : \mathbb{R} \to \mathbb{R}$ be a continuous-time signal and let $T > 0$. This signal, sampled at period T, is

$$\boxed{x^* = x\, \varpi_T} \tag{10.2}$$

where ϖ_T is the *Dirac comb* (see section 12.2.3). Denoting distributions like functions, we thus have

$$x^*(t) = x(t)\, \varpi_T(t) = x(t) \sum_{k=-\infty}^{+\infty} \delta(t - kT).$$

If x belongs to the space $\mathcal{O}_{\mathrm{M}}$ of indefinitely differentiable functions whose derivatives of all orders (including order 0) are slowly increasing, we obtain from expression (12.20) of section 12.2.3

$$\begin{aligned} x^*(t) &= \sum_{k=-\infty}^{+\infty} x(t)\, \delta(t - kT) = \sum_{k=-\infty}^{+\infty} x(kT)\, \delta(t - kT) \\ &= \sum_{k=-\infty}^{+\infty} x_d(k)\, \delta(t - kT). \end{aligned} \tag{10.3}$$

1. The signal x can be only defined in an interval of $\mathbb{R}$; in that case, x_d is defined in an interval of $\mathbb{Z}$.

REMARK 303.– *Relation* (10.3) *is exact for more general functions* x *than those belonging to* $\mathcal{O}_{\mathrm{M}}$. *Since the Dirac distribution is a measure with support reduced to* $\{0\}$, *it suffices that there exist a function* $y \in \mathcal{O}_{\mathrm{M}}$ *and, for all* $k \in \mathbb{Z}$, *a neighborhood* $\mathcal{N}_k$ *of* k, *such that* $x\,|_{\mathcal{N}_k} = y\,|_{\mathcal{N}_k}$. *In that case,* $x_d \in \mathbf{s}'$, *thus* $x^* \in \mathcal{S}'$ *(see section 12.3.2). In what follows, we denote by* $\mathbf{D}_T$ *the set of all functions* $x : \mathbb{R} \to \mathbb{R}$ *that satisfy the above condition. It is immediately clear that* $\mathbf{D}_T$ *is an* $\mathbb{R}$*-vector space, which we will call the space of "*T*-discretizable signals". Not all functions belonging to* $\mathbf{D}_T$ *are continuous.*

There exists a linear bijection $x_d \tilde{\mapsto} x^*$, and that is why these two quantities are often identified. Anyhow, we will distinguish in what follows the *discretized signal* x_d, which is a concrete entity, and the *sampled signal* x^*, which is purely a mathematical object, useful for many calculations. The (two-sided) Laplace transform of x^* satisfies

$$\hat{x}^*(s) = \sum_{k=-\infty}^{+\infty} x_d(k) \int_{-\infty}^{+\infty} \delta(t - kT) e^{-s\,t}\, dt = \sum_{k=-\infty}^{+\infty} x_d(k)\, e^{-s\,k\,T}$$

and hence according to (10.1), $\hat{x}^*(s) = X\left(e^{s\,T}\right)$, i.e.

$$\boxed{\hat{x}^*(s) = X(z), \quad z = e^{s\,T}}. \tag{10.4}$$

10.2.4. *Poisson summation formula*

Let $\mathcal{F}$ be the Fourier transform and $x \in \mathbf{D}_T$. According to equation (10.2) and the second Exchange theorem (see section 12.3.1), we have

$$\mathcal{F}x^* = \frac{1}{2\pi} (\mathcal{F}x) * (\mathcal{F}\varpi_T)$$

under the condition that the Fourier transform $\mathcal{F}x$ be with compact support. Under this hypothesis, according to relation (12.41) (section 12.41),

$$\begin{aligned}(\mathcal{F}x^*)(\omega) &= \frac{1}{2\pi}\left((\mathcal{F}x) * \frac{2\pi}{T}\varpi_{\omega_s}\right)(\omega) \\ &= \frac{1}{T}\sum_{k=-\infty}^{+\infty} (\mathcal{F}x)(\omega - k\omega_s).\end{aligned}$$

On the other hand, from equation (10.4), $(\mathcal{F}x^*)(\omega) = X\left(e^{i\omega T}\right)$, and by substituting $i\omega = s$, we have the equality

$$\boxed{X\left(e^{sT}\right) = \tfrac{1}{T} \sum_{k=-\infty}^{+\infty} \hat{x}(s - i\,k\,\omega_s)} \tag{10.5}$$

(where $\hat{x}$ stands for the bilateral Laplace transform of x), which is an expression of the *Poisson summation formula*.

10.2.5. *Sampling theorem*

Let $x \in \mathbf{D}_T$ be a (continuous-time) signal such that $\mathcal{F}x$ has a compact support and let

$$\omega_{\max} = \inf \{\omega \geq 0 : \operatorname{supp}(\mathcal{F}x) \subset [-\omega, \omega]\}$$

(where supp is the support).

This signal x is discretized at the period $T = 1/f_s$, and we write $x_d(k) = x(kT)$ $(k \in \mathbb{Z})$.

DEFINITION 304.– *The frequency $f_N = f_s/2$ and the angular frequency $\omega_N = 2\pi f_N$ are called the* Nyquist frequency *and the* Nyquist angular frequency, *respectively.*

We have the following result, called the "sampling theorem" or the "Shannon theorem". Shannon, indeed, pointed out the importance of this result in the problem of sampling.[2]

THEOREM 305.– *A sufficient condition under which we can reconstruct the signal x from the discretized signal x_d is: $\omega_N > \omega_{\max}$.*

PROOF. In order to make it more readable, the proof is carried out in a way that all distributions encountered are treated as functions. We have relation (10.5), which can also be written as

$$\boxed{X\left(e^{i\omega T}\right) = \tfrac{1}{T} \sum_{k=-\infty}^{+\infty} \hat{x}\left[i\left(\omega - k\omega_s\right)\right]}. \tag{10.6}$$

Assuming that $\omega_N > \omega_{\max}$, we have for all $\omega \in [0, \omega_N]$,

$$X\left(e^{i\omega T}\right) = \frac{1}{T}\hat{x}(i\omega). \tag{10.7}$$

Let the "normalized angular frequency" be

$$\theta = \omega T. \tag{10.8}$$

The "function" $\theta \mapsto X\left(e^{i\theta}\right)$ is the Fourier transform of the sequence $x_d \in \mathrm{s}'$ (see section 12.3.3). Therefore, knowing x_d, we know $X\left(e^{i\omega T}\right)$ for all $\omega \in [0, \omega_N]$, and

2. According to some authors, Cauchy already knew this result in 1841. This is questionnable and, more probably, the first to discover this condition was Whittaker in 1915.

thus the "function" $\omega \mapsto \hat{x}(i\omega)$, according to equation (10.7). This is the Fourier transform of x, and therefore $x \in \mathcal{S}'$ is known (see section 12.3.1). The elements of $\mathcal{S}'$ are distributions; *since the distribution defined by x is known, the function x is known almost everywhere.* ■

REMARK 306.– *(i)* A priori, *sampling can cause a loss of information. The sampling theorem shows that the condition $\omega_N > \omega_{\max}$ (called the "Shannon condition") is sufficient for all the information contained in x to be also contained in x_d. (ii) Equality* (10.7) *is the relation that exists between $\mathcal{F}x_d$ and $\mathcal{F}x$. Using the normalized angular frequency* (10.8), *we can write*

$$X\left(e^{i\theta}\right) = \frac{1}{T}\,\hat{x}(i\omega)$$

and $X\left(e^{i\theta}\right) = (\mathcal{F}x_d)(\theta)$, $\hat{x}(i\omega) = (\mathcal{F}x)(\omega)$ (see sections 12.3.3 and 12.3.4). Remember that $\mathcal{F}x_d$ is 2π-periodic. Since $(\mathcal{F}x_d)(-\theta) = \overline{(\mathcal{F}x_d)(\theta)}$, $\mathcal{F}x_d$ is determined by its restriction to $[0,\pi]$ and π is the value of θ when the value of ω is ω_N.

Let us now see in a more explicit manner how x can be reconstructed from x_d when the Shannon condition is satisfied. Let us define the "sine cardinal" function, denoted by sinc, as

$$\boxed{\operatorname{sinc}\varphi = \begin{cases} \frac{\sin\varphi}{\varphi} & \text{if } \varphi \neq 0 \\ 1 & \text{if } \varphi = 0 \end{cases}}.$$

The shape of this function is shown in Figure 10.1.

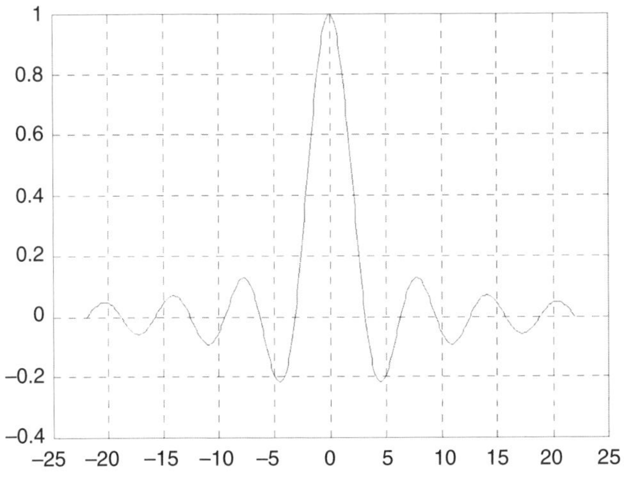

Figure 10.1. *Sine cardinal function*

PROPOSITION 307.– *Suppose the Shannon condition is satisfied (see Remark 306). Then, we have the* Shannon interpolation formula

$$\boxed{x(t) = \sum_{k=-\infty}^{+\infty} x_d(k)\,\mathrm{sinc}\left[\omega_N (t - kT)\right]}. \tag{10.9}$$

PROOF. The Fourier transform of x is given by (10.7). By inverse transform (section 12.3.1), we obtain

$$x(t) = \frac{1}{2\pi}\int_{-\infty}^{+\infty} \hat{x}(i\omega)\, e^{i\omega t} d\omega = \frac{T}{2\pi}\int_{-\omega_N}^{\omega_N} X\left(e^{i\omega T}\right) e^{i\omega t} d\omega.$$

According to (10.1), we have

$$x(t) = \frac{2}{\omega_N}\sum_{k=-\infty}^{+\infty} x_d(k)\int_{-\omega_N}^{\omega_N} e^{i\omega(t-kT)} d\omega$$

from which we deduce (10.9). ■

REMARK 308.– *Let us examine what will happen if the Shannon condition is not satisfied. (i) Suppose x is a sinusoidal signal, sampled at period T, and with angular frequency $3\omega_N/2$. Its spectrum consists of two rays at frequency $-3\omega_N/2$ and $3\omega_N/2$ (see relation* (12.42) *of section 12.3.1). According to equation* (10.6)*, the distribution $\omega \mapsto X\left(e^{i\omega T}\right)$ (denoted like a function, for convenience) is $2\omega_N$-periodic. The ray at angular frequency $-3\omega_N/2$ of the spectrum of x thus generates, by the summation in* (10.6)*, a ray at angular frequency $\omega_N/2$. As a result, the discretized signal x_d is identical to what we would obtain by discretization (at the same period T) of a sinusoidal signal with angular frequency $\omega_N/2$. Through the interpolation formula* (10.9)*, it is this last signal that we generate, instead of x. (This phenomenon was observed by Nyquist, and this is the reason why $f_N = f_s/2$ is called the Nyquist frequency.) (ii) Take a second example by considering the continuous-time signal x whose spectrum is shown in Figure 10.2. The spectrum of the discretized signal x_d is shown in Figure 10.3 when the Shannon condition is satisfied, and in Figure 10.4 when it is not. Sampling, in this last case, has destroyed the information which was contained in x, by a phenomenon we call* spectrum aliasing.

REMARK 309.– *The interpolation formula* (10.9) *is* non-causal, *because in order to calculate $x(t)$ according to this expression, one must know all the $x_d(k)$, including those for which $kT > t$ (Figure 10.5 shows a typical example of reconstruction of signal x from the discretized signal x_d). In some applications, where the signal is processed off-line, this does not present any difficulty, but the control engineer, who always has to work in real-time, does not have the appropriate crystal ball. It is therefore necessary to make a causal approximation of formula* (10.9)*; such an approximation is only valid if $\omega_N \geq \lambda\omega_{\max}$, where $\lambda > 1$. A factor $\lambda = 5$ or 10 is commonly used.*

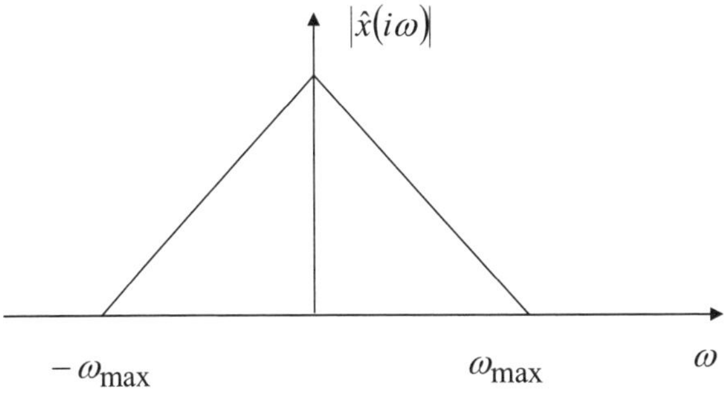

Figure 10.2. *Spectrum of the continuous-time signal*

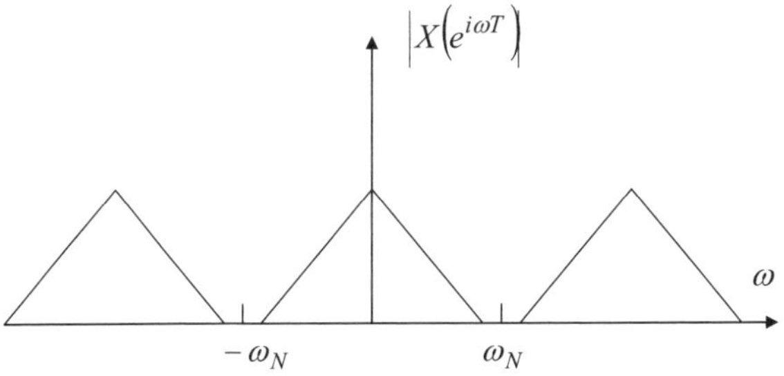

Figure 10.3. *Spectrum of the discretized signal (Shannon condition satisfied)*

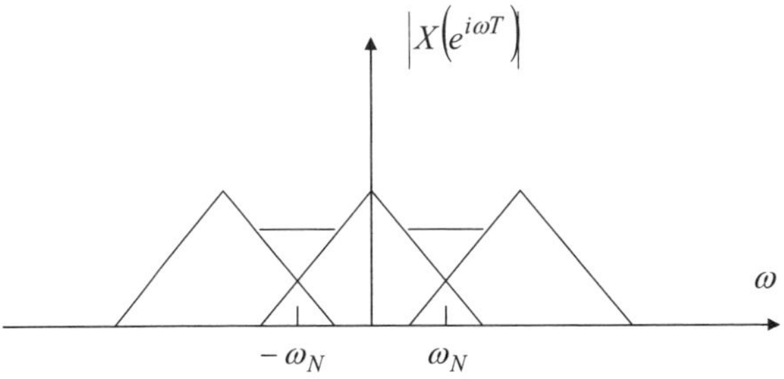

Figure 10.4. *Spectrum of the discretized signal (spectrum aliasing)*

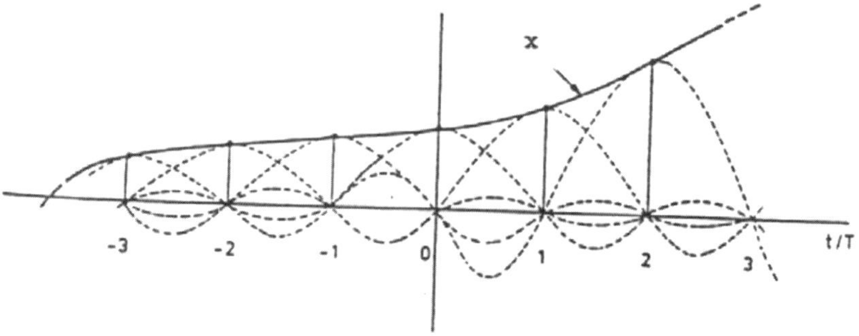

Figure 10.5. *Shannon interpolation*

10.2.6. *Hold*

A *hold* providing a solution to the problem was brought up in Remark 309 (section 10.2.5), i.e. the non-causal nature of the interpolation formula (10.9). One way to approximately reconstruct a signal x which has been discretized is to hold its value over each sampling period. This hold operation is obviously causal and consists ofan extrapolation.

A hold of order n $(n \geq 0)$ makes it possible to determine an estimate of $x(t)$, $t \in [kT, (k+1)T)$, from the values $x_d(k)$, ..., $x_d(k-n)$.

Zero-order hold

The zero-order hold (Z.O.H.) is the most commonly used because it requires the least calculations. Let x_d be a discretized signal $(x_d(k) = x(kT))$. The "sampled-and-held signal" (with Z.O.H.) is defined by

$$x_{h0}(t) = x_d(k), t \in [kT, (k+1)T).$$

The function x_{h0} is a staircase function (such a function is discontinuous in general). Suppose x is differentiable; then according to the mean value formula,

$$|x(t) - x_{h0}(t)| \leq T \sup_t |\dot{x}(t)|$$

and hence the interpolation error is small if x varies sufficiently slowly.

First-order hold

The sampled-and-held signal with a first-order hold is defined by

$$x_{h1}(t) = x_d(k) + \frac{t - kT}{T}(x_d(k) - x_d(k-1)), t \in [kT, (k+1)T).$$

Note that x_{h1} is a continuous function.

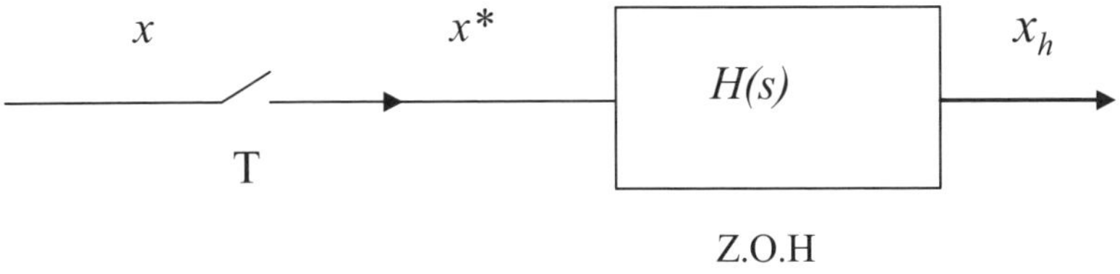

Figure 10.6. *Sample-and-hold*

Transfer function of zero-order hold

In everything that follows, we limit ourselves to the case of a zero-order hold. The sampled-and-held signal x_{h0} is denoted by x_h. Its Laplace transform is given by

$$\hat{x}_h(s) = \int_{-\infty}^{+\infty} x_h(t)\, e^{-s\,t}\, dt = \sum_{k=-\infty}^{+\infty} x(kT) \int_{k\,T}^{(k+1)\,T} e^{-s\,t}\, dt.$$

The integral on the right-hand side is equal to $e^{-s\,k\,T}\, H(s)$, where

$$\boxed{H(s) = \frac{1-e^{-s\,T}}{s}}. \tag{10.10}$$

We thus obtain $\hat{x}_h(s) = \sum_{k=-\infty}^{+\infty} x(kT)\, e^{-s\,k\,T}\, H(s)$, and we also have

$$\boxed{\hat{x}_h(s) = \hat{x}^*(s)\, H(s)}. \tag{10.11}$$

The transfer function $H(s)$ is thus that of the zero-order hold and the signal x_h can be represented as in Figure 10.6; it is said to be obtained from x by means of a *sample-and-hold*.

10.3. Discrete-time systems

10.3.1. *General description*

Let Σ be a continuous-time system with transfer matrix $G(s)$; Σ is assumed to be linear time-invariant throughout this chapter. By discretizing the output y of Σ at sampling period T, we obtain the discrete-time signal y_d. This signal can then be processed by a calculator in order to generate a discrete-time control u_d. This control signal can be converted, using a Z.O.H., into a continuous-time signal u_h (note that $u_h : t \mapsto u_h(t)$ is not a *continuous function*). The feedback by "discrete-time control" is achieved by using this signal u_h as input to Σ.

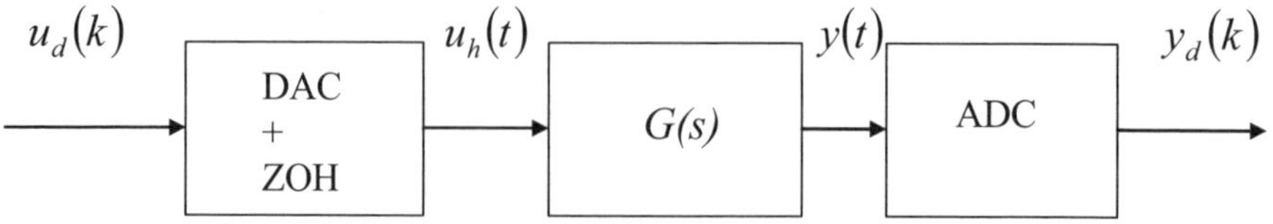

Figure 10.7. *Discretized system*

It is then necessary to convert y_d into a normalized signal, whose components vary between $-10\,V$ and $+10\,V$, for example; the combination of the discretization and this normalization is called the *analog-to-digital conversion* (ADC).

On the other hand, the signal u_d must be held, and then amplified. A *digital-to-analog* (DAC) conversion is thus necessary.

We thus arrive at the diagram of the *discretized system* Σ_d, with input u_d and output y_d, as represented in Figure 10.7.

The system Σ_d is a linear time-invariant discrete-time system. It admits a state-space representation (the proof of Theorem 129 of section 7.1.2 extends to the case of discrete-time systems without difficulty – see [15] or [22] for more details) and, as a result, a transfer matrix $F(z)$.

As a first step, we are going to show how to determine $F(z)$ from $G(s)$, and then we will show how to determine a state-space representation of Σ_d from that of Σ.

REMARK 310.– *The discretized output y_d is only representative of the output y if the Shannon condition is satisfied (see Theorem 305). As a result, it is essential to incorporate a low-pass filter with lower cutoff frequency ω_N into Σ, at the output of this system. This filter is called an* "anti-aliasing filter"*, since its purpose is to avoid spectrum aliasing (see section 10.2.5, Remark 308).*

10.3.2. *Sampled system*

The notion of the *sampled system* (as we will refer to in this text) is purely abstract, as well as that of a *sampled signal*. But it is useful to calculate $F(z)$ as discussed above.

Suppose Σ_0 is a continuous-time system, with transfer matrix $\hat{f}(s)$, receiving a sampled-signal $u^*(t)$ (with period T) at its input. The output $y(t)$ of this system is sampled at period T and it results in a sampled output $y^*(t)$ (see Figure 10.8).

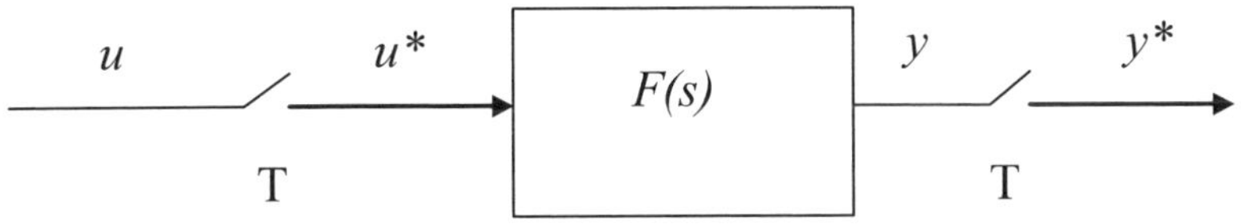

Figure 10.8. *Sampled system*

Let $f(t) = \mathcal{L}^{-1}\left\{\hat{f}(s)\right\}$ be the impulse response of Σ_0 (see section 2.5.2). Assuming that the initial conditions are zero, the output y is given by

$$y = f * u^* = f * (\varpi_T\, u)$$

and the sampled output y^* thus satisfies

$$y^* = \varpi_T\, f * (\varpi_T\, u) = f^* * u^*,$$

(these calculations are valid if both f and u belong to $\mathbf{D}_T \cap \mathcal{S}'(\Gamma)$, where Γ is a non-empty interval of $\mathbb{R}$: see sections 12.3.4 and 10.2.3). According to the Exchange theorem, we thus have

$$\hat{y}^*(s) = \hat{f}^*(s)\, \hat{u}^*(s)$$

and as a consequence of (10.4),

$$Y(z) = F(z)\, U(z) \tag{10.12}$$

where $Y(z), U(z)$, and $F(z)$ are the z-transforms of the sequences $(y_d(k)), (u_d(k))$, and $(f_d(k))$, respectively.

10.3.3. *Discretized system*

Consider the system in Figure 10.8, with input $u_d = (u_d(k))$ and output $y_d = (y_d(k))$. This is a discrete-time system, but it cannot be considered a discretized system yet (according to the definition given in section 10.3.1), because the input of the continuous-time system Σ_0 does not receive a sampled-and-held signal. We can decompose the system Σ_d, resulting from the discretization of Σ at period T, according to the diagram in Figure 10.9.

The discrete-time signal $u_d = (u_d(k))$, represented by the sampled signal u^* according to (10.3), is now held before entering Σ. The system in Figure 10.9 is identical to the one in Figure 10.8 when $\hat{f}(s) = G(s)\, H(s)$. We deduce the following theorem:

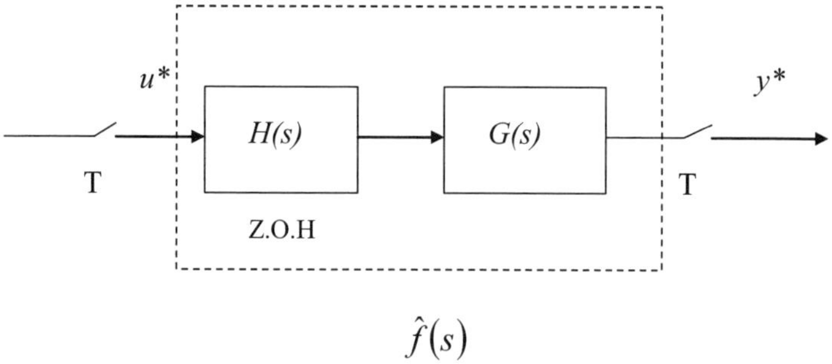

Figure 10.9. *Discretized system*

THEOREM 311.– *The transfer matrix* $G_d(z)$ *of the discretized system* Σ_d *can be expressed as a function of the transfer matrix* $G(s)$ *of the continuous-time system* Σ *by the relation*

$$\boxed{G_d(z) = \left(1 - z^{-1}\right) \mathcal{Z}\left[\mathcal{L}^{-1}\left\{\frac{G(s)}{s}\right\}\right]}. \tag{10.13}$$

PROOF. According to (10.12), $Y(z) = G_d(z)\, U(z)$, where $G_d(z) = \mathcal{Z}[f(t)]$,[3] $f(t) = \mathcal{L}^{-1}\{G(s)\, H(s)\} = \mathcal{L}^{-1}\left\{G(s)\, \frac{1-e^{-sT}}{s}\right\}$. Let $h(t) = \mathcal{L}^{-1}\left\{\frac{G(s)}{s}\right\}$. Then, $f(t) = h(t) - h(t - kT)$, and hence $G_d(z) = \left(1 - z^{-1}\right) H(z)$. ■

Some examples of transfer functions of discretized systems are given in Table (10.14).

$G(s)$	$G_d(z)$
$\frac{1}{s}$	$\frac{T}{z-1}$
$\frac{1}{s^2}$	$\frac{T^2\,(z+1)}{2\,(z-1)^2}$
e^{-Ts}	z^{-1}
$\frac{a}{s+a}$	$\frac{1-e^{-aT}}{z-e^{-aT}}$

(10.14)

EXAMPLE 312.– *Consider the third row of Table* (10.14). *Since* $G(s) = \frac{1}{s^2}$ *(double integrator),* $G_d(z) = \left(1 - z^{-1}\right) \mathcal{Z}\left[\mathcal{L}^{-1}\left\{\frac{1}{s^3}\right\}\right]$. *According to relation* (12.83) *of section 12.4.4,*

$$h(t) = \mathcal{L}^{-1}\left\{\frac{1}{s^3}\right\} = \frac{t^2}{2}\mathbf{1}(t).$$

3. We slightly abuse the language here. More precisely, one should write $G_d(z) = \mathcal{Z}[f_d(k)] = \mathcal{Z}[f(kT)]$.

As a result, $h_d(k) = h(kT) = \frac{T^2}{2} k^2 \mathbf{1}(k)$, *and according to table* (12.64) *of section 12.3.5,* $\mathcal{Z}\left[\mathcal{L}^{-1} k^2 \mathbf{1}(k)\right] = \frac{z(z+1)}{z-1}$. *Finally,* $G_d(z) = \frac{T^2(z+1)}{2(z-1)^2}$.

10.3.4. *State-space representation of a discrete-time system*

Discretization of a state-space system

Let there be a continuous-time system $\{A, B, C, D\}$:

$$\begin{cases} \dot{x} = A\,x + B\,u \\ y = C\,x + D\,u. \end{cases}$$

If the input u of this system is a sampled-and-held signal (with Z.O.H.), we have

$$u(t) = u_d(k), t \in [kT, (k+1)T).$$

The integration of the state equation between the instants kT and $(k+1)T$ provides, according to relation (12.112) of section 12.5.2,

$$\begin{aligned} x_d(k+1) &= e^{A\,T} x_d(k) + \int_{kT}^{(k+1)T} e^{A[(k+1)T-t]} B\,u(t)\; dt \\ &= e^{A\,T} x_d(k) + \int_{kT}^{(k+1)T} e^{A[(k+1)T-t]} dt\; B\,u_d(k) \end{aligned}$$

where $x_d(k) = x(kT)$. Putting $y_d(k) = y(kT)$ and

$$\boxed{A_d = e^{A\,T}, \quad B_d = \int_0^T e^{A\,t}\, B\, dt} \tag{10.15}$$

we obtain (changing t to $(k+1)T - t$ in the integral)

$$\begin{cases} x_d(k+1) = A_d\, x_d(k) + B_d\, u_d(k) \\ y_d(k) = C\, x_d(k) + D\, u_d(k). \end{cases} \tag{10.16}$$

State-space representation of a discrete-time system

A discrete-time system such as equation (10.16) is not always obtained by discretization of a continuous-time system.

Using the shift-forward operator q (see section 12.3.5), equation (10.16) is written as

$$\begin{cases} q\, x_d = A_d\, x_d + B_d\, u_d \\ y_d = C\, x_d + D\, u_d. \end{cases} \tag{10.17}$$

Equations (10.17) constitute a state-space representation of a *causal* discrete-time system. *From a purely formal point of view*, these equations have the same structure as equations (7.4) of section 7.1.2: we only need to replace the differential operator ∂ by the shift-forward operator q. The terminology introduced in section 7.1.2 (state matrix, control matrix, etc.) is therefore maintained.

A *not necessarily causal* discrete-time system has a state-space representation of the form

$$\boxed{\left\{\begin{array}{l} q\,x_d = A_d\,x_d + B_d\,u_d \\ y_d = C\,x_d + W_d\,(q)\,u_d \end{array}\right.} \tag{10.18}$$

where $W_d\,(q)$ is a polynomial matrix. This system is *causal* if and only if, $W_d\,(q)$ is a constant matrix D (possibly zero), i.e. if the future values of the input have no influence on the present output. The control engineer only has to do with causal systems (see section 10.2.5, Remark 309).

System (10.18) is said to be *strictly causal* if $W_d\,(q) = 0$. This means that the system input has no immediate influence on the output. The notions of causal system and strictly causal system, relative to discrete-time systems, correspond to the notions of proper system and strictly proper system, relative to continuous-time systems (see section 2.5.3).[4] Using the notion of transfer matrix, we are led to the following:

DEFINITION 313.– *A discrete-time system Σ_d is said to be* causal *(resp.,* strictly causal, bicausal*) if its transfer matrix $G(z)$ is proper (resp., strictly proper, biproper). (See section 13.6.1.)*

REMARK 314.– *It is also possible, and to some extent preferable, to represent a discrete-time system using the operator*

$$\delta = \frac{q-1}{T} \tag{10.19}$$

(where T is the sampling period if the system considered is a discretized system and where T is any real number > 0 – for example, $T = 1$ – otherwise) because δ is, like $\partial = \frac{d}{dt}$, a "differential operator". This remark only has full significance in the framework of linear time-varying systems: see [15] or [22] (where $T = 1$). Using this operator, a discrete-time system admits a state-space representation

$$\left\{\begin{array}{l} \delta\,x_d = \tilde{A}_d\,x_d + \tilde{B}_d\,u_d \\ y_d = C\,x_d + \tilde{W}_d\,(\delta)\,u_d \end{array}\right.$$

4. This is the reason why some authors call causal (resp., strictly causal) a continuous-time system which, in the terminology of this book, is called proper (resp., strictly proper). We should avoid this abuse of language in our opinion. For general considerations on causality, see [41].

where

$$\tilde{A}_d = \frac{A_d - I_n}{T}, \quad \tilde{B}_d = \frac{B_d}{T}.$$

Such a system is causal (resp., strictly causal) if and only if, $\tilde{W}_d(\delta) = D$ *(resp.,* $\tilde{W}_d(\delta) = 0$*).*

Calculations

Let us consider a discretized system and see how one can, in practice, calculate the matrices A_d and B_d defined by equation (10.15). The matrix A_d is an exponential matrix and thus can be calculated by one of the methods discussed in section 12.5.2. The same methods make it possible to calculate

$$\Psi = \int_0^T e^{A\tau} d\tau = I_n T + A \frac{T^2}{2} + A^2 \frac{T^3}{3!} + \ldots$$

and we then have

$$A_d = I_n + A\Psi, \quad B_d = \Psi B.$$

In the particular case where A is invertible, $\Psi = A^{-1}(A_d - I_n)$ and we thus have

$$B_d = A^{-1}(A_d - I_n) B.$$

In the general case, let $\Phi(t) = e^{At}$ and $\Gamma(t) = \Psi(t) B$. We have

$$\frac{d}{dt}\begin{pmatrix} \Phi(t) & \Gamma(t) \\ 0 & I_m \end{pmatrix} = \begin{pmatrix} \Phi(t) & \Gamma(t) \\ 0 & I_m \end{pmatrix}\begin{pmatrix} A & B \\ 0 & 0 \end{pmatrix},$$

and hence

$$\boxed{\begin{pmatrix} A_d & B_d \\ 0 & I_m \end{pmatrix} = \exp\left\{\begin{pmatrix} A & B \\ 0 & 0 \end{pmatrix} T\right\}}. \tag{10.20}$$

Equality (10.20) has the advantage of expressing very synthetically the relation that exists between the state and control matrices of the discretized system and those of the continuous-time system, but it does not generally constitute the most economic way to make calculations.

10.3.5. *Calculation of the state of a discretized system*

Consider the first equation of equation (10.16). Let there be an initial condition

$$x_d(0) = x_0.$$

We have

$$\begin{aligned} x_d(1) &= A_d\, x_0 + B_d\, u_d(0)\,, \\ x_d(2) &= A_d\, x_d(1) + B_d\, u_d(1) = A_d^2\, x_0 + \begin{bmatrix} B_d & A_d\, B_d \end{bmatrix} \begin{bmatrix} u_d(1) \\ u_d(0) \end{bmatrix} \end{aligned}$$

and by induction, we easily establish that

$$x_d(k) = A_d^k\, x_0 + \begin{bmatrix} B_d & A_d\, B_d & \ldots & A_d^{k-1}\, B_d \end{bmatrix} \begin{bmatrix} u_d(k-1) \\ \vdots \\ u_d(1) \\ u_d(0) \end{bmatrix}. \quad (10.21)$$

10.4. Structural properties of discrete-time systems

10.4.1. *Poles and zeros*

The definitions of the various kinds of poles and zeros of discrete-time systems are identical to those of continuous-time systems.[5] Let us recall some of the essential points regarding discrete-time systems in state-space forms:

– The *poles of the system* Σ_d described by (10.18) are the eigenvalues of A_d, and the *order* of Σ_d is equal to the number of its poles (see section 2.3.7). These poles are also the Smith zeros of the matrix $z\, I_n - A_d$, and this observation allows one to define their structural indices, their orders, and their degrees (see section 13.2.5).

– The *invariant zeros* of Σ_d are the Smith zeros of the "Rosenbrock matrix"

$$\begin{bmatrix} z\, I_n - A_d & -B_d \\ C & W_d(z) \end{bmatrix}.$$

– The *transmission poles* (resp., *zeros*) of Σ_d are the MacMillan poles (resp., zeros) of the transfer matrix

$$G_d(z) = C\, (z\, I_n - A_d)^{-1}\, B_d + W_d(z)$$

(see sections 2.4.2, 2.4.4 and 2.4.5).

– The *non-controllable modes* (also called the *input-decoupling zeros*, *i.d.z.s*) of Σ_d are the Smith zeros of

$$\begin{bmatrix} z\, I_n - A_d & -B_d \end{bmatrix}$$

5. A unitary definition of poles and zeros of continuous-time and discrete-time systems (possibly time-varying) has been given in [15] and [22] using module theory.

and its *non-observable poles* (also called its *output-decoupling zeros*, $o.d.z.s$) are the Smith zeros of

$$\begin{bmatrix} z\,I_n - A_d \\ C \end{bmatrix}$$

(see sections 7.2.2 and 7.2.3).

– The *hidden modes* can be determined by

$$\{\text{hidden modes}\} = \{\text{system poles}\} \setminus \{\text{transmission poles}\}$$

(see section 7.2.6, Theorem 179(i)) and according to Corollary 176 of section 7.2.5 we can determine the *input–output decoupling zeros* ($i.o.d.z.$) by

$$\{i.o.d.z.\} = \{i.d.z.\} \dot{\cup} \{o.d.z.\} \setminus \{\text{hidden modes}\}.$$

– The blocking zeros of $\mathbf{\Sigma}_d$ (if any) are the complex numbers z such that $G_d(z) = 0$ (see section 2.4.4, Definition 27) while the *zeros of* $\mathbf{\Sigma}_d$ remain defined according to Definition 178 (section 7.2.6).

In what follows, the discrete-time system $\mathbf{\Sigma}_d$ is assumed to be causal, defined (except when otherwise stated) by a state-space representation such as that of equation (10.17), and is denoted by $\{A_d, B_d, C, D\}$ (or $\{A_d, B_d, C\}$ if $D = 0$).

10.4.2. *Controllability*

Controllability of a discrete time state-space system

DEFINITION 315.– *A discrete time system $\{A_d, B_d, C, D\}$ is controllable (resp., 0-controllable) if there exists a control sequence that allows one to transfer its state x_d from any initial value $x_d(0)$ to any final value x_d^* (resp., to the origin) in finite time.*

REMARK 316.– *Numerous authors call* reachability *(resp.,* controllability*) the notion called controllability (resp., 0-controllability) above: see [64] for example. We use here again the terminology adopted in [22] in order to better reveal the unity between continuous-time and discrete-time.*

THEOREM 317.– *(i) The system $\mathbf{\Sigma}_d = \{A_d, B_d, C, D\}$ is controllable if, and only if,* $\operatorname{rk}\Gamma = n$*, where n is the order of $\mathbf{\Sigma}_d$ and where Γ_d is the* controllability matrix

$$\Gamma_d = \begin{bmatrix} B_d & A_d\,B_d & \dots & A_d^{n-1}\,B_d \end{bmatrix} \tag{10.22}$$

("Kalman criterion for controllability"). (ii) $\mathbf{\Sigma}_d$ is 0-controllable if, and only if, $\operatorname{im} A_d^n \subset \operatorname{im}\Gamma_d$.[6]

6. With a slight abuse of language: these symbols denote the images of the linear mappings represented by the matrices Γ_d and A_d^n in the canonical bases.

PROOF. (i) According to (10.21), $\mathbf{\Sigma}_d$ is controllable if and only if, there exists an integer N such that $\operatorname{rk}\Gamma_{d_N} = n$, where

$$\Gamma_{d_N} = \begin{bmatrix} B_d & A_d\,B_d & \cdots & A_d^{N-1}\,B_d \end{bmatrix}.$$

Now, according to the Cayley–Hamilton theorem (section 13.3.4, Theorem 537), $\operatorname{rk}\Gamma_{d_N} = \operatorname{rk}\Gamma_{d_n}$ for any $N \geq n$, since A_d^n is a linear combination with real coefficients of I_n, A_d, ..., A_d^{n-1}. (ii) If $\operatorname{rk} A_d^n \subset \operatorname{rk}\Gamma_d$, for any $x_d\,(0) \in \mathbb{R}^n$, there exists a finite sequence $(u\,(k))_{0\leq k\leq n}$ such that

$$A_d^n\,x_d\,(0) + \Gamma_d \begin{bmatrix} u\,(n-1) \\ \vdots \\ u\,(0) \end{bmatrix} = 0,$$

and hence $\mathbf{\Sigma}_d$ is 0-controllable. The converse also holds according to the Cayley–Hamilton theorem. ■

PROPOSITION 318.– *The following two conditions are equivalent: (a)* $\operatorname{im} A_d^n \subset \operatorname{im}\Gamma_d$*; (b) for any row* $v^T = \begin{bmatrix} v_1 & \ldots & v_n \end{bmatrix}$*, if* $v^T\,A_d^i\,B_d = 0$, $0 \leq i \leq n-1$*, then* $v^T\,A_d^n = 0$.

PROOF. Condition (a) means that for any $x \in \operatorname{im} A_d^n$, $x \in \operatorname{im}\Gamma$. Now, we have $v^T\,A_d^i\,B_d = 0$, $0 \leq i \leq n-1$, if and only if, for any $x \in \operatorname{im}\Gamma$, $v^T\,x = 0$. Also, we have $v^T\,A_d^n = 0$ if and only if, for any $x \in \operatorname{im} A_d^n$, $v^T\,x = 0$, from which we arrive at the equivalence as stated. ■

* *Intrinsic definitions of controllability and* 0*-controllability*

Definition 315 is, of course, only valid for a discrete-time system in *state-space form*. More generally, a discrete-time system $\mathbf{\Sigma}_d$ can be defined as a finitely presented $\mathbf{R}$-module of M, where $\mathbf{R} = \mathbb{R}\,[q]$ (see section 2.2.5, Remark 8). We are thus led to the following definition, in which the input of $\mathbf{\Sigma}_d$, its output, and the way it is represented (state-space form, Rosenbrock representation, etc.) do not play any role.

DEFINITION 319.– *The system* $\mathbf{\Sigma}_d$ *is controllable if the module* M *is free (or, in an equivalent manner, torsion-free, since* $\mathbf{R}$ *is a principal ideal domain).*

In the case where $\mathbf{\Sigma}_d$ is a state-space system, Definitions 319 and 315 are equivalent, according to Theorem 317. Indeed, all the proofs of section 7.1.3 can be transposed to the case of discrete-time systems. The canonical decomposition according to controllability, and Theorem 165 of section 7.2.2 (i.e. $\mathbf{\Sigma}_d$ is controllable if and only if, it has no $i.d.z.$) remain valid.

REMARK 320.– *The statement of Theorem 129 (section 7.1.2) remains valid* mutatis mutandis*: all discrete-time control systems admit a state-space representation (of the form of equation* (10.18)*).*

Let there be the multiplicative set $S = \{q^n, n \geq 0\}$ and let $\mathbf{A} = S^{-1}\mathbf{R}$ be the ring consisting of all elements of the form r/q^n, $r \in \mathbf{R}$, $n \geq 0$[7]; in addition, let $\breve{M}$ be the $\mathbf{A}$-module $\mathbf{A} \otimes_{\mathbf{R}} M$ consisting of all elements of the form m/q^n, $m \in M$, $n \geq 0$ (see section 13.6.5). Consider the following definition of 0-controllability, which is intrinsic like Definition 319 of controllability:

DEFINITION 321.– *The system $\mathbf{\Sigma}_d$ is 0-controllable if the $\mathbf{A}$-module $\breve{M}$ is free.*

REMARK 322.– *The ring $\mathbf{A}$ is a principal ideal domain and so is $\mathbf{R}$, and hence the module $\breve{M}$ is free if, and only if, it is torsion-free (see Corollary 555 of section 13.4.2).*

PROPOSITION 323.– *For a discrete-time* state-space *system $\mathbf{\Sigma}_d = \{A_d, B_d, C, D\}$, Definitions 315 and 321 of 0-controllability are equivalent.*

PROOF. The $\mathbf{A}$-module $\breve{M}$ is free if, and only if, the matrix $\begin{bmatrix} I_n - q^{-1} A_d & q^{-1} B_d \end{bmatrix}$ (which is a presentation matrix of $\breve{M}$) is right-invertible (see the proof of Theorem 141 in section 7.1.3). This condition means that if $\breve{v}^T = \begin{bmatrix} \breve{v}_1 & \dots & \breve{v}_n \end{bmatrix}$ is a row of elements of a right $\mathbf{A}$-module, the equality $\breve{v}^T \begin{bmatrix} I_n - q^{-1} A_d & q^{-1} B_d \end{bmatrix} = 0$ implies $\breve{v}^T = 0$. The first of these equalities is equivalent to (a) $\breve{v}^T = \breve{v}^T q^{-1} A_d$ and (b) $\breve{v}^T q^{-1} B_d = 0$. Right-multiplying (a) by q^{-1}, we obtain $\breve{v}^T q^{-1} = \breve{v}^T q^{-2} A_d$, and hence $\breve{v}^T q^{-2} A_d B_d = 0$ according to (b). This last equality, in turn, implies $\breve{v}^T q^{-3} A_d^2 B_d = 0$, etc. For any $i \in \{0, ..., n-1\}$, we can right-multiply the equality $\breve{v}^T q^{-i-1} A_d^i B_d = 0$ by q^{i+1} and (since q is an invertible element of $\mathbf{A}$) this equality is equivalent to $\breve{v}^T A_d^i B_d = 0$. The equality $\breve{v}^T \begin{bmatrix} I_n - q^{-1} A_d & q^{-1} B_d \end{bmatrix} = 0$ is thus equivalent to $\breve{v}^T A_d^i B_d = 0$, $0 \leq i \leq n-1$. As a result, if $\operatorname{im} A_d^n \subset \operatorname{im} \Gamma$, we have $\breve{v}^T A_d^n = 0$ according to Proposition 318, and hence from (a), we have $\breve{v}^T = 0$, and $\breve{M}$ is free. Conversely, if $\breve{M}$ is free, the equalities $\breve{v}^T A_d^i B_d = 0$, $0 \leq i \leq n-1$, imply $\breve{v}^T = 0$, and hence $\operatorname{im} A_d^n \subset \operatorname{im} \Gamma$. ■

From Definitions 319 and 321, the "*Popov–Belevitch–Hautus test for controllability* (resp., 0-*controllability*)" is stated as follows:

PROPOSITION 324.– *The discrete-time state-space system $\mathbf{\Sigma}_d = \{A_d, B_d, C, D\}$ is controllable (resp., 0-controllable) if, and only if,* $\operatorname{rk}_{\mathbb{C}} \begin{bmatrix} z I_n - A_d & B_d \end{bmatrix} = n$ *for any complex number z (resp., for any complex number $z \neq 0$).*

Controllability and discretization

Let $\mathbf{\Sigma}$ be a continuous-time control system, the input of which is an independent variable with m components (see section 2.3.1), discretized at a sampling period of T. Let $\mathbf{\Sigma}_d$ be the discretized system.

7. *The ring $\mathbf{A}$ is the ring of Laurent polynomials in q and is denoted by $\mathbb{R}\left[q, q^{-1}\right]$.*

THEOREM 325.– *For* $\mathbf{\Sigma}_d$ *to be controllable, it is* necessary *that* $\mathbf{\Sigma}$ *be controllable, and it is* sufficient *that* $\mathbf{\Sigma}$ *be controllable and has no poles* λ_1, λ_2 *such that* $\lambda_1 - \lambda_2 = 2\pi k i/T$, *where* k *is any non-zero integer and* $i = \sqrt{-1}$.

PROOF. (A) *Preliminary calculations.* According to Theorem 129 (section 7.1.2), we can assume that $\mathbf{\Sigma}$ is a state-space system $\{A, B, C, D\}$. Let

$$\varepsilon(s) = e^{s\,T}, \quad \psi(s) = \int_0^T e^{s\,t}\,dt$$

which are two entire functions (see section 12.4.2). According to equation (10.15), we have

$$A_d = \varepsilon(A), \quad B_d = \psi(A)\,B.$$

We get $\psi(s) = \left(e^{s\,T} - 1\right)/s$ for $s \neq 0$ and $\psi(s) = T$ for $s = 0$, and hence $\psi(s) = 0$ if and only if, $e^{s\,T} = 1$ with $s \neq 0$, i.e. $s = 2\pi k i/T$ where k is a non-zero integer. On the other hand, $\varepsilon(s_1) = \varepsilon(s_2)$ if and only if, $s_1 - s_2 = 2\pi k i/T$, where k is an integer. The pair (A, B) is non-controllable if and only if, there exists $v \in \mathbb{C}^n \setminus \{0\}$ and $\lambda \in \mathbb{C}$ such that

$$\text{(a)} \quad v^T A = v^T \lambda; \quad \text{(b)} \quad v^T B = 0$$

(see section 7.6, Exercise 222). Likewise, (A_d, B_d) is non-controllable if and only if, there exists $v \in \mathbb{C}^n \setminus \{0\}$ and $\lambda \in \mathbb{C}$ such that

$$\text{(c)} \quad v^T A_d = v^T e^{\lambda\,T}; \quad \text{(d)} \quad v^T B_d = 0.$$

Assuming that Condition (a) holds, $v^T \varepsilon(A) = v^T \varepsilon(\lambda)$ and $v^T \psi(A) = v^T \psi(\lambda)$ (see section 12.4.2, Proposition 439). As a result, $v^T A_d = v^T e^{\lambda\,T}$ and

$$v^T B_d = v^T \psi(A)\,B = v^T \psi(\lambda)\,B = \psi(\lambda)\,v^T B.$$

In addition, assuming that Condition (c) holds, $v^T e^{A\,T} = v^T e^{\lambda\,T}$, i.e. $v^T e^{(A - \alpha I_n)T} = v^T e^{(\lambda - \alpha)\,T}$, where $\alpha > \max\{\operatorname{Re}\beta : \beta \in \operatorname{Sp}(A)\}$. According to Propositions 439 and 441 (section 12.4.2), this is equivalent to $v^T A = v^T (\lambda + 2\pi k i/T)$, where k is an integer.

(B) *Proof by contradiction.* Let there be conditions (i), (ii) and (iii) below: (i) $\mathbf{\Sigma}_d$ is controllable; (ii) $\mathbf{\Sigma}$ is controllable; (iii) $\mathbf{\Sigma}$ has no poles λ_1, λ_2 such that $\lambda_1 - \lambda_2 = 2\pi k i/T$, $k \neq 0$. 1) Suppose that (ii) does not hold, and let $v \neq 0$, $\lambda \in \mathbb{C}$ be such that (a) and (b) hold. Then, $v^T A_d = v^T \varepsilon(\lambda)$ and $v^T B_d = v^T B\,\psi(\lambda) = 0$, and hence we have (c) and (d) and it follows that (i) does not hold. 2) Suppose (i) does not hold. All eigenvalues of A_d are of the form $e^{\lambda T}$, where λ is an eigenvalue of A. Thus, let λ be an eigenvalue of A and $v \neq 0$ be such that (c) and (d) hold. According to (c), $v^T\ A = v^T\ (\lambda + 2\pi ki/T)$, i.e. $\lambda + 2\pi ki/T$ is an eigenvalue of A and v^T is an associated left-eigenvector. As a result, either (iii) does not hold or $k = 0$. Suppose $k = 0$; according to (d), $v^T\ B_d = v^T\ B\,\psi(\lambda) = 0$, and hence $v^T\ B = 0$ or $\psi(\lambda) = 0$. In the first case, (ii) does not hold; in the second, $\lambda = 2\pi li/T$, $l \neq 0$, and (iii) does not hold. 3) Suppose (iii) does not hold and $m = 1$. There thus exist v_1^T and v_2^T such that $v_j^T\ A = \lambda_j\, v_j^T\ (j = 1, 2)$ with $\lambda_1 - \lambda_2 = 2\pi ki/T$, $k \neq 0$. Therefore, $v_1^T \neq v_2^T$ according to (13.29) (see section 13.3.3), and $e^{\lambda_1 T} = e^{\lambda_2 T} \triangleq \mu$, from which $v_j^T\ A_d = \mu\, v_j^T\ (j = 1, 2)$. As a result, the eigenvalue μ of A_d has geometric multiplicity at least 2 and $\operatorname{rk}\begin{bmatrix} \mu I_n - A_d & B_d \end{bmatrix} \leq n - 2 + m < n$, thus (i) does not hold. ■

We observe that if $\lambda = \pi ki/T$ ($k \in \mathbb{Z} \setminus \{0\}$) is an eigenvalue of A, so is $\bar{\lambda}$, and $\lambda - \bar{\lambda} = 2\pi ki/T$. We deduce the following:

Let $\omega_{\max} > 0$ and let $\mathcal{E}_{\omega_{\max}}$ be the set of all controllable continuous-time systems, the poles of which have an imaginary part $\leq \omega_{\max}$. Let $\mathcal{D}_{\omega_{\max}}$ be the set of systems obtained by discretization at sampling frequency f_s of all systems belonging to $\mathcal{E}_{\omega_{\max}}$; let $\omega_s = 2\pi f_s$ and $\omega_N = \omega_s/2$. The result below, which is applicable to *systems*, is a consequence of Theorem 325 and is analogous to the sampling theorem (section 10.2.5, Theorem 305), which is applicable to *signals*:

COROLLARY 326.– *A necessary and sufficient condition for all such systems belonging to* $\mathcal{D}_{\omega_{\max}}$ *to be controllable is* $\omega_N > \omega_{\max}$.

10.4.3. *Observability*

Observability of a discrete-time state-space system

The "behavioral" definition of observability of a discrete-time system is as follows:

DEFINITION 327.– *A discrete-time state-space system* $\Sigma_d = \{A_d, B_d, C, D\}$ *is observable (resp., 0-observable) if, from its inputs and outputs* $u_d(k)$ *and* $y_d(k)$, $k \in \{0, ..., N-1\}$, *for a sufficiently large* N, *we can determine the state* $x_d(0)$ *(resp.,* $x_d(N)$*).*

THEOREM 328.– *Let there be the "observability matrix"*

$$\Omega_d = \begin{bmatrix} C^T & A_d^T\, C^T & \dots & \left(A_d^T\right)^{n-1} C^T \end{bmatrix}^T$$

where n *is the order of* Σ_d. *(i) System* Σ_d *is observable if, and only if,* $\operatorname{rk}\Omega_d = n$ *("Kalman criterion for observability"). (ii)* Σ_d *is 0-observable if, and only if,* $\ker \Omega_d \subset \ker A_d^n$.[8]

8. With the same abuse of language as previously mentioned.

PROOF. According to expression (10.21) of section 10.3.5, we have

$$y_d(k) = C\,A_d^k\,x_0 + \begin{bmatrix} B_d & A_d\,B_d & \dots & A_d^{k-1}\,B_d \end{bmatrix} \begin{bmatrix} u_d(k-1) \\ \vdots \\ u_d(1) \\ u_d(0) \end{bmatrix} + D\,u_d(0).$$

By writing these equalities from $k = 0$ to $k = \kappa - 1$ $(\kappa \geq 1)$, we thus obtain an expression of the form

$$\begin{bmatrix} y_d(0) \\ y_d(1) \\ \vdots \\ y_d(\kappa-1) \end{bmatrix} = \Omega_{d_\kappa}\,x_0 + T_\kappa \begin{bmatrix} u_d(0) \\ u_d(1) \\ \vdots \\ u_d(\kappa-1) \end{bmatrix} \tag{10.23}$$

where

$$\Omega_{d_\kappa} = \begin{bmatrix} C \\ C\,A_d \\ \vdots \\ C\,A_d^{\kappa-1} \end{bmatrix}.$$

According to the Cayley–Hamilton theorem, $\operatorname{rk}\Omega_{d_N} = \operatorname{rk}\Omega_{d_n}$ for any $N \geq n$, and hence we can restrict to the case $\kappa = N = n$. Let $\Omega_d = \Omega_{d_n}$. (1) According to equation (10.23), we can determine x_0 as a function of $u_d(k)$ and $y_d(k)$, $k \in \{0, ..., n-1\}$, if and only if, Ω_d is left-invertible, i.e. of rank n. (2) According to (10.21), $x_d(n)$ is determined in a unique manner as a function of $u_d(k)$ and $y_d(k)$, $k \in \{0, ..., n-1\}$ if and only if, $A_d^n x_0$ is determined in a unique manner as a function of *the same quantities*. If $\ker\Omega_d$ is not included in $\ker A_d^n$, there exist x_0 and x_0' such that $x_0 - x_0' \in \ker\Omega_d$ and $x_0 - x_0' \notin \ker A_d^n$. We thus have $\Omega_d\,x_0 = \Omega_d\,x_0'$ and $A_d^n x_0 \neq A_d^n x_0'$, and therefore $u_d(k)$ and $y_d(k)$, $k \in \{0, ..., n-1\}$, do not allow one to determine $A_d^n x_0$, and $\mathbf{\Sigma}_d$ is not 0-observable. Conversely, suppose $\ker\Omega_d \subset \ker A_d^n$. Let $f : x \mapsto A_d^n x$ and let $\bar{f}$ be the linear mapping induced by f on $\mathbb{R}^n / \ker\Omega_d$ (see section 13.3.2, Remark 518). We have $\bar{f}(\bar{x}_0) = A_d^n x_0$, where $\bar{x}_0 = x_0 + \ker\Omega_d$, and hence $\mathbf{\Sigma}_d$ is 0-observable. ■

* *Intrinsic definitions of observability and 0-observability*

Suppose now that the discrete-time control system $\mathbf{\Sigma}_d$ is characterized by a finitely presented $\mathbf{R}$-module M (with $\mathbf{R} = \mathbb{R}[q]$), and that $\mathbf{\Sigma}_d$ has input u_d and output y_d (see section 7.1.1). Using an approach similar to that followed in section 10.4.2 (and preserving the notation introduced in this section), we are led to the following definition, which generalizes Definition 327:

DEFINITION 329.– *System $\mathbf{\Sigma}_d$ is observable (resp., 0-observable) if $M = [y_d, u_d]_{\mathbf{R}}$ (resp., $\mathbf{A} \otimes_{\mathbf{R}} M = \mathbf{A} \otimes_{\mathbf{R}} [y_d, u_d]_{\mathbf{R}}$).*

Popov–Belevitch–Hautus test

We deduce immediately from Definition 329 the following criterion ("Popov–Belevitch–Hautus test"):

PROPOSITION 330.– *The discrete-time state-space system* $\Sigma_d = \{A_d, B_d, C, D\}$ *is observable (resp., 0-observable) if and only if,* $\mathrm{rk}_{\mathbb{C}} \begin{bmatrix} z\, I_n - A_d^T & C^T \end{bmatrix} = n$ *for any complex number* z *(resp., for any complex number* $z \neq 0$*).*

COROLLARY 331.– *A discrete-time system* Σ_d *is observable (resp., 0-observable) if and only if, it has no o.d.z. (resp., if all its o.d.z.s – if any – are zero).*

Observability and discretization

Theorem 325 can be transposed without difficulty to the case of loss of observability due to discretization. The details are left to the reader.

EXAMPLE 332.– *Let there be a minimal continuous-time system with transfer function*

$$\frac{\pi^2}{s^2 + \pi^2} + \frac{1}{s+1}.$$

A "natural" state-space representation of this system is $\{A, B, C\}$ *with*

$$A = \begin{bmatrix} -1 & 0 & 0 \\ 0 & 0 & -\pi^2 \\ 0 & 1 & 0 \end{bmatrix}, \quad B = \begin{bmatrix} 1 \\ 1 \\ 0 \end{bmatrix}$$
$$C = \begin{bmatrix} 1 & 0 & \pi^2 \end{bmatrix}.$$

If this state-space system is discretized at period $T = 2$*, we obtain the discrete-time system* $\{A_d, B_d, C\}$ *where*

$$A_d = \begin{bmatrix} e^{-2} & 0 & 0 \\ 0 & 1 & 0 \\ 0 & 0 & 1 \end{bmatrix}, \quad B_d = \begin{bmatrix} 1 - e^{-2} \\ 0 \\ 0 \end{bmatrix}$$

(a suitable method for calculating A_d *is the use of the inverse Laplace transform – see section 12.5.2 – and we can also calculate* B_d *using the second equality of equation* (10.15)*. System* $\{A_d, B_d, C\}$ *is neither controllable nor observable (the rank of its controllability matrix is* 1 *and that of its observability matrix is* 2*). The step response of the continuous-time system (–) and the interpolated one of the discretized system (- -) are represented in Figure 10.10.*

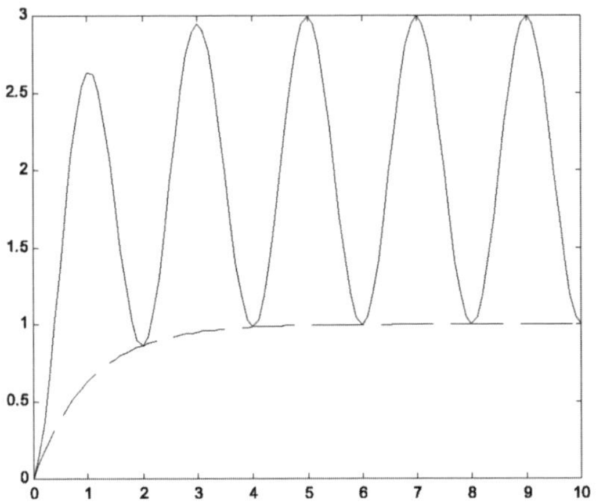

Figure 10.10. *Step responses*

10.4.4. *Rosenbrock representation*

The Rosenbrock representation, and, in particular, the left and right forms, detailed in section 2.3.5 for continuous-time systems, are identical in the case of discrete-time systems, replacing the ∂ operator by the q operator (see [15] or [22]). A left form is always observable while a right form is always controllable.

10.4.5. *Stability*

Stability of a discrete-time control system $\mathbf{\Sigma}_d$ is defined analogously to that of a continuous-time control system $\mathbf{\Sigma}$. Assume without loss of generality (according to Remark 320 of section 10.4.2) that $\mathbf{\Sigma}_d$ is a state-space system $\{A_d, B_d, C, D\}$. The *free behavior* of $\mathbf{\Sigma}_d$ is therefore, according to Definition 20 (section 2.3.8), the vector space spanned by its state x_d when its input u_d is zero. In this case, according to (10.21), x_d is the sequence $(x_d(k))$ defined as a function of the initial state x_0 by

$$x_d(k) = A_d^k x_0. \tag{10.24}$$

Remarks 181 and 184 of section 7.3 lead us to the following definition (from a "behavioral" point of view):

DEFINITION 333.– *A discrete-time linear time-invariant system* $\mathbf{\Sigma}_d$ *is* stable *(resp.,* marginally stable*) if all the variables of its free behavior tend to* 0 *(resp., are bounded) as* k *tends to* $+\infty$.

LEMMA 334.– *A discrete-time state-space system $\mathbf{\Sigma}_d = \{A_d, B_d, C, D\}$ is stable (resp., marginally stable) if and only if, $\lim_{k \to +\infty} A_d^k = 0$ (resp., the sequence $\left(A_d^k\right)$ is bounded).*

THEOREM 335.– *The discrete-time linear time-invariant system $\mathbf{\Sigma}_d$ is* stable *(resp., marginally stable) if and only if, all its poles lie in the open unit disk $|z| < 1$ (resp., in the closed unit disk $|z| \leq 1$, those that belong to the circle $|z| = 1$ – if any – having all their structural indices equal to 1).*

PROOF. Assume without loss of generality (according to Remark 320 of section 10.4.2) that $\mathbf{\Sigma}_d$ is a state-space system $\{A_d, B_d, C, D\}$. Changing the basis, if necessary, the study of A_d^k comes down to the case where A_d is in Jordan form,

$$A_d = \bigoplus_{\lambda, l} J_{\lambda, l}$$

(diagonal sum): see section 13.3.4, Theorem 530. We thus have

$$A_d^k = \bigoplus_{\lambda, l} \left(J_{\lambda, l}\right)^k .$$

For any $k \geq l - 1$, we have $J_{\lambda, l} = \lambda I + J_{0,l}$, and since the matrices λI and $J_{0,l}$ commute,

$$\left(J_{\lambda, l}\right)^k = \sum_{j=0}^{k} \binom{k}{j} \lambda^{k-j} \left(J_{0,l}\right)^j = \sum_{j=0}^{l-1} \binom{k}{j} \lambda^{k-j} \left(J_{0,l}\right)^j ,$$

where the last equality is due to the fact that $\left(J_{0,l}\right)^l = 0$. Thus, we obtain the stated result. ■

DEFINITION 336.– *The state matrix A_d of a discrete-time system is a* stability matrix *if all its eigenvalues belong to the open unit disk $|z| < 1$.*

Suppose now that $\mathbf{\Sigma}_d$ has been obtained by the discretization of a continuous-time system $\mathbf{\Sigma}$. We have the following:

THEOREM 337.– *The system $\mathbf{\Sigma}_d$ is stable (resp., marginally stable) if, and only if, $\mathbf{\Sigma}$ has the same property.*

PROOF. Suppose, without loss of generality (see Theorem 129, section 7.1.2) that $\mathbf{\Sigma}$ is a state-space system $\{A, B, C, D\}$. Let $T > 0$ be the sampling period. System $\mathbf{\Sigma}_d$ is thus a state-space system $\{A_d, B_d, C, D\}$, where $A_d = \exp(AT)$. Changing the basis (if necessary), the problem comes down to the case where A is in Jordan form, i.e.

$$A = \bigoplus_{\lambda,l} J_{\lambda,l}.$$

As a result,

$$A_d = \bigoplus_{\lambda,l} \exp\left(J_{\lambda,l}\, T\right) = \bigoplus_{\lambda,l} e^{\lambda\, T}\, e^{J_{0,l}\, T}$$

and $e^{J_{0,l}\, T}$ is given by expression (12.107) of section 12.5.2 (replacing t by T), which proves the theorem. ■

DEFINITION 338.– *Let* $\mathbf{\Sigma}_d$ *be a* minimal *discrete-time system, and suppose that this system is* bicausal *(Definition 313). The system* $\mathbf{\Sigma}_d$ *(or, abusing the language, its transfer matrix) is said to be* bistable *if its transmission poles and zeros all lie in the open unit disk.*

10.5. Pseudocontinuous systems

10.5.1. *Bilinear transform*

The "bilinear transform" is defined by

$$\boxed{w = \lambda\, \tfrac{z-1}{z+1}, \quad \lambda > 0} \tag{10.25}$$

Interpretation

This transform can be interpreted as an approximation to integration by trapezoidal rule. Indeed, consider the differential equation

$$\partial\, y = u, \tag{10.26}$$

and let $\hat{y}\,(s)$ and $\hat{u}\,(s)$ be the bilateral Laplace transforms of y and u, respectively. They satisfy the relation

$$\hat{y}\,(s) = \frac{1}{s}\,\hat{u}\,(s)\,. \tag{10.27}$$

On the other hand, by integrating equation (10.26) between the instants $k\,T$ and $(k+1)\,T$, we obtain

$$y\left[(k+1)\,T\right] - y\,(k\,T) = \int_{k\,T}^{(k+1)\,T} u\,(t)\;dt. \tag{10.28}$$

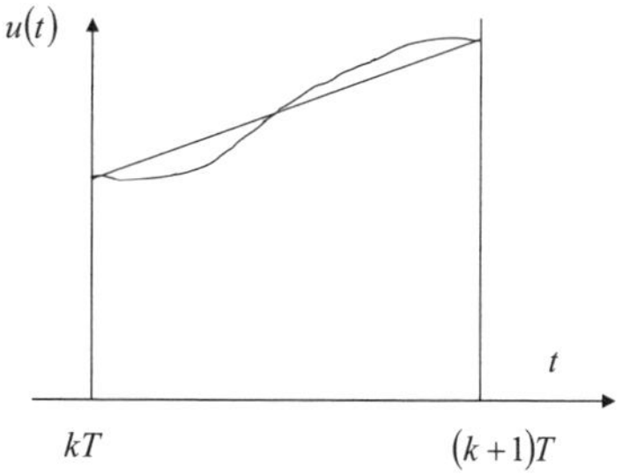

Figure 10.11. *Trapezoidal rule*

Suppose that the restriction of u to the interval $[k\,T,(k+1)\,T)$ has the graph represented by the curve in Figure 10.11. The integral figuring in the right-hand side of (10.28) is the area subtended by this graph. An approximation of this area is that subtended by the segment joining the points $(k\,T, u\,(k\,T))$ and $((k+1)\,T, u\,[(k+1)\,T])$, and hence we have

$$y\,[(k+1)\,T] - y\,(k\,T) \simeq T\,\frac{u\,[(k+1)\,T] + u\,(k\,T)}{2},$$

and therefore

$$(q-1)\;y_d \simeq \frac{T}{2}\;(q+1)\;u_d.$$

Let $Y\,(z)$ and $U\,(z)$ be the bilateral z-transforms of y_d and u_d, respectively. According to expression (12.67) of section 12.3.5, we obtain

$$Y\,(z) \simeq \frac{T}{2}\frac{z+1}{z-1}\,U\,(z)\,. \tag{10.29}$$

The trapezoidal rule leads therefore to approximation (10.29) of equation (10.27). Returning to the complex variable w defined by (10.25), we obtain

$$w \simeq s \text{ with } \lambda = \frac{2}{T}. \tag{10.30}$$

Properties

PROPOSITION 339.– *(i) The bilinear transform is a diffeomorphism*[9] *from the interior of the unit disk,* $|z| < 1$, *onto the left half-plane* $\operatorname{Re} w < 0$, *from the exterior of the unit disk,* $|z| > 1$, *onto the right half-plane* $\operatorname{Re} w > 0$, *except the point* λ, *and from*

9. That means it is a differentiable bijection whose inverse function is also differentiable.

the unit circle, $|z| = 1$, except the point -1, onto the imaginary axis. The inverse diffeomorphism is given by

$$z = \frac{1 + w/\lambda}{1 - w/\lambda}. \tag{10.31}$$

(ii) We have $z = e^{i\theta}$ $(-\pi < \theta < \pi)$ if and only if, $w = i\omega$ with $\omega = \lambda \tan \frac{\theta}{2}$.

PROOF. Let $z = \rho\, e^{i\theta}$. Then

$$w = \lambda \frac{\rho e^{i\theta} - 1}{\rho e^{i\theta} + 1}$$

and we easily see that

$$\operatorname{Re} w = \frac{\lambda\left(\rho^2 - 1\right)}{\rho^2 + 2\rho\cos\theta + 1}$$

which proves (i) according to (10.31). For $\rho = 1$,

$$w = \lambda \frac{e^{i\theta/2} - e^{-i\theta/2}}{e^{i\theta/2} + e^{-i\theta/2}} = i\lambda \tan \frac{\theta}{2}.$$

■

REMARK 340.– *For small values of $\frac{\theta}{2}$ (θ being expressed in radians), $\tan \frac{\theta}{2} \simeq \frac{\theta}{2}$, and hence the relation (10.8) of section 10.2.5 is satisfied with a good precision for $\lambda = \frac{2}{T}$ (with this value, the bilinear transform is called the "Tustin transform"). If we want this relation to be verified with a good precision in the neighborhood of $\theta_0 \neq 0$, $|\theta_0|$ being large enough for the approximation $\tan \frac{\theta_0}{2} \simeq \frac{\theta_0}{2}$ to be too rough, we take λ such that $\lambda \tan \frac{\theta_0}{2} = \omega_0 = \frac{\theta_0}{T}$, and hence*

$$\boxed{\lambda = \frac{\theta_0}{T \tan \frac{\theta_0}{2}}}. \tag{10.32}$$

With this value, the bilinear transform is called the "Tustin transform with prewarping". If we take θ_0 tending to 0 in equation (10.32), we get back to the classic value $\lambda = \frac{2}{T}$ in equation (10.30).

10.5.2. *Pseudocontinuous representations*

Let $\mathbf{\Sigma}_d$ be a discrete-time system with transfer matrix $G_d(z)$ and let

$$\breve{G}(w) = G_d\left(\frac{1 + w/\lambda}{1 - w/\lambda}\right).$$

The transfer matrix $\breve{G}$ is "analogous" to that of a continuous-time system, because, according to (10.30), the variable w is "analogous" to the Laplace variable s. We call $\breve{G}(w)$ the "pseudocontinuous form" of the transfer matrix $G_d(z)$.

It is likewise possible to determine a "pseudocontinuous state-space representation" $\breve{\mathbf{\Sigma}}$ of a discrete-time system $\mathbf{\Sigma}_d$, by using the operator

$$\Delta = \lambda \frac{q-1}{q+1}, \tag{10.33}$$

that is "analogous" to the differential operator ∂. Note that

$$q = \frac{1+\Delta/\lambda}{1-\Delta/\lambda}. \tag{10.34}$$

PROPOSITION 341.– *(i) If -1 is not a pole of $\mathbf{\Sigma}_d$, a pseudocontinuous representation of this system is $\breve{\mathbf{\Sigma}}$ given by*

$$\begin{cases} \Delta\, \eta_d = \breve{A}\, \eta_d + \breve{B}\, u_d \\ y_d = \breve{C}\, \eta_d + \breve{D}\, u_d \end{cases} \tag{10.35}$$

where

$$\eta_d = \frac{q+1}{\lambda T}\, x_d, \quad \Delta\, \eta_d = \delta\, x_d$$

(with $\delta = \frac{q-1}{T}$),

$$\begin{aligned} \breve{A} &= \lambda\,(A_d - I_n)\,(A_d + I_n)^{-1}, \quad \breve{B} = \frac{2}{T}\,(A_d + I_n)^{-1}\, B_d, \\ \breve{C} &= T\lambda C\,(A_d + I_n)^{-1}, \quad \breve{D} = \left[D - C\,(A_d + I_n)^{-1}\, B_d\right]. \end{aligned}$$

(ii) The transfer matrix of $\breve{\mathbf{\Sigma}}$ is $\breve{G}(w)$, where

$$\breve{G}(w) = \breve{C}\left(w\, I_n - \breve{A}\right)^{-1} \breve{B} + \breve{D}. \tag{10.36}$$

PROOF. (i) We have

$$\eta_d = \frac{1}{T\,\lambda}\,(A_d + I_n)\, x_d + \frac{1}{T\,\lambda}\, B_d\, u_d,$$

and hence x_d can be expressed as a function of η_d if and only if, -1 is not an eigenvalue of A_d, and in that case

$$x_d = T\,\lambda\,(A_d + I_n)^{-1}\, \eta_d - (A_d + I_n)^{-1}\, B_d\, u_d.$$

It follows that

$$\Delta \eta_d = \frac{1}{T} \left[(A_d - I_n) \, x_d + B_d \, u_d \right] = \check{A} \, \eta_d + \check{B} \, u_d.$$

We obtain $y_d = \check{C} \, \eta_d + \check{D} \, u_d$ by an analogous rationale. (ii) We have with zero initial conditions

$$\mathcal{Z} \left(\Delta \, \eta_d \right) = w \, \mathcal{Z} \left(\eta_d \right).$$

According to (10.35),

$$\mathcal{Z} \left(\Delta \, \eta_d \right) = \check{A} \, \mathcal{Z} \left(\eta_d \right) + \check{B} \, U \left(z \right),$$

and hence

$$\left(w \, I_n - \check{A} \right) \, \mathcal{Z} \left(\eta_d \right) = \check{B} \, U \left(z \right)$$

from which we deduce (10.36). ■

DEFINITION 342.– *Suppose that* -1 *is not a pole of* Σ_d. *Then,* $\check{\Sigma}$ *defined by* (10.35) *is a* pseudocontinuous state-space representation *of* Σ_d.

REMARK 343.– *(i) The above pseudocontinuous state-space system* $\check{\Sigma}$ *is proper but not strictly proper, even if* Σ_d *is strictly causal. (ii) If* Σ_d *is obtained by discretization of a continuous-time state-space system* $\Sigma = \{A, B, C, D\}$ *with sampling period* T, *the set of poles of* Σ_d *is* $\left\{ e^{s\,T}, s \in P \right\}$, *where* P *is the set of poles of* Σ, *and hence* -1 *is not a pole of* Σ_d. *Now, take* $\lambda = 2/T$ *and* $T \to 0^+$; *then*

$$A_d = e^{A\,T} = I_n + T\,A + o\,(T), \quad B_d = \int_0^T e^{A\,t} \, dt \, B = T\,B + o\,(T) \quad (10.37)$$

where $o\,(T)\,/T \to 0$. *As a result,*

$$\check{A} \to A, \quad \check{B} \to B, \quad \check{C} \to C, \quad \check{D} \to D.$$

10.5.3. **Intrinsic definition of a pseudocontinuous system*

The considerations that follow call upon notions of extension and restriction of the ring of scalars, and of functor (see section 13.6.5).

From the algebraic point of view, consider the discrete-time system Σ_d as associated with a finitely presented $\mathbf{R}$-module M ($\mathbf{R}$ denoting the principal ideal domain $\mathbb{R}\,[q]$: see section 10.4.2). According to equation (10.33), in order that Δ be able to act on the $\mathbf{R}$-module M, we need to be able to divide the elements of M by $q + 1$.

Thus, let there be the multiplicative set $Q = \{(q+1)^n, n \geq 0\} \subset \mathbf{R}$ and let $\mathbf{B} = Q^{-1}\mathbf{R}$ be the principal ideal domain consisting of all elements of the form $r/(q+1)^n$, $r \in \mathbf{R}$, $n \geq 0$ (see section 13.6.5). Since $\Delta \in \mathbf{B}$, the principal ideal domain $\mathbf{C} = \mathbb{R}[\Delta]$ is a subring of $\mathbf{B}$.

Conversely, according to (10.34), $1 - \Delta/\lambda = 2/(q+1)$. Let there be the multiplicative set $T = \{(1-\Delta/\lambda)^n, n \geq 0\} \subset \mathbf{C}$ and let $\mathbf{D} = T^{-1}\mathbf{C}$. According to equation (10.34), $q \in \mathbf{D}$, and hence $\mathbf{R}$ is a subring of $\mathbf{D}$.

We can thus go from an $\mathbf{R}$-module to a $\mathbf{C}$-module and conversely by means of two functors, denoted by $\mathcal{G}$ and $\mathcal{H}$, respectively, and defined as follows:

– Functor $\mathcal{G}$: extension of the ring of scalars from $\mathbf{R}$ to $\mathbf{B}$, then restriction of the ring of scalars from $\mathbf{B}$ to $\mathbf{C}$ (see section 13.6.5).

– Functor $\mathcal{H}$: extension of the ring of scalars from $\mathbf{C}$ to $\mathbf{D}$, then restriction of the ring of scalars from $\mathbf{D}$ to $\mathbf{R}$.

These two functors are exact, as they are the compositions of two exact functors, according to ([102], Corollary 3.74 and Theorem 9.31).

Let M be an $\mathbf{R}$-module. The function $\theta_M : M \to \mathbf{B} \otimes_{\mathbf{R}} M$ defined by $m \mapsto 1 \otimes m$ (where 1 is the unit element of $\mathbf{B}$) is $\mathbf{R}$-linear, and according to equation (13.66), section 13.6.5:

$$\ker \theta_M = \{m \in M : (q+1)^n \ m = 0 \text{ for sufficiently large } n\}.$$

Likewise, let $\breve{M}$ be a $\mathbf{C}$-module. The function $\psi_{\breve{M}} : \breve{M} \to \mathbf{D} \otimes_{\mathbf{C}} \breve{M}$ defined by $\breve{m} \mapsto 1 \otimes \breve{m}$ is $\mathbf{C}$-linear, and

$$\ker \psi_{\breve{M}} = \left\{\breve{m} \in \breve{M} : (1-\Delta/\lambda)^n \ \breve{m} = 0 \text{ for sufficiently large } n\right\}.$$

LEMMA 344.– *Let M be a finitely presented $\mathbf{R}$-module. (i) If M is free of rank n (resp., torsion), then $\mathcal{G}M$ is a finitely presented $\mathbf{C}$-module which is free of rank n (resp., torsion). (ii) Let N be a submodule of M. Then $\mathcal{G}N$ is a submodule of $\mathcal{G}M$ and $\mathcal{G}M/\mathcal{G}N = \mathcal{G}(M/N)$.*

PROOF. See ([10], sections II.1 and II.5) and ([102], Theorem 3.76). ■

DEFINITION 345.– *Let Σ_d be a discrete-time system, i.e. a finitely presented $\mathbf{R}$-module M. The* pseudocontinuous system *$\breve{\Sigma}$ associated with Σ_d is the finitely presented $\mathbf{C}$-module $\breve{M} = \mathcal{G}M$. Conversely, let $\breve{\Sigma}$ be a pseudocontinuous system, i.e. a finitely presented $\mathbf{C}$-module $\breve{M}$. The discrete-time system associated with $\breve{\Sigma}$ is the finitely presented $\mathbf{R}$-module $M = \mathcal{H}\breve{M}$.*

PROPOSITION 346.– *Let $\mathbf{\Sigma}_d$ be a discrete-time control system, with input $u_d = [u_{d_1} \, ... \, u_{d_m}]^T$ and output $y_d = [y_{d_1} \, ... \, y_{d_p}]^T$. Then, the pseudocontinuous system $\breve{\mathbf{\Sigma}}$ is a control system with input $\breve{u}_d = [\breve{u}_{d_1} \, ... \, \breve{u}_{d_m}]^T$ and output $\breve{y}_d = [\breve{y}_{d_1} \, ... \, \breve{y}_{d_p}]^T$, where $\breve{u}_{d_i} = \theta_M (u_{d_i})$ and $\breve{y}_{d_j} = \theta_M (y_{d_j})$ $(1 \leq i \leq m, 1 \leq j \leq p)$. If $\mathbf{\Sigma}_d$ is controllable, then so too is $\breve{\mathbf{\Sigma}}$.*

PROOF. Lemma 344 implies that Property (ii) of section 7.1.1 holds. If $\mathbf{\Sigma}_d$ is controllable, this system is a free $\mathbf{R}$-module M, and hence $\breve{\mathbf{\Sigma}}$ is the $\mathbf{C}$-module $\mathcal{G}M$ which is free according to Lemma 344. ■

We will illustrate Definition 345 through three examples, the first is elementary and the next two are "pathological":

EXAMPLE 347.– *Let there be the system $\mathbf{\Sigma}_d : q \, y_d = u_d$, and let M be the associated $\mathbf{R}$-module. The $\mathbf{B}$-module $\mathbf{B} \otimes_{\mathbf{R}} M$ is defined for any value of $n \geq 0$ by*

$$\frac{q}{(q+1)^n} \, \breve{y}_d = \frac{1}{(q+1)^n} \, \breve{u}_d$$

where $\breve{y}_d = \theta_M (y_d)$ and $\breve{u}_d = \theta_M (u_d)$. By taking $n = 1$, we obtain

$$(1 + \Delta/\lambda) \, \breve{y}_d = (1 - \Delta/\lambda) \, \breve{u}_d$$

which defines $\breve{M} = \mathcal{G}M$. In practice, we obtain $\breve{\mathbf{\Sigma}}$ by replacing the operator q by its expression (10.34) as a function of Δ.

EXAMPLE 348.– *Let there be the system $\mathbf{\Sigma}_d : (q+1) \, y_d = 0$. The module $\breve{M} = \mathcal{G}M$ is reduced to 0.*

EXAMPLE 349.– *Let there be the system $\mathbf{\Sigma}_d : (q^2 - 1) \, y_d = u_d$, which admits a strictly causal state-space representation of order 2, the state matrix of which has eigenvalues -1 and 1 (Lemma 341 is therefore not applicable). Using the functor $\mathcal{G}$, we obtain*

$$4 \, (\Delta/\lambda) \, \breve{y}_d = (1 - \Delta/\lambda)^2 \, \breve{u}_d.$$

To put $\breve{\mathbf{\Sigma}}$ in state-space form, we can set $\breve{x}_d = 4 \, \breve{y}_d + (2 - \Delta/\lambda) \, \breve{u}_d$ and we obtain

$$\begin{cases} \Delta \, \breve{x}_d = \breve{u}_d, \\ \breve{y}_d = \frac{1}{4} \, \breve{x}_d + \frac{1}{4} \, (\Delta/\lambda - 2) \, \breve{u}_d, \end{cases}$$

which is an improper first-order state-space representation.

10.5.4. *Structural properties of pseudocontinuous systems*

Consider a pseudocontinuous system $\check{\Sigma}$. Its structural properties (stability, controllability, observability, etc.) are defined in the same manner as those of a continuous-time system and they are thus made explicit by replacing (in a formal way) the operator Δ by ∂.

Let Σ_d be a discrete-time causal control system and let $\check{\Sigma}$ be the associated pseudocontinuous control system. Suppose that -1 is not a pole of Σ_d. We then have the following:

THEOREM 350.– *The pseudocontinuous system $\check{\Sigma}$ is stable (resp., controllable, observable, etc.) if and only if, Σ_d has the same property.*

PROOF. We can assume without loss of generality that Σ_d is defined by a state-space representation $\{A_d, B_d, C, D\}$ according to Remark 320 (section 10.4.2). Therefore, $\check{\Sigma}$ is the state-space system $\{\check{A}, \check{B}, \check{C}, \check{D}\}$ given by Proposition 341. The spectrum of $\check{A}$ is included in the left half-plane if and only if, that of A_d is included in the open unit disk according to Proposition 339 (section 10.5.1) and Proposition 439 (section 12.4.2). On the other hand, we show the equivalence between the controllability of (A_d, B_d) and that of $(\check{A}, \check{B})$ by applying the Popov–Belevitch–Hautus test. It is the same for the equivalence between the observability of (C, A_d) and that of $(\check{C}, \check{A})$. ■

10.6. Synthesis of discrete-time control

10.6.1. *Direct approaches*

The design methods of continuous-time controls studied in the preceding chapters (PID and RST controllers, state feedback control with integral action – or more generally "internal model", state feedback/observer synthesis, etc.) can be reformulated in the context of discrete time. In this way, we are led to syntheses of discrete-time controls which we can qualify as "direct approaches". These approaches are discussed in a rather complete manner in [4]; they will not be developed here.

10.6.2. *Discretization by approximation*

It is also possible to derive a digital controller from an analog controller by approximation. The *Euler approximation*,

$$\partial \simeq \frac{q-1}{T} \tag{10.38}$$

is the simplest one. It comes down to writing, if x is a differentiable continuous-time signal,

$$\dot{x}(t) \simeq \frac{x(t+T) - x(t)}{T}$$

and that $t = kT$. This approximation is of course only valid if x does not vary too rapidly between two sampling instants, and hence if the sampling period T is sufficiently small.

Consider, for example, an analog PID controller, whose transfer function is

$$K(s) = k\left(1 + \frac{1}{T_I\, s} + \frac{T_d\, s}{1 + \frac{T_d}{N}\, s}\right).$$

With zero initial conditions, the approximation (10.38) is equivalent to

$$s \simeq \frac{z-1}{T}.$$

We thus obtain the digital PID controller with transfer function

$$K_d(z) = k\left(1 + \frac{T}{T_I}\frac{1}{z-1} + \frac{T_d}{T}\frac{z-1}{1 + \frac{T_d}{N\,T}(z-1)}\right).$$

10.6.3. *Passage through a pseudocontinuous system*

The effectiveness of the approach detailed here below has been pointed out in a more general context [99].

General procedure

Let $\mathbf{\Sigma}_d$ be a discrete-time control system, not having a pole at -1 and having the appropriate structural properties (controllability, observability, etc.) and let $\breve{\mathbf{\Sigma}}$ be the associated pseudocontinuous system (which has the same structural properties according to Theorem 350). We apply one of the design methods of continuous-time control discussed in section 10.6.1 to $\breve{\mathbf{\Sigma}}$ (the operator ∂ being replaced by Δ). We thus obtain a suitable pseudocontinuous controller $\breve{\mathbf{\Theta}}$, for which the pseudocontinuous closed-loop system is stable. Now let $\mathbf{\Theta}_d$ be the discrete-time controller associated with $\breve{\mathbf{\Theta}}$ (i.e. $\mathbf{\Theta}_d = \mathcal{H}\breve{\mathbf{\Theta}}$, with the notation in section 10.5.3*). According to Theorem 350, $\mathbf{\Sigma}_d$, fed back by $\mathbf{\Theta}_d$, is stable.

REMARK 351.– *The pseudocontinuous control system $\breve{\mathbf{\Sigma}}$ is proper but not strictly proper (see section 10.5.2, Remark 343(i)). As a result, for the procedure we are proposing to be applicable, the methods used to design continuous-time controls have to apply to this type of system (see section 6.5, Exercises 121 and 128; section 8.4, Exercise 269; section 9.3.1, Remark 290; section 9.4, Exercise 302).*

Consideration of the computing time

The above controller Θ_d is causal but not strictly causal (even if $\breve{\Theta}$ is strictly proper): see Exercise 358. If the computing time of the control $u_d\,(k)$ is not negligible relative to the sampling period (which depends both on the controller complexity and the computing rapidity), it is necessary, when implementing the control law in real time, to only "send" the control $u_d\,(k)$ at instant $(k+1)\,T$, thus using a strictly causal controller.

EXAMPLE 352.– *Consider the continuous-time system* Σ *defined by the left form* (6.25) *(section 6.3.6). The (Z.O.H.) discretized system* Σ_d *with sampling period* $T = 0.1$ *is described by the left form* $A_d\,(q)\; y_d = B_d\,(q)\; u_d$*, where*

$$
\begin{aligned}
A_d\,(q) &= q^2 - 1.72\,q + 0.74 \\
B_d\,(q) &= -0.016\,q + 0.051.
\end{aligned}
$$

We wish to design a digital RST controller with equation

$$S_d\,(q)\; u_d = R_d\,(q)\; (r_d - y_d) \tag{10.39}$$

having characteristics similar to those of the continuous-time RST controller whose polynomials are defined by equalities (6.28) *and* (6.29)*.*

(i) Case of a non-strictly causal *controller. As mentioned above, the implementation of such a controller is possible if the computing time of the control* $u_d\,(k)$ *is small compared with the sampling period (up to about 20% of this period). The more powerful computers are, the more frequent this situation becomes. The pseudocontinuous system* $\breve{\Sigma}$ *associated with* Σ_d*, for* $\lambda = \frac{2}{T}$*, is described by the left form* $\breve{A}\,(\Delta)\;\breve{y} = \breve{B}\,(\Delta)\;\breve{u}$*, with*

$$
\begin{aligned}
\breve{A}\,(\Delta) &= \Delta^2 + 2.99\,\Delta + 1.99, \\
\breve{B}\,(\Delta) &= 0.019\,\Delta^2 - 0.587\,\Delta + 3.98.
\end{aligned}
$$

Note that the coefficients of these polynomials are close to those of the polynomials $A\,(\partial)$ *and* $B\,(\partial)$*. We choose, like in section 6.3.6,*

$$\breve{A}_{cl}\,(\Delta) = \breve{A}_c\,(\Delta)\,(\Delta + 10)^2$$

with $\breve{A}_c\,(\Delta) = (\Delta + 2)^2$*. The polynomials* $\breve{S}\,(\Delta)$ *and* $\breve{R}\,(\Delta)$ *which realize the desired pole placement while satisfying condition* $\breve{S}\,(0) = 0$ *(integrator) are*

$$
\begin{aligned}
\breve{S}\,(\Delta) &= \Delta^3 + 21.40\,\Delta^2 + 209.54\,\Delta, \\
\breve{R}\,(\Delta) &= 83.02\,\Delta^2 + 266.24\,\Delta + 200.83,
\end{aligned}
$$

and define the pseudocontinuous controller $\breve{\Theta}$*. This one here is associated with the digital controller* Θ_d *defined by the left form* (10.39)*, where*

$$
\begin{aligned}
S_d\,(q) &= q^3 - 1.367\,q^2 + 0.542\,q - 0.175, \\
R_d\,(q) &= 1.867\,q^3 - 1.315\,q^2 - 1.828\,q + 1.353.
\end{aligned}
$$

(ii) Case of a strictly causal *controller. If we want to obtain a strictly causal controller, we first replace* $A_d(q)$ *by*

$$\breve{A}_d(q) = q\,A(q),$$

thus the system $\mathbf{\Sigma}_d$ *by the system* $\breve{\mathbf{\Sigma}}_d$*, which has a supplementary unity delay. The pseudocontinuous system associated with* $\breve{\mathbf{\Sigma}}_d$ *is (again with* $\lambda = \frac{2}{T}$*) described by the left form* $\breve{A}_2(\Delta)\,\breve{y} = \breve{B}_2(\Delta)\,\breve{u}$ *where*

$$\begin{aligned}\breve{A}_2(\Delta) &= \Delta^3 + 22.99\,\Delta^2 + 61.84\,\Delta + 39.83,\\ \breve{B}_2(\Delta) &= -0.019\,\Delta^3 + 0.97\,\Delta^2 - 15.72\,\Delta + 79.67.\end{aligned}$$

Choose now $\breve{A}_{cl}(\Delta) = \breve{A}_c(\Delta)(\Delta+10)^2(\Delta+2/T)^2$*, where each root* $\Delta = -2/T$ *corresponds to a supplementary pole* $q = 0$ *introduced by the computing delay. We obtain the polynomials*

$$\begin{aligned}\breve{S}(\Delta) &= \Delta^4 + 44.65\,\Delta^3 + 539.05\,\Delta^2 + 4858.88\,\Delta,\\ \breve{R}(\Delta) &= 84.71\,\Delta^3 + 1963.91\,\Delta^2 + 5593.19\,\Delta + 4016.66\end{aligned}$$

that define a strictly proper pseudocontinuous controller $\tilde{\mathbf{\Theta}}$*, which is associated with a discrete-time controller* $\tilde{\mathbf{\Theta}}_d$*. The latter is defined by a left form whose polynomials can be denoted by* $\tilde{S}_d(q)$ *and* $R_d(q)$*. The digital controller we are looking for is defined by the left form* (10.39) *where* $S_d(q) = q\,\tilde{S}_d(q)$*. We finally obtain (after simplification of the root* $q = 0$*, common to* $S_d(q)$ *and* $R_d(q)$*)*[10]

$$\begin{aligned}S_d(q) &= q^4 - 1.398\,q^3 + 0.637\,q^2 - 0.145\,q - 0.095,\\ R_d(q) &= 1.903\,q^3 - 1.344\,q^2 - 1.863\,q + 1.383.\end{aligned}$$

(iii) In the two cases, $S_d(1) = 0$*, which means that the controller is an integrator system (*$S_d(1)$ *is the sum of the coefficients of the polynomial* $S_d(q)$*); in addition,* $R_d(-1) = 0$*; this is due to the fact that, since* each pseudocontinuous controller is strictly proper, *its zero at infinity (i.e. at* $w = \infty$*) is transformed into a zero at* $z = -1$ *through the inverse bilinear transform* (10.31) *(it is important that this condition be respected for the digital controller to have zero gain at the Nyquist frequency, which plays the role of "high frequencies" for discrete-time systems: see section 4.2.6). It now remains to verify the good behavior of the closed-loop system. The events are as follows: (i) unit step command at instant* $t = 0$*; (ii) disturbance step of amplitude* 0.3 *adding to* y *at instant* $t = 5$*. The output* y_d *and the control* u_d *are represented as functions of time in Figure 10.12, with the first (- -) and the second (–) controller – see below. The two responses are almost identical; the second controller nonetheless generates a slight delay compared with the first (which was predictable). These responses are also very similar to those in Figures 6.6 and 6.8 of section 6.3.6 (this time, however, the simulation is performed without measurement noise).*

10. The reader is requested to show that due to the factor $\Delta + 2/T$ in $\breve{A}_{cl}(\Delta)$, the polynomials $S_d(q)$ and $R_d(q)$ must have this root in common (the digital RST controller thus is 0-controllable, but not controllable, before the indicated simplification is performed).

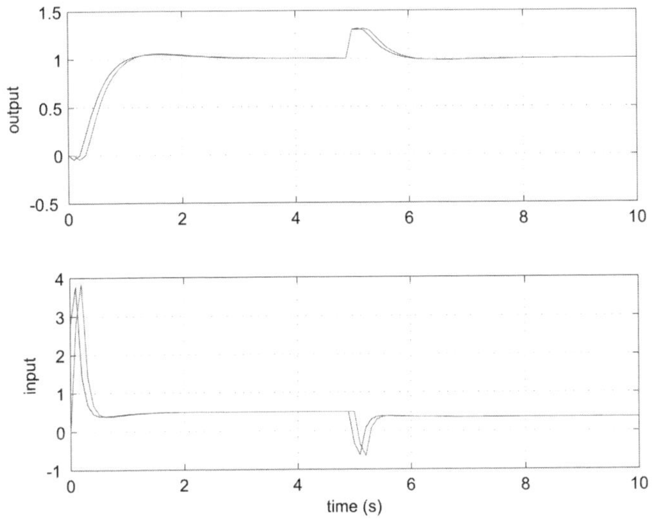

Figure 10.12. *Time responses (Example 352)*

EXAMPLE 353.– *Consider the same continuous-time system* $\mathbf{\Sigma}$ *as in Example 352, but in the situation specified in section 6.4.7, no.1. We must not only ensure a control without any static error, but also reject a sinusoidal disturbance with angular frequency* $\omega_0 = 1$ *rad/s. The notation used here is identical to that in the cited paragraph (*mutatis mutandis*). We assume that the computing time is small compared with the sampling period, and that it is therefore useless to design a strictly causal digital controller (otherwise, the method already detailed in Example 352(ii) can be used) The easiest way is to use the bilinear transform with prewarping at the normalized angular frequency* $\theta_0 = \omega_0 T$, *with coefficient* λ *given by the expression* (10.32) *(for another approach, see Exercise 361). Therefore,*

$$D_1(\Delta) = \Delta^2 + 2\varsigma\,\omega_0\,\Delta + \omega_0^2$$

with $\varsigma = 0.005$ *for example. The polynomials* $A_s(\Delta)$ *and* $A_{cl}(\Delta)$ *are, respectively,*

$$\begin{aligned} A_s(\Delta) &= (\Delta + 1 + i\,\omega_0)(\Delta + 1 - i\,\omega_0)(\Delta + 10)^2, \\ A_{cl}(\Delta) &= A_s(\Delta)(\Delta + 2)(\Delta + 1)^2, \end{aligned}$$

and the corresponding polynomials $\check{R}(\Delta)$, $\check{S}(\Delta)$, *and* $\check{T}(\Delta)$ *are*

$$\begin{aligned} \check{R}(\Delta) &= 99.58\,\Delta^4 + 360.90\,\Delta^3 + 437.50\,\Delta^2 + 276.91\,\Delta + 100.58, \\ \check{S}(\Delta) &= \Delta^5 + 21.08\,\Delta^4 + 221.38\,\Delta^3 + 23.27\,\Delta^2 + 220.17\,\Delta, \\ \check{T}(\Delta) &= 0.50\,\Delta^4 + 11.06\,\Delta^3 + 71.41\,\Delta^2 + 120.70\,\Delta + 100.58. \end{aligned}$$

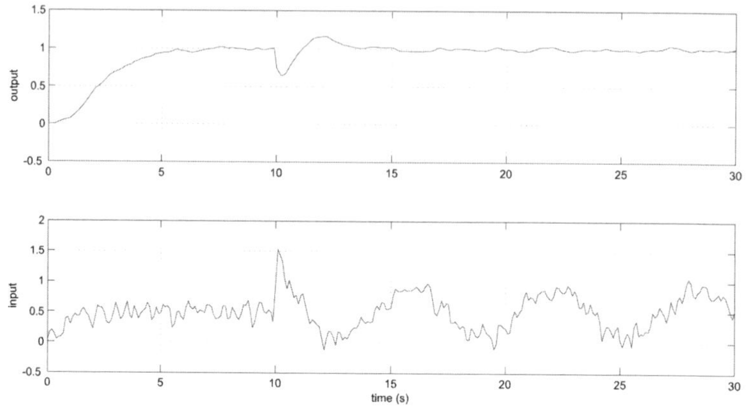

Figure 10.13. *Time responses (Example 353)*

Note that this pseudocontinuous controller is strictly proper (see Example 352 (iii)) and that $\check{R}(0) = \check{T}(0)$ *(see section 6.4.2). The polynomial* $\check{S}(\Delta)$ *can be put in the form*

$$\check{S}(\Delta) = \Delta\, D_1(\Delta)\, (\Delta + \beta)\left(\Delta + \bar{\beta}\right)$$

with $\beta = -10.53 + 10.45\, i$*, and*

$$\check{T}(\Delta) = A_s(\Delta)\, \frac{\check{R}(0)}{A_s(0)}.$$

This pseudocontinuous controller $\tilde{\mathbf{\Theta}}$ *is associated with a discrete-time controller* $\mathbf{\Theta}_d$ *using the transform* (10.33) *with the value of* λ *as specified above. This controller* $\mathbf{\Theta}_d$ *is a digital RST controller whose polynomials are*

$$\begin{aligned}
R_d(q) &= 2.274\, q^5 - 6.043\, q^4 + 3.081\, q^3 + 4.462\, q^2 - 5.356\, q + 1.581, \\
S_d(q) &= q^5 - 3.334\, q^4 + 4.208\, q^3 - 2.598\, q^2 + 0.9139\, q - 0.1905, \\
T_d(q) &= 10^{-2}\, (2.396\, q^5 - 3.514\, q^4 - 0.809\, q^3 + 3.315\, q^2 \\
&\quad -1.567\, q + 0.217).
\end{aligned}$$

We have $S_d(1) = 0$*,* $R_d(-1) = 0$ *(see Example 352(iii)), and* $R_d(1) = T_d(1)$ *since* $\check{T}(0) = \check{R}(0)$ *(this is the condition for the static error to be zero). The simulation of the closed-loop system yields the responses in Figure 10.13 (with the same conditions as in section 6.4.7, n°1). These responses are almost identical to those in Figure 6.12*

10.7. Exercises

EXERCISE 354.– *Establish all the results in table* (10.14).

EXERCISE 355.– *Let there be the sinusoid* $x(t) = \sin(\pi t)$. *(a) What is its frequency? (b) This sinusoid is being discretized at frequency* $f_e = 1$, *and hence the sampling instants are the rational integers. What is the obtained discretized signal* x_d*? (c) Can we replace the strict inequality by an inequality which is not strict in the statement of Theorem 305?*

EXERCISE 356.– *Consider the continuous-time state-space system* $\mathbf{\Sigma} = \{A, B, C\}$ *with*

$$A = \begin{bmatrix} -2 & 0 \\ 1 & -2 \end{bmatrix}, \quad B = \begin{bmatrix} 1 \\ 0 \end{bmatrix},$$
$$C = \begin{bmatrix} 0 & 1 \end{bmatrix}.$$

(i) Determine the poles of $\mathbf{\Sigma}$. *(ii) The system* $\mathbf{\Sigma}$ *is discretized with sampling period* $T = 0.2$; *determine the discretized system* $\mathbf{\Sigma}_d = \{A_d, B_d, C_d\}$. *(iii) Calculate the transfer function of* $\mathbf{\Sigma}_d$ *and its poles.*

EXERCISE 357.– *The altitude* θ *of a satellite satisfies the equation*

$$J\,\frac{d^2\theta}{dt^2} = \Gamma(t)$$

where J *is the moment of inertia and* $\Gamma(t)$ *is the torque exerted at instant* t. *We put* $y = \theta$ *and* $u = \Gamma/J$. *(i) Determine the observable canonical form* $\{A, B, C\}$ *of this system* $\mathbf{\Sigma}$. *(ii) The system* $\mathbf{\Sigma}$ *is discretized with period* T. *Determine the state-space realization* $\{A_d, B_d, C\}$ *of the discretized system* $\mathbf{\Sigma}_d$ *by applying relation* (10.15) *of section 10.3.4. Is it an observable canonical form? (iii) Determine the left form* $D(q)\, y_d = N(q)\, u_d$ *which governs the system* $\mathbf{\Sigma}_d$. *(iv) We now assume that the torque exerted to the satellite is no longer* u *(up to factor* J*) but* $u(t-\tau)$, *where* τ *is the transmission delay of the information from Earth. By putting* $\tau = 3T$, *write the left form which governs* $\mathbf{\Sigma}_d$, *taking this delay into account. (v) Write the observable canonical form of* $\mathbf{\Sigma}_d$ *in this situation. (vi) Determine the block diagram of* $\mathbf{\Sigma}_d$, *analogous to that in Figure 7.5 (section 7.4.2) but where the integrators are replaced by delays* q^{-1}. *(vii) Generalize the above to the case where* $\tau = nT$, $n \geq 1$. *(viii) Considering an RST controller calculated according to the methods presented in Examples 352 and 353, is it well adapted for the* control of a time-delay system *such as this one here? If yes, is it not a kind of* predictive control*?*

EXERCISE 358.– *Let* $\breve{\mathbf{\Sigma}}$ *be a pseudocontinuous system defined by a state-space representation* $\{\breve{A}, \breve{B}, \breve{C}, \breve{D}\}$. *Determine the corresponding discrete-time state-space* *system $\Sigma_d = \{A_d, B_d, C, D\}$ by making the appropriate hypothesis on the eigenvalues of $\breve{A}$. In the general case, is Σ_d strictly causal?*

EXERCISE 359.– *Let Σ be a continuous-time state-space system of order n, with equation $\dot{x} = A\,x + B\,u$, and discretized at period T. We denote the state and control matrices of the discretized system Σ_d by A and B, respectively. (i) Show that for $T \to 0^+$, $A_d \to I_n$, $B_d \to 0$ and that all the poles of Σ_d tend to 1. (ii) We denote the operator $(q-1)/T$ as δ. Determine the state and control matrices $\tilde{A}_d$ and $\tilde{B}_d$ of Σ_d when its state-space equation is expressed using the operator δ, according to $\delta\, x_d = \tilde{A}_d\, x_d + \tilde{B}_d\, u_d$. By abuse of language, we will call this equation the* "delta transform" *of the state equation of Σ_d [4]. (iii) Determine the limits of $\tilde{A}_d$ and $\tilde{B}_d$ as $T \to 0^+$; deduce that Σ_d is better represented using the delta operator (instead of the shift-forward operator q) after quantization when the sampling period is very small. (iv) In order to implement the control law, does the operator δ pose any problem? (Compare with the operator Δ.)*

EXERCISE 360.– *Let Σ be the continuous-time system defined by the left form $(\partial + 1)\, y = u$. (i) This system is discretized at period $T \ll 1$. Determine the transfer function $G_d(z)$of the discretized system Σ_d, and then the transfer function $\breve{G}(w)$ of the associated pseudocontinuous system $\breve{\Sigma}$ (with $\lambda = 2/T$). (ii) Determine $\lim_{T \to 0} \breve{G}(w)$. What is remarkable about this limit? (iii) We hypothesize that the computing time is negligible compared with the sampling period $T = 0.2$. Determine the pseudocontinuous RST controller having the following properties: (a) it is an integrator system; (b) $\breve{R}/\breve{S}$ has relative degree $\delta_0 = 1$; (c) all the poles of the pseudocontinuous closed-loop system are placed at $-10/3$; (d) the transfer function between the reference signal and the output is of order 1. (iv) Determine the corresponding discrete-time RST controller. Calculate $S_d(1)$, $R_d(1)$, $T_d(1)$, and $R_d(-1)$. Interpretation ? (vi) Examine the problem again in the case where we cannot neglect the computing time compared with the sampling period.*

EXERCISE 361.– *How do you treat Example 353 using a bilinear transform without prewarping?*

EXERCISE 362.– *(i) Can a discretized system have poles in the subset $(-\infty, 0)$ of the real line? (ii) Can it have poles at 0?*

EXERCISE 363.– * *Consider the discrete-time system defined by the left form*

$$(q+1)\, q\, y_d = (q+1)\; u_d.$$

*(i) Is this system stabilizable? (ii) Show that the associated pseudocontinuous system is controllable. Does this result contradict Theorem 350 or Proposition 346? (iii) Can this system be obtained by discretization of a continuous-time system?**

Chapter 11

Identification

A "visual" method of identification has already been discussed in this book (see Exercise 71, section 3.6). Other methods that are a little less rustic exist (Strejc's and Broïda's, for example [77]) whose results, however, are not significantly better. We present here one of the widely used and efficient identification approaches, i.e. the parametric identification. The idea is to first choose *a priori* a model $\mathcal{M}(\theta)$ of the system (for example, on the basis of physical considerations), where θ is the *parameter vector* to be estimated. The estimation of θ is done by minimizing a cost function (or "criterion"), for example the norm l_2 of the "prediction error" (see section 12.2.1 for the definition of this norm). To clarify ideas, consider the discrete–time system

$$\left(q^2 + a_1 q + a_2\right) y_d = \left(b_1 q + b_2\right) u_d.$$

This equation determines the model structure (which is a discrete–time left form of the second order). The coefficients to be estimated are the a_i's and b_i's $(1 \leq i \leq 2)$. Putting

$$\theta = \begin{bmatrix} a_1 & a_2 & b_1 & b_2 \end{bmatrix}^T,$$

the identification problem now comes down to estimating θ, by minimizing an appropriate criterion $J(\theta)$.

11.1. Random signals

The natural framework used to discuss parametric identification methods is that of *random signals* (also called *stochastic processes*). The theory of random signals is based on the theory of probabilities (see section 12.7). We will begin with the study of *pseudo-random signals* (in the sense of [5]), which share many properties with *ergodic random signals* (see section 11.1.5), and we will limit our discussion to the former (general stochastic processes are studied in, e.g. [8] and [37]).

11.1.1. *Moments of order 1 and 2*

Let $x = \{x(t)\}_{t\in\mathbb{Z}}$ be a discrete-time signal with values in $\mathbf{K} = \mathbb{R}$ or $\mathbb{C}$ (i.e. $x \in \mathbf{K}^{\mathbb{Z}}$). The *mean* (or *moment of order 1*) of this signal is

$$\boxed{\overline{x(t)} = \lim_{N\to+\infty} \tfrac{1}{2N+1} \sum_{t=-N}^{N} x(t)}.$$

(if it exists, and with an abuse of language because $\overline{x(t)}$ does not depend on t).

REMARK 364.– *The mean of a signal only defined for* $t \geq 0$ *is*

$$\overline{x(t)} = \lim_{N\to+\infty} \frac{1}{N} \sum_{t=0}^{N-1} x(t).$$

The *moment of order 2* of x (if it exists), also called its *mean energy*, is

$$\boxed{\overline{|x(t)|^2} = \lim_{N\to+\infty} \tfrac{1}{2N+1} \sum_{t=-N}^{N} |x(t)|^2}.$$

DEFINITION 365.– *A signal x is said to be* centered *if* $\overline{x(t)} = 0$.

11.1.2. *Correlation and cross-correlation*

DEFINITION 366.– *Let* $x \in \mathbf{K}^{\mathbb{Z}}$. *Its* correlation sequence *(or, abusing the language, its* correlation function*) is* $r_{xx} = \{r_{xx}(\tau)\}_{\tau\in\mathbb{Z}}$ *such that*

$$\boxed{r_{xx}(\tau) = \overline{x^*(t)\, x(t+\tau)} = \lim_{N\to+\infty} \tfrac{1}{2N+1} \sum_{t=-N}^{N} x^*(t)\, x(t+\tau)}$$

if it exists,[1] *in which case the signal x is said to be* correlatable. *Let* $y \in \mathbf{K}^{\mathbb{Z}}$. *The* cross-correlation sequence *(or* function*) of x and y is the sequence* $r_{xy} = \{r_{xy}(\tau)\}_{\tau\in\mathbb{Z}}$ *such that*

$$\boxed{r_{xy}(\tau) = \overline{x^*(t)\, y(t+\tau)} = \lim_{N\to+\infty} \tfrac{1}{2N+1} \sum_{t=-N}^{N} x^*(t)\, y(t+\tau)}$$

1. In this chapter, x^* denotes the conjugate of x.

if it exists. Two signals $x, y \in \mathbf{K}^{\mathbb{Z}}$ *are said to be* cross-correlatable *if they are correlatable and* r_{xy} *exists, and* uncorrelated *if* $r_{xy} = 0$.

PROPOSITION 367.– *(i)* r_{yx} *exists if and only if* r_{xy} *exists and* $r_{yx}(\tau) = r_{xy}^{*}(-\tau)$ *(in such a way that the relation "being cross-correlatable" is symmetric). (ii) If* x *and* y *are cross-correlatable, then they are of finite mean energy and*

$$|r_{xy}(\tau)| \leq \sqrt{r_{xx}(0)}\sqrt{r_{yy}(0)}.$$

PROOF. (i) is obvious and (ii) is a consequence of the Schwarz inequality. ■

We write

$$\boxed{r'_{x,y}(\tau) = \frac{r_{xy}(\tau)}{\sqrt{r_{xx}(0)}\sqrt{r_{yy}(0)}}}$$

and according to Proposition 367 we have

$$\boxed{\left|r'_{x,y}(\tau)\right| \leq 1}.$$

DEFINITION 368.– *The sequence* $r'_{x,x}$ *is the* normalized correlation sequence *(or* function*) of* x *and* $r'_{x,y}$ *is the* normalized cross-correlation sequence *(or* function*) of* x *and of* y.

PROPOSITION 369.– *A set* $\mathcal{A}$ *of pairwise cross-correlatable signals is a* $\mathbf{K}$*-vector space.*

PROOF. It suffices to notice that if $z = \mu x$ ($\mu \in \mathbf{K}$), $r_{zz} = |\mu|^2 r_{xx}$ and that $r_{x+y,x+y} = r_{xx} + r_{xy} + r_{yx} + r_{yy}$. ■

DEFINITION 370.– *Let* $f = \{f(t)\}_{t\in\mathbb{Z}}$ *be a discrete-time signal. This signal is said to be* of positive type *(written as* $f \gg 0$*) if for any sequence* (c_k) *of complex numbers and any sequence* (τ_k) *of elements of* $\mathbb{Z}$*, we have*

$$\sum_{(k,l)\in I\times I} c_k\, c_l^{*}\, f(\tau_k - \tau_l) \geq 0$$

for every finite set of indices I.

PROPOSITION 371.– *Let* x *be a correlatable signal. Then* $r_{xx} \gg 0$.

PROOF. Let (c_k) be a sequence of complex numbers and (τ_k) be a sequence of elements of $\mathbb{Z}$. For any finite set of indices I, we have

$$\sum_{(k,l)\in I^2} c_k c_l^* \sum_{t=-N}^{N} x^*(t)\, x(t+\tau_k-\tau_l)$$

$$= \sum_{(k,l)\in I^2} c_k c_l^* \sum_{t=-N+\tau_k}^{N+\tau_k} x^*(t-\tau_k)\, x(t-\tau_l)$$

and

$$\sum_{t=-N+\tau_k}^{N+\tau_k} = \sum_{t=-N}^{N} + \sum_{t=N+1}^{N+\tau_k} - \sum_{t=-N}^{-N+\tau_k-1}.$$

The two sums on the right-hand side, divided by $2N+1$, tend to 0 as $N \to +\infty$. On the other hand,

$$\sum_{(k,l)\in I\times I} \frac{c_k c_l^*}{2N+1} \sum_{t=-N}^{N} x^*(t-\tau_k)\, x(t-\tau_l)$$

$$= \frac{1}{2N+1} \sum_{t=-N}^{N} \left| \sum_{k\in I} c_k^* x(t-\tau_k) \right|^2 \geq 0.$$

■

*The following result is a variant of the *Bochner theorem* [5], the proof of which can be found in ([37], Chapter X).

THEOREM 372.– *Let x be a correlatable signal. (i) r_{xx} can be written in the form of a Fourier–Stieltjes integral*

$$r_{xx}(\tau) = \frac{1}{2\pi} \int_{-\pi}^{\pi} e^{i\omega\tau}\, d\Phi_{xx}(\omega),$$

where Φ_{xx} is a bounded non-decreasing function called the spectral function *of x.*[2] *(ii) $\overline{r_{xx}(\tau)} = \sigma(0) = \Phi(0^+) - \Phi(0^-) \geq 0$. (iii) $\overline{|r_{xx}(\tau)|^2} = \frac{1}{4\pi^2} \sum_{\omega} \sigma^2(\omega)$, where $\sigma(\omega) = \Phi(\omega^+) - \Phi(\omega^-) \geq 0$ (σ can only be non-zero in a countable set of points). (iv) As a result, if Φ is a continuous function, then $\overline{r_{xx}(\tau)} = \overline{|r_{xx}(\tau)|^2} = 0$.**

2. In this chapter, the normalized angular frequency (see (10.8), section 10.2.5) is denoted by ω and not θ, to avoid confusion with the parameter vector.

PROPOSITION 373.– *Let x be a correlatable signal. Then,* $\overline{r_{xx}(t)} = \left|\overline{x(t)}\right|^2 + \overline{r_{yy}(t)}$, *where* $y(t) = x(t) - \overline{x(t)}$.

PROOF. Writing $m = \overline{x(t)}$, we have

$$r_{xx}(\tau) = \overline{(m^* + y^*(t))(m + y(t+\tau))} = |m|^2 + r_{yy}(\tau)$$

because $\overline{y(t)} = 0$. On the other hand, the Bochner theorem shows that $\overline{r_{yy}(t)}$ exists. ■

11.1.3. *Pseudo-random signals*

DEFINITION 374.– *A pseudo-random signal is a correlatable signal whose spectral function Φ_{xx} is absolutely continuous (see section 12.2.3). The derivative φ_{xx} of this spectral function is called the* spectral density *of x.*

THEOREM 375.– *(i) A pseudo-random signal is a correlatable signal such that*

$$r_{xx}(\tau) = \frac{1}{2\pi}\int_{-\pi}^{\pi} e^{i\omega\tau}\varphi_{xx}(\omega)\,d\omega = \left(\mathcal{F}^{-1}\varphi_{xx}\right)(\tau);$$

its spectral density φ_{xx} is such that $\varphi_{xx}(\omega) = (\mathcal{F}r_{xx})(\omega) \geq 0$ and $\varphi_{xx} \in L_1[-\pi,\pi]$ (where $L_1[-\pi,\pi]$ denotes the space of integrable functions on $[-\pi,\pi]$). (ii) Its correlation sequence r_{xx} is such that $\overline{r_{xx}(\tau)} = \overline{|r_{xx}(\tau)|^2} = 0$. (iii) A pseudo-random signal is centered.

PROOF. (i) and (ii) are immediately clear from Definition 374 and the Bochner theorem. (iii) is derived from Proposition 373. ■

DEFINITION 376.– *Two pseudo-random signals x and y are said to be* absolutely cross-correlatable *if they are cross-correlatable and if*

$$r_{xy} = \mathcal{F}^{-1}\varphi_{xy}$$

where $\varphi_{xy} \in L_1[-\pi,\pi]$ is called the cross-spectral density *of x and of y.*

Let $R_{xy}(z)$ be the z-transform of r_{xy}.

PROPOSITION 377.– *If x and y are absolutely cross-correlatable, then y and x are also absolutely cross-correlatable, and $R_{yx}(z) = R_{xy}\left(z^{*\,-1}\right)^*$; in particular, $\varphi_{yx}(\omega) = \varphi_{xy}^*(\omega)$. If $\mathbf{K} = \mathbb{R}$, $R_{yx}(z) = R_{xy}\left(z^{-1}\right)$; in particular $\varphi_{xy}^*(\omega) = \varphi_{xy}(-\omega)$.*

PROOF. Suppose x and y are absolutely cross-correlatable. Then

$$\varphi_{yx}(\omega) = \sum_{\tau=-\infty}^{+\infty} r_{yx}(\tau)\, z^{-\tau} = \sum_{\tau=-\infty}^{+\infty} r_{xy}^{*}(-\tau)\, z^{-\tau}$$

(according to Proposition 367 (i)), thus $\varphi_{yx} \in L_1[-\pi, \pi]$ and $\varphi_{yx}(\omega) = \varphi_{xy}^{*}(\omega)$. The other assertions are clear. ■

THEOREM 378.– *A set $\mathcal{M}$ of pairwise absolutely cross-correlatable pseudo-random signals is a* **K**-*vector space.*

PROOF. According to the proof of Proposition 369, if $z = \mu\, x$ $(\mu \in \mathbf{K})$, then $\varphi_{zz} = |\mu|^2\, \varphi_{xx}$ and $\varphi_{x+y,x+y} = \varphi_{xx} + \varphi_{yy} + \varphi_{xy} + \varphi_{yx}$. ■

11.1.4. *Filtering and factorization*

We call a *digital filter* a discrete-time linear time-invariant system functioning from initial instant $t_0 \to -\infty$ with zero initial conditions. From the mathematical point of view, a digital filter is characterized by an input–output relation which is a convolution operator $u \mapsto y = g * u$, i.e.

$$y(t) = \sum_{\tau=-\infty}^{+\infty} g(t-\tau)\, u(\tau)\,.$$

The impulse response of this filter (i.e. its response to a unit impulse $\boldsymbol{\delta}_0$ – see section 12.3.5) is the sequence $g = (g(t))_{t\in\mathbb{Z}}$, and its transfer function is $G(z) = \mathcal{Z}\{g\}$.

If we assume that the function $G(z)$ is rational, this filter is *causal and stable* if and only if the rational function $G(z)$ is proper and all its poles lie in the open unit disc, which is equivalent to saying that g is positively supported and belongs to l_1 (see Theorem 446, section 12.4.4).

Interference formula

Let there be two *stable causal* digital filters with rational transfers functions $G_1(z)$ and $G_2(z)$. Let x_1 (resp., x_2) be the input to the first (resp., second) filter and y_1 (resp., y_2) be its output (see Figure 11.1).

THEOREM 379.– *If x_1 and x_2 are absolutely cross-correlatable, then so are y_1 and y_2, and*

$$\boxed{R_{y_1 y_2}(z) = G_1\left(z^{-1}\right) G_2(z)\, R_{x_1 x_2}(z)} \qquad (11.1)$$

where $R_{y_1 y_2}(z)$ and $R_{x_1 x_2}(z)$ are the z-transforms of $r_{y_1 y_2}$ and $r_{x_1 x_2}$ respectively. Relation (11.1) *is called the* interference formula.

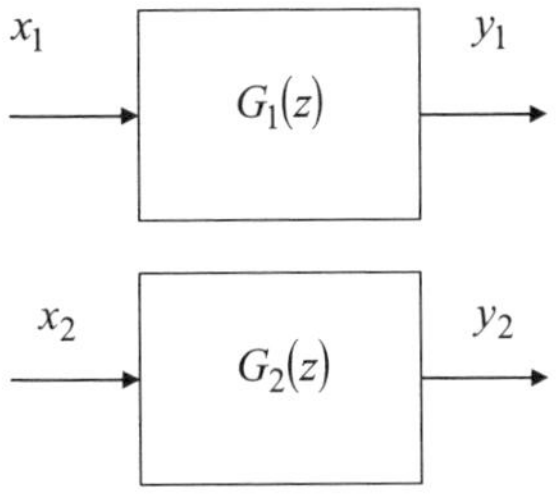

Figure 11.1. *Pseudo-random signal filtering*

PROOF. Let $g_i = \mathcal{Z}^{-1}\{G_i\}$ $(i = 1, 2)$. According to the Exchange theorem (section 12.3.5),

$$y_i(t) = \sum_{k=-\infty}^{+\infty} g_i(t-k)\, x_i(k),$$

as a result

$$y_1^*(t)\, y_2(t+\tau) = \sum_{k=-\infty}^{+\infty} \sum_{j=-\infty}^{+\infty} g_1(t-k)\, g_2(t+\tau-k+j)\, x_1^*(k)\, x_2(k+j)$$

from which we get

$$r_{y_1 y_2}(\tau) = \sum_{k=-\infty}^{+\infty} \sum_{j=-\infty}^{+\infty} g_1(t-k)\, g_2(t+\tau-k+j)\, r_{x_1 x_2}(j).$$

On the other hand,

$$R_{y_1 y_2}(z) = \sum_{\tau=-\infty}^{+\infty} r_{y_1 y_2}(\tau)\, z^{-\tau}$$

and we thus obtain (11.1) by writing $-\tau = (t-k) - (t-k+\tau-j) - j$. ■

Pseudo-white noise

A *pseudo-white noise* is a correlatable signal w such that $r_{ww} = \lambda\, \delta_0$. We thus have

$$\boxed{r_{ww}(\tau) = \begin{cases} \lambda \text{ if } \tau = 0 \\ 0 \text{ otherwise} \end{cases}}. \qquad (11.2)$$

Since $r_{ww}(0) = \overline{|w(t)|^2} \geq 0$, λ is a real non-negative number, called the *variance* of the pseudo-white noise w; $\sigma = \sqrt{\lambda}$ is its *standard deviation*.

DEFINITION 380.– *The pseudo-white noise with correlation sequence* (11.2) *is said to be* normalized *if its variance is equal to* 1.

The proof of the following theorem is obvious and its details are left to the reader:

THEOREM 381.– *A pseudo-white noise with variance* λ *is a pseudo-random signal with constant spectral density equal to* λ.

A *pseudo-random binary sequence* (PRBS) is a pseudo-random signal w such that $w(t_n) \in \{-\sigma, \sigma\}$, where (t_n) is a strictly increasing sequence of elements of $\mathbb{Z}$, chosen in such a way that $r_{ww}(\tau) \simeq 0$ if $\tau \neq 0$ (for more details, see [110], Chapter 5). Such a signal is therefore an approximation of a pseudo-white noise.

Spectral factorization

DEFINITION 382.– *A pseudo-random signal* x *is said to be* rational *if* $R_{xx}(z)$ *is a rational function. It is said to be* persistently exciting *if* $\varphi_{xx}(\omega) > 0$ *for any* $\omega \in [-\pi, \pi]$*, and to be* persistently exciting of order N $(N \in \{1, 2, ...\} \cup \{+\infty\})$ *if there exist* N *distinct values* $\omega_j \in [0, \pi]$ *for which* $\varphi_{xx}(\omega_j) > 0$.

THEOREM 383.– (Spectral factorization theorem). *Let* x *be a persistently exciting* real *rational pseudo-random signal (i.e.* $\mathbf{K} = \mathbb{R}$*). There exists a pseudo-white noise* w *with well-determined variance, as well as a unique bistable and bicausal filter, the impulse response and the transfer function of which are denoted by* g *and* $G(z)$ *respectively, such that* $x = g * w$ *and* $g(0) = 1$.

PROOF. According to Proposition 377, if z_k (resp., p_k) is a zero (resp., a pole) of $R_{xx}(z)$, then $1/z_k$, z_k^* and $1/z_k^*$ (resp., $1/p_k$, p_k^* and $1/p_k^*$) are again zeros (resp., poles) of $R_{xx}(z)$. On the other hand, $R_{xx}(z)$ has no poles on the unit circle because $\varphi_{xx} \in L_1[-\pi, \pi]$, and $R_{xx}(z)$ has no zeros on the unit circle because $\varphi_{xx}(\omega) > 0$ for all $\omega \in [-\pi, \pi]$. We can thus write

$$R_{xx}(z) = \lambda \frac{\prod_{k \in K} (z - z_k)(z - z_k^*)(z - 1/z_k)(z - 1/z_k^*)}{\prod_{j \in J} (z - p_j)(z - p_j^*)(z - 1/p_j)\left(z - 1/p_{kj}^*\right)},$$

$\lambda > 0$. We can assume without loss of generality that $|z_k| < 1$ and $|p_j| < 1$, $k \in K$, $j \in J$. Thus we have

$$G(z) = \frac{\prod_{k \in K} (z - z_k)(z - z_k^*)}{\prod_{j \in J} (z - p_j)(z - p_j^*)} z^{2n},$$

where $n = \text{card}(J) - \text{card}(K)$. Therefore, $G(z)$ is biproper and bistable. We have

$$g(0) = \lim_{|z| \to +\infty} G(z) = 1$$

according to the Initial value theorem (see section 12.3.5). This transfer function $G(z)$ is uniquely determined. There exists a signal w such that $x = g * w$; this signal w is given by

$$w = \mathcal{Z}^{-1}\left\{\frac{X(z)}{G(z)}\right\}.$$

It is a pseudo-random signal according to Theorem 379, and since

$$\boxed{R_{xx}(z) = G\left(z^{-1}\right) G(z) \lambda} \tag{11.3}$$

we have

$$R_{ww}(z) = \frac{1}{G\left(z^{-1}\right) G(z)} R_{xx}(z) = \lambda,$$

which proves that w is a pseudo-white noise. ■

Relation (11.3) expresses the "bicausal spectral factorization" that has been done.

Pseudo-colored noise

Just as white light is a superimposition of all colors and thus has a spectrum that contains all visible frequencies, a pseudo-white noise has a constant spectral density. If white light goes through a color filter (a piece of colored glass for example), we obtain colored light behind this filter. By analogy, a *pseudo-colored noise* is the output of a digital filter excited by a pseudo-white noise.

Theorem 383 shows that every rational pseudo-random signal x is a pseudo-colored noise.

11.1.5. *Ergodic random signals*

Ergodicity of up to second order

Let $(\Omega, \mathcal{F}, P)$ be a probability space. A discrete-time random signal x is a sequence of random variables $x(t) : \omega \in \Omega \to x(t, \omega) \in (\mathbf{K}, \mathcal{B})$ (with $t \in \mathbb{Z}$ and where $\mathcal{B}$ is the Borelian σ-algebra of $\mathbf{K}$). This signal is said to be *real* if $\mathbf{K} = \mathbb{R}$

and *stationary* if its law of probability $\nu_{x(t)}$ is independent of t. It is *ergodic of first order* if all random variables $x(t)$ are of first order and if for all $\omega \in \Omega$,[3]

$$\boxed{E[x(t)] = \overline{x(t,\omega)}.}$$

The signal x is *ergodic up to second order* if, in addition, all random variables $x(t)$ are of second order and (for all $\omega \in \Omega$)

$$\boxed{E[x^*(t)\, x(t+\tau)] = \overline{x^*(t,\omega)\, x(t+\tau,\omega)}}. \tag{11.4}$$

In what follows, "ergodic" means "ergodic up to second order". We further assume that, for any $\omega \in \Omega$, $x(.,\omega)$ is a pseudo-random signal in the way specified in section 11.1.3 (this implies, in particular, that $E[x(t)] = 0$, i.e. x is *centered*).

White noise and colored noise

A discrete-time white noise w is an ergodic random signal whose all realizations $w(.,\omega)$ are pseudo-white noises of the same *variance* $E\left[|w(t)|^2\right] = \lambda \geq 0$ (see relation (11.2), section 11.1.4).[4] We include as an additional hypothesis that the random variables $w(t)$ and $w(\tau)$ are *independent* whenever $t \neq \tau$.[5]

Let x be a random signal, ergodic up to the second order, such that $R_{xx}(z) = \mathcal{Z}\{r_{xx}\}$ is a rational function. The realizations of x are pseudo-colored noises (see section 11.1.4) and x is thus called a *colored noise*.

Temporal law

Let X be a real ergodic random signal. Its *temporal law* is the set of probability laws of all random variables

$$(X(t_1), X(t_2), ..., X(t_n))$$

where $\{t_1, ..., t_n\}$ spans the set of all strictly increasing finite sequences of elements of $\mathbb{Z}$.

In particular, a random signal (assumed to be real ergodic) is said to be *Gaussian* if its temporal law is Gaussian, that is all the variables discussed here are Gaussian.

3. More precisely, for almost every ω in the sense of the probability measure P, i.e. "almost surely". That is how the expression "for all $\omega \in \Omega$" should be interpreted, in what follows.

4. Some authors call a *pseudo-white noise* what we call here a discrete-time white noise.

5. These random variables are of course non-correlated. Two independent random variables are non-correlated, but the converse does not hold, except in particular cases (the Gaussian case, for example).

11.2. Open-loop identification

11.2.1. *Notation*

Here and in subsequent sections, all systems are linear discrete-time ones and are represented by means of the delay operator q^{-1} in place of the shift-forward operator q (we thus work over the ring $\mathbf{A}$ introduced in section 10.4.2).

Let $G\left(q^{-1}\right)$ be a rational function with indeterminate q^{-1}. The left multiplication by $G\left(q^{-1}\right)$ represents the convolution by the impulse response g of the filter with transfer function $G\left(z^{-1}\right)$. In other words, if

$$G\left(q^{-1}\right) = \sum_{\tau=-\infty}^{+\infty} g\left(\tau\right) q^{-\tau},$$

where the sequence $\left(g\left(\tau\right)\right)_{\tau\in\mathbb{Z}}$ is positively supported for causality of the filter, the output $y\left(t\right)$ is given by

$$y\left(t\right) = G\left(q^{-1}\right) u\left(t\right) = \sum_{\tau=-\infty}^{+\infty} g\left(\tau\right) u\left(t-\tau\right).$$

Note that in this formalism, the *Initial value theorem* can be written in the following particularly simple form:

$$g\left(0\right) = G\left(0\right).$$

REMARK 384.– *In this section, where open-loop systems are considered, we assume that these systems are stable in order to avoid the divergence of the output signal* $\{y\left(t\right)\}$.

11.2.2. *Least squares method*

Deterministic approach

Consider a stable system with the following model:

$$\begin{aligned} A\left(q^{-1}\right) y\left(t\right) &= q^{-r} B\left(q^{-1}\right) u\left(t\right) + e\left(t\right), \qquad (11.5) \\ A\left(q^{-1}\right) &= 1 + \sum_{k=1}^{n_A} a_i\, q^{-i}, \\ B\left(q^{-1}\right) &= \sum_{k=1}^{n_B} b_i\, q^{-i}. \end{aligned}$$

where u and y are respectively the input and the output of the system, and e is the error due to the uncertainty on the coefficients a_i and b_i which are to be identified, thus *a priori* are not precisely known (possibly totally unknown). Integer $r \geq 0$ is the *pure delay* (adding to the delay due to the discretization with Z.O.H.).

Let

$$\begin{aligned} \theta &= \begin{bmatrix} a_1 & \dots & a_{n_A} & b_1 & \dots & b_{n_B} \end{bmatrix}^T, \\ \phi^T(t-1) &= \begin{bmatrix} -y(t-1) & \dots & -y(t-n_A) & u(t-r-1) \\ & \dots & u(t-r-n_B) & \end{bmatrix}. \end{aligned}$$

The relation (11.5) is thus written as

$$y(t) = \phi^T(t-1)\,\theta + e(t). \tag{11.6}$$

DEFINITION 385.– *The vector $\phi(t-1)$ (that contains the data $y(\tau)$ and $u(\tau - r)$ up to instant $t-1$) is called the* regression vector.

The *least squares method* consists of minimizing the criterion

$$J(\theta,t) = \frac{1}{t}\sum_{\tau=1}^{t} e(\tau)^2. \tag{11.7}$$

where $t > 1$. Let

$$\Phi(t-1) = \begin{bmatrix} \phi^T(0) \\ \vdots \\ \phi^T(t-1) \end{bmatrix}, \tag{11.8}$$

$$\begin{aligned} Y(t) &= \begin{bmatrix} y(1) & \dots & y(t) \end{bmatrix}^T, \\ E(t) &= \begin{bmatrix} e(1) & \dots & e(t) \end{bmatrix}^T. \end{aligned} \tag{11.9}$$

Then

$$Y(t) = \Phi(t-1)\,\theta + E(t)$$

and

$$J(\theta,t) = \frac{1}{t}E^T(t)\,E(t) = \frac{1}{t}\left(Y(t) - \Phi(t-1)\,\theta\right)^T\left(Y(t) - \Phi(t-1)\,\theta\right).$$

For $J(.,t)$ to be minimum at point $\hat{\theta}(t)$, it is necessary that the Euler condition $\frac{\partial J}{\partial \theta}\left(\hat{\theta}(t),t\right) = 0$ be satisfied (see section 12.6.4, Proposition 453), with

$$\frac{\partial J}{\partial \theta}(\theta,t) = -\frac{2}{t}\Phi^T(t-1)\left(Y(t) - \Phi(t-1)\,\theta\right).$$

As a result, $\hat{\theta}(t)$ will be a solution of the so-called *normal equation*

$$\Phi^T(t-1)\,\Phi(t-1)\,\theta = \Phi^T(t-1)\,Y(t).$$

This equation always admits a solution [29], but its uniqueness is not guaranteed unless $\Phi^T(t-1)\,\Phi(t-1)$ is invertible, that is

$$\boxed{\operatorname{rk}\Phi(t-1) = n_A + n_B}. \tag{11.10}$$

Suppose this condition holds. We then have

$$HJ(\theta,t) = \frac{2}{t}\,\Phi^T(t-1)\,\Phi(t-1) > 0,$$

(where HJ is the Hessian matrix of J : see section 12.6.2), thus $J(.,t)$ is strictly convex (section 12.6.4, Proposition 454). According to Theorem 455 (section 12.6.4), $J(.,t)$ thus admits a *strict global minimum* at point $\hat{\theta}(t)$ such that

$$\hat{\theta}(t) = \left(\Phi^T(t-1)\,\Phi(t-1)\right)^{-1}\Phi^T(t-1)\,Y(t) = \Phi^{\dagger}(t-1)\,Y(t)$$

(for the second equality, see section 13.5.7, Proposition 583). This solution can be made more explicit using (11.8) and (11.9):

$$\boxed{\hat{\theta}(t) = \left(\sum_{\tau=1}^{t}\phi(\tau-1)\,\phi^T(\tau-1)\right)^{-1}\sum_{\tau=1}^{t}\phi(\tau-1)\,y(\tau)}. \tag{11.11}$$

We obtain the following theorem:

THEOREM 386.– *The least squares problem has a unique solution if and only if condition* (11.10) *holds (that is, if the data contain sufficient information). This solution is given by* (11.11).

Recursive least squares

Let

$$P(t) = \left(\sum_{\tau=1}^{t}\phi(\tau-1)\,\phi^T(\tau-1)\right)^{-1}.$$

The inconvenient of solution (11.11) is that it is necessary to calculate $P(t)$, which is the inverse of a matrix of size $(n_A+n_B)\times(n_A+n_B)$. We have the recurrence formula

$$P^{-1}(t) = P^{-1}(t-1) + \phi(t-1)\,\phi^T(t-1).$$

Applying the "Inversion lemma" (section 13.1.4, Lemma 493) with $A = P^{-1}(t-1)$, $B = \phi(t-1)$, $C = 1$ and $D = \phi^T(t-1)$, we get

$$\boxed{\begin{cases} P(t) = P(t-1) - K(t-1)\,\phi^T(t-1)\,P(t-1) \\ K(t-1) = \frac{P(t-1)\,\phi(t-1)}{1+\phi^T(t-1)\,P(t-1)\,\phi(t-1)} \end{cases}} \tag{11.12}$$

On the other hand, according to (11.11),

$$\hat{\theta}(t) = P(t)\sum_{\tau=1}^{t}\phi(\tau-1)\,y(\tau). \tag{11.13}$$

Using (11.12), we obtain

$$\boxed{\hat{\theta}(t) = \hat{\theta}(t-1) + K(t-1)\left[y(t) - \phi^T(t-1)\,\hat{\theta}(t-1)\right]}. \tag{11.14}$$

Relations (11.12) and (11.14) define the *recursive least squares algorithm*, which does not require any matrix inversion.

Let us study this from a stochastic point of view.

Formulation based on prediction error

All random signals are assumed to be real ergodic. The input and output of the system are now considered to be such signals, and vector θ is a *random variable* with values in $\mathbb{R}^n$ ($n = n_A + n_B$). Consider the "stochastic version" of model (11.5), known as the "ARX model"[6]

$$\boxed{A\left(q^{-1},\theta\right)y(t) = q^{-r}B\left(q^{-1},\theta\right)\,u(t) + w(t)\,,} \tag{11.15}$$

where

$$y(t) = \phi^T(t-1)\,\theta + w(t)\,. \tag{11.16}$$

Let $\mathcal{F}_{t-1}$ be the σ-algebra generated by the random variables $y(t-i)$, $u(t-i)$, $i \geq 1$ and θ (see section 12.7.2). (Regarding the input, only the values $u(t-r-i)$, $i \geq 1$, are actually involved, but we do not take this into account for the sake of clarity in the following discussion.)

DEFINITION 387.– *We call* one-step optimal prediction *of $y(t)$ the random variable given by*

$$\boxed{\hat{y}(t \mid t-1) = E\left[y(t) \mid \mathcal{F}_{t-1}\right]} \tag{11.17}$$

---

6. *ARX* stands for *AutoRegressive* (which is the part $A\left(q^{-1}\right)y(t) = w(t)$) and *X* for the eXternal part $q^{-r}B\left(q^{-1}\right)u(t)$ (u is indeed an external signal).

and prediction error *the random variable*

$$\boxed{\varepsilon(t) = y(t) - \hat{y}(t \mid t-1)}. \tag{11.18}$$

Let us take the following two hypotheses, the first of which makes more explicit Remark 384 (section 11.2.1).

(H0) : The co-domain of the random variable θ is a subset $\mathcal{D}$ of $\mathbb{R}^n$ such that for any $\theta_0 \in \mathcal{D}$, $A\left(z^{-1}, \theta_0\right) \neq 0$ for any $|z| \geq 1$ and the polynomials $A\left(q^{-1}, \theta_0\right)$ and $B\left(q^{-1}, \theta_0\right)$ are co-prime.

(H1) : w is a *white noise* with variance $E\left[w(t)^2\right] = \lambda$ and is such that for every instant t, $w(t)$ is independent of the σ-algebra $\mathcal{F}_{t-1}$.

THEOREM 388.– *Under Hypotheses (H0) and (H1), the one-step optimal prediction of* $y(t)$ *is given by:* $\hat{y}(t \mid t-1) = \phi^T(t-1)\,\theta$.

PROOF. Let $z(t) = \phi^T(t-1)\,\theta$ and let $y^\bullet(t)$ be a prediction of $y(t)$, that is (like $z(t)$) an element of $L^2(\Omega, \mathcal{F}_{t-1}, P)$. We have

$$E\left[|y(t) - y^\bullet(t)|^2\right] = E\left[|z(t) - y^\bullet(t) + w(t)|^2\right].$$

The random variables $z(t) - y^\bullet(t)$ and $w(t)$ are independent, thus we have, according to Proposition 470(ii) (section 12.7.2),

$$E\left[|y(t) - y^\bullet(t)|^2\right] = E\left[|z(t) - y^\bullet(t)|^2\right] + \lambda \geq \lambda.$$

The minimum of the left-hand side is attained when $y^\bullet(t) = z(t)$, as a result $z(t) = E\left[y(t) \mid \mathcal{F}_{t-1}\right]$ according to Definition 474 (section 12.7.3). ■

REMARK 389.– *(i) The above optimal prediction* $\hat{y}(t \mid t-1)$ *can be expressed linearly as a function of* θ, *thus it is called a* linear regression *in statistics. (ii) According to Theorem 388, the* least squares criterion (11.7) *is also written as*

$$\boxed{J(\theta, t) = \frac{1}{t}\sum_{\tau=1}^{t} (\varepsilon(\tau))^2}. \tag{11.19}$$

According to the ergodic hypothesis (11.4), *as* $t \to +\infty$, $J(\theta, t) \to E\left[(\varepsilon(\tau))^2\right]$.

Bias of the estimator

Let we take Hypothesis (H2):

(H2): There exists a unique value $\breve{\theta} \in \mathcal{D}$ of θ such that the data y and u satisfy the relation

$$y(t) = \phi^T(t-1)\,\breve{\theta} + w(t). \tag{11.20}$$

REMARK 390.– *(i) Hypothesis (H2) means that the identified system actually has the ARX structure* (11.15) *and that $\breve{\theta}$ is the "true value" of parameter θ (see section 11.2.5 for more details). (ii) To be rigorous, one should distinguish between the noise $\{w(t)\}$ of the model (which is only fictitious) and the noise $\{\breve{w}(t)\}$ of the actual system (and to write $y(t) = \phi^T(t-1)\,\breve{\theta} + \breve{w}(t)$ in place of* (11.20)*). But this complicates the derivation without changing the conclusion (see Exercise 428); for the sake of simplicity, w and $\breve{w}$ are not distinguished in what follows.*

Let

$$\boxed{\hat{\theta}(t) = \arg\min_{\theta} J(t,\theta)} \tag{11.21}$$

(assuming that $J(t,.)$ admits a strict global minimum, that is condition (11.10) holds).

DEFINITION 391.– *The estimator that provides the estimate $\hat{\theta}(t)$ (or, abusing the language, the estimator $\hat{\theta}(t)$) is said to be non-biased if $E\left[\hat{\theta}(t)\right] = \breve{\theta}$, and* asymptotically non-biased *if $\lim_{t\to+\infty} E\left[\hat{\theta}(t)\right] = \breve{\theta}$.*

THEOREM 392.– *Under Hypotheses (H0) to (H2), the least squares estimator is non-biased.*

PROOF. According to (11.13) and Hypothesis (H2), we have, by putting $\Gamma(t) = tP(t)$

$$\begin{aligned}\hat{\theta}(t) &= \Gamma(t)\frac{1}{t}\sum_{\tau=1}^{t}\phi(\tau-1)\left(\phi^T(\tau-1)\,\breve{\theta} + w(\tau)\right)\\ &= \breve{\theta} + \Gamma(t)\frac{1}{t}\sum_{\tau=1}^{t}\phi(\tau-1)\,w(\tau).\end{aligned} \tag{11.22}$$

The random variables $\phi(\tau-1)$ and $w(\tau)$ are independent according to Hypothesis (H1) and $w(\tau)$ is centered, thus according to Proposition 470(i) (section 12.7.3)

$$E[\phi(\tau-1)\,w(\tau)] = E[\phi(\tau-1)]\;E[w(\tau)] = 0,$$

therefore $E\left[\hat{\theta}(t)\right] = \breve{\theta}$. ■

Consistency of the estimator

DEFINITION 393.– *The estimator in Definition 391 is said to be* consistent *if*

$$\lim_{t\to+\infty} \hat{\theta}(t) = \breve{\theta} \text{ almost surely.} \tag{11.23}$$

REMARK 394.– *A consistent estimator is* asymptotically non-biased.

Let us take the two following hypotheses:

(H3) : The input $\{u(t)\}$ is a persistently exciting signal of infinite order.

(H4) : For any instants t and τ, the noise $w(t)$ is independent of the input $u(\tau)$.

REMARK 395.– *Hypothesis (H4) is realistic in the context of open-loop identification which is the subject of this section. If the identification is made in closed-loop, the feedback introduces a cross-correlation between $u(t)$ and the noise $w(\tau)$, $\tau \leq t$.*

THEOREM 396.– *Under Hypotheses (H0) to (H4), the least squares estimator is consistent.*

PROOF. We have

$$\Gamma(t)^{-1} = \frac{1}{t} P^{-1}(t) = \frac{1}{t} \sum_{\tau=1}^{t} \phi(\tau-1)\, \phi^T(\tau-1)$$

(see section 11.1.1, Remark 364). As a result, if $E\left[\phi(t-1)\, \phi^T(t-1)\right]$ is invertible, we have as $t \to +\infty$

$$\Gamma(t) \to \left\{ E\left[\phi(t-1)\, \phi^T(t-1)\right]^{-1} \right\}.$$

The invertibility of $E\left[\phi(t-1)\, \phi^T(t-1)\right]$ is ensured by Hypotheses (H3) and (H4) ([110] Chapter 5, Complement 5.2); we will come back to this point in the more general context of section 11.2.5. On the other hand,

$$\frac{1}{t} \sum_{\tau=1}^{t} \phi(\tau-1)\, w(\tau) \to E\left[\phi(t-1)\, w(t)\right] = 0$$

(see proof of Theorem 392). The theorem is thus a consequence of (11.22). ■

Residue analysis

THEOREM 397.– *Under Hypotheses (H0) to (H4), if $\hat{\theta}(t) = \breve{\theta}$, then the prediction error $\{\varepsilon(t)\}$ is a white noise such that for any instants t and τ, $\varepsilon(t)$ is independent of the input $u(\tau)$.*

PROOF. It suffices to note that, for $\theta = \breve{\theta}$, $\varepsilon(t) = w(t)$. ■

EXAMPLE 398.– *Let us take the continuous-time minimal system with transfer function*

$$G(s) = \frac{s-2}{s^2 + 0.2\,s + 1}. \tag{11.24}$$

Discretizing this system (with Z.O.H.) at sampling period $T_s = 0.5$ *s (to clarify ideas, we can indeed assume that the unit of time chosen is the second), we obtain the discrete-time system with transfer function*

$$G_d(z) = \frac{0.2193\,z - 0.6853}{z^2 - 1.6718\,z + 0.9048} = \frac{0.2193\,z^{-1} - 0.6853\,z^{-2}}{1 - 1.6718\,z^{-1} + 0.9048\,z^{-2}} \triangleq \frac{\breve{B}(z^{-1})}{\breve{A}(z^{-1})}.$$

The system input is a PRBS with amplitude 1 *and the "equation noise"* w *is a Gaussian white noise with standard deviation* 1*. The identification is done over 1000 points. (The sampling period* T_s *chosen can seem quite large – taking into account the undamped characteristic frequency* $\omega_0 = 1$ *rad/s of the system –, but this order of magnitude is essential to obtain a correct identification. With respect to the control,* T_s *can be much smaller, and we can conceive two different sampling periods: a large enough one for the identification, another smaller one for the control, so that the latter will quickly react to a disturbance.) The input and the output, over the first 200 points (thus the first 100 seconds), are shown in Figure 11.2 (–),*[7] *as well as the simulated output based on the identified model and in the absence of noise (- -); this makes it possible to get a first appreciation of the quality of the identification and also of the noise level. Residue analysis is shown in Figure 11.3.*[8] *According to some statistic criteria that we are not going to detail here, normalized correlation sequence* $r'_{\varepsilon\varepsilon}$ *and normalized cross-correlation sequence* $r'_{\varepsilon u}$ *can be considered to be zero (except* $r'_{\varepsilon\varepsilon}(0)$*) if they stay inside the confidence intervals represented by darker areas, which is the case here. The transfer function* $\hat{G}(s)$ *of the continuous-time identified system is obtained by inverting formula* (10.13) *of section 10.3.3, based on the transfer function of the discrete-time identified system (in practice, this inversion is done through a state-space realization, by inverting relation* (10.20) *of section 10.3.4). In this way we obtain*

$$\hat{G}(s) = \frac{1.163\,s - 1.952}{s^2 + 0.2161\,s + 0.962}$$

which is very close to (11.24)*. The step responses of the actual system (–) and of the identified system (- -) are shown in Figure 11.4. The curves are almost merge.*

EXAMPLE 399.– *This example is the same as Example 398, except in the way that the white noise* $w(t)$ *affects the system: here it is no longer an "equation noise", as in*

7. Figures 11.2 to 11.7 are at the end of section 11.2.

8. Where "lag" stands for the argument τ of $r_{yu}(\tau)$.

relation (11.15)*, but a measurement noise. This means that the system satisfies*

$$\left\{ \begin{array}{c} y(t) = y_0(t) + w(t), \\ \breve{A}(q^{-1}) \, y_0(t) = q^{-r} \, \breve{B}(q^{-1}) \, u(t). \end{array} \right. \tag{11.25}$$

This is the only change compared with the previous case. Note that this situation is more realistic. The input and output, for the first 200 points, are shown in Figure 11.5 (–), as well as the output simulated based on the identified model and in the absence of noise (- -). According to the residue analysis, shown in Figure 11.6, the identification is a priori *poor. The transfer function of the continuous-time identified system is indeed*

$$\hat{G}(s) = \frac{6.099\, s - 9.491}{s^2 + 4.354\, s + 4.001}$$

and thus is quite different from (11.24)*. The step responses of the actual system (–) and of the identified system (- -), shown in Figure 11.7, confirm the poor quality of the identification.*

Bias analysis

THEOREM 400.– *If the noise* w *is a* measurement noise, *the least squares estimator is in general biased.*

PROOF. Let us follow again the approach used in the proof of Theorem 392. Recall first that according to (11.25)

$$\breve{A}(q^{-1})\, y(t) = \breve{B}(q^{-1})\, u(t-r) + \breve{A}(q^{-1})\, w(t). \tag{11.26}$$

Polynomial $\breve{A}(q^{-1})$ is of the form $\breve{A}(q^{-1}) = 1 + q^{-1}\breve{A}^{\bullet}(q^{-1}, \theta)$. As a result, if $\breve{\theta}$ is the true value of the parameter, we have from (11.13), writing $v(t) = A^{\bullet}(q^{-1}, \theta)\, w(t)$,

$$E\left[\hat{\theta}(t)\right] = \breve{\theta} + \Gamma(t) \frac{1}{t} \sum_{\tau=1}^{t} E\left[\phi^T(\tau-1)\, v(\tau-1)\right],$$

thus $E\left[\hat{\theta}(t)\right] \neq \breve{\theta}$ because (except for a special case) the components of $\phi(\tau-1)$ are correlated with $v(\tau-1)$. ■

Theorem 400 – as well as Example 11.15 – shows that the least squares method, based on an ARX model, does not make it possible to correctly resolve all identification problems. For this reason, we shall study more complex methods in the following sections.

11.2.3. *Models and prediction*

Consider the identification problem of a system in the presence of measurement noise (see Example 399). Assuming that this measurement noise is white, the input $u(t)$, the output $y(t)$ and the white noise $w(t)$ satisfy the relation (11.26). This expression has a structure that can be put into several different forms.

Models

ARMAX model[9]

This model is

$$\boxed{A\left(q^{-1},\theta\right)\, y(t) = q^{-r}\, B\left(q^{-1},\theta\right)\, u(t) + C\left(q^{-1},\theta\right)\, w(t)}. \tag{11.27}$$

Adding the constraint

$$A\left(q^{-1},\theta\right) = C\left(q^{-1},\theta\right), \tag{11.28}$$

equation (11.26) has this structure.

In general, we can assume that polynomial $C\left(z^{-1},\theta\right)$ of an ARMAX model is such that $C\left(z^{-1},\theta\right) \neq 0$ for $|z| \geq 1$, according to the spectral factorization theorem (section 11.1.4, Theorem 383).

Output error model

The *OE* (*Output Error*) model is

$$\boxed{y(t) = \frac{q^{-r}\, B\left(q^{-1},\theta\right)}{F\left(q^{-1},\theta\right)}\, u(t) + w(t)}. \tag{11.29}$$

This is the structure that corresponds to the problem in Example 11.15 (with $A\left(q^{-1},\theta\right)$ replaced by $F\left(q^{-1},\theta\right)$).

Box and Jenkins model

We can consider the more general case where the measurement noise n is a *colored noise*.[10] According to the spectral factorization theorem, n can be assumed to be generated by the model

$$n(t) = \frac{C\left(q^{-1},\theta\right)}{D\left(q^{-1},\theta\right)}\, w(t),$$

9. *ARMA* stands for *AutoRegresive Moving Average* and *X* is from eXternal.

10. Do not confuse the noise n with the integer n in Hypothesis (H0).

where the transfer function

$$H\left(q^{-1},\theta\right)=\frac{C\left(q^{-1},\theta\right)}{D\left(q^{-1},\theta\right)}$$

is bicausal and bistable and such that $H(0,\theta)=1$. We thus obtain the *Box and Jenkins model* (BJ)

$$\boxed{y(t)=\frac{q^{-r}\,B\left(q^{-1},\theta\right)}{F\left(q^{-1},\theta\right)}\,u(t)+\frac{C\left(q^{-1},\theta\right)}{D\left(q^{-1},\theta\right)}\,w(t)}. \tag{11.30}$$

Prediction error model

We gather all models obtained above into the general structure

$$\boxed{A\left(q^{-1}\right)\,y(t)=\frac{q^{-r}\,B\left(q^{-1},\theta\right)}{F\left(q^{-1},\theta\right)}\,u(t)+\frac{C\left(q^{-1},\theta\right)}{D\left(q^{-1},\theta\right)}\,w(t)} \tag{11.31}$$

called the *prediction error model* (PEM). We are now going to see why.

One-step optimal prediction

General system (11.31) (in which polynomials $A\left(q^{-1},\theta\right)$, $B\left(q^{-1},\theta\right)$, $C\left(q^{-1},\theta\right)$, $D\left(q^{-1},\theta\right)$ and $F\left(q^{-1},\theta\right)$ appear explicitly; the coefficients of these polynomials are the components of the parameter-vector θ to be identified) can be put in the more concise form

$$\boxed{y(t)=G\left(q^{-1},\theta\right)\,u(t)+H\left(q^{-1},\theta\right)\,w(t)\,.} \tag{11.32}$$

The following hypothesis, which generalizes Hypothesis (H0) (section 11.2.3), is in force; n is the number of components of θ.

(H0'): The random variable θ takes its values in a subset $\mathcal{D}$ of $\mathbb{R}^n$ such that for any $\theta_0\in\mathcal{D}$, $G\left(q^{-1},\theta_0\right)$ is the transfer function of a stable and strictly causal filter, $H\left(q^{-1},\theta_0\right)$ is the transfer function of a bistable and bicausal filter, and $H(0,\theta_0)=1$ (see section 11.1.4, Theorem 383, and section 11.2.1); G and H are rational functions with respect to the indeterminate q^{-1} and the components of θ_0.

REMARK 401.– *More concisely, $G\left(q^{-1},\theta\right)$ and $H\left(q^{-1},\theta\right)$ can be denoted in what follows as $G\left(q^{-1}\right)$ and $H\left(q^{-1}\right)$, respectively, when this is not ambiguous.*

Let $\mathcal{F}_{t-1}$ be the σ-algebra generated by the random variables $y(t-i)$, $u(t-i)$, $i\geq 1$ and θ. The one-step optimal prediction $\hat{y}(t\mid t-1)$ and the prediction error $\varepsilon(t)$ are defined as in Definition 387 (section 11.2.2).

THEOREM 402.– *Under Hypothesis (H0') and Hypothesis (H1) of section 11.2.2, the one-step optimal prediction of* $y(t)$ *is given by*

$$\boxed{\hat{y}(t \mid t-1) = \left[1 - \frac{1}{H(q^{-1})}\right] y(t) + \frac{G(q^{-1})}{H(q^{-1})} u(t)}. \tag{11.33}$$

PROOF. Let $z(t)$ be the left-hand side of (11.33) and $y^{\bullet}(t)$ be a prediction of $y(t)$, i.e. (like $z(t)$) an element of $L^2(\Omega, \mathcal{F}_{t-1}, P)$. We have $y(t) - y^{\bullet}(t) = z(t) - y^{\bullet}(t) + w(t)$, thus we can conclude as in the proof of Theorem 388. ■

11.2.4. *Output Error method and ARMAX method*

The identification method now consists of determining $\hat{\theta}(t)$ defined by (11.21), thus of minimizing criterion (11.19) (see section 11.2.2, Remark 389). This minimization is usually carried out using the Levenberg–Marquardt algorithm (see section 12.6.5), which requires the knowledge of the gradient of the criterion with respect to the parameters to be identified. Let θ_i be the ith component of θ; we have

$$\frac{\partial J}{\partial \theta_i}(t, \theta) = \frac{2}{t} \sum_{\tau=1}^{t} \varepsilon(\tau) \frac{\partial \varepsilon}{\partial \theta_i}(\tau).$$

What needs to be done next is to evaluate the partial derivatives $\frac{\partial \varepsilon}{\partial \theta_i}$.

To illustrate this point, we will calculate these partial derivatives for the Output Error model and for the ARMAX model (see also Exercise 422).

Case of an Output Error model

We have

$$G(q^{-1}) = q^{-r} \frac{B(q^{-1})}{F(q^{-1})}, \quad H(q^{-1}) = 1.$$

As a result, according to (11.33),

$$\varepsilon(t) = y(t) - q^{-r} \frac{B(q^{-1})}{F(q^{-1})} u(t).$$

Therefore,

$$\frac{\partial \varepsilon}{\partial b_i}(t) = -\frac{q^{-r-i}}{F(q^{-1})} u(t), \quad \frac{\partial \varepsilon}{\partial f_i}(t) = \frac{q^{-r-i} B(q^{-1})}{F(q^{-1})^2} u(t).$$

Case of an ARMAX model

We have

$$G\left(q^{-1}\right)=q^{-r}\frac{B\left(q^{-1}\right)}{A\left(q^{-1}\right)},\quad H\left(q^{-1}\right)=\frac{C\left(q^{-1}\right)}{A\left(q^{-1}\right)},$$

and therefore, from (11.33),

$$\varepsilon\left(t\right)=\frac{A\left(q^{-1}\right)}{C\left(q^{-1}\right)}y\left(t\right)-q^{-r}\frac{B\left(q^{-1}\right)}{C\left(q^{-1}\right)}u\left(t\right).$$

We deduce that

$$\begin{aligned}\frac{\partial\varepsilon}{\partial a_i}\left(t\right) &= \frac{q^{-i}}{C\left(q^{-1}\right)}y\left(t\right),\quad \frac{\partial\varepsilon}{\partial b_i}\left(t\right)=-\frac{q^{-r-i}}{C\left(q^{-1}\right)}u\left(t\right),\\ \frac{\partial\varepsilon}{\partial c_i}\left(t\right) &= -\frac{q^{-i}}{C\left(q^{-1}\right)}\varepsilon\left(t\right).\end{aligned}$$

EXAMPLE 403.– *This example is identical to Example 399 (section 11.2.2), but this time the system is identified using the Output Error method. The input and the output, for the first 200 points, are shown in Figure 11.8 (–), as well as the simulated output based on the identified model in the absence of a noise (- -). The latter clearly better "follows" the actual output than in Figure 11.5. The residue analysis, shown in Figure 11.9, is good. For verification, the transfer function of the identified continuous-time system is*

$$\hat{G}\left(s\right)=\frac{0.9465\,s-1.991}{s^2+0.1933\,s+0.9941}$$

and is thus very close to (11.24)*. The step responses of the actual system (–) and of the identified system (- -), shown in Figure 11.10, confirm the good quality of the identification. The (very important) problem of identification in the presence of a white measurement noise is thus correctly resolved by the Output Error method.*

EXAMPLE 404.– *This example is identical to Example 403 but the system is identified using the ARMAX method. The input and output, for the first 200 points, are shown in Figure 11.11 (–), as well as the simulated output based on the identified model in the absence of a noise (- -). The residue analysis, shown in Figure 11.12, is correct. The polynomials of the identified discretized system are the following:*

$$\begin{aligned}\hat{A}\left(q^{-1}\right) &= 1-1.6750\,q^{-1}+0.9052\,q^{-2},\\ \hat{B}\left(q^{-1}\right) &= 0.1907\,q^{-1}-0.6495\,q^{-2},\\ \hat{C}\left(q^{-1}\right) &= 1-1.6886\,q^{-1}+0.9237\,q^{-2}.\end{aligned}$$

The fact that $\hat{C}\left(q^{-1}\right)$ *is very close to* $\hat{A}\left(q^{-1}\right)$ *makes it possible to conclude that the noise is definitely a measurement noise (according to relation* (11.28)*). The step responses of the actual system (–) and of the identified system (- -), shown in Figure 11.13, confirm the excellent quality of the identification. The problem of identification in the presence of a white measurement noise is thus also resolved by the ARMAX method.*

EXAMPLE 405.– *The difference between this example and Example 403 is that, this time, the measurement noise is very much colored. It is represented in Figure 11.14 over the first 100 points. The identification is carried out using the Output Error method. The input and the output, over the first 200 points, are shown in Figure 11.15 (–), as well as the simulated output based on the identified model in the absence a of noise (- -). The residue analysis, shown in Figure 11.16, demonstrates that the Output Error model does not correspond to the case treated. Indeed, the prediction error was not whitened at the end of the optimization. Nevertheless, the step responses of the actual system (–) and the identified system (- -) in Figure 11.17, show that the system was correctly identified. Analysis of this phenomenon is the subject of the following section.*

11.2.5. *Consistency of the estimator and residues*

Consistency

Consider the general model (11.32) and suppose that the actual system is governed by the equation

$$y(t) = \breve{G}\left(q^{-1}\right) u(t) + \breve{H}\left(q^{-1}\right) w(t), \tag{11.34}$$

where the filter $\breve{G}\left(q^{-1}\right)$ is stable and strictly causal, and the filter $\breve{H}\left(q^{-1}\right)$ is bicausal, bistable and such that $\breve{H}(0) = 1$.

Hypothesis (H2) (section 11.2.2) is a hypothesis of *identifiability* according to the following:

DEFINITION 406.– *The system is said to be* identifiable *if there exists a* unique value $\breve{\theta} \in \mathcal{D}$ *of* θ *such that* $G\left(q^{-1}, \breve{\theta}\right) = \breve{G}\left(q^{-1}\right)$ *and* $H\left(q^{-1}, \breve{\theta}\right) = \breve{H}\left(q^{-1}\right)$.[11]

Suppose that now model (11.32) can be written as

$$\boxed{y(t) = G\left(q^{-1}, \theta_s\right) u(t) + H\left(q^{-1}, \theta_\nu\right) w(t)} \tag{11.35}$$

11. There are several notions of identifiability in the literature. The notion presented here is a variant of the so-called *structural identifiability*. Another concept of identifiability is presented in [110].

so that the parameter θ to be determined is separated into two *independent* parts θ_s and θ_ν, the first part characterizing only the deterministic part of the system to be identified and the second part characterizing only the stochastic part. We have

$$\theta = (\theta_s, \theta_\nu) . \tag{11.36}$$

Let

$$\hat{\theta}(t) = \left(\hat{\theta}_s(t), \hat{\theta}_\nu(t)\right),$$

where $\hat{\theta}(t)$ is defined by (11.21) (see section 11.2.2).

DEFINITION 407.– *The deterministic part of the system is said to be* identifiable *if there exists a unique value* $\breve{\theta}_s$ *of* θ_s *such that* $G\left(q^{-1}, \breve{\theta}_s\right) = \breve{G}\left(q^{-1}\right)$.

THEOREM 408.– *Let Hypothesis (H0') (section 11.2.3) be in force, as well as Hypotheses (H1),(H3) and (H4) (section 11.2.2). (i) If the deterministic part of the system is identifiable, then the estimator of this part is consistent, i.e.*

$$\lim_{t\to+\infty} \hat{\theta}_s(t) = \breve{\theta}_s \quad \textit{almost surely.}$$

(ii) If the whole system is identifiable and if the variance $\lambda = \sigma^2$ *of the white noise* w *is non-zero, then the estimator of the system is consistent, i.e.*

$$\lim_{t\to+\infty} \hat{\theta}(t) = \breve{\theta} \quad \textit{almost surely.}$$

PROOF. The one-step optimal prediction can be written as

$$\hat{y}(t \mid t-1) = \left[1 - \frac{1}{H\left(q^{-1}, \theta_\nu\right)}\right] y(t) + \frac{G\left(q^{-1}, \theta_s\right)}{H\left(q^{-1}, \theta_\nu\right)} u(t)$$

whereas the actual system is defined by (11.34). The prediction error $\varepsilon(t) = y(t) - \hat{y}(t \mid t-1)$ thus satisfies

$$\varepsilon(t) = \frac{\Delta G\left(q^{-1}, \theta_s\right)}{H\left(q^{-1}, \theta_\nu\right)} u(t) + \frac{\breve{H}\left(q^{-1}\right)}{H\left(q^{-1}, \theta_\nu\right)} w(t), \tag{11.37}$$

where $\Delta G\left(q^{-1}, \theta_s\right) = \breve{G}\left(q^{-1}\right) - G\left(q^{-1}, \theta_s\right)$. As $t \to +\infty$, $J(t, \theta) \to E\left[\varepsilon(t)^2\right]$ (see section 11.2.2, Remark 389) and according to Hypothesis (H4), which implies that the right-hand side of (11.37) is the sum of two independent random variables, we have

$$E\left[\varepsilon(t)^2\right] = E\left[\left(\frac{\Delta G\left(q^{-1}, \theta_s\right)}{H\left(q^{-1}, \theta_\nu\right)} u(t)\right)^2\right] + E\left[\left(\frac{\breve{H}\left(q^{-1}\right)}{H\left(q^{-1}, \theta_\nu\right)} w(t)\right)^2\right].$$

Let Q_1 be the first quantity on the right-hand side of the above equality and Q_2 the be second quantity. We have

$$E\left[\varepsilon(t)^2\right] = r_{\varepsilon\varepsilon}(0) = \frac{1}{2\pi}\int_{-\pi}^{\pi} \varphi_{\varepsilon\varepsilon}(\omega)\, d\omega.$$

Therefore, according to the Interference formula (see section 11.1.4, Theorem 379),

$$\begin{aligned} Q_1 &= \frac{1}{2\pi}\int_{-\pi}^{\pi} \left|\frac{\Delta G\left(e^{-i\omega}, \theta_s\right)}{H\left(e^{-i\omega}, \theta_\nu\right)}\right|^2 \varphi_{uu}(\omega)\, d\omega, \\ Q_2 &= \frac{1}{2\pi}\int_{-\pi}^{\pi} \left|\frac{\breve{H}\left(e^{-i\omega}\right)}{H\left(e^{-i\omega}, \theta_\nu\right)}\right|^2 \varphi_{ww}(\omega)\, d\omega, \end{aligned}$$

where $\varphi_{uu}(\omega) > 0$ for an infinite number of distinct angular frequencies $\omega \in [-\pi, \pi]$ according to Hypothesis (H3) and $\varphi_{ww}(\omega) = \lambda$. (i) Suppose that the deterministic part is identifiable. According to Hypothesis (H3), Q_1 is minimum (and equal to 0) if and only if $\Delta G\left(e^{-i\omega}, \theta_s\right) = 0$ for an infinite number of distinct frequencies $\omega \in [-\pi, \pi]$, that is $\theta_s = \breve{\theta}_s$; indeed, a non-zero rational function can only have a finite number of zeros. (ii) Assume that the previous condition is satisfied and $\lambda > 0$. We then have

$$E\left[\varepsilon(t)^2\right] = \frac{\lambda}{2\pi}\int_{-\pi}^{\pi} \left|\frac{\breve{H}\left(e^{-i\omega}\right)}{H\left(e^{-i\omega}, \theta_\nu\right)}\right|^2 d\omega.$$

Let

$$\tilde{H}\left(q^{-1}, \theta_\nu\right) = \frac{\breve{H}\left(q^{-1}\right)}{H\left(q^{-1}, \theta_\nu\right)}.$$

This is the transfer function of a bicausal bistable filter such that $\tilde{H}(0, \theta_\nu) = 1$; it can be written as

$$\tilde{H}\left(q^{-1}, \theta_\nu\right) = 1 + \sum_{\tau=1}^{+\infty} \tilde{h}(\tau)\, q^{-\tau}.$$

According to the Parseval equality (12.45) of section 12.3.3,

$$E\left[\varepsilon(t)^2\right] = \lambda\left(1 + \sum_{\tau=1}^{+\infty} \tilde{h}(\tau)^2\right).$$

Since $\lambda > 0$, $E\left[\varepsilon(t)^2\right]$ is minimum if and only if the l_2 norm of $\left(\tilde{h}(\tau)\right)$ is also minimum. Now, by assuming that the system is entirely identifiable, the $\tilde{h}(\tau)$, $\tau \geq 1$, are all zero if and only if $H\left(q^{-1}, \theta_\nu\right) = \breve{H}\left(q^{-1}\right)$, that is to say $\theta = \breve{\theta}$ (taking into account the fact that $\theta_s = \breve{\theta}_s$). ■

REMARK 409.– *The statement of Theorem 408(ii) remains valid if we do not have the separation* (11.36) *of* θ *into two parts* θ_s *and* θ_ν *(as the reader can check by slightly modifying the proof of this theorem), and Example 404 is an illustration of this remark.*

Residue analysis

The prediction error has the important property stated in the theorem below (which generalizes Theorem 397).

THEOREM 410.– *Under the hypotheses of Theorem 408, if the system is identifiable and if the parameter* θ *has its true value* $\breve{\theta}$*, then the prediction error* $\{\varepsilon(t)\}$ *is a white noise such that for any instants* t *and* τ*,* $\varepsilon(t)$ *is independent of the input* $u(\tau)$*.*

PROOF. We have

$$\varepsilon(t) = w(t),$$

and so the theorem is a consequence of Hypothesis (H4). ■

11.2.6. *Filtering of data*

It is rare for the "deterministic part" of the system, as defined in section 11.2.5, to be identifiable in the absolute. Indeed, this subsystem (whose model is, e.g. to design a control law) is in fact very complex (see section 4.2.1). Therefore, it makes sense to identify only the "useful part" of this subsystem, in a bounded interval of frequencies (that consists of frequencies at which the control has energy). Let $[\omega_{\min}, \omega_{\max}]$ be such an interval. If we restrict the behavior of the deterministic part to $[\omega_{\min}, \omega_{\max}]$, then the above-mentioned "useful part" becomes identifiable if both this interval and the model structure are correctly chosen.

In practice, we therefore filter the data by a *bandpass filter* with transfer function $H_{bp}(z^{-1})$, assumed to be bicausal and bistable for reasons discussed below. In other words, the data provided to the algorithm are, instead of the input/output $u(t)$ and $y(t)$, the *filtered* input/output

$$u_f(t) = H_{bp}(q^{-1})\, u(t), \quad y_f(t) = H_{bp}(q^{-1})\, y(t).$$

Suppose the transfer function of the entire deterministic part is

$$G(q^{-1}) = G_u(q^{-1}) + G_n(q^{-1}),$$

where $G_u(q^{-1})$ is the "useful part" and $G_n(q^{-1})$ is the part to be neglected. The bandpass filter is chosen in a way that

$$\begin{aligned} H_{bp}(e^{-i\omega})\, G_u(e^{-i\omega}) &\simeq G_u(e^{-i\omega}) \\ H_{bp}(e^{-i\omega})\, G_n(e^{-i\omega}) &\simeq 0. \end{aligned}$$

In absence of noise, the inputs and outputs satisfy the relation

$$y(t) = \left[G_u\left(q^{-1}\right) + G_n\left(q^{-1}\right)\right] u(t).$$

Multiplying the two sides of this equality by $H_{bp}\left(q^{-1}\right)$, we obtain the relation satisfied by the filtered inputs and outputs

$$y_f(t) = \left[G_u\left(q^{-1}\right) + G_n\left(q^{-1}\right)\right] u_f(t),$$

which yields

$$y_f(t) \simeq G_u\left(q^{-1}\right) u_f(t).$$

Let us now examine how this is formalized in the presence of a noise. Consider the general model (11.32) of section 11.2.3 and let

$$\boxed{H\left(q^{-1}, \theta\right) = \frac{H_i\left(q^{-1}, \theta\right)}{H_{bp}\left(q^{-1}\right)}}, \tag{11.38}$$

where $H_i\left(q^{-1}, \theta\right)$ is that part of the stochastic term which is *to be identified*; in $H\left(q^{-1}, \theta\right)$, this part is multiplied by $1/H_{bp}\left(q^{-1}\right)$ which is the transfer function of a *bandstop filter*. According to Theorem 402, the prediction error is given by

$$\varepsilon(t) = \frac{1}{H_i\left(q^{-1}, \theta\right)} y_f(t) - \frac{G\left(q^{-1}, \theta\right)}{H_i\left(q^{-1}, \theta\right)} u_f(t).$$

As a result, the heuristic method that consists of supplying the filtered data to the algorithm can be interpreted as a particular choice of the noise model (according to relation (11.38)).

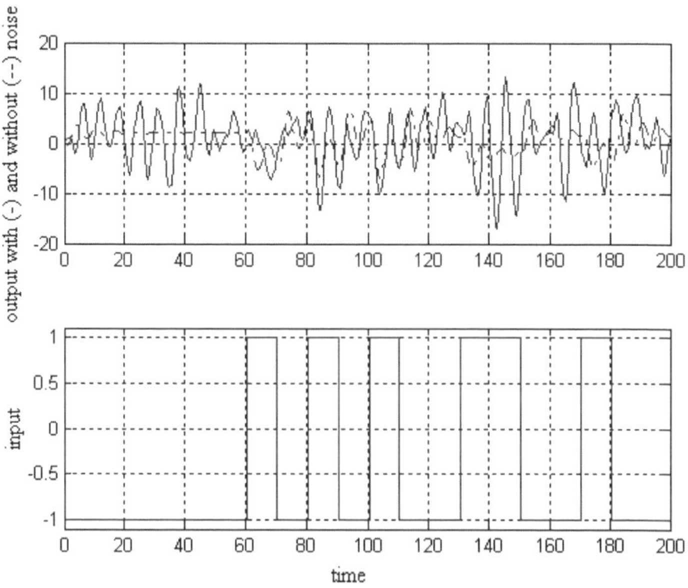

Figure 11.2. *Simulated data and output – Example 398*

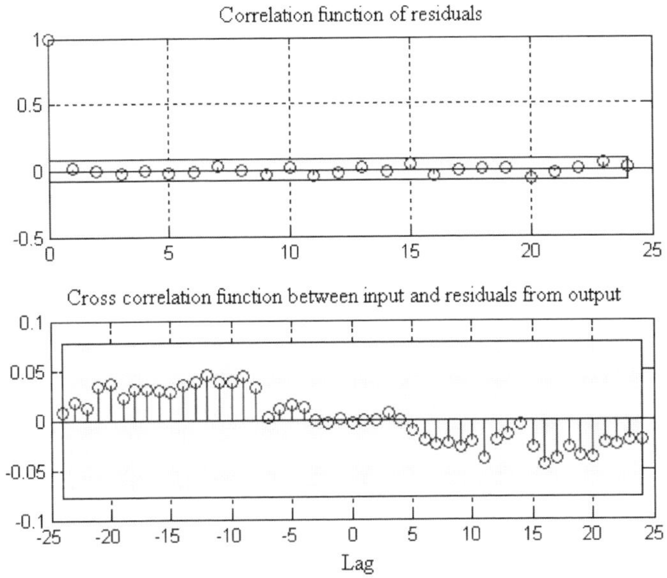

Figure 11.3. *Residue analysis – Example 398*

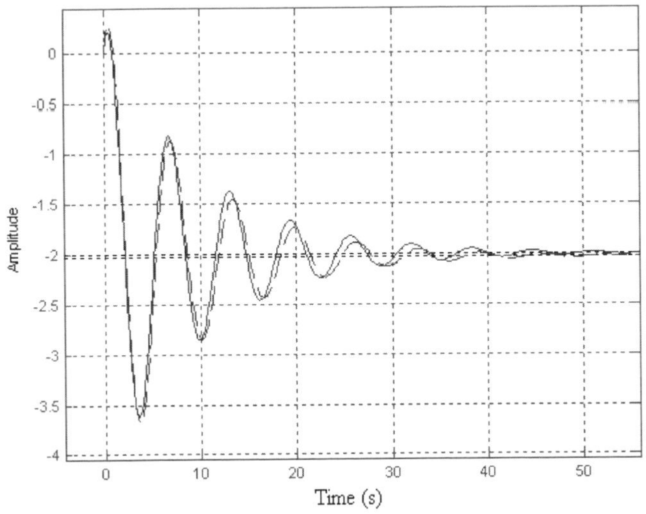

Figure 11.4. *Step responses – Example 398*

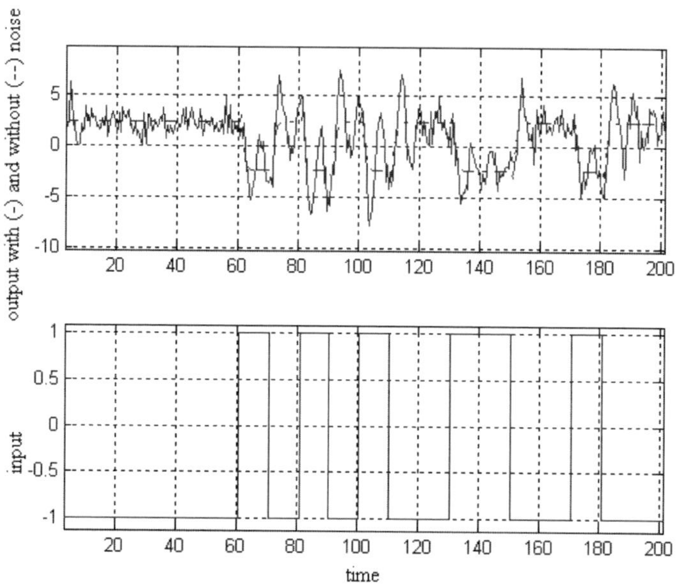

Figure 11.5. *Simulated data and output – Example 399*

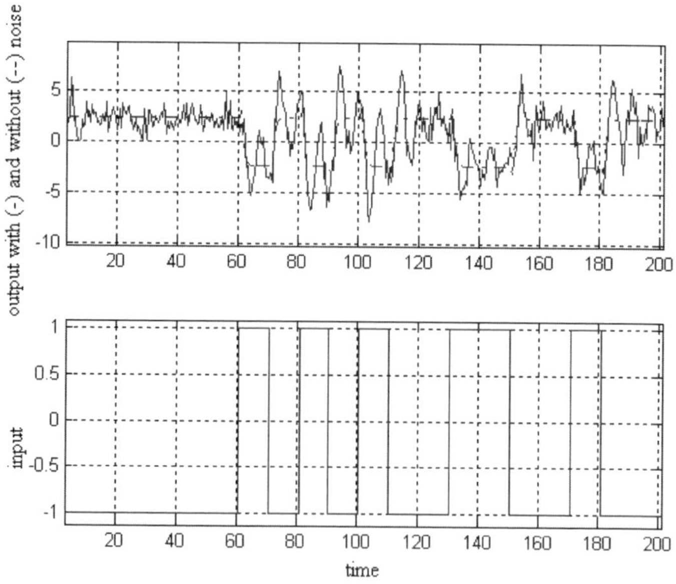

Figure 11.6. *Residue analysis – Example 399*

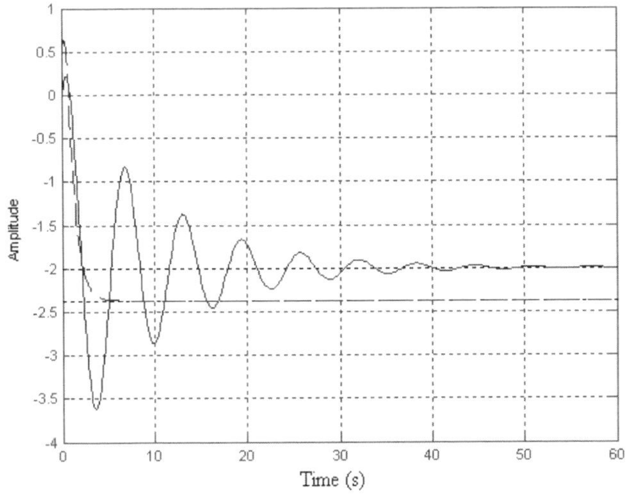

Figure 11.7. *Step responses – Example 399*

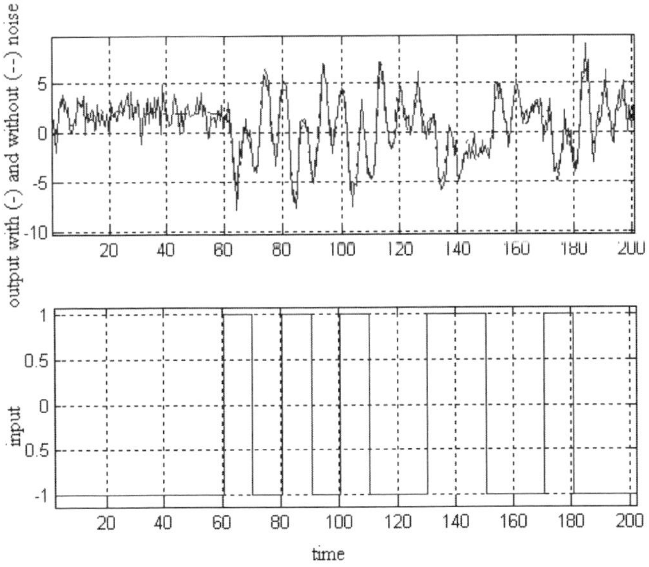

Figure 11.8. *Simulated data and output – Example 403*

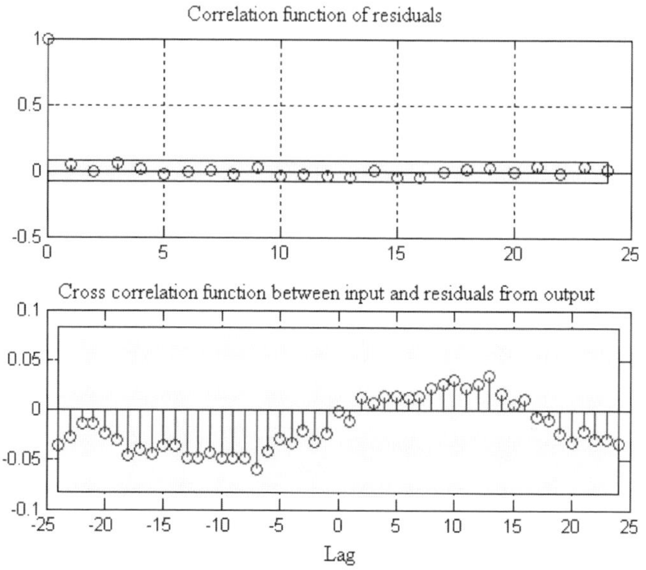

Figure 11.9. *Residue analysis – Example 403*

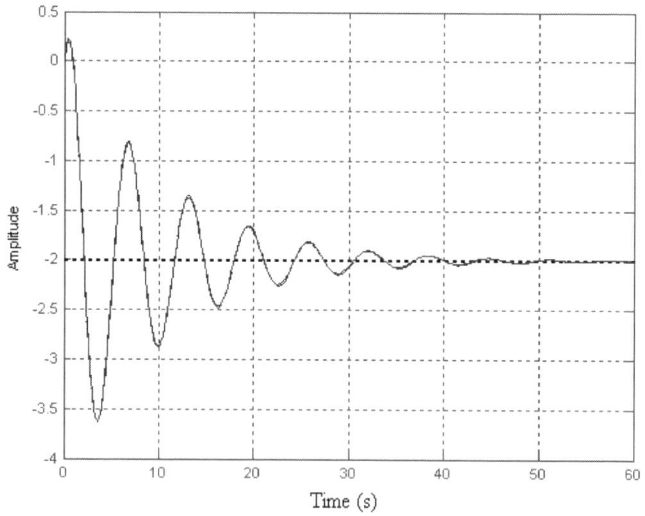

Figure 11.10. *Step responses – Example 403*

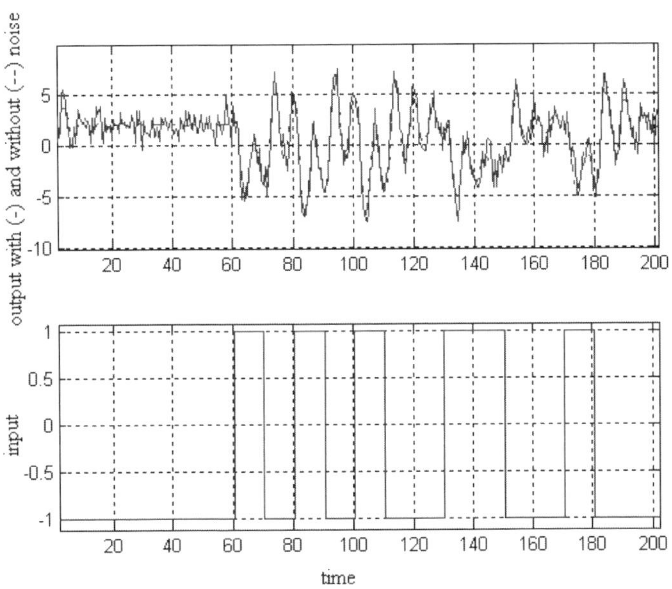

Figure 11.11. *Simulated data and output – Example 404*

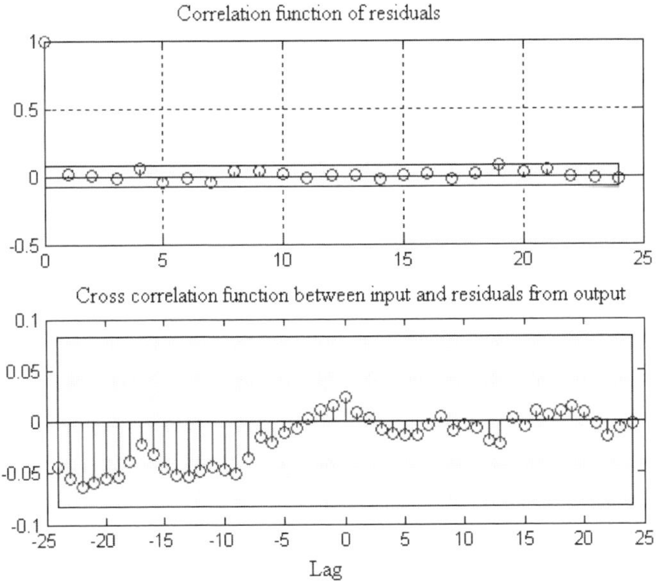

Figure 11.12. *Residue analysis – Example 404*

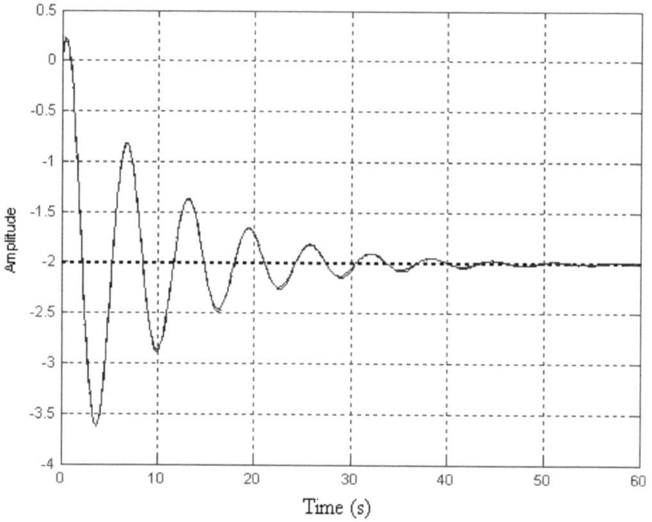

Figure 11.13. *Step responses – Example 404*

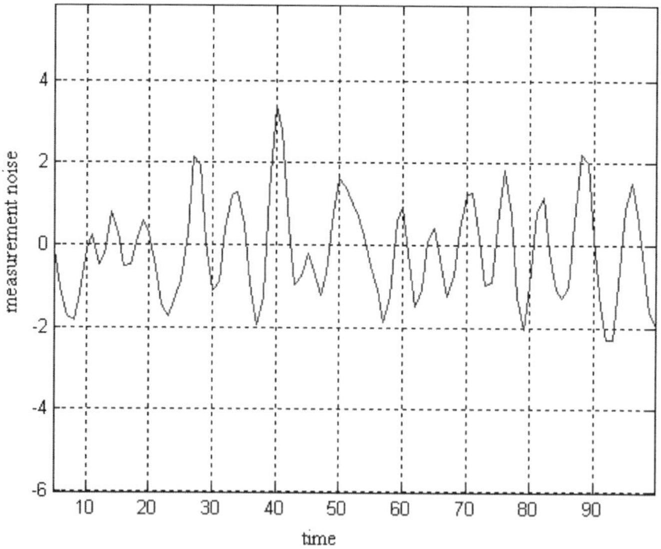

Figure 11.14. *Measurement noise – Example 405*

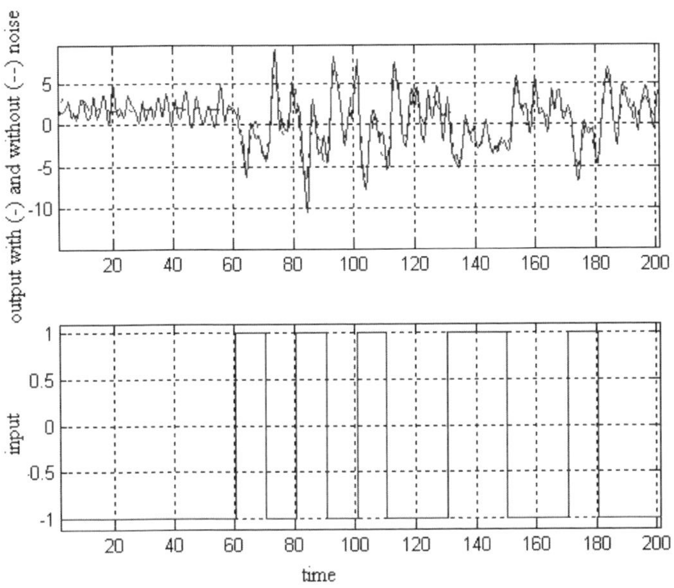

Figure 11.15. *Simulated data and output – Example 405*

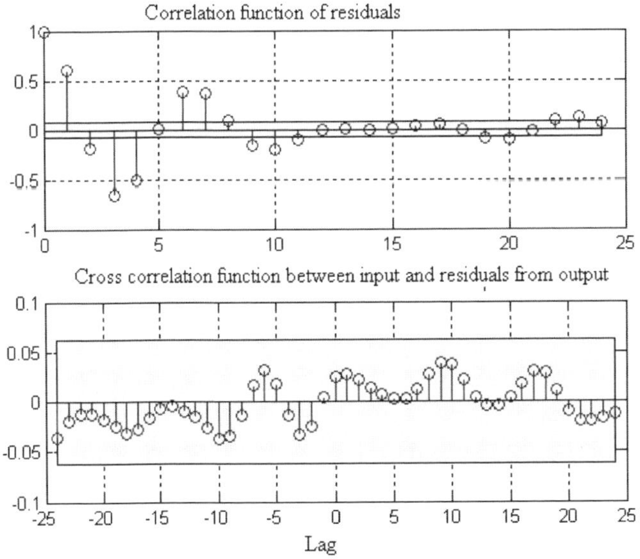

Figure 11.16. *Residue analysis – Example 405*

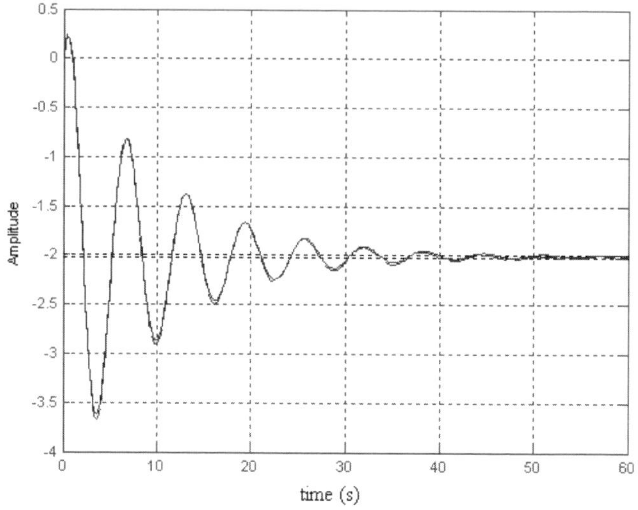

Figure 11.17. *Step responses – Example 405*

11.3. Closed-loop identification

11.3.1. *Direct and indirect approach*

Consider a system with equation (11.34) where the filter $\breve{G}\left(q^{-1}\right)$ is strictly causal (but may be unstable) and the filter $\breve{H}\left(q^{-1}\right)$ is bicausal, bistable and such that $\breve{H}(0) = 1$. The case where the variance λ of the noise is zero is trivial, therefore we will assume $\lambda > 0$ in what follows. In the present section, we will discuss the problem of identification of the system when it is fed back by a causal regulator, according to the equation

$$u(t) = r(t) - K\left(q^{-1}\right) y(t) \; ; \tag{11.39}$$

the feedback system is assumed to be stable and the reference signal $\{r(t)\}$ satisfies Hypothesis (H5) below:

(H5): For any instants t and τ, $r(t)$ is independent of the noise $w(\tau)$. Hypothesis (H0') of section 11.2.3 is replaced by the hypothesis (H0") hereafter (where n denotes the number of components of θ).

(H0"): The random variable θ takes its values in a subset $\mathcal{D}$ of $\mathbb{R}^n$ such that for any $\theta_0 \in \mathcal{D}$, $G\left(q^{-1}, \theta_0\right)$ is the transfer function a strictly causal filter stabilized by the controller (11.39) ; $H\left(q^{-1}, \theta_0\right)$ is the transfer function of a bistable and bicausal filter such that $H(0, \theta_0) = 1$; G and H are rational functions with respect to q^{-1} and to the components of θ_0.

Hypothesis (H1) of section 11.2.2, which is still realistic, is in force.

Let us now look at what difficulty arises in the identification process due to the feedback loop. Since $y(t)$ is correlated with the $w(\tau)$, $\tau \leq t$ (due to equation (11.32)), relation (11.39) introduces a cross-correlation between $u(t)$ and the $w(\tau)$, $\tau \leq t$ (see section 11.2.2, Remark 395). Therefore, Hypothesis (H4) of section 11.2.2 is not satisfied, which invalidates the proof of Theorem 408 (section 11.2.5). It is therefore necessary to reconsider the problem of the consistency of the estimator.

Direct approach

The direct approach of closed-loop identification consists in identifying the system using the input and output signals $\{u(t)\}$ and $\{y(t)\}$ while the feedback loop is ignored. Besides, this is unavoidable when the controller is unknown. The analysis of the direct approach is carried out in section 11.3.2.

Indirect approach

The indirect approach, which is only possible when the controller is known, consists of two steps.

1) The *first step* consists in identifying the feedback system. To do this, various methods are of course possible, in particular the one that consists of minimizing the l_2 norm of the prediction error.

For the following discussion, we will assume that this is the method being used; the model of the feedback system, deduced from (11.32) and (11.39), can then be written as

$$y(t) = G_c\left(q^{-1}, \theta\right) r(t) + H_{c1}\left(q^{-1}, \theta\right) w(t) \tag{11.40}$$

where

$$G_c = \frac{G}{1+GK}, \quad H_{c1} = \frac{H}{1+GK}.$$

Let $\theta_0 \in \mathcal{D}$; we can carry out the bicausal spectral factorization of the signal $H_{c1}\left(q^{-1}, \theta_0\right) w(t)$ (see section 11.1.4, Theorem 383) if this signal is persistently exciting, that is if $\lambda \triangleq \varphi_{ww} > 0$ and Hypothesis (H6) below is in force:

(H6) : For any $\theta_0 \in \mathcal{D}$, the filters $G\left(q^{-1}, \theta_0\right)$ and $K\left(q^{-1}\right)$ have no poles on the unit circle.

Under this hypothesis, model (11.40) can be put in the form

$$y(t) = G_c\left(q^{-1}, \theta\right) r(t) + H_c\left(q^{-1}, \theta\right) w'(t) \tag{11.41}$$

where, according to Hypothesis (H0"), for any $\theta_0 \in \mathcal{D}$, $G_c\left(q^{-1}, \theta_0\right)$ is the transfer function of a stable and strictly causal filter, $H_c\left(q^{-1}, \theta_0\right)$ is the transfer function of a bicausal bistable filter such that $H_c(0,\theta_0) = 1$, and w' is a white noise, with variance same as w; w and w' are not distinguished in the sequel. In other words, the model (11.41) satisfies Hypothesis (H0') (section 11.2.3).

According to Hypothesis (H1) mentioned above, we can determine the optimal one-step prediction $\hat{y}(t \mid t-1)$ *based on model* (11.41); this prediction is given (according to Theorem 402, section 11.2.3) by

$$\hat{y}(t \mid t-1) = \left[1 - \frac{1}{H_c(q^{-1})}\right] y(t) + \frac{G_c(q^{-1})}{H_c(q^{-1})} r(t). \tag{11.42}$$

Hypothesis (H5) means that Hypothesis (H4) is also in force *for model* (11.41) (*mutatis mutandis*). Let

$$\breve{G}_c = \frac{\breve{G}}{1+\breve{G}K}, \qquad \breve{H}_{c1} = \frac{\breve{H}}{1+\breve{G}K}$$

and, assuming that $\breve{G}$ and K do not have poles on the unit circle, let $\breve{H}_c(q^{-1})$ be a bistable bicausal filter such that

$$\breve{H}_c(q^{-1})\,\breve{H}_c(q) = \breve{H}_{c1}(q^{-1})\,\breve{H}_{c1}(q)$$

and $\breve{H}_c(0) = 1$. We have $G_c = \breve{G}_c$ and $H_c = \breve{H}_c$ if and only if $G = \breve{G}$ and $H = \breve{H}$; consequently:

LEMMA 411.– *The feedback system is identifiable with model* (11.41) *if and only if the open-loop system is identifiable with model* (11.32).

Now, let us consider Hypothesis (H3') below:

(H3'): The signal $\{r(t)\}$ persistently exciting of infinite order.

REMARK 412.– *(i) The system input u is given by*

$$u(t) = \frac{1}{1+\breve{G}(q^{-1})\,K(q^{-1})} r(t) - \frac{K(q^{-1})\,\breve{H}(q^{-1})}{1+\breve{G}(q^{-1})\,K(q^{-1})} w(t). \tag{11.43}$$

Let the sensitivity function be

$$\breve{S}_o(z^{-1}) = \frac{1}{1+\breve{G}(z^{-1})\,K(z^{-1})}.$$

We have according to Hypothesis (H5)

$$\varphi_{uu} = \left|\breve{S}_o\right|^2 \varphi_{rr} + \left|K\,\breve{H}\,\breve{S}_o\right|^2 \lambda \tag{11.44}$$

(where the transfer functions are evaluated on the unit circle). Therefore, if Hypothesis (H3') is in force or if $K \neq 0$, $\{u(t)\}$ *is a persistently exciting signal of infinite order.*
(ii) According to (11.43), *the transfer function between* w *and* u *is* $G_1 = -K\breve{H}/\left(1+\breve{G}K\right)$. *Applying the Interference formula* (11.1) *(section 11.1.4) with* $x_1 = x_2 = w$ *and* $G_2 = 1$, *we get* $R_{uw}(z) = G_1\left(z^{-1}\right)\lambda$; *thus, according to Proposition 377 (section 11.1.3)*

$$\varphi_{uw}\,\varphi_{wu} = \left|K\,\breve{H}\,\breve{S}_o\right|^2 \lambda^2. \tag{11.45}$$

Let $\hat{\theta}(t)$ be the estimator defined by (11.21), where J is defined by (11.19) and (11.18), and $\hat{y}(t \mid t-1)$ is given by (11.42). The result below follows immediately from the above and from Theorem 408:

COROLLARY 413.– *Suppose Hypotheses (H0"), (H1), (H3'), (H5) and (H6) are in force. If the open-loop system* (11.34) *is identifiable with model* (11.32)*, then the estimator* $\hat{\theta}(t)$ *of the feedback system (designed based on model* (11.41)*) is consistent.*

REMARK 414.– *Suppose model* (11.32) *can be put in the form* (11.35). *Then* G_c *depends only on* θ_s *(i.e. on that part of the parameter* θ *which only characterizes the deterministic part of the open-loop system) and therefore can be written as* $G_c\left(q^{-1},\theta_s\right)$. *On the other hand, even though* H *only depends on* θ_ν *(i.e. that part of the parameter* θ *that only characterizes the stochastic part of the open-loop system),* H_c *depends on both* θ_ν *and* θ_s*, thus entirely on parameter* θ. *As a result, if we replace the hypothesis of system identifiability in the statement of Corollary 413 by the hypothesis of identifiability of the* deterministic part *of the open-loop system (with the other hypotheses remaining unchanged),* we cannot guarantee the consistency of the estimator of this part *(see the proof of Theorem 408). But the hypothesis of total system identifiability (including the stochastic part) is sometimes too strong, thus we will revisit this subject later (see section 11.3.3).*

2) The *second step* consists of determining, using the estimate $\hat{G}_c\left(q^{-1}\right)$ of the actual closed-loop transfer function $\breve{G}_c\left(q^{-1}\right)$, an estimate $\hat{G}\left(q^{-1}\right)$ of the actual open-loop transfer function $\breve{G}\left(q^{-1}\right)$. We have the equality

$$\breve{G} = \frac{\breve{G}_c}{1-\breve{G}_c K}.$$

As a rcsult,

$$\hat{G} = \frac{\hat{G}_c}{1-\hat{G}_c K} \tag{11.46}$$

is an estimate of $\breve{G}\left(q^{-1}\right)$, and it is this one that is used in the *classic* indirect approach. If the estimate $\hat{G}_c\left(q^{-1}\right)$ of $\breve{G}_c\left(q^{-1}\right)$ were perfect (that is, $\hat{G}_c\left(q^{-1}\right) = \breve{G}_c\left(q^{-1}\right)$), (11.46) would certainly be a perfect estimate of $\breve{G}\left(q^{-1}\right)$. However, in practice, (11.46) might well be a *poor estimate* of $\breve{G}\left(q^{-1}\right)$ due to the fact that the estimate $\hat{G}_c\left(q^{-1}\right)$ of $\breve{G}_c\left(q^{-1}\right)$ is inevitably spoilt by errors. Thus, $\hat{G}_c\left(q^{-1}\right)$ may have too high an order, generate instabilities, not correspond to anything real, etc.

11.3.2. *Consistency of estimator in the direct approach*

Identification of the global system

In the direct approach in a strict sense, as was already been mentioned, the controller is not supposed to be known. Under Hypotheses (H0") and (H1), the one-step optimal prediction $\hat{y}\left(t \mid t-1\right)$ and the prediction error are given by (11.33) and (11.37), respectively. This last expression can be written in the form

$$\varepsilon\left(t\right) = \frac{1}{H\left(q^{-1}, \theta\right)} z\left(t\right) + w\left(t\right),$$

$$z\left(t\right) = \Delta G\left(q^{-1}, \theta\right) u\left(t\right) - \Delta H\left(q^{-1}, \theta\right) w\left(t\right), \tag{11.47}$$

where

$$\Delta G\left(q^{-1}, \theta\right) = \breve{G}\left(q^{-1}\right) - G\left(q^{-1}, \theta\right), \quad \Delta H\left(q^{-1}, \theta\right) = \breve{H}\left(q^{-1}\right) - G\left(q^{-1}, \theta\right).$$

Let $\hat{\theta}\left(t\right)$ be the estimator defined by (11.21), where J is defined by (11.19) and (11.18).

LEMMA 415.– *Suppose Hypotheses (H0") and (H1) are in force; if the integral in expression* (11.48) *below admits a minimum, then* $\hat{\theta}\left(t\right) \rightarrow \theta_{\bullet}$ *as* $t \rightarrow +\infty$, *where, with* $\Delta G = \Delta G\left(e^{-i\omega}, \theta\right)$, $\Delta H = \Delta H\left(e^{-i\omega}, \theta\right)$, $(.)^{*}$ *= conjugate of* $(.)$,

$$\theta_{\bullet} \in \arg\min_{\theta \in \mathcal{D}} \frac{1}{2\pi} \int_{-\pi}^{\pi} \frac{1}{\left|H\left(e^{-i\omega}, \theta\right)\right|^2} \varphi_{zz}\left(\omega\right) d\omega, \tag{11.48}$$

$$\varphi_{zz}\left(\omega\right) = \begin{bmatrix} \Delta G & -\Delta H \end{bmatrix} \varphi_{\varsigma\varsigma}\left(\omega\right) \begin{bmatrix} \Delta G^{*} \\ -\Delta H^{*} \end{bmatrix}, \tag{11.49}$$

$$\varphi_{\varsigma\varsigma}\left(\omega\right) = \begin{bmatrix} \varphi_{uu}\left(\omega\right) & \varphi_{uw}\left(\omega\right) \\ \varphi_{wu}\left(\omega\right) & \lambda \end{bmatrix}. \tag{11.50}$$

PROOF. According to Hypothesis (H1), the random variables $z\left(t\right)$ and $w\left(t\right)$ are independent, thus $E\left[\varepsilon\left(t\right)^2\right] = \frac{1}{|H|^2} E\left[z\left(t\right)^2\right] + \lambda$, from which we derive (11.48) if a minimum such as that mentioned exists (see section 12.6.4); for this it is *sufficient* that system (11.34) is identifiable based on the model (11.32)). The expression of $\varphi_{zz}(\omega)$ is deduced from (11.47). ■

If system (11.34) is identifiable based on the model (11.32), Lemma 415 shows that its estimator $\hat{\theta}(t)$ given by (11.21) is consistent if $\varphi_{\varsigma\varsigma}(\omega) > 0$ for an infinite number of values of ω in $[-\pi, \pi]$. Let

$$\varphi_{uu}^{r} \triangleq \left|\breve{S}_o\right|^2 \varphi_{rr}, \quad \varphi_{uu}^{w} \triangleq \left|K\,\breve{H}\,\breve{S}_o\right|^2 \lambda, \tag{11.51}$$

where, according to (11.44),

$$\varphi_{uu} = \varphi_{uu}^{r} + \varphi_{uu}^{w};$$

φ_{uu}^{r} (resp., φ_{uu}^{w}) is the contribution of signal r (resp., of noise w) to the spectral density φ_{uu} of u. From (11.45),

$$\varphi_{uu}^{w} = \varphi_{uw}(\omega)\,\varphi_{wu}(\omega)/\lambda.$$

PROPOSITION 416.– *If* $\lambda > 0$, *the following conditions are equivalent: (i) Hypothesis (H3') is satisfied. (ii)* $\varphi_{uu}^{r}(\omega) > 0$ *for an infinite number of values of* ω *in* $[-\pi, \pi]$. *(iii)* $\varphi_{\varsigma\varsigma}(\omega) > 0$ *for an infinite number of values of* ω *in* $[-\pi, \pi]$.

PROOF. (i) is equivalent to (ii) according to (11.51); (ii) is equivalent to (iii) according to Proposition 576 (section 13.5.6), (11.50), and the equality $\varphi_{uu}^{r} = \varphi_{uu} - \varphi_{uw}(\omega)\,\varphi_{wu}(\omega)/\lambda$. ■

The result below is thus clear and consistent with Corollary 413 (section 11.3.1):

THEOREM 417.– *Under Hypotheses (H0"), (H1), (H3'), (H5) and* $\lambda > 0$, *the estimator* $\hat{\theta}(t)$ *is consistent if the open-loop system* (11.34) *is identifiable based on model* (11.32).

Identification of the deterministic part

In practice, we most often seek to identify the deterministic part of the system (which is the subject of interest of Theorem 408(i) in the context of open-loop identification). The hypothesis of the identifiability of the global system (i.e. including the stochastic part) is often too strong (see section 11.3.1, Remark 414). Consider now the situation where only the deterministic part of the system is identifiable. In the following section, the signal $\{u(t)\}$ is assumed to be persistently exciting (see Remark 412(i), section 11.3.1).

According to (11.49), (11.50) and Proposition 576 (section 13.5.6), writing

$$\varphi_{ww}^{r}(\omega) \triangleq \lambda - \varphi_{wu}(\omega)\,\varphi_{uw}(\omega)/\varphi_{uu}(\omega),$$

$$B\left(e^{-i\omega}\right) \triangleq \Delta H\left(e^{-i\omega}\right) \varphi_{uw}(\omega) / \varphi_{uu}(\omega), \tag{11.52}$$

we get

$$\begin{aligned} & \left[\begin{array}{cc} \Delta G & -\Delta H \end{array}\right] \varphi_{\varsigma\varsigma}(\omega) \left[\begin{array}{c} \Delta G^* \\ -\Delta H^* \end{array}\right] \\ = & \left[\begin{array}{cc} \Delta G + B & -\Delta H \end{array}\right] \left[\begin{array}{cc} \varphi_{uu} & 0 \\ 0 & \varphi_{ww}^r \end{array}\right] \left[\begin{array}{c} (\Delta G + B)^* \\ -\Delta H^* \end{array}\right]. \end{aligned}$$

Note that according to (11.44), (11.45) and (11.51),

$$\varphi_{ww}^r = \lambda \frac{\varphi_{uu}^r}{\varphi_{uu}}. \tag{11.53}$$

From (11.48), we get

$$\theta_\bullet \in \arg\min_{\theta \in \mathcal{D}} \frac{1}{2\pi} \int_{-\pi}^{\pi} \frac{\varphi_{uu}}{|H|^2} \left\{ |\Delta G + B|^2 + |\Delta H|^2 \frac{\varphi_{ww}^r}{\varphi_{uu}} \right\} d\omega \tag{11.54}$$

(i) If the stochastic part is not identifiable, we cannot have $\Delta H = 0$, which implies $B \neq 0$ according to (11.52). It follows that we will not have $\Delta G \to 0$, and thus *the estimator of the deterministic part is asymptotically biased. A fortiori*, it is thus not consistent according to Remark 394 (section 11.2.2).

(ii) Nevertheless, note that according to (11.52) and (11.53)

$$|B|^2 = |\Delta H|^2 \frac{\lambda}{\varphi_{uu}} \frac{\varphi_{uu}^w}{\varphi_{uu}}, \quad |\Delta H|^2 \frac{\varphi_{ww}^r}{\varphi_{uu}} = |\Delta H|^2 \frac{\lambda}{\varphi_{uu}} \left(1 - \frac{\varphi_{uu}^w}{\varphi_{uu}}\right),$$

thus, if $|\Delta H|$ bounded, if $\lambda / \varphi_{uu} \to 0$, and if $\varphi_{uu}^w / \varphi_{uu} \to 0$ (in other words, if the signal-to-noise ratio becomes infinitely large), then the integrand of (11.54) tends to

$$\frac{\varphi_{uu}}{|H|^2} |\Delta G|^2.$$

Thus, if the deterministic part of the open-loop system is identifiable, the integral in (11.54) becomes minimum when $\Delta G = 0$, and the estimator of the deterministic part becomes consistent (but this is almost a tautology).

Therefore, the direct approach leads to the same conclusion as the indirect approach (see Remark 414): except for the situation where the signal-to-noise ratio is so large that we can consider that the identification is realized in the absence of a noise, *in order to identify the deterministic part of the open-loop system, it is necessary to identify this system globally, thus this system needs to be entirely identifiable.*

11.3.3. *A third path*

Modified direct approach

The purpose of this approach is to resolve the difficulties outlined in the previous section; even though it is based on the direct approach, it utilizes explicitly the knowledge of the controller. The model of the open-loop system (11.34) is assumed to be of the form (11.32) with $G = G\left(q^{-1}, \theta_s\right)$ (on the other hand, H is assumed to be entirely dependent on the parameter θ, which makes the difference with the model (11.35)). In addition, we assume that $\theta = \left(\theta_s, \theta_\beta\right)$, where the random variables θ_s and θ_β are *independent*, and that the set $\mathcal{D}$ defined in Hypothesis (H0") is of the form $\mathcal{D}_s \times \mathcal{D}_\beta$.

According to (11.37), under Hypotheses (H0") and (H1), the prediction error is given by

$$\varepsilon(t) = \frac{1}{H\left(q^{-1}, \theta\right)} \left[\Delta G\left(q^{-1}, \theta_s\right) u(t) + v(t)\right],$$

where $v(t) = \breve{H}\left(q^{-1}\right) w(t)$ and $u(t)$ satisfies (11.39). Using this last expression, we obtain

$$\varepsilon(t) = \frac{\breve{S}_o\left(q^{-1}\right)}{H\left(q^{-1}, \theta\right)} \left[\Delta G\left(q^{-1}, \theta_s\right) r(t) + \left(1 + G\left(q^{-1}, \theta_s\right) K\left(q^{-1}\right)\right) v(t)\right].$$

Therefore, under Hypothesis (H5), the spectral density $\varphi_{\varepsilon\varepsilon}$ of $\{\varepsilon(t)\}$ satisfies (with $\Delta G_{\theta_s} = \Delta G\left(e^{-i\omega}, \theta_s\right)$, $H_\theta = H\left(e^{-i\omega}, \theta\right)$, etc.)

$$\varphi_{\varepsilon\varepsilon} = \frac{1}{\left|H_\theta\right|^2} \left|\Delta G_{\theta_s}\right|^2 \left|\breve{S}_o\right|^2 \varphi_{uu}^r + \frac{1}{\left|H_\theta\right|^2} \left|1 + G_{\theta_s} K\right|^2 \left|\breve{S}_o\right|^2 \varphi_{vv}. \quad (11.55)$$

The result below can be proven the same way as Theorem 383 (section 11.1.4):

LEMMA 418.– *Under Hypothesis (H6), there exists a unique rational function $R\left(z^{-1}, \theta_s\right)$ such that for any $\theta_s \in \mathcal{D}_s$, the filter $R\left(q^{-1}, \theta_s\right)$ is bicausal, bistable, such that $R(0, \theta_s) = 1$, and $\left|R_{\theta_s}\right|^2 = \left|1 + G_{\theta_s} K\right|^2$.*[12]

The criterion to be minimized becomes as $t \to +\infty$

$$J(\theta) = \frac{1}{2\pi} \int_{-\pi}^{\pi} \varphi_{\varepsilon\varepsilon}(\omega)\, d\omega.$$

Choose $H\left(q^{-1}, \theta_s, \theta_\beta\right)$ of the form

$$H\left(q^{-1}, \theta_s, \theta_\beta\right) = R\left(q^{-1}, \theta_s\right) \mathcal{H}\left(q^{-1}, \theta_\beta\right), \quad (11.56)$$

12. The variable θ_{0s} is denoted by θ_s, for more simplicity.

where for every $\theta_\beta \in \mathcal{D}_\beta$, the filter $\mathcal{H}\left(q^{-1}, \theta_\beta\right)$ is bicausal, bistable and such that $\mathcal{H}(0, \theta_\beta) = 1$. We thus have

$$J(\theta) = J_1(\theta_s, \theta_\beta) + J_2(\theta_\beta), \tag{11.57}$$

$$J_1(\theta_s, \theta_\beta) = \frac{1}{2\pi}\int_{-\pi}^{\pi} \frac{1}{|H_\theta|^2} |\Delta G_{\theta_s}|^2 \left|\breve{S}_o\right|^2 \varphi_{uu}^r \, d\omega,$$

$$J_2(\theta_\beta) = \frac{1}{2\pi}\int_{-\pi}^{\pi} \frac{1}{\left|\mathcal{H}_{\theta_\beta}\right|^2} \left|\breve{S}_o\right|^2 \varphi_{vv} \, d\omega.$$

THEOREM 419.– *Suppose Hypotheses (H0"), (H1), (H3'), (H5) and (H6) are in force and the* deterministic part *of the open-loop system* (11.34) *is identifiable. Assume also that the model chosen for the identification is* (11.32) *with H of the form* (11.56). *Then, the estimator $\hat{\theta}_s(t)$ of the deterministic part of the system is consistent.*

PROOF. According to (H0") and (H1), the optimal one-step prediction $\hat{y}(t \mid t-1)$ is given by (11.33). According to (H5), $\varphi_{\varepsilon\varepsilon}$ is of the form (11.55) and from (H3'), $\varphi_{uu}^r(\omega) > 0$ for an infinite number of values of ω in $[-\pi, \pi]$ (see Proposition 416). Hypothesis (H6) allows one to write (11.56), and since the deterministic part of the system is identifiable, $J_1(\theta_s, \theta_\beta)$ is minimum (and equal to 0) if and only if $\Delta G\left(q^{-1}, \theta_s\right) = 0$, i.e. $\theta_s = \breve{\theta}_s$. (On the other hand, as $t \to +\infty$, $\hat{\theta}_\beta(t) \to \hat{\theta}_{\beta\bullet}$, where $\hat{\theta}_{\beta\bullet} \in \arg\min_{\theta_\beta \in \mathcal{D}_\beta} J_2(\theta_\beta)$, if $J_2(\theta_\beta)$ admits a minimum.) ■

Relation with the indirect approach

The one-step predictor (11.33), fed back by the controller (11.39), can be written (in closed-loop)

$$\hat{y}(t \mid t-1) = \left[1 - \frac{1 + G\left(q^{-1}, \theta_s\right) K\left(q^{-1}\right)}{H\left(q^{-1}, \theta\right)}\right] y(t) + \frac{G\left(q^{-1}, \theta_s\right)}{H\left(q^{-1}, \theta\right)} r(t). \tag{11.58}$$

1) Suppose that for every $\theta_s \in \mathcal{D}_s$, $G\left(q^{-1}, \theta_s\right)$ is stable, and $K\left(q^{-1}\right)$ as well. Then $R\left(q^{-1}, \theta_s\right) = 1 + G\left(q^{-1}, \theta_s\right) K\left(q^{-1}\right)$ and, according to (11.56), (11.58) can be written as

$$\hat{y}(t \mid t-1) = \left[1 - \frac{1}{\mathcal{H}\left(q^{-1}, \theta_\beta\right)}\right] y(t) + \frac{G_c\left(q^{-1}, \theta_s\right)}{\mathcal{H}\left(q^{-1}, \theta_\beta\right)} r(t),$$

an expression that becomes identical to (11.42) if, in this last equation, we replace $H_c\left(q^{-1}, \theta\right)$ by $\mathcal{H}\left(q^{-1}, \theta_\beta\right)$.

2) Conversely, if we make use of the indirect approach of section 11.3.1 with (i) $G = G\left(q^{-1}, \theta_s\right)$, and thus $G_c = G_c\left(q^{-1}, \theta_s\right)$, and (ii) $H_c\left(q^{-1}, \theta\right)$ replaced by

$\mathcal{H}\left(q^{-1},\theta_\beta\right)$ (in other words, provided that the deterministic part and the stochastic part of the *closed-loop system* are characterized by *independent parameters*), Theorem 408(i) (section 11.2.5) shows that under Hypotheses (H0”), (H1), (H3’) and (H5)[13], we obtain for the deterministic part of the closed-loop system a consistent estimator $\hat{\theta}_s(t)$ (the problem mentioned in Remark 414 is thus resolved).

On the other hand, the difficulty inherent to the second step of the indirect approach, in its classic formulation, is avoided. Indeed, $\hat{\theta}_s(t)$ is an estimator of the *deterministic part of the open-loop system* as well as of the *deterministic part of the closed-loop system*.

EXAMPLE 420.– *Let us take Example 399 (section 11.2.2) again. This time, the system is fed back by a proportional controller with gain $k = 0.4$ (according to the relation $u = k(r-y)$). The system is identified in a direct approach by the ARMAX method (with $n_A = n_B = n_C = 2$, like in Example 404, where n_P denotes the number of unknown parameters of polynomial $P = A$, B or C). The reference signal is a PRBS of amplitude 1, the measurement noise w is a Gaussian white noise of standard deviation 1 and the identification is carried out over 1000 points. Theorem 417 is applicable. The input and the output, over the first 200 points, are represented in Figure 11.18 (–), as well as the output simulated based on the identified model in the absence of a noise (- -). The residue analysis is shown in Figure 11.19. We see that the prediction error $\{\varepsilon(t)\}$ is whitened correctly; on the other hand, $\varepsilon(t)$ is not correlated with the input $u(\tau)$, $\tau > t$ (but is correlated with the $u(\tau)$, $\tau \leq t$, in accordance with the observation made at the beginning of section 11.3.1). For verification purposes, the transfer function of the identified continuous-time system is*

$$\hat{G}(s) = \frac{1.139\,s - 1.974}{s^2 + 0.1495\,s + 0.9825}$$

and thus is close to (11.24). The step responses of the actual system (–) and of the identified system (- -), shown in Figure 11.20, confirm the good quality of the identification. Nevertheless, the comparison with the results obtained in Example 404 shows that the closed-loop identification is (due to the cross-correlation between the input and the output) an unfavorable circumstance.

11.4. Exercises

EXERCISE 421.– *Consider the ARX model*

$$y(t) + a\,y(t-1) = b\,u(t-1) + w(t).$$

13. Hypothesis (H6) thus becomes useless.

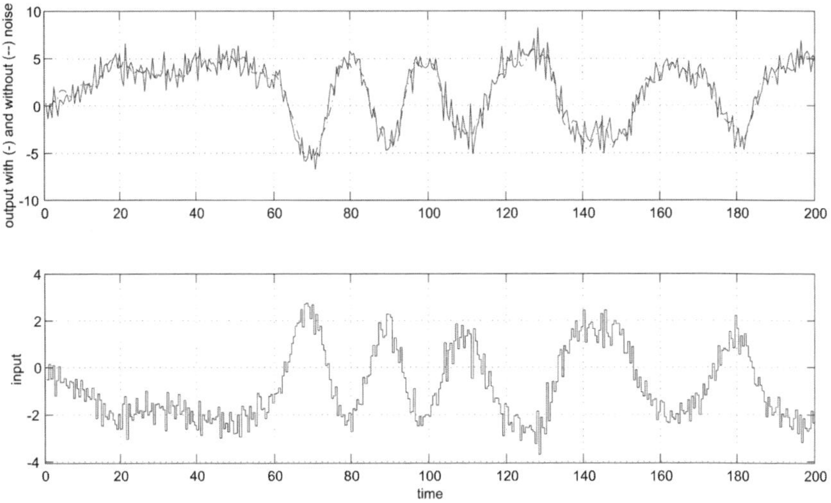

Figure 11.18. *Simulated data and output – Example 420*

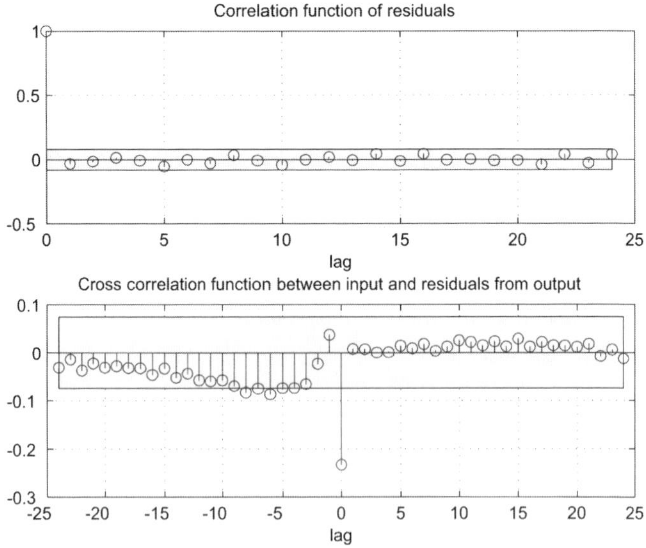

Figure 11.19. *Residue analysis – Example 420*

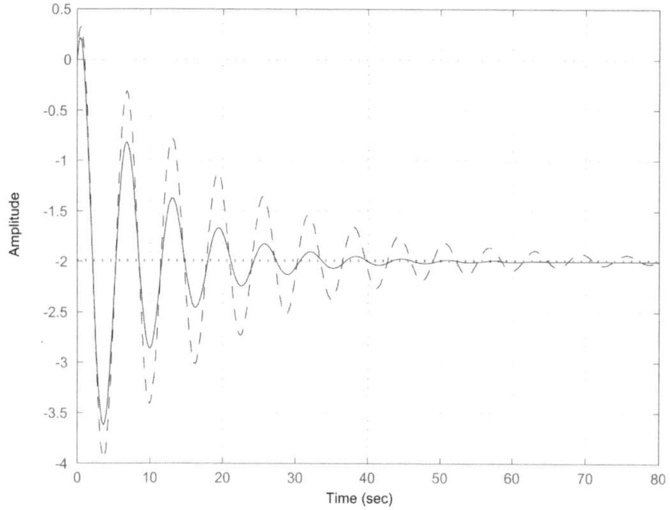

Figure 11.20. *Step responses – Example 420*

(i) Let $\theta = \begin{bmatrix} a & b \end{bmatrix}$*. Express* $\hat{\theta}(t)$ *as a function of* r_{yy}*,* r_{yu} *and* $r_{u\,u}$ *when* $t \to +\infty$*. (ii) We obtain experimentally* $r_{yy}(0) = 0.03$*,* $r_{yy}(1) = -10^{-3}$*,* $r_{u\,u}(0) = 0.05$*,* $r_{yu}(0) = -0.02$ *and* $r_{yu}(-1) = -0.036$*. Calculate* $r'_{yy}(1)$*,* $r'_{yu}(0)$ *and* $r'_{yu}(-1)$*. Are the results coherent? (iii) Determine the identified coefficients* $\hat{a}$ *and* $\hat{b}$*. (iv) Is the identified system stable? What is its static gain?*

EXERCISE 422.– *Consider the "PEM model"* (11.31)*. Determine the prediction error as a function of the data and its various polynomials, and then the partial derivatives of this error with respect to the coefficients of these polynomials.*

EXERCISE 423.– *We consider the system of Example 398 and we try to identify this system using the Output Error method. Assuming that the deterministic part is identifiable, is the estimator of this part consistent when Hypotheses (H1) and (H3) of section 11.2.2 are in force?*

EXERCISE 424.– *We consider the system of Example 405 and we try to identify this system using the ARMAX method. (i) Assume that the deterministic part is identifiable (the vector* θ_s *of Definition 407 being composed of the coefficients of the polynomials A and B of the ARMAX model). Are the hypotheses of Theorem 408(i) satisfied? *(ii) Is it still possible to obtain a correct estimation of the "deterministic part"? (If yes, indicate how).**

EXERCISE 425.– *Consider a system having the structure*

$$A\left(q^{-1}\right)\, y\left(t\right) = q^{-r}\, B\left(q^{-1}\right)\, u\left(t\right) + \frac{C\left(q^{-1}\right)}{D\left(q^{-1}\right)}\, w\left(t\right) \tag{11.59}$$

with the usual conventions. (a) Show that

$$A\left(q^{-1}\right)\, y\left(t\right) = q^{-r}\, B\left(q^{-1}\right)\, u\left(t\right) + \frac{1}{D_1\left(q^{-1}\right)}\, w\left(t\right), \tag{11.60}$$

where $D_1\left(q^{-1}\right)$ *is a power series in* q^{-1}. *(b) Justify the fact that* (11.60) *is a correct approximation of* (11.59) *when* $D_1\left(q^{-1}\right)$ *is a* polynomial of sufficiently high degree *such that* $\left|D_1\left(z^{-1}\right)\right| > 0$ *for* $|z| \geq 1$. *(c) Multiplying the two sides of* (11.60) *by* $D_1\left(q^{-1}\right)$, *show that we can identify with a good precision* $\tilde{A}\left(q^{-1}\right) \triangleq A\left(q^{-1}\right)\, D_1\left(q^{-1}\right)$ *and* $\tilde{B}\left(q^{-1}\right) \triangleq B\left(q^{-1}\right)\, D_1\left(q^{-1}\right)$ *using the least squares method (and, possibly,* recursive *least squares). (This procedure is one of the variants of the* generalized least squares method *called the* indirect procedure.*)*

EXERCISE 426.– *Let us take system* (11.32) *where*

$$\begin{aligned} G\left(q^{-1},\theta\right) &= G_c\left(q^{-1}\right)\, G_i\left(q^{-1},\theta\right) \\ H\left(q^{-1},\theta\right) &= H_c\left(q^{-1}\right)\, H_i\left(q^{-1},\theta\right) \end{aligned}$$

and where $G_c\left(z^{-1}\right)$ *and* $H_c\left(z^{-1}\right)$ *are known transfer functions,* $G_i\left(z^{-1},\theta\right)$ *and* $H_i\left(z^{-1},\theta\right)$ *are to be identified (usual hypotheses are in force:* $G_c\left(q^{-1}\right)$ *and* $G_i\left(q^{-1},\theta\right)$ *are causal and stable,* $H_c\left(q^{-1}\right)$ *and* $H_i\left(q^{-1},\theta\right)$ *are bicausal bistable and such that* $H_c\left(0\right) = H_i\left(0,\theta\right) = 1$*). How can we estimate* θ*?*

EXERCISE 427.– *(i) Is it possible to treat Example 420 correctly using the Output Error method? (ii) According to the theory discussed in section 11.3.3, is it possible to treat correctly Example 420 with an ARMAX model whose stochastic part has less coefficients (*$n_C = 1$ *for example)?*

EXERCISE 428.– *Go over the development of this chapter again and carefully distinguish between noises* $w\left(t\right)$ *and* $\breve{w}\left(t\right)$ *of the model and of the system respectively, and thus justify Remark 390(ii).*

Chapter 12

Appendix 1: Analysis

What is presented here is a brief summary of a few elements of analysis which are very useful in control theory. The need for conciseness is not compatible with the rigor required in pure mathematics. More precise presentations including proofs can be found in numerous books ([35], [103], [83], among others). The theory of measure and integration is only touched upon in Remark 431 below; this should not be an inconvenience for the reader who has not studied this subject, which is developed in great detail in ([35], vol. II), or in [103], using a different approach. *Following are some of the simplifications made: we are only concerned with Hausdorff topological spaces with a countable base, whose topology is completely described by the convergence of sequences[1]; two functions that are equal almost everywhere in Lebesgue's sense are considered equal (in other words, a function is "identified" with its Lebesgue class); a distribution is expressed the same way as a function and the "integral notation" (12.14) used systematically below.*

12.1. Topology

12.1.1. *Topological spaces*

Open sets

A topological space E is a set, embedded with a *topology* consisting of *open sets*. The open sets satisfy the following axioms: $\varnothing$ and E are open, any union of open sets is open, as well as any *finite* intersection of open sets. If A is a subset of E, the

1. In the case of topological spaces with uncountable base – in particular, in spaces of distributions – one has only to replace sequences by nets, or equivalently, by filters.

topology of A said to be "induced by that of E" if its open sets are the $\Omega \cap A$ where the Ω's are the open sets of E.

Some fundamental concepts

A *closed subset* of E is the complement of an open subset of E. If A is a non-empty subset of E (possibly reduced to one point), a *neighborhood* of A in E is a subset of E containing an open set containing A. The *closure* of a subset A of E is the smallest closed subset containing A, that is the intersection of all closed subsets containing A.

A topological space E is said to be *compact* if from any covering of E by a family of open sets, one can extract a finite subcovering; one can show that a compact subset of E is necessarily closed in E.

The topological space E is *connected* if the only subsets of E which are both open and closed are $\varnothing$ and E; roughly speaking, this means that E is "all in one piece". Let E be any topological space and $x \in E$; the union of all connected subsets of E containing x is a connected set $C(x)$, called the *connected component* of x in E. The intersection of two distinct components of E is obviously empty and E is the union of its connected components.

Convergence and continuity

We say that a sequence (x_n) in the topological space E converges to $y \in E$ if for any neighborhood V of $\{y\}$ (also called a neighborhood of y), there exists an integer N such that $x_n \in V$ whenever $n \geq N$. A subset A of E is said to be *dense* in E if every point of E is a limit of a sequence of points of A.

Let E and F be two topological spaces and let $f : E \to F$ be a function. This function is said to be *continuous* if for any open subset Ω_F of F, the inverse image $f^{-1}(\Omega_F)$ is open in E. In this case, for any point $y \in E$ and for any sequence (x_n) which converges to y in E, the sequence $(f(x_n))$ converges to $f(y)$ in F; and conversely, if this condition holds (*with the simplifications mentioned above*), f is continuous.

If $f : E \to F$ is continuous and E is compact, then $f(E) \subset F$ is compact.

Metric spaces

Metric spaces are among the most important topological spaces. A metric space is a set E equipped with a function $d : E \times E \to \mathbb{R}^+$, called a *distance*, such that (i) $d(x, y) = d(y, x)$ and $d(x, x) = 0$, (ii) $d(x, y) + d(y, z) \leq d(x, z)$ ("triangle inequality"), (iii) if $d(x, y) = 0$, then $x = y$.[2] (If d only satisfies (i) and (ii) and if its

2. Quantifiers are understood when there is no ambiguity.

codomain is $\mathbb{R}^+ \cup \{+\infty\}$, this function is called a *pseudometric*.) If (E, d) is a metric space, the open (resp., closed) ball with center x and radius r ($x \in E, r \geq 0$) is the set of all $y \in E$ such that $d(x, y) < r$ (resp., $d(x, y) \leq r$). Of course, an open ball with zero radius is the empty set. Let $x \in E$ and $V_{n,x} = \{y \in E : d(x, y) \leq 1/n\}$. The set of all $V_{n,x}$, $n \geq 1$, is a fundamental system of neighborhoods of x, in other words a set W_x is a neighborhood of x if and only if there exists $n \geq 1$ such that $V_{n,x} \subset W_x$. Therefore, an open set of E is a union of open balls.

Completeness

A sequence (x_n) in a metric space (E, d) is called a Cauchy sequence if $d(x_n, x_m)$ tends to 0 as n and m both tend to infinity. All convergent sequences are Cauchy, but the converse is false in general. A metric space in which every Cauchy sequence converges is said to be *complete*.

12.1.2. *Topological vector spaces*

General definition

Let E be a vector space defined over the field $\mathbf{K} = \mathbb{R}$ or $\mathbb{C}$ ($\mathbb{R}$ denotes the field of real numbers and $\mathbb{C}$ the field of complex numbers). We will equip this space with a *topology* which is *compatible* with its algebraic structure, i.e. such that the following two conditions hold:

– if (λ_n) is a sequence of numbers converging to λ in $\mathbf{K}$, and if (x_n) is a sequence of elements of E converging to x in E, then the sequence $(\lambda_n x_n)$ converges to λx in E;

if (y_n) is another sequence of elements of E converging to a point y in E, then the sequence $(x_n + y_n)$ converges to $x + y$.

Note that according to the above compatibility conditions, (x_n) converges to x if and only if $x_n - x$ converges to 0. A $\mathbf{K}$-vector space equipped with such a topology is called a *topological vector space*.

If E and F are two topological vector spaces, an *isomorphism* from E *onto* F is a bijective linear function of $E \to F$, which is continuous as well as its inverse function.

Normed vector space

A *seminorm* on E is a function $\|.\| : E \to \mathbb{R}^+$ such that the following two conditions hold:

– $\|\lambda x\| = |\lambda| \, \|x\|$ (where λ is a scalar, i.e. an element of $\mathbf{K}$);

– $\|x + y\| \leq \|x\| + \|y\|$ ("triangle inequality").

Therefore, $d(x,y) = \|x-y\|$ is a pseudometric. The function $\|.\|$ is a *norm* if the following supplementary condition holds:

– if $\|x\| = 0$, then $x = 0$

(that is if the above pseudometric is a distance). The topology of a normed vector space can be characterized as follows: a sequence (x_n) of elements of E converges to 0 in E if and only if the sequence $(\|x_n\|)$ tends to 0 in $\mathbb{R}$. (However, the topology of a topological vector space is not always defined by a norm nor by a distance. In particular, this is the case of distribution spaces which will be studied later.)

A *bounded* subset of E is a set included in a ball. A Cauchy sequence in E is bounded.[3] A vector x such that $\|x\| = 1$ is said to be *unitary* and the set of unitary vectors forms the "unit sphere".

Let E be a normed vector space. A compact subset of E is necessarily closed and bounded (but if E is infinite-dimensional, there exist closed and bounded subsets of E which are not compact). If E is complete, it is called a *Banach space*.

Case of $\mathbb{R}$ *and* $\mathbb{C}$

In particular, an open ball of $\mathbb{R}$ is an *open interval* of the form (a,b), $-\infty < a \leq b < +\infty$. In the complex plane $\mathbb{C}$, the open ball with center z and radius $r \geq 0$ is nothing but the *open disc* with center z and radius r, which is the set of $s \in \mathbb{C}$ such that $|z-s| < r$.

Duality

A linear form f on E is a linear function from E into $\mathbf{K}$. For any $x \in E$, the scalar $f(x)$ is often denoted by $\langle f, x \rangle$ ("duality bracket"). The set of all linear forms (resp., of all continuous linear forms) on E is a $\mathbf{K}$-vector space, which we call the *algebraic dual* (resp., the topological dual) of E and we denote by E^* (resp., E'). If E is finite-dimensional, E^* and E' coincide.

If E is a normed vector space, we can define a norm on E' (see section 12.1.3). The topology defined by this norm is called the "strong topology" of E'.

We can define on E' another topology, called the *weak* topology*, in the following manner: let (x'_n) be a sequence of elements of E'; we say that this sequence weakly* converges to 0 in E' if, for any $y \in E$, the sequence $(\langle x'_n, y \rangle)$ tends to 0.

Hilbert spaces

Let $\mathbf{K} = \mathbb{R}$ or $\mathbb{C}$. A *pre-Hilbert* space E over the field $\mathbf{K}$ is a $\mathbf{K}$-vector space equipped with a *scalar product* denoted as $\langle .,. \rangle$. This scalar product is a positive

3. This does not hold for a Cauchy net.

definite Hermitian form, i.e. a function from $E \times E$ into $\mathbf{K}$ for which the following conditions hold:

– $\langle x + x', y \rangle = \langle x, y \rangle + \langle x', y \rangle$ and $\langle x, y + y' \rangle = \langle x, y \rangle + \langle x, y' \rangle$;

– $\langle x, \lambda y \rangle = \lambda \langle x, y \rangle$; thus, the function $\langle x, . \rangle : y \mapsto \langle x, y \rangle$ is linear;[4]

– $\langle x, y \rangle = \overline{\langle y, x \rangle}$ (Hermitian symmetry if $\mathbf{K} = \mathbb{C}$, symmetry if $\mathbf{K} = \mathbb{R}$);

– $\langle x, x \rangle \in \mathbb{R}^+$ and $\langle x, x \rangle = 0$ if and only if $x = 0$.

One can easily show that the function $x \mapsto \sqrt{\langle x, x \rangle} \triangleq \|x\|$ is a norm on E, called a *pre-Hilbert norm*. A pre-Hilbert space is thus a particular type of normed $\mathbf{K}$-vector space.

We have the *Schwarz inequality*:

$$\boxed{|\langle x, y \rangle| \leq \|x\| \, \|y\|}. \tag{12.1}$$

Note that if $\mathbf{K} = \mathbb{C}$, the function $x \rightarrow \langle x, y \rangle$ is *antilinear*[5], whereas it would be linear if $\langle ., . \rangle$ were a duality bracket.

A complete pre-Hilbert $\mathbf{K}$-space is called a *Hilbert space*. A pre-Hilbert norm on a Hilbert space is called a Hilbert norm. The following is the "orthogonal projection theorem in Hilbert spaces":

THEOREM 429.– *Let E be a pre-Hilbert space and $F \neq \{0\}$ be a complete subspace of E. For any $x \in E$, there exists one and only one point of F, denoted by $\hat{x}$, such that $\|x - \hat{x}\| = \min_{y \in F} \|x - y\|$. This point is characterized by the relation $\langle y, x - \hat{x} \rangle = 0$ for all $y \in F$, and is thus the* orthogonal projection *of x onto F. The function $x \mapsto \hat{x}$ is $\mathbf{K}$-linear (it is also continuous with norm* 1: *see section 12.1.3).*

Finite-dimensional spaces

According to the *Riesz theorem*, if E is a *finite-dimensional* $\mathbf{K}$-vector space, there exists a unique topology on E compatible with its structure of $\mathbf{K}$-vector space. In addition, if F is a topological vector space and if f is a linear function from E into F, f is necessarily continuous. In the elements of algebra in Chapter 13, we are only concerned with finite dimension, therefore the notion of continuity (which is not algebraic) is never discussed; from a certain point of view, we can consider that it is understood. A $\mathbf{K}$-vector space of dimension $n < +\infty$ is identified with

4. According to an equivalent but different convention, the function $\langle ., y \rangle : x \mapsto \langle x, y \rangle$ is assumed to be linear.

5. A function $f : x \mapsto f(x)$ is antilinear if $f(x_1 + x_2) = f(x_1) + f(x_2)$ and $f(\lambda x) = \bar{\lambda} f(x)$.

$\mathbf{K}^n$ once a basis is chosen (see section 13.3.1). The topology of this space can thus be defined by one of the following norms (among others): $\|x\|_1 = \sum_{i=1}^n |x_i|$, $\|x\|_\infty = \max_{1 \le i \le n} |x_i|$ (with $x = (x_1, ..., x_n)$).

A subset of a finite-dimensional $\mathbf{K}$-vector space is compact *if and only if* it is closed and bounded.

Every finite-dimensional $\mathbf{K}$-vector space is complete and is characterized by the fact that it is *locally compact*, that is to say there exist compact neighborhoods of 0 (the unit ball if the space is normed, for example).

Euclidean and Hermitian spaces

Let E be a finite-dimensional pre-Hilbert space over $\mathbf{K}$. If $\mathbf{K} = \mathbb{R}$, E is called an *Euclidean* space, and a pre-Hilbert norm on such a space is said to be Euclidean. If $\mathbf{K} = \mathbb{C}$, E is called a *Hermitian space* and a pre-Hilbert norm on such a space is said to be *Hermitian*. Obviously, an Euclidean space or a Hermitian space is a Hilbert space; conversely, a finite-dimensional Hilbert space is a Euclidean space or a Hermitian space.

A Hilbert $\mathbf{K}$-space of dimension n is identified with $\mathbf{K}^n$ in an *orthonormal basis*; the vectors of this space also identified with the column vectors with n entries in $\mathbf{K}$; then the "standard scalar product" is $\langle x, y \rangle = x^* y$ (see above as well as section 13.5.1). The pre-Hilbert norm $\|x\|_2$ associated with this scalar product is called the "standard Euclidean norm" if $\mathbf{K} = \mathbb{R}$, and is called the "standard Hermitian norm" if $\mathbf{K} = \mathbb{C}$.

12.1.3. *Continuous linear operators*

Let E and F be two normed $\mathbf{K}$-vector spaces and let $\mathbf{u}$ be a $\mathbf{K}$-linear function from E into F ($\mathbf{u}$ is also called a linear operator from E into F). We say that $\mathbf{u}$ is continuous (in the norm topology) if $\|\mathbf{u}\| < +\infty$, where

$$\boxed{\|\mathbf{u}\| \triangleq \sup_{x \neq 0} \frac{\|\mathbf{u}(x)\|}{\|x\|} = \sup_{\|x\| \le 1} \|\mathbf{u}(x)\| = \sup_{\|x\| = 1} \|\mathbf{u}(x)\|} \tag{12.2}$$

(the reader can prove the two equalities on the right-hand side as an exercise).

The continuous linear operators from E into F form a $\mathbf{K}$-vector space denoted by $\mathcal{L}(E, F)$ and the quantity (12.2) defines a norm on this space. This norm is sometimes called the "operator norm induced by the norms of E and F" (and the "operator norm induced by the norm of E" if $E = F$). For the explicit calculation of this norm in the case where E and F are finite-dimensional Hilbert spaces, see section 13.5.7.

If $\mathbf{u} \in \mathcal{L}(E,F)$ and $\mathbf{v} \in \mathcal{L}(F,G)$, where E, F and G are normed $\mathbf{K}$-vector spaces, we have

$$\|\mathbf{v}\,\mathbf{u}\| \leq \|\mathbf{v}\|\ \|\mathbf{u}\|$$

and we express this property by saying that an operator norm is *multiplicative*.

One can show that $\mathcal{L}(E,F)$ is complete, thus a Banach space, if F is a Banach space. In particular, $E' = \mathcal{L}(E,\mathbf{K})$ is a Banach space equipped with the norm (12.2); this norm defines the "strong topology" of E'. If a sequence (x'_n) of elements of E' converges to zero in this topology, this sequence is said to *strongly* converge to zero.

If E is finite-dimensional, all linear operator of E into F are continuous (this remains true if F is any topological vector space).

12.2. Sequences, functions and distributions

12.2.1. *Sequences*

l_p spaces

– Case $p \in [1, +\infty)$

We denote by l_p the $\mathbf{K}$-vector space consisting of all sequences $(x_n)_{n\in\mathbb{Z}}$ of elements of $\mathbf{K}$ such that $\sum_{n=-\infty}^{+\infty} |x_n|^p < \infty$. Then

$$\boxed{\|x\|_p \triangleq \left(\sum_{n=-\infty}^{+\infty} |x_n|^p\right)^{1/p}} \tag{12.3}$$

is a norm on l_p and that space, equipped with the norm (12.3), is a Banach space.

In the case $p = 2$, we define on the space l_2 the scalar product

$$\langle x, y\rangle_2 \triangleq \sum_{n-\infty}^{+\infty} \bar{x}_n\, y_n \tag{12.4}$$

where $\bar{x}_n$ is the conjugate of x_n. The space l_2 is a Hilbert space.

– Case $p = +\infty$

The space l_∞ is that of all bounded sequences; this space is equipped with the norm

$$\boxed{\|x\|_\infty \triangleq \sup_n |x_n|}. \tag{12.5}$$

This is a Banach space.

The following result can easily be shown:

THEOREM 430.– *If* $1 \leq q \leq p \leq +\infty$, *then* $l_q \subset l_p$.

Sequences with positive support

The *support* of a sequence (x_n) is the set of all integers n for which $x_n \neq 0$. This sequence (x_n) is said to be *with positive support*, or *positively supported*, if $x_n = 0$ for $n < 0$.

Convolution of two sequences

General case

Let h and u be two sequences of elements of $\mathbf{K}$. Their *convolution product* is the sequence y, denoted $y = h * u$ and defined by

$$\boxed{y_n = \sum_{k=-\infty}^{+\infty} h_{n-k}\, u_k = \sum_{k=-\infty}^{+\infty} h_k\, u_{n-k}} \tag{12.6}$$

under the condition that the series converges for all values of n.

Note that the function $(h, u) \mapsto h*u$ is bilinear (like every product) and symmetric $(h * u = u * h)$.

Convolution and l_p spaces

As easily shown, if $h \in l_1$, then the above convolution product is well-defined whenever $u \in l_p$, $p \in [1, +\infty]$, and in this case $h * u \in l_p$. In addition, we have the *Young inequality*

$$\|h * u\|_p \leq \|h\|_1 \|u\|_p \,. \tag{12.7}$$

The linear operator H defined by $H\,u = h * u$ is called the *convolution operator* with *kernel* h. The Young inequality shows that if $h \in l_1$, then H is continuous from l_p into l_p and that $\|H\| \leq \|h\|_1$.

Convolution of sequences with positive support

If h and u are two positively supported sequences of complex numbers, their convolution product y must exist and is also positively supported since

$$y_n = \sum_{k=0}^{n} h_{n-k}\, u_k \,.$$

12.2.2. *Functions*

Lebesgue spaces

REMARK 431.– *Let S be a subset of $\mathbb{R}$ and x be a function from S into $\mathbf{K}^n$ or from S into $[-\infty, +\infty]$. The following notions are used below: (a) the measurability of S (in Lebesgue's sense); (b) the measurability of x (in Lebesgue's sense); the integral $\int_S x(t)\,dt$ (in Lebesgue's sense); the Lebesgue class of x.*
(i) If S is an interval of $\mathbb{R}$, the measurability of x is not a strong assumption. Indeed, there is no explicit example of a function which is not Lebesgue-measurable, although using the axiom of choice one can prove that such functions exist [103].
(ii) The set S is Lebesgue-measurable if and only if so is its characteristic function *$\varkappa_S : \mathbb{R} \to \mathbb{R}$ defined by $\varkappa_S(t) = 1$ if $t \in S$ and $\varkappa_S(t) = 0$ otherwise. The set S is said to be of (Lebesgue) measure zero if it is of "zero length". This happens, in particular, if S is countable. For example, the subset $[0,1] \cap \mathbb{Q}$ of $[0,1]$ is of measure zero (although $\mathbb{Q}$ is dense in $\mathbb{R}$), thus its complement $[0,1] \cap (\mathbb{R} \backslash \mathbb{Q})$ is of measure* 1.
(iii) Let $S \subset \mathbb{R}$ be Lebesgue-measurable. Two functions x and y from S into $\mathbf{K}^n$ or from S into $[-\infty, +\infty]$ are said to be equal almost everywhere *if $Z = \{t \in S : x(t) \neq y(t)\}$ is of (Lebesgue) measure zero. The* Lebesgue class *of x consists of all functions y which are equal to x almost everywhere.*
(iv) Let S be an interval of $\mathbb{R}$ and let x be a function from S into $\mathbf{K}^n$ or from S into $[-\infty, +\infty]$. If there exists in the Lebesgue class of x a function y such the usual Riemann integral $\int_S y(t)\,dt$ exists, then the Lebesgue integral $\int_S x(t)\,dt$ exists (i.e. x is Lebesgue-integrable on S) and is equal to $\int_S y(t)\,dt$. On the other hand, let $x : S \to \mathbf{K}^n$; the Lebesgue integral $\int_S x(t)\,dt$ exists if and only if x is measurable and $\int_S \|x(t)\|\,dt < +\infty$.

The Lebesgue spaces L_p ($p \in [1, +\infty]$) are spaces of functions which have properties very similar to those of l_p spaces.

– Case $p \in [1, +\infty)$

We denote by $\mathcal{L}_p$ the space of all functions $x : \mathbb{R} \to \mathbf{K}$ which are Lebesgue-measurable and such that $\int_{-\infty}^{+\infty} |x(t)|^p\,dt < \infty$. Provided that x is not distinguished from its Lesbesgue class, $\mathcal{L}_p$ is written L_p and the function $x \mapsto \|x\|_p$, as defined below, is a norm on L_p :

$$\boxed{\|x\|_p = \left(\int_{-\infty}^{+\infty} |x(t)|^p\,dt\right)^{1/p}}. \tag{12.8}$$

In the case $p = 2$, we define on space L_2 the scalar product

$$\boxed{\langle x, y\rangle_2 \triangleq \int_{-\infty}^{+\infty} \bar{x}(t)\,y(t)dt} \tag{12.9}$$

where $\bar{x}(t)$ is the conjugate of the complex number $x(t)$. We have $\|x\|_2 = \sqrt{< x, x >_2}$. The functions belonging to L_2 are said to be *square integrable*.

In signal theory, the quantity $\|x\|_2^2$ is called the *energy* of the signal x. As a result, L_2 can be interpreted as the space of signals with bounded energy.

– Case $p = +\infty$

A function $x : \mathbb{R} \to \mathbf{K}$ is said to be *essentially bounded* if it is measurable and there exists a finite constant $M > 0$ such that $|x(t)| \leq M$ almost everywhere. The greatest lower bound of all the constants M satisfying the above condition is denoted as

$$\boxed{\|x\|_\infty \triangleq \operatorname*{ess.sup}_{t\in\mathbb{R}} |x(t)|}. \tag{12.10}$$

The space $\mathcal{L}_\infty$ is that of all essentially bounded functions and the function $\mathcal{L}_\infty \ni x \mapsto \|x\|_\infty \in \mathbb{R}^+$ is a seminorm. Again, provided that x is not distinguished from its Lesbesgue class, $\mathcal{L}_\infty$ is written L_∞, and the above seminorm is then a norm on L_∞. From the point of view of signal theory, L_∞ can be interpreted as the space of signals with bounded absolute value.

According to the *Fisher–Riesz theorem*, all L_p spaces ($p \in [1, +\infty]$) are Banach spaces, and L_2 is a Hilbert space. Theorem 430 does not have an analogue in the case of functions.

Functions with positive support

The *support* of a function is the closure of the set of points at which it is non-zero. In control theory, we are essentially interested in functions with positive support, that is in functions x such that $x(t) = 0$ for $t < 0$. Here is why.

Consider a control system and its input u, defined in $\mathbb{R}$; it is thus a function from $\mathbb{R}$ into $\mathbf{K}^m$ ($\mathbf{K} = \mathbb{R}$ or $\mathbb{C}$). If we are only concerned with the behavior of the system from an initial instant t_0, which we can assume to be zero (by translation of the origin of time), then we also are only concerned with the restriction of u to $\mathbb{R}^+$. The knowledge of this is equivalent to that of the product $u^+ = u\,\mathbf{1}$, where $\mathbf{1}$ is the unit step, defined as follows:

$$\mathbf{1}(t) = \begin{cases} 0, & t < 0 \\ 1, & t \geq 0. \end{cases}$$

This function u^+ is with positively supported.

Convolution of two functions

General case

Let h and u be two functions from $\mathbb{R}$ into $\mathbf{K}$. Their *convolution product* is the function y from $\mathbb{R}$ into $\mathbf{K}$, denoted as $y = h * u$ and defined by

$$\boxed{y(t) = \int_{-\infty}^{+\infty} h(t-\tau)u(\tau)d\tau = \int_{-\infty}^{+\infty} h(\tau)u(t-\tau)d\tau} \tag{12.11}$$

under the condition that the integral converges for all values of t.

The function $(h, u) \to h * u$ is bilinear and symmetric.

Convolution and Lebesgue spaces

One can show that if $h \in L_1$, then the above convolution product is well defined whenever $u \in L_p$, $p \in [1, +\infty]$ and in this case $h * u \in L_p$. In addition, we have the *Young inequality*

$$\|h * u\|_p \leq \|h\|_1 \|u\|_p . \tag{12.12}$$

Let H be the linear operator defined by $H u = h * u$, called the convolution operator with *kernel* h. The Young inequality shows that if $h \in L_1$, then H is continuous from L_p into L_p and that $\|H\| \leq \|h\|_1$. To calculate $\|H\|$ in the case $p = 2$, see section 13.6.2, Theorem 589.

Algebra $\mathcal{K}_+$

We denote by $\mathcal{K}$ the set of all *locally integrable functions* from $\mathbb{R}$ into $\mathbf{K}$ (i.e. of those functions which are integrable on all compact intervals) and by $\mathcal{K}_+$ the set consisting of all positively supported functions belonging to $\mathcal{K}$. For example, the function $t \to e^{\alpha t}\,\mathbf{1}(t)$ belongs to $\mathcal{K}_+$, and the function $t \mapsto e^{e^t}\,\mathbf{1}(t)$ as well. For any h and u belonging to $\mathcal{K}_+$, integral (12.11) is written as

$$y(t) = \int_0^t h(t-\tau)\, u(\tau)d\tau,$$

thus it converges; in addition, $h * u \in \mathcal{K}_+$. The set $\mathcal{K}_+$ is a commutative $\mathbf{K}$-algebra (see section 13.1.1); indeed, it is both a $\mathbf{K}$-vector space and a commutative ring with convolution as a product. We thus call it a *convolution algebra*. Another example of convolution algebra is L_1. All these algebras are commutative since the convolution product is commutative.

Continuity of a convolution product

If h and u belong to $\mathcal{K}_+$ and if, in addition, h is a locally bounded function (that is, h is bounded on any compact interval), one can show that their convolution product $y = h * u$ is a continuous function.

12.2.3. *Distributions*

Introduction to the notion of distribution

Let $\mathcal{T}$ be a vector space of functions ϕ, from $\mathbb{R}$ into $\mathbb{C}$, which are very regular in the following sense:

– they are *indefinitely differentiable*;

– either they have a compact support, and in this case $\mathcal{T}$ is denoted as $\mathcal{D}$; or and they decrease very rapidly toward 0 at infinity, more precisely ϕ and its derivatives of all orders decrease more rapidly in the neighborhood of infinity (in absolute value) than any function of the form $\frac{1}{|t^k|}$ (where k is any positive integer): in this case $\mathcal{T}$ is denoted by $\mathcal{S}$. Space $\mathcal{S}$ is called the "*space of rapidly decaying functions at infinity*". We call $\mathcal{T}$ a space of *test functions*. Now let f be a function from $\mathbb{R}$ into $\mathbb{C}$. Let us form the integral

$$\langle T_f, \phi \rangle \triangleq \int_{-\infty}^{+\infty} f(t)\phi(t)dt. \tag{12.13}$$

For $\mathcal{T} = \mathcal{D}$, this integral converges whenever $f \in \mathcal{K}$.

For $\mathcal{T} = \mathcal{S}$, it converges for any function f in $\mathcal{K}$ that is not increasing (in absolute value) faster than any polynomial in the neighborhood of infinity; we denote the set of these functions by $\mathcal{O}$, and we call this the space of "slowly increasing functions at infinity". For example, all polynomials in t belong to $\mathcal{O}$.

Therefore, we associate with a function f the *linear form* $T_f : \phi \to \langle T_f, \phi \rangle \in \mathbb{C}$ which belongs to the dual $\mathcal{T}'$ of $\mathcal{T}$ ($f \in \mathcal{K}$ or $\mathcal{O}$, depending on whether $\mathcal{T} = \mathcal{D}$ or $\mathcal{S}$). If two functions f and g are such that $T_f = T_g$, one can show that $f = g$ almost everywhere, thus f and g are identified (by abuse of language). The knowledge of f is equivalent then to that of T_f, which makes it possible to identify f with T_f.

In the integral of (12.13), f and ϕ play a symmetric role if these functions belong to L_2 according to (12.9): the dual of L_2 is L_2 itself. But $\mathcal{T}$ is a space much smaller than L_2, therefore its dual $\mathcal{T}'$ is much larger (indeed, the stronger the conditions upon ϕ, the larger the set of functions f for which the integral (12.13) converges). In particular, $\mathcal{D}'$ is larger than $\mathcal{K}$, and $\mathcal{S}'$ is larger than $\mathcal{O}$. Therefore, we have embedded the spaces of functions $\mathcal{K}$ and $\mathcal{O}$ in the larger spaces, $\mathcal{D}'$ and $\mathcal{S}'$, respectively. To sum up

$$\boxed{\begin{array}{l} \mathcal{O} \subset \mathcal{S}' \subset \mathcal{D}' \\ \mathcal{O} \subset \mathcal{K} \subset \mathcal{D}' \end{array}}.$$

Spaces of distributions

The space $\mathcal{D}'$ is called the space of *distributions*. The space $\mathcal{S}'$ is called the space of *tempered distributions*.

A distribution is thus a linear form $T : \phi \mapsto \langle T, \phi \rangle$ $(T \in \mathcal{T}')$.

The spaces of distributions are equipped with the following topological structure: a sequence (T_n) of $\mathcal{T}'$ converges to $T \in \mathcal{T}'$ if for any test function $\phi \in \mathcal{T}$, $\langle T_n, \phi \rangle$ converges to $\langle T, \phi \rangle$.[6]

With this definition, $\mathcal{T}$ can be proved to be dense in $\mathcal{T}'$ and in L_p, $p \in [1, +\infty)$.

Taking into account relation (12.13), we find it convenient to denote the duality bracket $< T, \phi >$ by an integral, i.e.

$$\langle T, \phi \rangle = \int_{-\infty}^{+\infty} T(t)\, \phi(t) dt \tag{12.14}$$

where the distribution T thus appears the same way as a function $t \mapsto T(t)$. But it is a "generalized function"[7] which only has meaning under a summation, as in (12.14). A distribution which is not a function is said to be "singular".

Support of a distribution

We say that a distribution T is zero in an open subset Ω of $\mathbb{R}$ if $\langle T, \phi \rangle = 0$ for any function $\phi \in \mathcal{T}$, the support of which is included in Ω. The union of all open subsets in which T is zero is the largest open set in which T vanishes. Its complement S, which is closed, is by definition the *support* of T and is denoted by $\operatorname{supp} T$.

Now let Ω be an open neighborhood of $S = \operatorname{supp} T$ and ϕ_1, ϕ_2 be two functions of $\mathcal{T}$ that are equal in Ω. Then $\phi = \phi_1 - \phi_2$ is zero in Ω, therefore $\operatorname{supp} \phi$ is included in the complement of Ω which is itself included in S. As a result, $\langle T, \phi \rangle = 0$, which implies that $\langle T, \phi_1 \rangle = \langle T, \phi_2 \rangle$. This shows that $\langle T, \phi \rangle$ only depends on the restriction of ϕ to any open neighborhood of S.

For example, if T has a support included in $[t_1, t_2]$ (resp., $[t_1, +\infty)$), we can write

$$\langle T, \phi \rangle = \int_{t_1^-}^{t_2^+} T(t)\, \phi(t) dt \text{ (resp., } \langle T, \phi \rangle = \int_{t_1^-}^{+\infty} T(t)\, \phi(t) dt \text{)} \tag{12.15}$$

---

6. *This is the definition weak* topology (see section 12.1.2). But due to the very specific topological properties of $\mathcal{T}$ ($\mathcal{T}$, and thus $\mathcal{T}'$, are *Montel spaces*), weak* convergence and strong convergence of sequences coincide in $\mathcal{T}'$ [106].*

7. So are also Sato's hyperfunctions [22].

where by convention

$$\int_{t_1^-}^{t_2^+} \triangleq \lim_{\varepsilon_1 \to 0^+, \varepsilon_2 \to 0^+} \int_{t_1 - \varepsilon_1}^{t_2 + \varepsilon_2} .$$

This generalizes in the case of distributions the notion of support defined for functions, except that in the case of a function, t_1^- and t_2^+ can be replaced by t_1 and t_2, respectively.

As easily seen, every compactly supported distribution is tempered.

We denote by $\mathcal{T}'_+$ the subspace of $\mathcal{T}'$ consisting of positively supported distributions and for any such distribution T we thus have $\langle T, \phi \rangle = \int_{0^-}^{+\infty} T(t)\phi(t)dt$ for any $\phi \in \mathcal{T}$.

Differentiation of distributions

Definition

Let $T \in \mathcal{T}'$; we can define formally the derivative of this distribution by applying the integration by parts to (12.14):

$$\int_{-\infty}^{+\infty} \dot{T}(t)\phi(t)dt = -\int_{-\infty}^{+\infty} T(t)\dot{\phi}(t)dt$$

is a well-defined expression because $\dot{\phi} \in \mathcal{T}$. As a result,

$$\boxed{< \dot{T}, \phi > \triangleq - < T, \dot{\phi} >}. \tag{12.16}$$

We deduce that all distributions are indefinitely differentiable.

Derivative of a function in the sense of distributions

Let f be a locally integrable function from $\mathbb{R}$ into $\mathbb{C}$ (that is $f \in \mathcal{K}$). This function is generally not differentiable at all points of $\mathbb{R}$ in the usual sense. For example, the function $f(t) = t\mathbf{1}(t)$ is differentiable in $(-\infty, 0)$ (with a zero derivative) and in $(0, +\infty)$ (with a derivative of 1) but not at 0. On the other hand, if we deal with f the same way as with a distribution, f becomes differentiable. Its derivative is thus taken "in the sense of distributions". This derivative can be a function or a singular distribution.

In the example presented here, we have $\dot{f} = \mathbf{1}$, thus $\dot{f}$ is a function.

Absolutely continuous function

Let f be a locally integrable function and let $\dot{f}$ be its derivative in the sense of distributions. If $\dot{f}$ is a locally integrable *function*, one can show that

$$\boxed{f(t) - f(t_0) = \int_{t_0}^{t} \dot{f}(\tau)\, d\tau}$$

(see for example [35], section XVII. 5). Such a function f is said to be *absolutely continuous*.

It is immediate that an absolutely continuous function is continuous. An example of absolutely continuous function is the function $f(t) = t\,\mathbf{1}(t)$ considered above. This function is not differentiable in the usual sense.

The unit step, the Dirac distribution and its derivatives

Derivative of the unit step

The unit step is not differentiable in the usual sense of functions, due to its discontinuity at 0. In addition, this function is not absolutely continuous since it is not continuous. Its derivative in the sense of distributions is thus a singular distribution which we are going to determine.

We have for any $\phi \in \mathcal{T}$

$$\langle \dot{\mathbf{1}}, \phi \rangle = -\left\langle \mathbf{1}, \dot{\phi} \right\rangle = -\int_0^{+\infty} \dot{\phi}(t)\, dt = \phi(0)$$

Definition of δ

The derivative of the unit step is the *Dirac distribution* δ defined by

$$\boxed{\langle \delta, \phi \rangle = \phi(0).} \tag{12.17}$$

Using the "integral notation", we have

$$\int_{-\infty}^{+\infty} \delta(t)\phi(t)dt = \phi(0). \tag{12.18}$$

Obviously, the support of δ is $\{0\}$ and this distribution is therefore tempered.

Interpretation of δ

According to (12.18), we can interpret δ as a generalized function being everywhere zero, except at 0 where it is $+\infty$ and such that

$$\int_{-\infty}^{+\infty} \delta(t)\, dt = \lim_{\epsilon \to 0^+} \int_{-\epsilon}^{+\epsilon} \delta(t)dt = 1. \tag{12.19}$$

We can justify (12.19) in the following manner: let $\epsilon > 0$; we have according to (12.19) and (12.15): $\phi(0) = \int_{-\epsilon}^{+\epsilon} \delta(t)\phi(t)dt$. Since δ is a "non-negative (generalized) function", we have $\int_{-\epsilon}^{+\epsilon} \delta(t)\phi(t)\, dt = \phi(\eta) \int_{-\epsilon}^{+\epsilon} \delta(t)dt$ for some $\eta \in [-\epsilon, \epsilon]$, according to the mean value formula. We thus obtain the second equality of (12.19) by taking $\epsilon \to 0^+$, and the first by using the fact that $\delta(t) = 0$ for $t \neq 0$.

Note that here we repeatedly abuse the language: specifically, a function which is everywhere equal to zero except at 0 has a zero integral (Remark 431). Thus (12.19) would be impossible if δ were actually a function, as the notation and the rationale imply it; but the calculations which have just been made right here are exact and they can be given a precise mathematical sense.

Convergence to δ

In order to give a precise meaning of the above, let us state the following result:

THEOREM 432.– *Let (f_n) be a sequence of measurable functions from $\mathbb{R}$ to $\mathbb{R}^+$ such that $\int_{-\infty}^{+\infty} f_n(t)\,dt = 1$ and for every neighborhood V of 0 in $\mathbb{R}$, $\int_{\mathbb{R}\setminus V} f_n(t)\,dt$ converges to 0. Then (f_n) converges to δ in $\mathcal{D}'$.*

Derivatives of δ

According to (12.16), we have $\left\langle \dot{\delta}, \phi \right\rangle = -\dot{\phi}(0)$ and by iteration $\left\langle \delta^{(n)}, \phi \right\rangle = (-1)^n \phi^{(n)}(0)$.

Product of distributions

One cannot in general define the product of two distributions T and U. One can define the product of a distribution T by a function f according to the formula

$$\langle T f, \phi \rangle = \langle T, f\phi \rangle$$

under the condition that $f\phi$ belongs to $\mathcal{T}$ for any function ϕ of $\mathcal{T}$. If $\mathcal{T} = \mathcal{D}$, this holds whenever f is an indefinitely differentiable function; if $\mathcal{T} = \mathcal{S}$, this holds whenever f is a function belonging to the space $\mathcal{O}_{\mathrm{M}}$ of indefinitely differentiable functions whose derivatives of all orders (zeroth order included)[8] belong to $\mathcal{O}$.

If $T \in \mathcal{S}'_+$, this formula is meaningful, according to (12.15), whenever f is an indefinitely differentiable function whose derivatives of all orders are slowly increasing as $t \to +\infty$ (one such function does not generally belong to $\mathcal{O}_{\mathrm{M}}$ because we are not concerned with its behavior as $t \to -\infty$).

In particular, let ε_α be the function defined by $\varepsilon_\alpha(t) = e^{-\alpha t}$, $\alpha > 0$. It satisfies the above-mentioned property, thus the product $\varepsilon_\alpha T$ exists. This will be useful later to define the Laplace transform.

For $T = \delta$, we have

$$\boxed{\delta f = f(0)\delta}. \qquad (12.20)$$

8. The zeroth order derivative of f is f itself, of course.

Convolution of distributions

The *convolution product* of two distributions generalizes the convolution product of two functions. For this product to be meaningful, it is necessary that the supports of these distributions satisfy certain properties. In particular, if h is a compactly supported distribution, the product $h * u$ can be defined and the map $u \mapsto h * u$ is continuous *convolution operator* (with kernel h) from $\mathcal{D}'$ into $\mathcal{D}'$. We can calculate $h * u$ when $u \in \mathcal{D}$, and then when $u \in \mathcal{D}'$ by using the density of $\mathcal{D}$ in $\mathcal{D}'$.

Using the "integral notation" of distributions, we can formally write that (when $h * u$ is well-defined)

$$\boxed{(h * u)(t) = \int_{-\infty}^{+\infty} h(t - \tau)\, u(\tau)\, d\tau}$$

as if both h and u were functions.

For example, if $u \in \mathcal{D}$, $\int_{-\infty}^{+\infty} \delta(\tau)\, u(t - \tau) d\tau = u(t)$ according to (12.18), thus

$$\boxed{\delta * u = u} \tag{12.21}$$

and this result can now be extended to the case $u \in \mathcal{D}'$. As a result, the Dirac distribution is the "unit element" for the convolution product.

Convolution algebras of $\mathcal{D}'_+$ and $\mathcal{S}'_+$

If both h and u belong to $\mathcal{T}'_+$ ($\mathcal{T} = \mathcal{D}$ or $\mathcal{S}$), then the convolution product $h * u$ is well-defined and belongs to $\mathcal{T}'_+$. This makes $\mathcal{D}'_+$ and $\mathcal{S}'_+$ convolution algebras (these algebras are *commutative*, like $\mathcal{K}_+$). Since δ belongs to these two sets, $\mathcal{D}'_+$ and $\mathcal{S}'_+$ are *unitary algebras* (see section 13.1.1), contrary to $\mathcal{K}_+$, for $\delta \notin \mathcal{K}_+$.

Using the "integral notation", we obtain, when h and u belong to $\mathcal{T}'_+$:

$$(h * u)(t) = \int_{0^-}^{t^+} h(t - \tau)\, u(\tau)\, d\tau. \tag{12.22}$$

Derivative and convolution

We have also

$$\frac{d}{dt}(h * u) = \frac{dh}{dt} * u = h * \frac{du}{dt}. \tag{12.23}$$

In particular, taking $h = \delta$

$$\dot{u} = \dot{\delta} * u$$

In other words, differentiating a distribution is just as taking its convolution product with $\dot{\delta}$.

Translation of a distribution

Consider first a function $f : t \mapsto f(t)$. The function $f_{(\tau)}$ defined by $f_{(\tau)}(t) = f(t-\tau)$ is a translation (or shift) of f of time τ. For $\tau > 0$, this shift is a *delay*, or *lag*; for $\tau < 0$ it is an advance, or a *lead*.

Let us now define the "Dirac distribution at τ": it is the tempered distribution $\delta_{(\tau)}$, with support $\{\tau\}$, such that

$$\langle \delta_{(\tau)}, \phi \rangle = \phi(\tau), \phi \in \mathcal{S}.$$

The integral expression of this definition is $\phi(\tau) = \int_{-\infty}^{+\infty} \delta_{(\tau)}(t)\phi(t)dt$. Thus, $\phi(\tau) = \int_{-\infty}^{+\infty} \delta(t)\phi(t+\tau)\,dt = \int_{-\infty}^{+\infty} \delta(t-\tau)\phi(t)dt$ (after a change of variable), thus (abusing the language as we already did)

$$\boxed{\delta_{(\tau)}(t) = \delta(t-\tau)}. \tag{12.24}$$

Let $u \in \mathcal{D}$; we get

$$(\delta_{(\tau)} * u)(t) = \int_{-\infty}^{+\infty} \delta_{(\tau)}(\varsigma)u(t-\varsigma)d\varsigma = u(t-\tau).$$

In other words, the shift operator $u \mapsto u_{(\tau)}$ is nothing but the convolution operator $u \mapsto \delta_{(\tau)} * u$. We can now extend this definition to the case where $u \in \mathcal{D}'$ by density of $\mathcal{D}'$ in $\mathcal{D}$. As a result, for any distribution $T \in \mathcal{D}'$, the shifted distribution $T_{(\tau)}$ is defined by

$$\boxed{T_{(\tau)} = \delta_{(\tau)} * T}. \tag{12.25}$$

Dirac comb

Let $T > 0$ be a real number which we interpret as a period. The Dirac comb ϖ_T is defined by

$$\varpi_T = \sum_{n=-\infty}^{+\infty} \delta_{(nT)}.$$

Therefore, according to (12.24)

$$\boxed{\varpi_T(t) = \sum_{n=-\infty}^{+\infty} \delta(t-nT)}. \tag{12.26}$$

One can easily show that this series converges in $\mathcal{S}'$, thus defining a tempered distribution.

12.3. Fourier, Laplace and z transforms

12.3.1. *Fourier transforms of distributions*

Fourier transform in L_1

Let $u \in L_1$; thus we can calculate for every "angular frequency" ω

$$\boxed{\mathcal{F}u(\omega) = \int_{-\infty}^{+\infty} u(t)e^{-i\omega t}dt}. \tag{12.27}$$

The function $\mathcal{F}u$ (from $\mathbb{R}$ into $\mathbb{C}$) is called the *Fourier transform* of u, and $\mathcal{F}$ is the Fourier transform. This transform is $\mathbb{C}$-linear.

We have $|(\mathcal{F}u)(\omega)| \leq \int_{-\infty}^{+\infty} \left|u(t)\, e^{-i\omega t}\right| dt = \int_{-\infty}^{+\infty} |u(t)|\, dt$, and thus $\mathcal{F}u \in L_\infty$ and

$$\|\mathcal{F}u\|_\infty \leq \|u\|_1 .$$

REMARK 433.– *There exists various definitions of the Fourier transform. For example, we can define it as a function of frequency $\nu = \omega/2\pi$ instead of the angular frequency ω; this leads us to replace* (12.27) *by*

$$\mathcal{F}u(\nu) = \int_{-\infty}^{+\infty} u(t)e^{-i2\pi\nu t}dt. \tag{12.28}$$

Fourier transform in $\mathcal{S}$

Since $\mathcal{S} \subset L_1$, what we have discussed so far applies. But one can also show the following remarkable properties:

– The image of $\mathcal{S}$ by $\mathcal{F}$ is $\mathcal{S}$, and more specifically $\mathcal{F}$ is an isomorphism from the space $\mathcal{S}$ onto itself (i.e. an automorphism of $\mathcal{S}$).

– For any function $\breve{u}$ belonging to $\mathcal{S}$ ($\breve{u} : \omega \mapsto \breve{u}(\omega)$), we write

$$(\overline{\mathcal{F}}\breve{u})(t) = \int_{-\infty}^{+\infty} \breve{u}(\omega)e^{i\omega t}d\omega.$$

Then we have the *reciprocity formula*, valid for any function $u \in \mathcal{S} : \frac{1}{2\pi}\overline{\mathcal{F}}\mathcal{F}u = u$, in other words:

$$\boxed{\mathcal{F}^{-1} = \tfrac{1}{2\pi}\overline{\mathcal{F}}}. \tag{12.29}$$

REMARK 434.– *If* (12.28) *is used as the definition of the Fourier transform, we obtain the simpler relation $\mathcal{F}^{-1} = \overline{\mathcal{F}}$.*

Fourier transform in $\mathcal{S}'$

If both u and ϕ belong to $\mathcal{S}$, we immediately get

$$\boxed{\langle \mathcal{F}u, \phi \rangle = \langle u, \mathcal{F}\phi \rangle}. \tag{12.30}$$

This relation is now taken as a definition in the case where $u \in \mathcal{S}'$; this makes $\mathcal{F}$ an automorphism of $\mathcal{S}'$.

Using this definition, it is easily shown that

$$(\mathcal{F}\delta)(\omega) = 1; (\mathcal{F}\delta_{(\tau)})(\omega) = e^{-i\omega\tau} \tag{12.31}$$

and it follows that

$$\boxed{(\mathcal{F}\varpi_T)(\omega) = \sum_{n=-\infty}^{+\infty} e^{-i\omega nT}}. \tag{12.32}$$

The reciprocity formula (12.29) is valid in $\mathcal{S}'$. By applying it to (12.31), we get

$$\int_{-\infty}^{+\infty} e^{i\omega(t-\tau)}\, d\omega = 2\pi\delta(t-\tau), \tag{12.33}$$

an expression that becomes clearer from the point of view of physics if we exchange the roles played by the time and the angular frequency and by changing t to $-t$:

$$\boxed{\int_{-\infty}^{+\infty} e^{it(\omega_0 - \omega)}\, dt = 2\pi\delta\left(\omega - \omega_0\right)}. \tag{12.34}$$

The Fourier transform of the sinusoidal signal $t \mapsto e^{i\omega_0 t}$ is thus the distribution $\omega \mapsto 2\pi\delta(\omega - \omega_0)$ (denoted here as a function, with the usual abuse of language).

The support of the Fourier transform of a signal is called its *spectrum*. The spectrum of the sinusoidal signal considered above is the point $\{\omega_0\}$. Such a spectrum is called a "ray spectrum" (with a unique ray at ω_0).

Fourier transform in L_2

Let x and y be two functions belonging to $\mathcal{S}$. We have

$$\langle \mathcal{F}u, \mathcal{F}y \rangle_2 = \int_{-\infty}^{+\infty} \int_{-\infty}^{+\infty} \int_{-\infty}^{+\infty} \bar{u}\,(t)\, y\,(\tau)\, e^{i\omega(t-\tau)} dt\, d\tau$$

and according to (12.33) we obtain the *Plancherel–Parseval formula*:

$$\boxed{\langle \mathcal{F}u, \mathcal{F}y \rangle_2 = 2\pi \left\langle u, y \right\rangle_2}.$$

This equality allows us to extend the Fourier transform to L_2 (for $\mathcal{S}$ is dense in L_2), and the Plancherel–Parseval formula is still valid in that space. This makes $\mathcal{F}$ an automorphism of L_2.

REMARK 435.– *With Definition* (12.28) *of the Fourier transform, the Plancherel–Parseval relation becomes* $\langle \mathcal{F}u, \mathcal{F}y \rangle_2 = \langle u, y \rangle_2$.

Exchange theorem

According to (12.12), L_1 is a convolution algebra and we know that the Fourier transform is defined in L_1. Let h and u be two functions of L_1; then

$$\mathcal{F}\,(h * u)\,(\omega) = \int_{-\infty}^{+\infty}\int_{-\infty}^{+\infty}\int_{-\infty}^{+\infty} h\,(\tau)\,u\,(t-\tau)\,e^{-i\omega\tau}e^{-i\omega(t-\tau)}\,dt\,d\tau\,d\omega$$

and thus

$$\boxed{\mathcal{F}\,(h * u) = \mathcal{F}h\,\mathcal{F}u}\,. \tag{12.35}$$

This is the *Exchange theorem*, which is one of the fundamental properties of the Fourier transform (transformation of the convolution product into an ordinary product).

We see that this theorem also applies to the following cases:

– both h and u belong to $\mathcal{S}'_+$;

– h is a compactly supported distribution and u is a tempered distribution.

The Second Exchange theorem

Let $\tilde{h} = \mathcal{F}h$ and $\tilde{u} = \mathcal{F}u$. According to (12.35) and (12.29) we obtain the *Second Exchange theorem* (which is valid when $\tilde{h}$ and $\tilde{u}$ satisfy the same hypotheses as h and u above, respectively) :

$$\boxed{\mathcal{F}\left(\tilde{h}\,\tilde{u}\right) = \tfrac{1}{2\,\pi}\mathcal{F}\tilde{h} * \mathcal{F}\tilde{u}}\,. \tag{12.36}$$

12.3.2. *Fourier series*

Periodic distributions

A function u, defined in $\mathbb{R}$, is periodic with period $T > 0$ (or, in abbreviation, is T-periodic), if $u(t + nT) = u(t)$ for every real number t and every rational integer n, in other words if all shifted functions $u_{(nT)} = \delta_{(nT)} * u$ are equal to u. This definition, with this last formulation, can be extended to the case of a distribution $u \in \mathcal{D}'$. Every periodic distribution is tempered.

Trigonometric series

Let (a_n) be a sequence of complex numbers. We say that this sequence is *slowly increasing* if there exist a constant c and an integer k such that

$$|a_n| \leq c\,|n|^k \tag{12.37}$$

for all $n \neq 0$. The set of all slowly increasing sequences is a $\mathbb{C}$-vector space denoted by s′ (for a justification of this notation, see [95]).

For every sequence $(a_n) \in \mathrm{s}'$, the series

$$\boxed{u(t) = \sum_{n=-\infty}^{+\infty} a_n \, e^{-i \frac{2\pi}{T} n \, t}} \tag{12.38}$$

converges in $\mathcal{S}'$. (In the expression (12.38) we abuse the language, and the correct formulation is $u = \sum_{n=-\infty}^{+\infty} a_n \, e^{-i \frac{2\pi}{T} n \, \bullet}$.)

Indeed, let $\phi \in \mathcal{S}$; then $\left\langle e^{-i \frac{2\pi}{T} n \, \bullet}, \phi \right\rangle \triangleq \int_{-\infty}^{+\infty} e^{-i \frac{2\pi}{T} n \, t} \, \phi(t) \, dt = \mathcal{F} \phi \left(\frac{2\pi}{T} n \right)$. We know that $\mathcal{F} \phi \in \mathcal{S}$, thus the sequence with general term $a_n \, \mathcal{F} \phi \left(\frac{2\pi}{T} n \right)$ decreases faster than $\frac{1}{n^2}$ at infinity, for example, and thus the series $\sum_{n=-\infty}^{+\infty} a_n \left\langle e^{-i \frac{2\pi}{T} n \, \bullet}, \phi \right\rangle$ is absolutely convergent. We thus can now define the sum of series (12.38) by: $\langle u, \phi \rangle = \sum_{n=-\infty}^{+\infty} a_n \left\langle e^{-i \frac{2\pi}{T} n \, \bullet}, \phi \right\rangle$, as usual.

We immediately see that u is a T-periodic distribution.

One can show that any T-periodic distribution u is of this form and series (12.38) is called the *Fourier series expansion* of u.

Fourier coefficients

The above coefficients a_n are the *Fourier coefficients* of u. The difficulty in calculating them is that series (12.38) is in general not convergent in the "classic" sense.

1) Consider first the case where the T-periodic distribution u is written as

$$u = \sum_{n \in \mathbb{Z}} U_{(nT)}$$

(with $U_{(nT)} = \delta_{(nT)} * U$) where the distribution U has its support included in an open interval of the form $(\alpha, \alpha + T)$; we then have for every n

$$\boxed{a_n = \frac{1}{T} \int_{\alpha}^{\alpha+T} u(t) \, e^{i \frac{2\pi}{T} n \, t} \, dt}. \tag{12.39}$$

2) In the general case, we can proceed as follows: integrate the series $u - a_0$ term by term $k + 2$ times (where k satisfies (12.37)). We then obtain a quantity f, which is *a priori* a tempered distribution and is written as

$$f = \sum_{n \neq 0} b_n \, e^{-i \frac{2\pi}{T} n \, \bullet}$$

where $b_n = \frac{a_n}{\left(-i\frac{2\pi}{T}n\right)^{k+2}}$. The sequence (b_n) tends to 0 as fast as $\frac{1}{n^2}$ as n tends to infinity, thus the series f is *normally convergent* (that is to say, if we write $g_n = b_n\, e^{-i\frac{2\pi}{T}n\,\bullet}$, the series $\sum_n \|g_n\|_\infty = \sum_n |b_n|$ is convergent) and we can write for any real number α

$$\int_\alpha^{\alpha+T} f(t)\, e^{i\frac{2\pi}{T}m\,t}\, dt = \sum_{n\neq 0} b_n \int_\alpha^{\alpha+T} e^{-i\frac{2\pi}{T}(n-m)\,t}\, dt = T\, b_m$$

therefore the Fourier coefficients b_n of f can be calculated according to

$$b_n = \frac{1}{T}\int_\alpha^{\alpha+T} f(t)\, e^{i\frac{2\pi}{T}n\,t}\, dt.$$

Note that f is a continuous T-periodic function. Once the b_n's are known, we have from the above $a_n = \left(-i\frac{2\pi}{T}n\right)^{k+2} b_n$, $n \neq 0$.

Case of Dirac comb; Poisson summation formula

The Dirac comb ϖ_T is obviously a T-periodic distribution. We can apply formula (12.39) with $\alpha = -T/2$ and we obtain

$$\begin{aligned} a_n &= \frac{1}{T}\int_{-\frac{T}{2}}^{\frac{T}{2}} \sum_{k=-\infty}^{+\infty} \delta(t-kT)\, e^{i\frac{2\pi}{T}n\,t}\, dt \\ &= \frac{1}{T}\int_{-\frac{T}{2}}^{\frac{T}{2}} \sum_{k=-\infty}^{+\infty} \delta(t-kT)\, e^{i2\pi nk}\, dt = \frac{1}{T}\int_{-\frac{T}{2}}^{\frac{T}{2}} \delta(t)\, dt \end{aligned}$$

thus

$$\boxed{a_n = \tfrac{1}{T} \quad \text{for every } n}.$$

As a result,

$$\sum_{n=-\infty}^{+\infty} \delta(t-nT) = \frac{1}{T}\sum_{n=-\infty}^{+\infty} e^{-i\frac{2\pi}{T}n\,t}. \qquad (12.40)$$

Thus, replacing t by ω and T by $\frac{2\pi}{T}$:

$$\sum_{n=-\infty}^{+\infty} e^{-i\omega nT} = \frac{2\pi}{T}\sum_{n=-\infty}^{+\infty} \delta(\omega - n\frac{2\pi}{T}).$$

Consequently, according to (12.32) we have the *Poisson summation formula*:

$$\boxed{\mathcal{F}\left\{\sum_{n=-\infty}^{+\infty} \delta(t-nT)\right\}(\omega) = \tfrac{2\pi}{T}\sum_{k=-\infty}^{+\infty} \delta(\omega - k\tfrac{2\pi}{T})}. \qquad (12.41)$$

In other words, the Fourier transform of the (time domain) T-periodic Dirac comb is, up to a factor of $\frac{2\pi}{T}$, the (frequency domain) $\frac{2\pi}{T}$-periodic Dirac comb (because we work with angular frequencies: if we work with frequencies $\nu = \frac{\omega}{2\pi}$, this periodicity becomes $\frac{1}{T}$).

REMARK 436.– *The Poisson summation formula is often expressed in a different but equivalent manner. According to* (12.30), *we have for any function* $\phi \in \mathcal{S}$, $\langle \mathcal{F}\varpi_T, \phi \rangle = \langle \varpi_T, \mathcal{F}\phi \rangle$, *an equality which makes it possible to determine* $\mathcal{F}\varpi_T$ *from* ϖ_T. *(i) With* $T = 2\pi$, *it can be written according to* (12.41) *as*

$$\sum_{k \in \mathbb{Z}} \phi(k) = \sum_{n \in \mathbb{Z}} \mathcal{F}\phi(2n\pi).$$

(ii) If the Fourier transform is defined by (12.28), *this summation formula is simpler (and this is the most classic form):*

$$\sum_{k \in \mathbb{Z}} \phi(k) = \sum_{n \in \mathbb{Z}} \mathcal{F}\phi(n).$$

It remains valid more generally in cases than $\phi \in \mathcal{S}$ *: see ([35], section XXII.12).*

Fourier series expansion and Fourier transform

The T-periodic distribution u defined by (12.38) admits a Fourier transform and according to (12.33) we have

$$\begin{aligned}(\mathcal{F}u)(\omega) &= \sum_{n=-\infty}^{+\infty} a_n \int_{-\infty}^{+\infty} e^{-i\left(\omega - \frac{2\pi}{T} n\right)t}\, dt \\ &= \sum_{n=-\infty}^{+\infty} 2\pi a_n \delta\left(\omega - n\frac{2\pi}{T} t\right) \qquad (12.42)\end{aligned}$$

where $(a_n) \in \mathrm{s}'$. Conversely, if $(a_n) \in \mathrm{s}'$, the right-hand side of (12.42) is the Fourier transform of a T-periodic distribution u, thus this Fourier transform belongs to $\mathcal{S}'$. The formula (12.42), which generalizes (12.41), shows the close relationship that exists between the Fourier series expansion and the Fourier transform, when we deal with distributions. We have indeed obtained the following result:

THEOREM 437.– *A T-periodic distribution (which we can interpret as a time-domain signal) has a* ray spectrum, *these rays are, in the angular frequency domain, separated by* $\frac{2\pi}{T}$ *and weighted with the Fourier coefficients* a_n.

12.3.3. *Fourier transforms of sequences*

Definition

Let $a = (a_n) \in \mathrm{s}'$. Its Fourier transform is the sum of the series

$$\boxed{(\mathcal{F}a)(\theta) = \sum_{n=-\infty}^{+\infty} a_n e^{-in\theta}}. \tag{12.43}$$

This is series (12.38) where one replaces $\frac{2\pi t}{T}$ with θ. According to the above, $\mathcal{F}a$ is a 2π-periodic distribution. It thus admits a Fourier series expansion of form (12.43). As a result, $\mathcal{F}$ is an isomorphism from s' onto the space of 2π-periodic distributions.

The calculation of the inverse Fourier transform is nothing but the calculation of the Fourier coefficients of a 2π-periodic distribution. In the case where the sequence (a_n) converges to 0 as fast as $\frac{1}{n^2}$ for n tending to infinity, we have shown that

$$\boxed{a_n = \frac{1}{2\pi} \int_{-\pi}^{\pi} (\mathcal{F}a)(\theta) e^{in\theta} d\theta}. \tag{12.44}$$

Fourier transform in l_2

If both $a = (a_n)$ and $b = (b_n)$ belong to l_2, we can calculate according to (12.4) $\langle a, b \rangle_2 = \sum_{n-\infty}^{+\infty} \overline{a_n}\, b_n$. By (12.44), this quantity is, writing $\alpha = \mathcal{F}a$ and $\beta = \mathcal{F}b$:

$$\sum_{n-\infty}^{+\infty} \overline{a_n}\, b_n = \frac{1}{2\pi} \int_{-\pi}^{\pi} \bar{\alpha}(\theta)\, \beta(\varsigma) \sum_{n-\infty}^{+\infty} e^{in(\varsigma-\theta)} d\theta d\varsigma.$$

In addition, according to (12.40),

$$\sum_{n-\infty}^{+\infty} e^{in(\varsigma-\theta)} = \sum_{n=-\infty}^{+\infty} \delta(\theta - \varsigma - 2\pi n).$$

We finally obtain the Plancherel–Parseval formula

$$\boxed{\langle a, b \rangle_2 = \frac{1}{2\pi} \int_{-\pi}^{\pi} \overline{(\mathcal{F}a)}(\theta)\, (\mathcal{F}b)(\theta)\, d\theta}. \tag{12.45}$$

One can show that the Fourier transform is an isomorphism from l_2 onto the space of 2π-periodic functions which are square-integrable on $[-\pi, \pi]$, when this space is equipped with the scalar product in the right-hand side of (12.45).

The convolution algebra s'_+

Let s'_+ be the space of all positively supported sequences belonging to $\mathbf{s}'$. One easily shows that, for any two elements a and b of s'_+, the convolution product $a * b$ is well-defined and belongs to s'_+ ; thus s'_+ is a convolution algebra having the sequence $\boldsymbol{\delta}_0$ as a unit element; this one is defined by $(\boldsymbol{\delta}_0)_n = 1$ if $n = 0$ and $(\boldsymbol{\delta}_0)_n = 0$ if not.

Fourier transform of a convolution product

Consider two sequences $a = (a_n)$ and $b = (b_n)$, the convolution product and the Fourier transform of which are well-defined; a and b can be elements of s'_+, for example. We have

$$\begin{aligned}[\mathcal{F}(a * b)](\theta) &= \sum_{n=-\infty}^{+\infty} \sum_{k=-\infty}^{+\infty} a_k b_{n-k} e^{-i n \theta} \\ &= \sum_{k=-\infty}^{+\infty} \sum_{n=-\infty}^{+\infty} a_k e_k^{-in\theta} b_{n-k} e^{-i(n-k)\theta} = (\mathcal{F}a)(\theta)(\mathcal{F}b)(\theta);\end{aligned}$$

as a result we get the *Exchange theorem* (in the same form as (12.35)):

$$\boxed{\mathcal{F}(a * b) = \mathcal{F}a\,\mathcal{F}b}. \tag{12.46}$$

12.3.4. *Laplace transform*

Introduction

The Laplace transform (of functions or of distributions) is an extension of the Fourier transform. As a first step, we will discuss the *unilateral* (or *one-sided*) Laplace transform, applicable to positively supported functions or distributions. The *bilateral* (or *two-sided*) Laplace transform, whose theory is more difficult, will only be briefly mentioned at the end of this section (for a complete discussion of this question, see [106], Chapter VIII).

We have defined in section 12.2.3 the function $\varepsilon_\alpha : t \to e^{-\alpha t}$. Let $T \in \mathcal{D}'_+$ and suppose a real number α_0 exists such that $\varepsilon_{\alpha_0} T \in \mathcal{S}'_+$. Also let $\alpha \geq \alpha_0$. We have $\varepsilon_\alpha = \varepsilon_{\alpha_0} \varepsilon_{\alpha - \alpha_0}$ and it is obvious that $\varepsilon_{\alpha_0} \varepsilon_{\alpha-\alpha_0} T \in \mathcal{S}'_+$. Thus, $\varepsilon_\alpha T \in \mathcal{S}'_+$ for any $\alpha \geq \alpha_0$.

Thus we can define the set $\mathcal{A}_+$ of all distributions $T \in \mathcal{D}'_+$ for which there exists a real number α_0 such that $\varepsilon_\alpha T \in \mathcal{S}'_+$ whenever $\alpha \geq \alpha_0$. It is clear that $\mathcal{A}_+$ is a commutative convolution algebra. It is unitary, because $\delta \in \mathcal{A}_+$; furthermore,

$$\boxed{\mathcal{S}'_+ \subset \mathcal{A}_+ \subset \mathcal{D}'_+}.$$

As we will see, $\mathcal{A}_+$ is the space of distributions admitting a unilateral Laplace transform.

Definition

Let $u \in \mathcal{A}_+$ and

$$\gamma = \inf \left\{ \alpha \in \mathbb{R} : \varepsilon_\alpha u \in \mathcal{S}'_+ \right\}.$$

The real number γ is called the *abscissa of convergence* of the Laplace transform of u. The set of those $\alpha \in \mathbb{R}$ for which $\varepsilon_\alpha u \in \mathcal{S}'_+$ is therefore the interval $|\gamma, +\infty)$, an interval that can be either open or closed at γ.

The Laplace transform $\hat{u} = \mathcal{L}u$ is a function of the complex variable defined for $\alpha \in |\gamma, +\infty)$ by

$$\hat{u}(\alpha + i\omega) = \mathcal{F}\{\varepsilon_\alpha u\}(\omega). \tag{12.47}$$

Therefore, according to (12.27) and (12.15)

$$\boxed{\hat{u}(s) = \int_{0^-}^{+\infty} u(t)\, e^{-st} dt, \quad \mathrm{Re}(s) \in |\gamma, +\infty)}. \tag{12.48}$$

Strictly speaking, this expression only makes sense when u is a function, but, using the "integral notation" (12.14), it can be extended to the case where u is a distribution.

One can show that $\hat{u}$ is holomorphic (see section 12.4.1) in the open half-plane $\mathrm{Re}(s) > \gamma$.

It is clear that the Laplace transformation $\mathcal{L}$ is $\mathbb{C}$-linear.

Also note that if $u \in \mathcal{S}'_+$, then $\hat{u}(i\omega) = (\mathcal{F}u)(\omega)$

Useful Laplace transforms

u	$\hat{u}(s)$	γ
$\mathbf{1}$	$\frac{1}{s}$	0
δ	1	$-\infty$
$\delta_{(\tau)}$	$e^{-\tau s}$	$-\infty$
$\dot{\delta}$	s	$-\infty$
$\delta^{(n)}$	s^n	$-\infty$
$e^{-\alpha t}\,\mathbf{1}(t)$	$\frac{1}{s+\alpha}$	$-\alpha$
$t^m\, e^{-\alpha t}\,\mathbf{1}(t)$	$\frac{m!}{(s+\alpha)^{m+1}}$	$-\alpha$
$e^{-\alpha t}\sin(\omega t)\,\mathbf{1}(t)$	$\frac{\omega}{(s+\alpha)^2+\omega^2}$	$-\alpha$
$e^{-\alpha t}\cos(\omega t)\,\mathbf{1}(t)$	$\frac{s+\alpha}{(s+\alpha)^2+\omega^2}$	$-\alpha$

Exchange theorem

We easily extend the "Exchange theorem" to Laplace transforms: if h and u both belong to $\mathcal{A}_+$, and if their Laplace transforms have abscissae of convergence γ_1 and γ_2 respectively, then $h * u \in \mathcal{A}_+$ and this convolution product has a Laplace transform whose convergence abscissa is $\gamma = \max(\gamma_1, \gamma_2)$; furthermore,

$$\boxed{\mathcal{L}(h * u) = \mathcal{L}h\,\mathcal{L}u}. \tag{12.49}$$

Laplace transforms of derivatives

Case of a positively supported distribution

Let $u \in \mathcal{A}_+$. We know that $\dot{u} = \dot{\delta} * u$, thus

$$\mathcal{L}(\dot{u})(s) = s\,\hat{u}(s) \tag{12.50}$$

By induction, we obtain for any non-negative integer n

$$\boxed{\mathcal{L}\left(u^{(n)}\right)(s) = s^n\,\hat{u}(s)}. \tag{12.51}$$

Case of a differentiable function

Consider now a function u, defined and differentiable in an open neighborhood of $[0, +\infty)$ and the derivative of which is locally integrable.

By definition, the (unilateral) Laplace transform of u is the Laplace transform of the positively supported distribution $u_+ = \mathbf{1}\,u$ if u_+ belongs to $\mathcal{A}_+$. As a result, $\mathcal{L}u \triangleq \mathcal{L}u_+$. This Laplace transform is thus given by the formula (12.48) (where 0^- can be replaced by 0).

We have $\frac{d\,u_+}{dt} = \dot{u}\,\mathbf{1} + u\,\delta$ (Leibniz formula), from which, according to (12.20),

$$\frac{du_+}{dt} = \dot{u}\,\mathbf{1} + \delta\,u(0).$$

According to (12.50), $\mathcal{L}\left(\frac{d\,u_+}{dt}\right)(s) = s(\mathcal{L}u_+)(s)$, thus $s(\mathcal{L}u_+)(s) = \mathcal{L}(\dot{u})(s) + u(0)$, i.e.

$$\mathcal{L}(\dot{u})(s) = s\,\hat{u}(s) - u(0). \tag{12.52}$$

Extension: case of the sum of an n times continuously differentiable function and of a positively supported distribution

We are often led, in control theory, to consider signals of the form

$$u = f + T \tag{12.53}$$

where f is a function that is defined and n times continuously differentiable in an open neighborhood of $[0, +\infty)$ ($n \geq 1$)[9] and where $T \in \mathcal{A}_+$.

As in the above, the (unilateral) Laplace transform of u can be defined as the Laplace transform of the positively supported distribution $u_+ = f_+ + T$ and is thus given by the formula (12.48), but this time it is essential use 0^- as the lower limit of integration.

Now, u and f have the same restriction to every open interval $(-\varepsilon, 0)$,[10] where $\varepsilon > 0$ is a sufficiently small real number. Therefore, the restriction of u to such an interval is an n times continuously differentiable function and furthermore:

$$u^{(i)}\left(0^-\right) = f^{(i)}\left(0\right), \quad 0 \leq i \leq n-1. \tag{12.54}$$

Writing (12.53) and then taking the Laplace transform of the obtained expression, we get

$$\mathcal{L}\left(\dot{u}\right)\left(s\right) = \mathcal{L}\left(\dot{T}\right)\left(s\right) + \mathcal{L}\left(\dot{f}\right)\left(s\right).$$

As a result, according to (12.50) and (12.52)

$$\mathcal{L}\left(\dot{u}\right)\left(s\right) - s\,\hat{T}\left(s\right) + s\,\hat{f}\left(s\right) - f\left(0\right)$$

thus from (12.54)

$$\mathcal{L}\left(\dot{u}\right)\left(s\right) = s\,\hat{u}\left(s\right) - u\left(0^-\right).$$

By induction, we can establish that:

$$\mathcal{L}\left(u^{(n)}\right)\left(s\right) = s^n\,\hat{u}\left(s\right) - \sum_{i=0}^{n-1} s^{n-1-i} u^{(i)}\left(0^-\right).$$

9. More precisely, it suffices to assume that u is n times differentiable in an open neighborhood of $[0, +\infty)$ and that its nth derivative is locally integrable.

10. This is intuitively clear since T is positively supported. For a general definition of the notion of restriction of a distribution to an open set (see [35], section 17.13).

Denote the quantity $u^{(i)}(0^-)$ $(i \geq 0)$ by $\partial_0^i u_0$; then, in particular, $\partial_0^0 = 1$ and $u_0 = u^{(0)}(0^-)$ (∂_0 can viewed as a differential operator on the values at instant 0^-).[11] We have:

$$\sum_{i=0}^{n-1} s^{n-1-i} u^{(i)}\left(0^-\right) = \sum_{i=0}^{n-1} s^{n-1-i} \partial_0^i\, u_0 = \frac{s^n - \partial_0^n}{s - \partial_0} u_0.$$

therefore

$$\boxed{\mathcal{L}\left(u^{(n)}\right)(s) = s^n \hat{u}(s) - \tfrac{s^n - \partial_0^n}{s - \partial_0} u_0}. \tag{12.55}$$

Inverse Laplace transform

Bromwich–Schwartz theorem

This theorem characterizes the Laplace transforms of distributions and is stated in the following manner ([106], sectionVIII.4): consider a function of the complex variable $s \mapsto f(s)$, which is holomorphic in the half-plane $\mathrm{Re}(s) > \gamma$; this function is the Laplace transform of a distribution $u \in \mathcal{A}_+$ if and only if there exists a rational integer n and a half-plane $\mathrm{Re}(s) \geqslant \beta > \gamma$ such that the function $s \to (1 + |s|)^{-n} f(s)$ is bounded in this half-plane.

Inversion of a Laplace transform

Case 1 : Bromwich formula

Suppose that, in the above, $n = -2$, thus $f(s)$ decreases at the infinity at least as fast as $\frac{1}{s^2}$ (if $f(s)$ is a rational function, this means that it is of relative degree of at least 2 : see section 13.6.1). Thus there exists a constant c such that

$$|f(\alpha + i\omega)| \leq \frac{c}{\alpha^2 + \omega^2}$$

for $\alpha > \min(\beta, 0)$. Therefore the function of the real variable $f_\alpha : \omega \to f(\alpha + i\omega)$ belongs to L_1 and we can calculate

$$\frac{1}{2\pi}\left(\overline{\mathcal{F}}\, f_\alpha\right)(t) = \frac{1}{2\pi}\int_{-\infty}^{+\infty} f_\alpha(i\omega)\, e^{i\omega t} d\omega.$$

After the change of variable $s = \alpha + i\omega$ we get:

$$\frac{1}{2\pi}\left(\overline{\mathcal{F}} f_\alpha\right)(t) = e^{-\alpha t} u(t)$$

11. This notion is improper because, for example, $\dot{u}\left(0^-\right)$ cannot be derived from $u\left(0^-\right)$. It would be more rigorous – but misleading – to denote the above quantity $\partial_0^i u_0$ as $\partial_0^i u$.

where

$$\boxed{u(t) = \frac{1}{2\pi i} \int_{\alpha - i\infty}^{\alpha + i\infty} f(s)\, e^{st}\, ds}. \tag{12.56}$$

Thus, $\mathcal{F}\{\varepsilon_\alpha u\}(\omega) = f(\alpha + i\omega)$, which shows that $f(s) = \hat{u}(s)$ and (12.56) is the formula of the inverse Laplace transform, called the *Bromwich formula*. We will see in section 12.4.4 that this function u is continuous and positively supported.

Case 2 : Schwartz's method

Suppose now that $n > -2$. We have $s^{-n} f(s) = s^2 g(s)$ where $g(s) = s^{-(s+2)} f(s)$. The function $s \mapsto s^2\, g(s)$ is bounded in the half-plane $\operatorname{Re}(s) \geqslant \alpha > \min(\beta, 0)$, thus we can apply the Bromwich formula to the function g : the positively supported function v defined by

$$v(t) = \frac{1}{2\pi i} \int_{\alpha - i\infty}^{\alpha + i\infty} g(s)\, e^{st}\, ds$$

admits a Laplace transform $g(s)$. But since $f(s) = s^{n+2} g(s)$, $f(s)$ is the Laplace transform of the distribution $v^{(n+2)}$.

This shows that every distribution belonging to $\mathcal{A}_+$ is a finite order derivative of a continuous function.

This also shows that the Laplace transform is a bijection from $\mathcal{A}_+$ onto the set of functions of the complex variable characterized by the Bromwich–Schwartz theorem, a set which we can thus denote as $\mathcal{LA}_+$. Since $\mathcal{A}_+$ is a convolution algebra and the Laplace transform transforms the convolution product into the ordinary product, $\mathcal{LA}_+$ is an algebra for the ordinary product (as can also be directly verified).

Initial value theorem and final value theorem

Let u be a function admitting a Laplace transform $\hat{u}$. One can prove the following results [95] (be careful about the hypotheses!):

1) If u is such that $\lim_{t \to +\infty} u(t)$ exists, then the abscissa of convergence γ of $\hat{u}$ satisfies $\gamma \leq 0$ and we have the *final value theorem*:

$$\boxed{\lim_{s \in \mathbb{R},\, s \to 0^+} s\, \hat{u}(s) = \lim_{t \to +\infty} u(t)}.$$

2) If u admits a limit at 0 from the right, denoted as $u(0^+)$, we have the *initial value theorem*:

$$\boxed{\lim_{s \in \mathbb{R},\, s \to +\infty} s\hat{u}(s) = u(0^+)}.$$

Bilateral Laplace transform

Let us now generalize the method used in the beginning of this section. Let $u \in \mathcal{D}'$; the set of real numbers α such that $\varepsilon_\alpha u \in \mathcal{S}'$ is an interval $\Gamma = |\gamma^-, \gamma^+|$ of $\mathbb{R}$ (which can be, depending on the case, either open or closed at γ^- or at γ^+). If $\Gamma \neq \emptyset$, the set $B_c = \{s \in \mathbb{C} : \mathrm{Re}\,(s) \in \Gamma\}$ is the *band of convergence* of the Laplace transform of u. This transform, denoted as $\hat{u} = \mathcal{L}u$, is a function of the complex variable defined for $\alpha \in |\gamma^-, \gamma^+|$ by relation (12.47) or also by (with the usual abuse of language)

$$\boxed{\hat{u}\,(s) = \int_{-\infty}^{+\infty} u\,(t)\; e^{-st}\,dt, \qquad s \in B_c}$$

which generalizes (12.48) ; $\hat{u}\,(s)$ is holomorphic in the interior $\mathring{B}_c$ of B_c (if $\mathring{B}_c \neq \emptyset$). If $u \in \mathcal{D}'_+$, then $\gamma^+ = +\infty$ and then we come back to the unilateral Laplace transform.

Extension of the Exchange theorem

Let Γ be a *non-empty open* interval of $\mathbb{R}$ and let $\mathcal{S}'\,(\Gamma)$ be the set of all distributions $u \in \mathcal{D}'$ such that $\varepsilon_\alpha u \in \mathcal{S}'$ for every $\alpha \in \Gamma$. One can show that $\mathcal{S}'\,(\Gamma)$ is a commutative convolution algebra.

The "Exchange theorem" (12.49) is still valid when both h and u belong to $\mathcal{S}'\,(\Gamma)$. It follows that (12.51) is also still valid because $\delta^{(n)} \in \mathcal{S}'\,(\Gamma)$ for every non-empty open interval Γ of $\mathbb{R}$ containing zero.

*Paley–Wiener–Schwartz theorem

We denote by $\mathcal{E}$ the space of indefinitely differentiable functions from $\mathbb{R}$ into $\mathbb{C}$. This space is equipped with the following topology: a sequence (φ_n) converges to φ in $\mathcal{E}$ if for every integer $k \geq 0$, $\varphi_n^{(k)}$ converges uniformly to $\varphi^{(k)}$ on every compact set. One can show that $\mathcal{E}$ is a Fréchet space. Its dual $\mathcal{E}'$ is the space of compactly supported distributions. Using the Paley–Wiener–Schwartz theorem ([106], section VII.8), ([35], section XXII.18) one can characterize the elements of $\mathcal{E}'$ using their Laplace transforms. For an entire function $f\,(s)$ (see section 12.4.1 below) to be the Laplace transform of a distribution with support included in $[-a, a]$ $(a \geq 0)$, it is necessary and sufficient that there exists an integer $n \geq 0$ and a constant $c \geq 0$ such that for every $s \in \mathbb{C}$,

$$|f\,(s)| \leq c\,(1 + |s|)^n\; e^{a|\mathrm{Re}\,s|}.$$

(compare with the Bromwich–Schwartz theorem.)

12.3.5. *z-transform*

The z-transform plays the same role for sequences as the Laplace transform for distributions. Like the Laplace transform, it is a generalization of the Fourier transform. We first will consider the "unilateral" z-transform. The "bilateral" z-transform is discussed at the end of this section.

Definition

Let $x = (x_n)_{n\in\mathbb{Z}}$ be a sequence of complex numbers, assumed to be *positively supported* (section 12.2.1). Let

$$\rho = \inf\left\{r > 0 : \left(x_n r^{-n}\right) \in \mathrm{s}'\right\}$$

and suppose $\rho < +\infty$. For every $r \in |\rho, +\infty)$ (where the interval $|\rho, +\infty)$ is open or closed at ρ depending on the situation), the sequence $(x_n r^{-n})$ belongs to s'. This sequence thus admits a Fourier transform

$$\mathcal{F}\left\{x_n r^{-n}\right\}(\theta) = \sum_{n=0}^{+\infty} x_n r^{-n} e^{-in\theta} = \sum_{n=0}^{+\infty} x_n \left(re^{i\theta}\right)^{-n}.$$

Let us take the power series in z^{-1}

$$\boxed{X(z) = \sum_{n=0}^{+\infty} x_n z^{-n}}. \quad (12.57)$$

According to the above, this series converges for every z such that $|z| \in |\rho, +\infty)$, since

$$X\left(re^{i\theta}\right) = \mathcal{F}\left\{x_n r^{-n}\right\}(\theta). \quad (12.58)$$

The function $X : z \mapsto X(z)$ of the complex variable z is called the (unilateral) z-transform of x, it is defined for $|z| \in |\rho, +\infty)$ and is holomorphic in the open set $|z| > \varrho$. In this open set, the series (12.57) is absolutely convergent. The real number ϱ is called the *radius of convergence* of X.

The z-transform $\mathcal{Z} : x \mapsto X$ is $\mathbb{C}$-linear.

z-transform of an advanced sequence

We define the *advance operator* q, also called the *shift-forward operator*, acts on sequences of complex numbers in the following manner:

$$qx_n = x_{n+1}. \quad (12.59)$$

The z-transform of the sequence qx is $\mathcal{Z}\{qx\}(z) = \sum_{n=0}^{+\infty} x_{n+1} z^{-n} = z[X(z) - x_0]$. Generalizing this rationale:

$$\boxed{\mathcal{Z}\{q^k x\}(z) = z^k \left[X(z) - \sum_{n=0}^{k-1} x_n z^{-n}\right]}. \tag{12.60}$$

Exchange theorem

Let $x = (x_n)$ and $y = (y_n)$ be two positively supported sequences, the z-transforms of which have a radius of convergence of ϱ_1 and ϱ_2, respectively. Then, the positively supported sequence $x * y$ has a z-transform whose radius of convergence is $\rho = \max(\varrho_1, \rho_2)$ and

$$\boxed{\mathcal{Z}\{x * y\}(z) = X(z)\, Y(z)}. \tag{12.61}$$

In particular, let $\boldsymbol{\delta}_k$ ($k \geq 0$) be the sequence defined by $\boldsymbol{\delta}_k(n) = 1$ if $n = k$, $\boldsymbol{\delta}_k(n) = 0$ if not[12]. The sequence $\boldsymbol{\delta}_k$ is positively supported, its z-transform is z^{-k}. For every sequence $x = (x_n)$, $\boldsymbol{\delta}_k * x$ is the *delayed sequence* $q^{-k} x_n = x_{n-k}$. As a result,

$$\mathcal{Z}\{q^{-k} x\}(z) = \mathcal{Z}\{\boldsymbol{\delta}_k * x\}(z) = z^{-k} X(z). \tag{12.62}$$

Inverse z-transform

According to equation (12.58) which establishes the relationship between the Fourier transform and the z-transform, we have for any $r > \varrho$

$$x_n r^{-n} = \mathcal{F}^{-1} X\left(re^{i\theta}\right) = \frac{1}{2\pi} \int_{-\pi}^{\pi} X\left(re^{i\theta}\right) e^{in\theta} d\theta$$

from which, after the change of variable $z = re^{i\theta}$

$$\boxed{x_n = \frac{1}{2\pi i} \oint_{|z|=r} X(z)\, z^{n-1} dz}. \tag{12.63}$$

12. In order not to complicate the notation, we write the sequence $\boldsymbol{\delta}_k$ here as a function of n.

Useful z-transforms

We define the *unit step* $\mathbf{1}$ as the sequence defined by $\mathbf{1}_n = 1$ if $n \geq 0$, $\mathbf{1}_n = 0$ if $n < 0$.

(x_n)	$X(z)$	ϱ
$\boldsymbol{\delta}_0$	1	0
$\boldsymbol{\delta}_k$	z^{-k}	0
$\mathbf{1}$	$\frac{z}{z-1}$	1
$(a^n \mathbf{1}_n)$	$\frac{z}{z-a}$	$\lvert a \rvert$
$(n \mathbf{1}_n)$	$\frac{z}{(z-1)^2}$	1
$(n^2 \mathbf{1}_n)$	$\frac{z(z+1)}{(z-1)^3}$	1
$\sin(n\omega T)\, \mathbf{1}_n$	$\frac{z \sin \omega T}{z^2 - 2z \cos \omega T + 1}$	1
$\cos(n\omega T)\, \mathbf{1}_n$	$\frac{z^2 - z \cos \omega T}{z^2 - 2z \cos \omega T + 1}$	1

(12.64)

Initial value theorem and final value theorem

One can show the following results [95]:

1) $\lim_{|z| \to +\infty} X(z) = x_0$ (*initial value theorem*).

2) If $\lim_{n \to +\infty} x_n$ exists, then $X(z)$ has a radius of convergence $\varrho \leq 1$ and

$$\lim_{z \in \mathbb{R},\, z \to 1^+} (z-1)\, X(z) = \lim_{n \to +\infty} x_n$$

(*final value theorem*).

Bilateral z-transform

Let $x = (x_n)$ be a sequence of complex numbers, let

$$\left|\rho^-, \rho^+\right| = \left\{ r > 0 : \left(x_n r^{-n}\right) \in \mathbf{s}' \right\} \tag{12.65}$$

and suppose $|\rho^-, \rho^+| \neq \emptyset$. For every $r \in |\rho^-, \rho^+|$ the sequence $(x_n r^{-n})$ belongs to s' and thus admits a Fourier transform

$$\mathcal{F}\left\{x_n r^{-n}\right\}(\theta) = \sum_{n=-\infty}^{+\infty} x_n r^{-n} e^{-in\theta} = \sum_{n=-\infty}^{+\infty} x_n \left(re^{i\theta}\right)^{-n}.$$

The series

$$\boxed{X(z) = \sum_{n=-\infty}^{+\infty} x_n z^{-n}} \tag{12.66}$$

converges for all z such that $|z| \in |\rho^-, \rho^+|$ and satisfies the relation (12.58). For $z = re^{i\theta}$ with r equal to ρ^- or ρ^+ (if one of these elements belongs to $|\rho^-, \rho^+|$), the series (12.66) convergences in $\mathcal{S}'$.

Let $C_c = \{z \in \mathbb{C} : |z| \in |\rho^-, \rho^+|\}$. This set C_c is called the *annulus of convergence* of X. The function of the complex variable $X : z \mapsto X(z)$ is called the (bilateral) z-transform of x and is defined in C_c. In the interior $\overset{\circ}{C}_c$ of C_c (if it is non-empty), the series (12.57) is absolutely convergent and $X(z)$ is holomorphic. If this sequence x is positively supported, (12.66) becomes identical to (12.57) and $\rho^+ = +\infty$.

Extension of the Exchange theorem

We are now going to extend the Exchange theorem (12.61) to the case of the bilateral z-transform, as we have done at section12.3.4 for the bilateral Laplace transform.

Let Γ be a non-empty *open* interval included in $[0, +\infty)$ and let $\mathrm{s}'(\Gamma)$ (resp., $l_1(\Gamma)$) be the set of sequences of complex numbers (x_n) such that $(r^{-n} x_n) \in \mathrm{s}'$ (resp., $(r^{-n} x_n) \in l_1$) for every $r \in \Gamma$. As easily shown, $\mathrm{s}'(\Gamma) = l_1(\Gamma)$, and since l_1 is a convolution algebra (section 12.2.1), $\mathrm{s}'(\Gamma)$ is one too. The Exchange theorem (12.61) is also valid when x and y both belong to $\mathrm{s}'(\Gamma)$. We now deduce that

$$\boxed{\mathcal{Z}\{q^k x\}(z) = z^k X(z)} \tag{12.67}$$

where $\mathcal{Z}$ denotes the bilateral z-transform (of course, this equality does not contradict (12.60), and the reader is requested to wonder why).

12.4. Functions of one complex variable

As seen above, a Laplace transform and a z-transform are functions of one complex variable, and it is thus important to study some of the properties of these functions.

12.4.1. *Holomorphic functions*

Definition of a holomorphic function

A function $s \rightarrow f(s)$ is *holomorphic* in a non-empty open subset Ω of $\mathbb{C}$ if for any $s_0 \in \Omega$

$$\dot{f}(s_0) = \lim_{s \to s_0, s \neq s_0} \frac{f(s) - f(s_0)}{s - s_0}$$

exists. The set of all holomorphic functions in Ω is denoted by $\mathcal{O}(\Omega)$.

Analytic function

A function of the complex variable $s \rightarrow f(s)$ is *analytic* in Ω if, for any point z of $\mathbb{C}$, there exists a non-empty open disc $D_{z,r} = \{s \in \mathbb{C} : |s - z| < r\}$ contained in Ω such that f is the sum of a power series in $s - z$ which converges in $D_{z,r}$, that is to say f admits a Taylor series expansion in the neighborhood of z:

$$f(s) = \sum_{k=0}^{+\infty} \frac{(s-z)^k}{k!} f^{(k)}(z). \tag{12.68}$$

Identity of the two notions

Let p be a complex number and let j be a positive integer. We have

$$\frac{1}{(s-p)^j} = \frac{1}{(-p)^j} \frac{1}{\left(1 - \frac{s}{p}\right)^j}.$$

This function is the sum of a power series in s with radius of convergence $|p|$. By changing the origin of the plane, it follows that if $p \neq z$, this same function is the sum of a power series in $s - z$ with radius of convergence $|p - z|$.

Now let $f(z)$ be any rational function and let P be the set of its distinct poles, so that $f(s)$ is holomorphic in the open set $\mathbb{C} \backslash P$ (where $\mathbb{C} \backslash P$ designates the complement of P in $\mathbb{C}$). From (13.60), (13.61) and the above, it follows that $f(s)$ is analytic in $\mathbb{C} \backslash P$. As a result, holomorphy and analyticity are synonymous as far as rational functions are concerned. According to *Goursat's theorem* ([35], section IX.10, Problem 2) this holds true for all functions of one complex variable.

In addition, let $z \in \mathbb{C} \backslash P$ and

$$r = \min_{p \in P} |z - p|.$$

Power series (12.68) has a radius of convergence of r, thus it converges uniformly in the closed disc $|s - z| \leq \varrho$ for any $\rho < r$.

Analytic continuation principle

The following is proved in, e.g., ([25], IV.2.3) and ([35], (9.4.3)):

THEOREM 438.– *Let Ω be an open connected subset of the complex plane and $f \in \mathcal{O}(\Omega)$. The following conditions are equivalent:*
(i) there exists a point $z \in \Omega$ such that for all integers $n \geq 0$, $f^{(n)}(z) = 0$;
(ii) there exist a point $z \in \Omega$ and an open neighborhood $\mathcal{N} \subseteq \Omega$ of z such that $f|_{\mathcal{N}} = 0$;
(iii) there exists a compact infinite subset H of Ω such that $f(z) = 0$ for all $z \in H$.
(iv) $f = 0$ (i.e., $f(s) = 0$ for all $s \in \Omega$).

Entire functions

If a function of one complex variable is holomorphic in $\mathbb{C}$, then series (12.68) converges uniformly in any bounded subset of $\mathbb{C}$. Such a function is said to be entire. Thus, the set of all entire functions is $\mathcal{O}(\mathbb{C})$.

Meromorphic functions

Let Ω be an open connected subset of $\mathbb{C}$. A complex function h is said to be *meromorphic* in Ω if there exist $f, g \in \mathcal{O}(\Omega)$, $g \neq 0$, such that $h = f/g$. Let $z \in \Omega$; there exists a rational integer q such that $h(s) = (s-z)^q \, l(s)$ where $l(z) \neq 0$, and this representation is unique. If $q < 0$, h has a *pole* of order $-q$ at z; and if $q > 0$, h is said to have a *zero* of order q at z. There exists a real $r > 0$ such that

$$h(s) = \sum_{\nu \geq q} a_\nu (s-z)^\nu, \quad a_q \neq 0$$

for any $s \in \Omega$ such that $|s-z| < r$; the series in the right-hand member of the above equality is called a *Laurent series* (and is the *Laurent expansion* of h at z). This Laurent series is absolutely convergent for every s in the open disk $\Delta(z;r)$ with center z and radius r, and is normally –thus uniformly – convergent in any closed disk $\bar{\Delta}(z;\rho)$ with center z and radius ρ, $0 < \rho < r$. If $q < 0$, i.e., if z is a pole of h, then a_{-1} is called the *residue* of h at z and is denoted as $\operatorname{Res}(h;z)$.

12.4.2. *Functions of a matrix*

Let f be an analytic function in the neighborhood of 0 and let ρ be the radius of convergence of the power series

$$f(s) = \sum_{k=0}^{+\infty} a_k s^k, \quad a_k = \frac{f^{(k)}(0)}{k!}. \tag{12.69}$$

In other words,

$$\rho = \sup \left\{ c \geq 0 : \sum_{k=0}^{+\infty} |a_k| \, c^k < +\infty \right\}.$$

Let $A \in \mathbb{C}^{n \times n}$. We call *spectral radius* of matrix A the quantity

$$r(A) = \max \{ |\lambda| : \lambda \in \operatorname{Sp}(A) \}$$

where $\operatorname{Sp}(A)$ is the spectrum of A (section 13.3.3). One can show that the series

$$f(A) = \sum_{k=0}^{+\infty} a_k A^k$$

is convergent if $r(A) < \rho$ ([52], section V.4). Assuming that this condition holds, we have the following result:

PROPOSITION 439.– *Let $\lambda \in \mathbb{C}$ be an eigenvalue of A and let $x \in \mathbb{C}^n$ be an eigenvector of A associated with λ. Then, $f(\lambda)$ is an eigenvalue of $f(A)$ and x is an eigenvector of $f(A)$ associated with $f(\lambda)$.*

PROOF. We have $Ax = \lambda x$, thus $f(A)x = \sum_{k=0}^{+\infty} a_k A^k x = \sum_{k=0}^{+\infty} a_k \lambda^k x = f(\lambda)x$. ■

Exponential of matrix

The exponential function $f(s) = e^s$ can be expanded as follows:

$$e^s = \sum_{k=0}^{+\infty} \frac{s^k}{k!},$$

with an infinite radius of convergence (i.e. it is an entire function); as a result, if $A \in \mathbb{C}^{n \times n}$, the *exponential* of this matrix is defined by the series

$$e^A = \sum_{k=0}^{+\infty} \frac{A^k}{k!} \tag{12.70}$$

which converges for any square matrix A.

We have in particular

$$e^{\lambda I_n} = e^{\lambda} I_n$$

and thus

$$e^{0_{n \times n}} = I_n. \tag{12.71}$$

On the other hand, one can easily show that if two square matrices B *and* C are of the same order and *commute*, then

$$e^{B+C} = e^B e^C. \tag{12.72}$$

In particular, B and $-B$ commute, therefore $e^{0_{n \times n}} = e^B e^{-B}$, which proves that

$$e^{-B} = \left(e^B\right)^{-1}.$$

Finally, if B and C are two square matrices, consider their diagonal sum $B \oplus C$ (see section 13.1.4). It is immediate that

$$e^{B \oplus C} = e^B \oplus e^C \tag{12.73}$$

Logarithm of matrix

The function $f(s) = \ln(1+s)$ $(s \in \mathbb{R})$ can be expanded as follows:

$$\ln(1+s) = \sum_{k=1}^{+\infty} (-1)^{k-1} \frac{s^k}{k} \tag{12.74}$$

with radius of convergence $\rho = 1$.

DEFINITION 440.– *The power series in the right-hand side of* (12.74), *which converges for all complex numbers s such that $|s| < 1$,*[13] *is the principal branch of* $\ln(1+s)$.

In the complex plane, there are several "branches" of the logarithm, "inverse" of the exponential function, as shown here below (for more details, see [25]).

PROPOSITION 441.– *Let $f(s)$ be the principal branch of $\ln(1+s)$. All other determinations of $\ln(1+s)$ are of the form $f(s) + 2k\pi i$ (where k is an integer).*

PROOF. Consider two *complex* numbers x and y such that $e^x = e^y$. This equality is equivalent to $e^{x-y} = 1$, i.e. to $x - y = 2k\pi i$ $(k \in \mathbb{Z})$. ∎

If $A \in \mathbb{C}^{n \times n}$, we can thus define $\ln(I_n + A)$ by the expression

$$\ln(I_n + A) = \sum_{k=1}^{+\infty} (-1)^{k-1} \frac{A^k}{k}$$

provided that $r(A) < 1$.

12.4.3. *Integration in the complex plane*

Integration along a path

Path

A *path* is a piecewise continuously differentiable function from an interval $[a, b]$ of $\mathbb{R}$ into the complex plane $\mathbb{C}$. It is thus a "sufficiently regularly" parametrized curve of the complex plane, which we can denote as $\gamma : [a, b] \ni t \mapsto \gamma(t) \in \mathbb{C}$. It is oriented, because when the variable t traverses $[a, b]$ from a to b, the point $\gamma(t)$ traverses this curve in a well-determined sense. If there exists an open subset Ω of $\mathbb{C}$ such that $\gamma([a, b]) \subset \Omega$, then γ is called a path in Ω.

This path γ is *closed* if $\gamma(a) = \gamma(b)$.

13. It converges also for $s = 1$.

Integration

Let $f(s)$ be a meromorphic function in the open set $\Omega \subset \mathbb{C}$ and let P be the set of its distinct poles. Let also γ be a path in Ω and which does not pass through any point of P. Then we define the integral of f along the path γ: this is the quantity

$$\boxed{\int_{\gamma} f(s)\,ds = \int_{a}^{b} f(\gamma(t))\,\dot{\gamma}(t)\,dt}.$$

Index of a point with respect to a closed path

Let p be a point of the complex plane and let $\gamma : [a, b] \ni t \mapsto \gamma(t) \in \mathbb{C}$ be a closed path not passing through p. The quantity

$$\boxed{j(p;\gamma) = \frac{1}{2\pi i}\int_{\gamma} \frac{ds}{s-p}} \tag{12.75}$$

is called the *index* of p with respect to γ.

This index is a rational integer and is interpreted as the number of turns that γ goes around p (in the direct sense); equivalently, we can say that $j(p;\gamma)$ is the number of turns that s goes around p, in the direct sense, by following the closed path γ.

Indeed, let n be the number of turns as defined above and write

$$h(t) = \int_{a}^{t} \frac{\dot{\gamma}(\tau)}{\gamma(\tau) - p}\,d\tau.$$

Write

$$\gamma(\tau) - p = \varrho(\tau)\,e^{i\theta(\tau)},\ \varrho(\tau) > 0$$

(polar form of $\gamma(\tau) - p$). Then $\frac{\dot{\gamma}(\tau)}{\gamma(\tau)-p} = \frac{\dot{\varrho}(\tau)}{\varrho(\tau)} + i\theta(\tau)$, and thus $h(b) = i\Delta\theta$, where $\Delta\theta$ is the variation of the argument of $s - p$ when s traverses the closed path γ. We have $\Delta\theta = 2\pi n$.

Notice that if $\gamma_1, \gamma_2 : I \to \mathbb{C}$ are two closed paths which do not pass through 0, then $\gamma_1\gamma_2 : I \ni t \mapsto \gamma_1(t)\,\gamma_2(t) \in \mathbb{C}$ is again a closed path which does not pass through 0, and

$$j(0;\gamma_1\gamma_2) = j(0;\gamma_1) + j(0;\gamma_2). \tag{12.76}$$

Integration of elementary functions along a closed path

Let $k \geq 0$. Then $\int s^k \, ds = \frac{s^{k+1}}{k+1} + const$. Therefore, for any closed path γ, $\int_\gamma s^k \, ds = 0$. We deduce that for any polynomial $Q(s)$, $\int_\gamma Q(s) \, ds = 0$.

On the other hand, let

$$h(s) = \frac{1}{(s-p)^k}, \quad k \geq 2$$

We have

$$\int h(s) \, ds = \frac{1}{1-k} (s-p)^{1-k} + const.$$

As a result, for any closed path γ not passing through p, $\int_\gamma h(s) \, ds = 0$.

Residue theorem

This theorem is one of the most important ones in the theory of functions of one complex variable.

As a result of (13.60), (13.61), (12.75) and of the above, if $f(s)$ is a rational function and if γ is a closed path not passing through any pole of $f(s)$, we have

$$\boxed{\int_\gamma f(s) \, ds = 2\pi i \sum_{p \in P} j(p;\gamma) \operatorname{Res}(f;p)}. \tag{12.77}$$

A subset Ω of $\mathbb{C}$ is *convex* if whenever two points a and b belong to Ω, the segment $[a,b] = \{(1-t)\,a + tb : 0 \leq t \leq 1\}$ is included in Ω.

The equality (12.77) still holds when f is a meromorphic function in an open *convex*[14] set $\Omega \subset \mathbb{C}$ and when the closed path γ is in Ω (*Cauchy's residue theorem*) [103].

Cauchy's integral formula

Let f be a holomorphic function in a non-empty open *convex*[15] subset Ω of the complex plane. Let z be a point of Ω and γ be a closed path in Ω, which does not pass through z.

14. See Remark 442.

15. See Remark 442.

The function $\frac{f(s)}{s-z}$ is meromorphic in Ω and has a unique pole z; the residue of this function at z is $f(z)$. By applying the residue theorem (12.77) to $\frac{f(s)}{s-z}$, we thus obtain "Cauchy's integral formula":

$$j(z;\gamma) f(z) = \frac{1}{2\pi i} \int_{\gamma} \frac{f(s)}{s-z} ds \tag{12.78}$$

We can generalize this formula using the expansion (12.68), which converges uniformly in the closed disk $|s| \leq r$ if $r > 0$ is small enough for this disk to be included in Ω. The residue of $\frac{f(s)}{(s-z)^{k+1}}$ at z is $\frac{f^{(k)}(z)}{k!}$; therefore, we get:

$$\boxed{j(z;\gamma) f^{(k)}(z) = \frac{k!}{2\pi i} \int_{\gamma} \frac{f(s)}{(s-z)^{k+1}} ds, \quad k \geq 0}. \tag{12.79}$$

REMARK 442.– * *An open connected subset Ω of the complex plane is called* simply connected *if, roughly speaking, every closed path in Ω can be continuously shrunk until to be reduced to one point; such a subset Ω is an "open connected subset without holes" (so is any convex open set). In the above, "convex" can be everywhere replaced by "simply connected".* *

Maximum modulus principle

LEMMA 443.– *If $f(s)$ is holomorphic in $\Omega = \{s : |s - s_0| < r\}$ and if $|f(s)| \leq |f(s_0)|$ for all $s \in \Omega$, then $f(s)$ is constant in Ω.*

PROOF. We may assume that $f(s_0) \neq 0$. According to (12.78), whenever $0 < \rho < r$

$$f(s_0) = \frac{1}{2\pi} \int_0^{2\pi} f\left(s_0 + \rho e^{i\theta}\right) d\theta$$

or equivalently

$$\int_0^{2\pi} \left(1 - f\left(s_0 + \rho e^{i\theta}\right) / f(s_0)\right) d\theta = 0.$$

The real part of the integrand is ≥ 0 and is zero only when $f(s_0) = f\left(s_0 + \rho e^{i\theta}\right)$. ■

Now we can state the maximum modulus principle:

THEOREM 444.– *Let Ω be an open subset of the complex plane, let $\bar{\Omega}$ be its closure, and let f be a function which is both continuous in $\bar{\Omega}$ and holomorphic in Ω. If $|f(s)|$ admits a maximum in $\bar{\Omega}$, this maximum is attained on the frontier $\partial\Omega = \bar{\Omega} \backslash \Omega$.*

PROOF. *Assume that the maximum is attained at an interior point z. The connected component $C(z)$ of z in Ω is open for Ω is locally connected, and $\partial C(z) \subset \partial\Omega$. Hence, by Theorem 438 and Lemma 443, f is constant in Ω, thus f is constant in $\bar{\Omega}$ by continuity, and therefore $|f(s_0)| = |f(z)|$ at some point s_0 of $\partial\Omega$.* ■

12.4.4. *Applications to inverse transforms*

Calculations and properties of the inverse Laplace transform

Let us take a rational function $f(s)$ and suppose that $f(s)$ is of relative degree at least 2 (without loss of generality, according to section 12.3.4). Let P be the set of its distinct poles and let α be a real number such that $\alpha > \mathrm{Re}\,(p)\,, \forall p \in P$. Then we know, according to (12.56), that $f\,(s)$ is the Laplace transform of a function u and that

$$u(t) = \frac{1}{2\pi i} \int_{\alpha - i\infty}^{\alpha + i\infty} f(s) e^{st} ds.$$

We will show on the one hand how we can easily carry out this calculation using residues and on the other hand that this function u is positively supported.

Let us take the integral $J(t, A) = \int_{\alpha - iA}^{\alpha + iA} f(s) e^{st} ds$, $A > 0$.

1) First let $t \geq 0$.

We complete the path $s = \alpha + i\theta, \theta \in [-A, A]$ by the semi-circle $C_{\alpha -}$ defined by $s = \alpha + Ae^{i\,\varphi}, \varphi \in \left[\frac{\pi}{2}, \frac{3\pi}{2}\right]$, such that a closed path γ_{A-} is formed (see Figure 12.1).

There exists a constant $c > 0$ such that $|f(s)| \leq \frac{c}{A^2}$ for any $s \in C_{\alpha -}$ provided that A is sufficiently large. On the other hand, for any $s \in C_{\alpha -}$,

$$\left|e^{st}\right| = \left|e^{\mathrm{Re}(st) + i\,\mathrm{Im}(st)}\right| = \left|e^{\mathrm{Re}(st)}\right| \leq e^{\alpha t}, \tag{12.80}$$

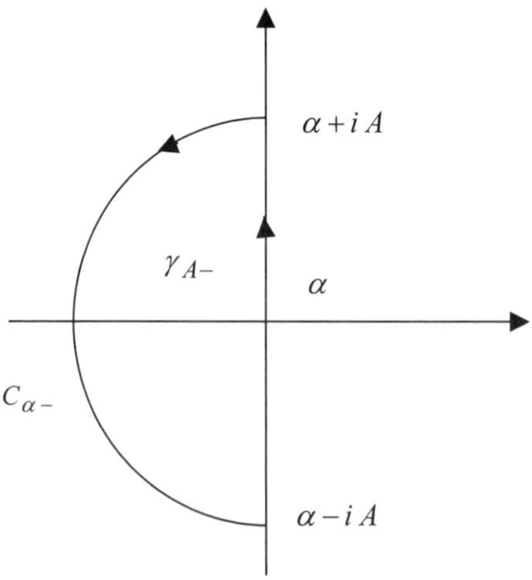

Figure 12.1. *Closed path*

therefore $\left|\int_{C_{\alpha-}} f(s)e^{st}ds\right| \leq \frac{ce^{\alpha t}}{A^2}\int_{C_{\alpha-}} ds = \frac{ce^{\alpha t}}{A^2}\pi A$. This quantity tends to 0 as $A \to +\infty$. Thus, $\lim_{A\to+\infty} J(t,A) = \lim_{A\to+\infty}\int_{\gamma_{A-}} f(s)e^{st}ds$. In addition, the closed path γ_{A-} turns in the direct sense one time around all the poles of $f(s)$ when A is sufficiently large, and thus also around the poles of $f(s)e^{st}$ since e^{st} is entire. Applying the residue theorem (12.77), we then obtain

$$\boxed{u(t) = \sum_{p\in P} \operatorname{Res}\left(f(s)e^{st};p\right),\ t \geq 0}. \tag{12.81}$$

For example, let $f(s) = \frac{1}{(s-p)^n}, n \geq 2$. We have $e^{st} = e^{pt}e^{(s-p)t} = e^{pt}\sum_{k=0}^{+\infty}\frac{(s-p)^k}{k!}t^k$. This function has a unique pole $s = p$ and its residue at this pole is $\frac{t^{n-1}e^{pt}}{(n-1)!}$. As a result

$$\mathcal{L}^{-1}\left\{\frac{1}{(s-p)^n}\right\}(t) = \frac{t^{n-1}e^{pt}}{(n-1)!}, \quad t \geq 0. \tag{12.82}$$

Note that this formula remains valid for $n = 1$.

2) Now suppose $t < 0$.

The upper bound (12.80) is no longer valid in $C_{\alpha-}$, which we will replace by the complementary semi-circle $C_{\alpha+}$ defined by $s = \alpha + Ae^{i\varphi}, \varphi \in \left[-\frac{\pi}{2}, \frac{\pi}{2}\right]$. We now integrate $f(s)e^{st}$ along the closed path γ_{A+}, defined as the union of the path $s = \alpha + i\theta, \theta \in [-A, A]$ and of $C_{\alpha+}$. This closed path does not enclose any of the poles of $f(s)$ and thus $\frac{1}{2\pi i}\int_{\alpha-i\infty}^{\alpha+i\infty} f(s)e^{st}ds = 0$ for $t < 0$. Therefore, we can specify (12.82) by writing

$$\boxed{\mathcal{L}^{-1}\left\{\frac{1}{(s-p)^n}\right\}(t) = \frac{t^{n-1}e^{pt}}{(n-1)!}\mathbf{1}(t)}. \tag{12.83}$$

THEOREM 445.– *Let $\hat{f}(s)$ be a rational function, let P be the set of its distinct poles, and let $\rho \in \mathbb{Z}$ be its relative degree (see section 13.6.1).*

i) The function $\hat{f}(s)$ is the Laplace transform of a distribution $f \in \mathcal{A}_+$, the abscissa of convergence of this Laplace transform is

$$\gamma = \max_{p\in P} \operatorname{Re}(p)$$

and this distribution is of the form

$$f = \sum_{i=0}^{-\rho} q_i\,\delta^{(i)} + g\,\mathbf{1}$$

(by convention, the sum is zero if $\rho > 0$), where the function g is analytic and of the form $\sum_{k,j} t^k e^{p_j t}$.

ii) This distribution f is tempered (thus admits a Fourier transform) if and only if $\gamma \leq 0$.

iii) The rational function $\hat{f}(s)$ is proper (in other words, we have $\rho \geq 0$) if and only if the distribution f is of the form $q_0 \delta + g$ where $q_0 \triangleq \lim_{|s| \to \infty} f(s)$ (the case where $\hat{f}(s)$ is strictly proper corresponds to the case where $q_0 = 0$).

iv) For every $p \in [1, +\infty)$, the following property holds: the function g belongs to L_p if and only if $\gamma < 0$; this is also a necessary and sufficient condition for g to admit a bounded Fourier transform.

Proof

i) According to (13.60) we have $\hat{f}(s) = \hat{q}(s) + \hat{g}(s)$, where $\hat{q}(s)$ is a polynomial and where $\hat{g}(s)$ is a strictly proper rational function. The polynomial $\hat{q}(s)$ is non-zero if and only if $\rho \leq 0$ and it is then of the form

$$\hat{q}(s) = \sum_{i=0}^{-\rho} q_i s^i.$$

Its inverse Laplace transform is thus

$$q = \sum_{i=0}^{-\rho} q_i \delta^i$$

which is a tempered distribution. We can decompose the strictly proper rational function $\hat{g}(s)$ into simple elements; each of these simple elements is the Laplace transform of a function of the form (12.83) and the corresponding abscissa of convergence is $\operatorname{Re}(p)$. We deduce from there the expression of γ. In addition, each of these functions is continuous (and even analytic) except at the origin; the same is true for $g = \mathcal{L}^{-1}\{\hat{g}\}$.

ii) The function g is slowly increasing – thus f is a tempered distribution – if and only if $\gamma \leq 0$.

iii) The rational function $\hat{f}(s)$ is proper (resp., strictly proper) if and only if the polynomial $\hat{q}(s)$ is a constant (resp., is zero).

iv) It is again sufficient to think of the decomposition of $\hat{g}(s)$ into its simple elements and then make use of their inverse Laplace transforms.

Note that if $\gamma < 0$, then $g \in L_\infty$, but the converse is not true. For example, $\mathcal{L}^{-1}\left\{\frac{1}{s}\right\} = \mathbf{1}(t)$: this function is bounded but $\gamma = 0$. Moreover, one can have $\gamma = 0$ and $g \notin L_\infty$: for example, $\mathcal{L}^{-1}\left\{\frac{1}{s^2}\right\} = t\,\mathbf{1}(t)$.

Consider the following example: $\mathcal{L}^{-1}\left\{\frac{s+1}{s}\right\} = \delta + \mathbf{1}$: this is a tempered distribution. The function g, in this case, is defined by $g(t) = 1, t \geq 0$; it is a slowly increasing function. We have $(\mathcal{F}g)(\omega) = (\mathcal{L}g)(i\omega) = \frac{1}{i\omega}$ if $\omega \neq 0$. This Fourier transform is not bounded (since $\gamma = 0$) and is not even a function: this is the distribution $\frac{1}{i}\mathcal{P}\frac{1}{\omega}$, where $\mathcal{P}\frac{1}{\omega}$ is the "principal value" of $\frac{1}{\omega}$ [106].

Calculation of the inverse z-transform

Application of the residue theorem

Let $f(z)$ be a rational function. According to (12.57), $f(z)$ is the z-transform of a positively supported sequence (x_n) if and only if $f(z)$ is a proper rational function. Let P be the set of distinct poles of $f(z)$ and let $\varrho = \max_{p \in P} |p|$. Let $r > \varrho$ and let γ be the closed path $z = re^{i\theta}, \theta \in [0, 2\pi]$. Formula (12.63) can be written as:

$$x_n = \frac{1}{2\pi i} \oint_\gamma f(z) z^{n-1} dz.$$

(We can also directly establish this formula from the residue theorem after the change of variable $s = z^{-1}$; $f(z)$ admits a power series expansion in $s = z^{-1}$.)

According to the residue theorem (12.77):

$$\boxed{x_n = \sum_{p \in P} \operatorname{Res}\left(f(z) z^{n-1}; p\right), \quad n \geq 0}. \tag{12.84}$$

Decomposition into simple elements and power series expansion

Let us now discuss another method which is often faster for the calculation of inverse z-transforms.

The proper rational function $f(z)$ is also a rational function (not necessarily proper) in $s = z^{-1}$. Indeed, $f(z)$ is an irreducible rational function $\frac{N(z)}{D(z)}$. Let $n = d^\circ(D)$. We have

$$\frac{N(z)}{D(z)} = \frac{\frac{N(z)}{z^n}}{\frac{D(z)}{z^n}} = \frac{N^*\left(z^{-1}\right)}{D^*\left(z^{-1}\right)} \tag{12.85}$$

where N^* and D^* are obviously polynomials. We can now apply the decomposition (13.60), (13.61) to $f^*\left(z^{-1}\right) = f(z)$. Write this decomposition in the form

$$f^*\left(z^{-1}\right) = Q^*\left(z^{-1}\right) + \sum_{p \in P} \sum_{1 \leq j \leq n_P} \frac{\alpha_{pj}}{\left(1 - pz^{-1}\right)^j}$$

where P is the set of distinct poles of $f(z)$ and n_P is the multiplicity of the pole $p \in P$.

The polynomial $Q^*\left(z^{-1}\right)$ is of the form

$$Q^*\left(z^{-1}\right) = \sum_{k=0}^{m} c_k z^{-k}$$

and its inverse z-transform is the finite sequence $c = \{c_0, ..., c_m\}$.

We thus know how to find the inverse z-transform of $f(z)$ if we know how to calculate the inverse z-transform of a term of the form $\frac{1}{(1-pz^{-1})^j}$. We have

$$\frac{1}{(1-pz^{-1})^j} = \sum_{k=0}^{+\infty} \binom{j+k-1}{k} p^k z^{-k}$$

where $\binom{j+k-1}{k}$ is the binomial coefficient $\frac{(j+k-1)!}{(j-1)!\,k!}$. As a result, for any $k \geq 0$,

$$\mathcal{Z}^{-1}\left\{\frac{1}{(1-pz^{-1})^j}\right\}(k) = \binom{j+k-1}{k} p^k. \tag{12.86}$$

Consequences

Essential consequences of the above, in particular of (12.86), are gathered in the following theorem:

THEOREM 446.– *A rational function $f(z)$, whose set of distinct poles is P, is the z-transform of a positively supported sequence $x = (x_n)$ if and only if $f(z)$ is proper. The radius of convergence of this z-transform is then*

$$\rho = \max_{p \in P} |p|.$$

The sequence (x_n) is slowly increasing, thus admitting a Fourier transform, if and only if $\rho \leq 1$. In addition, the following conditions are equivalent: (i) $\rho < 1$; (ii) $x \in l_1$; (iii) the Fourier transform $\mathcal{F}x$ is bounded.

12.4.5. *Argument principle*

The *argument principle* plays a key role in the study of the stability of closed-loop systems. Let us take the irreducible rational function

$$f(s) = \frac{\prod_{k=1}^{m}(s-z_k)}{\prod_{k=1}^{n}(s-p_k)} \tag{12.87}$$

where the z_k's and p_k's are the zeros and poles of $f(s)$ respectively. When they are multiple, they appear several times in (12.87).

The *logarithmic derivative* of $f(s)$ is $\frac{d\,f(s)}{f(s)}$ (this is the derivative of $\ln f(s)$). We have

$$\frac{d\,f(s)}{f(s)} = \sum_{k=1}^{m} \frac{d\,s}{s-z_k} - \sum_{k=1}^{n} \frac{d\,s}{s-p_k}. \tag{12.88}$$

Suppose now $\gamma : [a, b] \to \mathbb{C}$ a closed path not passing through any of the z_k's nor p_k's. The image of γ by f is the closed path η defined by $\eta(t) = f(\gamma(t))$.

Integrating the left-hand member of (12.88) along the path γ, we have

$$\int_\gamma \frac{df(s)}{f(s)} = \int_a^b \frac{\dot{f}(\gamma(t))\,\dot{\gamma}(t)}{f(\gamma(t))} dt.$$

Now let us integrate $\frac{ds}{s}$ along η :

$$\int_\eta \frac{ds}{s} = \int_a^b \frac{\dot{\eta}(t)}{\eta(t)} dt = \int_a^b \frac{\dot{f}(\gamma(t))\,\dot{\gamma}(t)}{f(\gamma(t))} dt.$$

Therefore, by integrating (12.88) along γ and by using (12.75) we obtain

$$j(0;\eta) = \sum_{k=1}^{m} j(z_k;\gamma) - \sum_{k=1}^{m} j(p_k;\gamma).$$

Let us now choose the closed path γ in such a way that, more specifically, γ encircles clockwise (i.e. in the indirect sense) one time n_z zeros and n_p poles of $f(s)$ (taking into account multiplicities), whereas it does not enclose the other poles and zeros of this rational function (and does not pass through any of these poles and zeros). We then have $j(z_k;\gamma) = j(p_k;\gamma) = -1$ for the zeros and poles which are enclosed and $j(z_k;\gamma) = j(p_k;\gamma) = 0$ for those that are not. As a result:

$$j(0;\eta) = n_p - n_z. \tag{12.89}$$

We can now state *Cauchy's argument principle:*

THEOREM 447.– *The number of times the image $\gamma\{f(s)\}$ turns around the origin in the direct sense is equal to $n_p - n_z$, where n_p and n_z are the number of poles and zeros respectively (multiplicities included) enclosed by γ in the clockwise sense.*

12.5. Differential equations

Below the reader will find a summary of the theory of differential equations, as well as some complements. The nature of the problems encountered by the control engineer makes it necessary to study linear differential equations with constant coefficients under hypotheses which are more general than in the usual approach.

12.5.1. *Generalities*

Classical notion of a solution

Let be the differential equation

$$\dot{x} = F(t, x) \tag{12.90}$$

where F is a function from $I \times \mathbf{K}^n$ into $\mathbf{K}^n$ (with $\mathbf{K} = \mathbb{R}$ or $\mathbb{C}$) and where I is an interval of $\mathbb{R}$ of the form $|t_1, +\infty)$, $-\infty \leq t_1 < +\infty$. From a concrete point of view, we always have $\mathbf{K} = \mathbb{R}$; however, for calculations, it is often more convenient to embed $\mathbb{R}$ in $\mathbb{C}$, thus to assume $\mathbf{K} = \mathbb{C}$. And this is what we are going to do here.

A function $x : I \ni t \mapsto x(t) \in \mathbb{C}^n$ is a solution (in the classic sense) of equation (12.90) if x is absolutely continuous and its derivative in the sense of distributions satisfies (12.90) almost everywhere.

The Cauchy problem

Let $(t_0, x_0) \in I \times \mathbb{C}^n$. The Cauchy problem consists of having to determine a function x which is a solution of (12.90) and which satisfies the *initial condition*

$$x(t_0) = x_0. \tag{12.91}$$

This obviously equivalent to the problem of determining a solution to the integral equation

$$x(t) = x_0 + \int_{t_0}^{t} F(\tau, x(\tau))\, d\tau.$$

For the Cauchy problem to have one and only one solution on the entire interval I, function F must satisfy certain conditions, specified here after. (For more general conditions, under which the uniqueness of the solution is not guaranteed, see [30], Chapter 2.)

The initial condition (12.91) includes the *initial instant* t_0 and the *initial state* x_0 (this terminology is only classic when equation (12.90) represents a *system*).

The Cauchy-Lipschitz theorem

The following, called the *Cauchy-Lipschitz theorem*, is fundamental, but its proof is outside the scope of this book (see [111], Proposition C.3.8).

Suppose that the function F satisfies the following hypotheses:

i) The function $t \mapsto F(t, \eta)$ is locally integrable on I, for any $\eta \in \mathbb{C}^n$;

ii) The function $\eta \mapsto F(t, \eta)$ is continuous in $\mathbb{C}^n$, for any $t \in I$.

iii) There exists a locally integrable function $\alpha : I \to \mathbb{R}^+$ such that for any $\eta, \varsigma \in \mathbb{C}^n$ and any $t \in I$,

$$\| F(t, \eta) - F(t, \varsigma) \| \leq \alpha(t) \| \eta - \varsigma \|. \tag{12.92}$$

Then the differential equation (12.90) admits a unique solution on $[t_0, +\infty)$ which satisfies the initial condition (12.91).

A function F satisfying condition $iii)$ is said to be *globally Lipschitz* (the adverb "globally" refers to the fact that, in (12.92), η and ς can vary in the whole $\mathbb{C}^n$ for a unique function α). In particular, if the function $(t, \eta) \mapsto F(t, \eta)$ is continuous and differentiable with respect to the second variable, and if the function $\eta \mapsto \frac{\partial F}{\partial \eta}(t, \eta)$ is locally uniformly bounded, then all the above conditions hold.

Differential equation of higher order

Consider the differential equation of order n

$$y^{(n)} = G\left(t, y^{(n-1)}, ..., y\right) \tag{12.93}$$

where G is a function from $I \times \mathbb{C}^p \times ... \times \mathbb{C}^p$ into $\mathbb{C}^p$, and where I is an interval of $\mathbb{R}$ defined as before. In order not to make the presentation cumbersome, we assume in what follows that $p = 1$ whenever we are dealing with differential equations of this type, except when otherwise stated.

We come back to the previous case by writing

$$\begin{cases} x_1 = y^{(n-1)}, \\ x_2 = y^{(n-2)}, \\ \quad \vdots \\ x_n = y. \end{cases} \tag{12.94}$$

Indeed, the differential equation (12.93) is equivalent to the system of differential equations

$$\begin{cases} \dot{x}_1 = G(t, x_1, ..., x_n), \\ \dot{x}_{k+1} = x_k, \ 1 \leq k \leq n-1 \end{cases}$$

which is of form (12.90).

The initial state at instant t_0 is, according to (12.94),

$$x(t_0) = \left[y^{(n-1)}(t_0) \ ... \ y(t_0)\right]^T. \tag{12.95}$$

Linear differential equation

The differential equation (12.90) is said to be linear if the mapping $\eta \mapsto F(t, \eta)$ is affine, i.e. if $F(t, \eta)$ is of the form

$$F(t, \eta) = A(t)\eta + f(t).$$

The differential equation can thus be written as

$$\dot{x} = A(t)x + f(t). \tag{12.96}$$

In this case, the conditions of the Cauchy-Lipschitz theorem are satisfied if both functions $t \mapsto A(t)$ and $t \mapsto f(t)$ are locally integrable on I. This is the case, for example, if these functions are piecewise continuous * (or, more generally, regulated [35]) *.

Linear differential equations of higher order

Consider a linear differential equation of order n of the form

$$\left(\partial^n + \sum_{i=1}^{n} \partial^{n-i} a_i\right) y = \left(\sum_{i=1}^{n} \partial^{n-i} b_i\right) u \tag{12.97}$$

where $\partial = d/dt$ and for any $i \in \{1, ..., n\}$, the coefficients a_i and b_i are functions of t defined on the interval I and satisfy the following property (P):

(P) The functions a_i and b_i are $n-i$ times differentiable in the interval I and their derivative of order $n-i$ is piecewise continuous * (or, more generally, regulated) * in this interval.

On the other hand, u is a function of t, defined in I, and is assumed to satisfy Property (P') we are going to specify below. Instead of proceeding as in (12.94), we define the variable x as follows:

$$\begin{cases} x_1 = y, \\ \partial x_1 = -a_1 x_1 + b_1 u + x_2, \\ \cdots \\ \partial x_{n-1} = -a_n x_1 + b_n u + x_n. \end{cases} \tag{12.98}$$

Equality (12.97) can be simplified as

$$\partial x_n = -a_n x_1 + b_n u. \tag{12.99}$$

We thus obtain the differential equation

$$\dot{x} = \begin{bmatrix} -a_1 & 1 & 0 & \cdots & 0 \\ -a_2 & 0 & \vdots & \ddots & \vdots \\ \vdots & 0 & \vdots & \ddots & 0 \\ \vdots & \vdots & \vdots & 0 & 1 \\ -a_n & 0 & 0 & \cdots & 0 \end{bmatrix} x + \begin{bmatrix} b_1 \\ \vdots \\ \vdots \\ \vdots \\ b_n \end{bmatrix} u, \tag{12.100}$$

$$y = \begin{bmatrix} 1 & 0 & \cdots & \cdots & 0 \end{bmatrix} x. \tag{12.101}$$

System (12.100) is of form (12.96) with the function $t \mapsto A(t)$ which is obviously locally integrable in I, and the function $t \mapsto f(t)$ will have this same property if the function $t \mapsto u(t)$ satisfies the following property (P'):

(P') The function u is piecewise continuous * (or, more generally, regulated) * in the interval I.

12.5.2. *Linear differential equations: constant coefficients*

Matrix form

Consider the differential equation (12.96), where the matrix $A(t) \in \mathbb{C}^{n \times n}$ is *constant*. This equation is thus written as

$$\dot{x} = Ax + f(t) \tag{12.102}$$

By a translation of the origin of time, the problem comes down to the case where the initial instant is $t_0 = 0$; thus we can assume that $I = [0, +\infty)$, and the function $f : I \ni t \mapsto f(t) \in \mathbb{C}^n$ is assumed to be locally integrable on I.

Homogenous equation

The homogenous equation associated with (12.102) is the one obtained when $f = 0$. It can thus be written as

$$\dot{x} = Ax. \tag{12.103}$$

By replacing A by At in (12.70) (see section 12.4.2), we obtain

$$e^{At} = \sum_{k=0}^{+\infty} \frac{t^k A^k}{k!} = I_n + tA + \frac{t^2 A^2}{2!} + \frac{t^3 A^3}{3!} \cdots \tag{12.104}$$

We can differentiate this power series term by term and we get

$$\begin{aligned} \frac{d}{dt}\left(e^{At}\right) &= A + tA^2 + \ldots = A\left(I_n + tA + \frac{t^2 A^2}{2!} + \ldots\right) \\ &= Ae^{At}, \end{aligned} \tag{12.105}$$

therefore the function $t \mapsto e^{At}$ is a solution of (12.103).

Calculation of e^{At}

There exist numerous ways to calculate e^{At}. One of them consists in reducing A to its Jordan form (section 13.3.4, Theorem 530). According to the Jordan theorem, there exists a change of basis matrix P such that $J = P^{-1}AP$ is a diagonal sum of Jordan blocks $J_{\lambda,k}$; each of these blocks appears in the diagonal sum a number of times that is equal to the geometric multiplicity of the eigenvalue λ and is associated with the elementary divisor $(s - \lambda)^k$.

It is immediate that $A = PJP^{-1}$ and that according to (12.104),

$$e^{At} = Pe^{Jt}P^{-1}.$$

According to (12.73), we are led to calculate $e^{J_{\lambda,k}t}$. We have $J_{\lambda,k} = \lambda I_k + J_{0,k}$ where I_k and $J_{0,k}$ commute; thus, by (12.72)

$$e^{J_{\lambda,k}t} = e^{\lambda t} e^{J_{0,k}t}. \tag{12.106}$$

The Jordan block $J_{0,k}$ is the matrix of dimension $k \times k$ defined by

$$J_{0,k} = \begin{bmatrix} 0 & 0 & \cdots & 0 \\ 1 & \ddots & \ddots & \vdots \\ \vdots & \ddots & \ddots & \vdots \\ 0 & \cdots & 1 & 0 \end{bmatrix};$$

as a result, according to (12.104),

$$e^{J_{0,k}t} = \begin{bmatrix} 1 & 0 & \cdots & \cdots & 0 \\ t & 1 & \ddots & & 0 \\ \frac{t^2}{2!} & t & \ddots & \ddots & \vdots \\ \vdots & \vdots & \ddots & \ddots & 0 \\ \frac{t^{k-1}}{(k-1)!} & \frac{t^{k-2}}{(k-2)!} & \cdots & t & 1 \end{bmatrix}. \tag{12.107}$$

Of course, in the case where the matrix A is diagonalizable, these calculations are simpler. Indeed, we then have

$$J = \text{diag}(\lambda_1, ..., \lambda_n)$$

where the λ_i's, $1 \leq i \leq n$, are the eigenvalues (not necessarily distinct) of A. We then obtain

$$e^{Jt} = \text{diag}\left(e^{\lambda_1 t}, ..., e^{\lambda_n t}\right).$$

Use of the inverse Laplace transform

Definition (12.48) of the Laplace transform extends without difficulty to the case of vector-valued functions (i.e. from $\mathbb{R}$ into $\mathbb{C}^n$) or matrix-valued functions (from $\mathbb{R}$ into $\mathbb{C}^{n\times m}$). We can calculate e^{At} using this generalized Laplace transform. We have indeed (12.52) with $u(t) = e^{At}$, thus $u(0) = I_n$ according to (12.71). And by (12.105), we get $s\hat{u}(s) - I_n = A\hat{u}(s)$, from which

$$\boxed{\mathcal{L}\left(e^{At}\,\mathbf{1}(t)\right) = (sI_n - A)^{-1}}. \tag{12.108}$$

A method to calculate e^{At} consists of calculating all entries of the matrix $(s\,I_n - A)^{-1}$, and then determining their inverse Laplace transforms, which are the entries of $e^{At}\,\mathbf{1}(t)$.

Solutions of the homogenous equation

If φ is any vector in $\mathbb{C}^n$, the function

$$x(t) = e^{At}\varphi \tag{12.109}$$

is the unique solution of (12.103) satisfying the initial condition $x(0) = \varphi$.

All the solutions of (12.103) are thus of the form (12.109), where φ spans $\mathbb{C}^n$. This set of solutions is thus a $\mathbb{C}$-vector space of dimension n, generated by the n functions $\eta_i(t) = e^{At}\varepsilon_i, 1 \leq i \leq n$, where $(\varepsilon_i)_{1\leq i\leq n}$ is the canonical basis of $\mathbb{C}^n$. (These n functions $\eta_i(t), 1 \leq i \leq n$, are linearly independent because e^{At} is invertible.)

Let $\lambda_1, ..., \lambda_q$ be the distinct eigenvalues of A. Each eigenvalue λ_i $(1 \leq i \leq q)$ corresponds to one or several Jordan blocks; let m_i be the maximal order of these Jordan blocks, i.e. the order of multiplicity of λ_i as a root of the minimal polynomial of A. According to (12.106) and (12.107), all solutions of (12.103) are of the form

$$x(t) = \sum_{i=1}^{q} e^{\lambda_i t} p_i(t) \tag{12.110}$$

where $p_i(t)$ is a polynomial in the indeterminate t, with coefficients in $\mathbb{C}^n$, and of degree $m_i - 1$.

Solutions of the complete equation

If z is a particular solution of equation (12.102), it is immediate that all solutions of (12.102) are of the form $x = y + z$, where y spans the set of solutions of the homogenous equation (12.103). These solutions thus form an affine $\mathbb{C}$-space of dimension n.

We determine a particular solution of (12.102) by applying the method of "variation of the constant" to the solutions of the homogenous equation (12.103). We saw that any solution of (12.103) is of the form (12.109). The method of variation of the constant consists in replacing the constant vector φ by an absolutely continuous function $t \mapsto \varphi(t) \in \mathbb{C}^n$. In other words, we look for a solution x of (12.102) in the form

$$x(t) = e^{At}\varphi(t). \tag{12.111}$$

By using this in (12.102), we get

$$Ae^{At}\varphi(t) + e^{At}\dot{\varphi}(t) = Ae^{At}\varphi(t) + f(t)$$

from which

$$\dot{\varphi}(t) = e^{-At}f(t)$$

and thus

$$\varphi(t) = x_0 + \int_0^t e^{-A\tau} f(\tau)\, d\tau, \quad x_0 \in \mathbb{C}^n$$

According to (12.111), we obtain

$$\boxed{x(t) = e^{At} x_0 + \int_0^t e^{A(t-\tau)} f(\tau)\, d\tau}. \tag{12.112}$$

The function in the right-hand side of (12.112) is the unique solution of (12.102) satisfying the initial condition $x(0) = x_0$.

Use of the Laplace transform

We recognize in (12.112) a convolution product. This is not surprising, for we can solve (12.102) using the Laplace transform. We then obtain from (12.102) and (12.52)

$$s\hat{x}(s) - x_0 = A\hat{x}(s) + \hat{f}(s)$$

from which

$$\hat{x}(s) = (sI_n - A)^{-1} x_0 + (sI_n - A)^{-1} \hat{f}(s)$$

and we obtain (12.112) according to (12.108).

Linear differential equation of higher order

Decomposition of the problem

In the case where the coefficients are constant, we can consider a linear differential equation of a more general form than (12.97), that is

$$\mathbf{a}(\partial)\, y = \mathbf{b}(\partial)\, u, \tag{12.113}$$

$$\begin{aligned} \mathbf{a}(\partial) &= \partial^n + a_1 \partial^{n-1} + \ldots + a_n, \\ \mathbf{b}(\partial) &= b_k \partial^{n-k} + \ldots + b_n, \quad b_k \neq 0, \quad k \in \mathbb{Z} \end{aligned}$$

where u is given and y is the unknown.

DEFINITION 448.– *The differential equation* (12.113) *is (i)* strictly proper *if* $k \geq 1$, *(ii)* proper *if* $k \geq 0$, *(iii)* improper *if* $k < 0$.

We can adopt the following approach [120]: the variable u is assumed to be of the form $u = f + T$, where f is an indefinitely differentiable function in I and where $T \in \mathcal{A}_{+}$ (see (12.53)). By performing the Euclidean division of $\mathbf{b}(\partial)$ by $\mathbf{a}(\partial)$ (see section 13.1.3), we get

$$\mathbf{b}(\partial) = \mathbf{a}(\partial)\,\mathbf{q}(\partial) + \mathbf{r}(\partial)$$

where $d^{\circ}(\mathbf{r}) < d^{\circ}(\mathbf{a})$, and where $d^{\circ}(\mathbf{q}) = -k$ if $k \leq 0$ and $\mathbf{q} = 0$ if not. Equation (12.113) is equivalent to

$$\mathbf{a}(\partial)\,(y - \mathbf{q}(\partial)\ u) = \mathbf{r}(\partial)\,u$$

as well as, by writing $z = y - \mathbf{q}(\partial)\ u$, to

$$\mathbf{a}(\partial)\,z = \mathbf{r}(\partial)\,u, \tag{12.114}$$

$$y = z + \mathbf{q}(\partial)\,u. \tag{12.115}$$

Putting $w = y - z$, we obtain

$$w = \mathbf{q}(\partial)\,u. \tag{12.116}$$

Solutions of the homogenous differential equation

We can proceed with the differential equation (12.114) as we have done with (12.97), replacing in this last equation y by z and $\mathbf{b}(\partial)$ by $\mathbf{r}(\partial)$. We then obtain the differential equation (12.100) with the coefficients b_i $(1 \leq i \leq n)$ replaced by the coefficients r_i of $\mathbf{r}(\partial)$. All coefficients are constant. The matrix A is a companion of the polynomial

$$\mathbf{a}(s) = s^n + a_1 s^{n-1} + ... + a_n$$

(see section 13.4.3); $\mathbf{a}(s)$ is thus the characteristic polynomial of A and is also called the *characteristic polynomial* of the differential equation (12.113). Since A is cyclic, $\mathbf{a}(s)$ is also the minimal polynomial of this matrix (section 13.4.3, Corollary 568).

Therefore, the distinct roots λ_i $(1 \leq i \leq q)$ of $\mathbf{a}(s)$ are the distinct eigenvalues of A. Moreover, the multiplicity order m_i of λ_i as a root of the polynomial $\mathbf{a}(s)$ is the integer m_i already encountered when analyzing equation (12.110); to λ_i, as an eigenvalue of A, there corresponds a unique Jordan block which has order m_i. Thus we can state the following result:

THEOREM 449.– *The solutions of the homogenous differential equation constitute a $\mathbb{C}$-vector space of dimension n. A basis of this vector space is formed by the n $\mathbb{C}$-linearly independent functions*

$$t \mapsto t^{k-1} e^{\lambda_i t},\ 1 \leq k \leq m_i,\ 1 \leq i \leq q.$$

Solutions of the complete equation

Taking the Laplace transform of (12.114) and using (12.55), we obtain an expression of the form

$$\hat{z}(s) = \frac{\mathbf{r}(s)}{\mathbf{a}(s)}\hat{u}(s) + \frac{\tilde{\mathbf{r}}(s,\partial_0)\,u_0 - \tilde{\mathbf{a}}(s,\partial_0)\,z_0}{\mathbf{a}(s)}$$

where $\tilde{\mathbf{r}}(s,\partial_0)$ (resp., $\tilde{\mathbf{a}}(s,\partial_0)$) is a polynomial with respect to the two variables s and ∂_0, of degree $d^\circ(\mathbf{r}) - 1$ or $-\infty$ (resp., $n-1$ or $-\infty$) *with respect to each of these variables*; this degree is denoted as $\mathrm{d}^\circ(\tilde{\mathbf{r}})$ (resp., $d^\circ(\tilde{\mathbf{a}})$) in what follows. In the same manner, taking the Laplace transform of (12.116) we obtain

$$\hat{w}(s) = \mathbf{q}(s)\;\hat{u}(s) + \tilde{\mathbf{q}}(s,\partial_0)\;u_0$$

where $\tilde{\mathbf{q}}(s,\partial_0)$ is a polynomial with respect to the two variables s and ∂_0, of degree equal to $d^\circ(\mathbf{q}) - 1$ or $-\infty$ *with respect to each of these variables*; this degree is denoted as $d^\circ(\tilde{\mathbf{q}})$ in the sequel. Finally, we have, according to (12.115)

$$\hat{y}(s) = \frac{\mathbf{b}(s)}{\mathbf{a}(s)}u(s) + \frac{\tilde{\mathbf{r}}(s,\partial_0)\;u_0 - \tilde{\mathbf{a}}(s,\partial_0)\,z_0}{\mathbf{a}(s)} + \tilde{\mathbf{q}}(s,\partial_0)\,u_0. \tag{12.117}$$

Let

$$\hat{y}_f(s) = \frac{\mathbf{b}(s)}{\mathbf{a}(s)}u(s).$$

This quantity is obtained from (12.117) by putting $z^{(i)}(0^-) = 0$ $(0 \le i \le d^\circ(\tilde{\mathbf{a}}))$ and $u^{(i)}(0^-) = 0$ $(0 \le i \le \max(d^\circ(\tilde{\mathbf{r}}), d^\circ(\tilde{\mathbf{q}})))$, ("zero initial conditions"). Its inverse Laplace transform is, according to (12.22) and (12.49), for $t \ge 0$

$$y_f(t) = \int_{0^-}^{t^+} g(t-\tau)\,u(\tau)\,d\tau, \quad \text{where } g = \mathcal{L}^{-1}\left\{\frac{\mathbf{b}(s)}{\mathbf{a}(s)}\right\}. \tag{12.118}$$

This is an element of $\mathcal{A}_+$, called the *forced response*.

We now write

$$\hat{y}_{lr}(s) = \frac{\tilde{\mathbf{r}}(s,\partial_0)\,u_0 - \tilde{\mathbf{a}}(s,\partial_0)\,z_0}{\mathbf{a}(s)}. \tag{12.119}$$

This is a strictly proper rational function with respect to s; it thus admits, according to Theorem 445, an inverse Laplace transform which is an indefinitely differentiable function y_{lr}. When the initial conditions $z^{(i)}(0^-)$ $(0 \le i \le d^\circ(\tilde{\mathbf{a}}))$ and $u^{(i)}(0^-) = 0$ $(0 \le i \le d^\circ(\tilde{\mathbf{r}}))$ vary, y_{lr} spans, according to Theorem 449, a $\mathbb{C}$-vector space of dimension n, having a basis consisting of the n functions specified in the above-cited theorem. The functions belonging to this vector space are referred to as the *regular free responses*.

Lastly, we write

$$\hat{y}_{li}(s) = \tilde{\mathbf{q}}(s, \partial_0)\, u_0. \tag{12.120}$$

If $\tilde{\mathbf{q}}(s, \partial_0) \neq 0$, the inverse Laplace transform y_{li} spans, as the initial conditions vary, a $\mathbb{C}-$vector space of dimension equal to ρ, where $\rho = 1 + d^\circ(\tilde{\mathbf{q}})$, having as a basis the distributions δ, $\dot{\delta}, ..., \delta^{(\rho-1)}$. The distributions belonging to this vector space are called the *irregular free responses*.

We can now gather the results we have obtained in the following theorem, where $\rho = 1 + d^\circ(\tilde{\mathbf{q}}) = d^\circ(\mathbf{q})$.

THEOREM 450.– *i) Suppose u is of the form $u = f + T$, where f is an indefinitely differentiable function in I and where $T \in \mathcal{A}_+$. Then the solutions y of the differential equation* (12.113) *are of the same form. They can be decomposed according to $y = y_{lr} + y_{li} + y_f$, where y_f is the forced response, obtained with zero initial conditions and given by* (12.118), *y_{lr} is the regular free response whose Laplace transform is given by* (12.119), *and y_{li} is the irregular free response whose Laplace transform is given by* (12.120).

ii) The regular free responses form a $\mathbb{C}$-vector space of dimension n, a basis of which is formed by the n functions in Theorem 449.

iii) The irregular free responses are zero if $k \geq 0$ and they form a $\mathbb{C}$-vector space of dimension $-k$ if $k < 0$, a basis of which is formed by the distributions δ, $\dot{\delta}, ..., \delta^{(-k-1)}$.

iv) As a result, the free responses of the differential equation (12.113) *form a $\mathbb{C}$-vector space of dimension* $\max(n, n-k)$,[16] *and are all "regular" (that is they are all locally integrable functions) if and only if $k \geq 0$.*

REMARK 451.– *Theorem 450 becomes simpler when $k \geq 0$ (case of a "proper differential equation") since the irregular free responses are then reduced to 0 (in other words, all free responses are regular). Suppose this is the case and, in addition, u is a locally integrable function; then, according to Theorem 445 and the properties of the convolution product (section 12.2.2), the forced response y_f is a* locally integrable *function if $k \geq 0$ and a* continuous function *if $k > 0$.*

16. In [120], this quantity $\max(n, n-k)$ is referred to as the *order* of the differential equation (12.113), for obvious reasons.

12.6. Functions of several variables; optimization

The presentation on functions of several variables made here is extremely succinct (with only a few proofs, and limited to the case of real-valued functions). For a general presentation, see ([35], Chapter VIII).

12.6.1. *Functions of class* C^1

Let Ω be a non-empty open subset of $\mathbb{R}^n$ and $x^* \in \Omega$. We say that a function $J : \Omega \to \mathbb{R}$ admits a *partial derivative* $\frac{\partial J}{\partial x_i}(x^*)$ with respect to the variable x_i at the point x^* if the partial mapping

$$J_i : x_i \mapsto J\left(x_1^*, ... x_{i-1}^*, x_i, x_{i+1}^*, ..., x_n^*\right)$$

is differentiable at x_i^*; and in this case we have by definition

$$\frac{\partial J}{\partial x_i}(x^*) = \frac{dJ_i}{dx_i}(x_i^*) .$$

The function J is said to be *differentiable* at the point x^* if there exists a linear form, denoted by $dJ(x^*)$ and called the *derivative* of J at x^*, such that

$$\boxed{J(x^* + h) - J(x^*) = dJ(x^*)\, h + o(\|h\|)} \tag{12.121}$$

where $o(\|h\|)$ is a function, defined in the neighborhood of 0 and such that

$$\lim_{\substack{\|h\| \to 0 \\ \|h\| \neq 0}} \frac{o(\|h\|)}{\|h\|} = 0.$$

In the canonical basis of $\mathbb{R}^n$, $dJ(x^*)$ is represented by the row matrix

$$\boxed{DJ(x^*) = \left[\begin{array}{ccc} \frac{\partial J}{\partial x_1}(x^*) & \dots & \frac{\partial J}{\partial x_n}(x^*) \end{array}\right]}$$

called the *Jacobian matrix* of J at x^*. If we identify the vector $h = (h_1, ..., h_n)$ with the column matrix $\left[\begin{array}{ccc} h_1 & \dots & h_n \end{array}\right]^T$, we have

$$dJ(x^*)\, h = DJ(x^*)\, h,$$

in a way that leads us to identify the derivative $dJ(x^*)$ with the Jacobian matrix $DJ(x^*)$.

The column vector $\nabla J(x^*) = DJ(x^*)^T$ is the *gradient* of J at x^*.

The function J is said to be of class C^1 if it is differentiable at any point of Ω and if its derivative $dJ : x \to dJ(x)$ is continuous. One can show that J is of class C^1 if, and only if it admits partial derivatives $\frac{\partial J}{\partial x_i}$ with respect to all its variables at any point of Ω and if these partial derivatives are continuous.

12.6.2. *Functions of class* C^2

Let $J : \Omega \rightarrow \mathbb{R}$ be a function admitting a partial derivative $\frac{\partial J}{\partial x_i}(x)$ at any point $x \in \Omega$ and consider the partial mapping

$$(DJ)_{i,j} : x_j \mapsto \frac{\partial J}{\partial x_i}\left(x_1^*, ...x_{j-1}^*, x_j, x_{j+1}^*, ..., x_n^*\right).$$

If it is differentiable at the point $x_j^* \in \Omega$, we say that J admits a *second order partial derivative* $\frac{\partial^2 J}{\partial x_i \partial x_j}(x^*)$ with respect to the variables x_i and x_j at x^*, and by definition

$$\frac{\partial^2 J}{\partial x_i \partial x_j}(x^*) = \frac{d\,(DJ)_{i,j}}{dx_j}\left(x_j^*\right).$$

The function J is said to be *twice differentiable* at the point x^* if it is of class C^1 and if its derivative $dJ : \Omega \longrightarrow (\mathbb{R}^n)'$ is differentiable at x^* (where $(\mathbb{R}^n)'$ is the dual of $\mathbb{R}^n$: see section 12.1.2; $(\mathbb{R}^n)'$ is identified with the $\mathbb{R}$-vector space of row matrices with n entries belonging to $\mathbb{R}$, i.e. with $\mathbb{R}^{1 \times n}$). The derivative of dJ at x^* is called the *second derivative* of J at that point and is denoted as $d^2 J(x^*)$. It is an element of $\mathcal{L}(\mathbb{R}^n, \mathcal{L}(\mathbb{R}^n, \mathbb{R}))$, represented in the canonical bases by the square matrix of partial derivatives of order 2,

$$\boxed{HJ(x^*) = \left(\frac{\partial^2 J}{\partial x_i \partial x_j}(x^*)\right)_{1 \leq i \leq n, 1 \leq j \leq n}}$$

called the *Hessian matrix* of J at x^*. This matrix is symmetric ($\frac{\partial^2 J}{\partial x_i \partial x_j} = \frac{\partial^2 J}{\partial x_j \partial x_i}$). Let h_1, h_2 be two vectors of $\mathbb{R}^n$; we have, based on the above identifications

$$\boxed{d^2 J(x^*)(h_1, h_2) = h_1^T HJ(x^*)\, h_2}. \tag{12.122}$$

The function J is said to be of class C^2 if it is twice differentiable at any point of Ω and if its second derivative $d^2 J : x \rightarrow d^2 J(x)$ is continuous. One can show that J is of class C^2 if and only if it admits second order partial derivatives $\frac{\partial^2 J}{\partial x_i \partial x_j}$ with respect to all its variables at any point of Ω and if these partial derivatives are continuous.

We leave it to the reader to define by induction the pth derivative $d^p J(x^*)$, as well as a function of class C^p.

12.6.3. *Taylor's formula*

Set $h^1 = h$ and $h^p = \left(h^{p-1}, h\right)$ for $p > 1$.

Taylor's formula with Young's remainder

Let $J : \Omega \to \mathbb{R}$ be a function which is $p-1$ times differentiable in Ω and p times differentiable at point x^*. Therefore

$$\boxed{J(x^*+h) = J(x^*) + \sum_{k=1}^{p} \tfrac{1}{k!} d^k J(x^*) h^k + o(\|h\|^p)} \tag{12.123}$$

(For $p = 1$, this formula is nothing but the definition of the derivative $dJ(x^*)$.)

Taylor's formula with Lagrange's remainder

Let $J : \Omega \to \mathbb{R}$ be a function which has a derivative or order $p-1$ in Ω. Suppose the closed segment $[x^*, x^*+h]$ is contained in Ω and J admits a derivative of order p at any point of the open segment (x^*, x^*+h).[17] Then there exists $\theta \in (0,1)$ such that

$$\boxed{J(x^*+h) = J(x^*) + \sum_{k=1}^{p-1} \tfrac{1}{k!} d^k J(x^*) h^k + \tfrac{1}{p!} d^p J(x^*+\theta h) h^p} \tag{12.124}$$

12.6.4. *Convexity, coercivity, ellipticity*

For more details on what follows, the interested reader may consult [29].

Convexity

A non-empty subset Ω of $\mathbb{R}^n$ is said to be *convex* if for any points a and b of Ω, the segment $[a, b]$ is included in Ω.

Let $J : \Omega \to \mathbb{R}$, where Ω is convex. The function J is said to be *convex* (resp., *strictly convex*) if for any distinct points a and b of Ω and any $\lambda \in (0,1)$, $J((1-\lambda)a + \lambda b) \leq (1-\lambda) J(a) + \lambda J(b)$ (resp., $J((1-\lambda)a + \lambda b) < (1-\lambda) J(a) + \lambda J(b)$).

Let us recall the following definitions, where Ω denotes a non-empty part of $\mathbb{R}^n$:

– The function $J : \Omega \to \mathbb{R}$ admits a global (resp., strict global) minimum at $x^* \in \Omega$ if $J(x^*) \leq J(x)$ for any $x \in \Omega$ (resp., $J(x^*) < J(x)$ for any $x \in \Omega$, $x \neq x^*$); we then write $x^* \in \arg\min_{x \in \Omega} J(x)$ (resp., $x^* = \arg\min_{x \in \Omega} J(x)$).

17. The segment $[a, b]$ (resp., (a, b)) is the set of all points of the form $(1-\lambda)a + \lambda b$, $\lambda \in [0,1]$ (resp., $\lambda \in (0,1)$). See section 12.4.3.

– The function $J : \Omega \to \mathbb{R}$ admits a *local* (resp., *strict local) minimum* at $x^* \in \Omega$ if there exists a neighborhood V of x^* in Ω such that the restriction of J to V admits a global (resp., strict global) minimum at x^*.

Convexity plays a key role in the theory of optimization for the following reason:

THEOREM 452.– *Let Ω be a convex subset of $\mathbb{R}^n$ and $J : \Omega \to \mathbb{R}$. If J admits a local minimum at x^* and is* convex *(resp.,* strictly convex*), then this minimum is* global *(resp.,* strict global*).*

Note that a strictly convex function does not necessarily admit a minimum. This is the case, for example, of $J(x) = 1/x$ in $\Omega = (0, +\infty)$. A necessary condition for a function to admit a local minimum is the *Euler condition*, an immediate consequence of (12.121) :

PROPOSITION 453.– *Let Ω be a non-empty open subset of $\mathbb{R}^n$ and $J : \Omega \to \mathbb{R}$ be a differentiable function. For J to admit a local minimum at point x^*, the* Euler condition *$dJ(x^*) = 0$ must hold.*

The following is a consequence of (12.123) and (12.124) :

PROPOSITION 454.– *Let Ω be a convex open subset of $\mathbb{R}^n$ and let $J : \Omega \to \mathbb{R}$ be function of class C^2. The function J is convex if and only if $HJ(x) \geq 0$ for every $x \in \Omega$. If $HJ(x) > 0$ for every $x \in \Omega$, then J is strictly convex.*

On the other hand, the following is a classic result:

THEOREM 455.– *Let Ω be a convex open subset of $\mathbb{R}^n$ and let $J : \Omega \to \mathbb{R}$ be a differentiable function. If J is* convex *and there exists a solution to the Euler equation $dJ(x^*) = 0$, then $J(x^*)$ is a* global minimum *of J. If in addition J is* strictly convex*, the solution x^* here above is* unique *and J admits a* strict global minimum *at this point.*

Coercivity

Let Ω be an *unbounded* part of $\mathbb{R}^n$ and $J : \Omega \to \mathbb{R}$ be continuous function. This function is said to be *coercive* if

$$\lim_{\substack{\|x\| \to +\infty \\ x \subset \Omega}} J(x) = +\infty.$$

THEOREM 456.– *Let Ω be an* unbounded closed *part of $\mathbb{R}^n$; every coercive function $J : \Omega \to \mathbb{R}$ admits a global minimum.*

PROOF. Let $\tilde{x} \in \Omega$ and $\Phi = \{x \in \Omega : J(x) \leq J(\tilde{x})\}$. The set Φ is closed (because J is continuous) and bounded, thus compact. As a result, the restriction of J to Φ admits a minimum (see the proof of Theorem 444). This here is obviously a global minimum of J. ■

Ellipticity

Let $J : \mathbb{R}^n \to \mathbb{R}$ be a function of class C^2. This function is said to be *elliptic* if there exists a real $\delta > 0$ such that for every $x \in \mathbb{R}^n$, $HJ(x) - \delta I_n \geq 0$.

THEOREM 457.– *Let $J : \mathbb{R}^n \to \mathbb{R}$ be an elliptic function; this function is strictly convex and coercive.*

PROOF. It is obvious that J is strictly convex. In addition, according to (12.124), for any points x and x^* such that $x \neq x^*$

$$J(x) = J(x^*) + dJ(x^*)(x - x^*) + \frac{1}{2} d^2 J(y)(x - x^*)^2$$

for some $y \in (x^*, x)$. According to (12.122) and the condition of ellipticity,

$$J(x) \geq J(x^*) + dJ(x^*)(x - x^*) + \frac{\delta}{2} \|x - x^*\|^2,$$

thus $J(x) \to +\infty$ as $\|x\| \to +\infty$. ■

COROLLARY 458.– *Let $J : \mathbb{R}^n \to \mathbb{R}$ be an elliptic function; this function admits a strict global minimum, attained at a point x^* which is the unique solution of the Euler equation $dJ(x^*) = 0$.*

12.6.5. *Optimization algorithms*

The principal optimization algorithms (for problems without constraints and except for the conjugate gradients algorithm) are briefly described below.

Gradient method

Let Ω be a non-empty open subset of $\mathbb{R}^n$ and $J : \Omega \to \mathbb{R}$ be a differentiable function. Let $\theta \in \Omega$ and $h \in \mathbb{R}^n$ be such that $\theta + h \in \Omega$. We have (see section 12.6.3)

$$J(\theta + h) - J(\theta) = dJ(\theta)\, h + o(\|h\|) \tag{12.125}$$

and $dJ(\theta) = \nabla J(\theta)^T$.

LEMMA 459.– *Let h be of the form $-\rho \nabla J(\theta)$, $\rho > 0$. Then $J(\theta - \rho \nabla J(\theta)) < J(\theta)$ if $\nabla J(\theta) \neq 0$ and if $\rho > 0$ is sufficiently small.*

PROOF. According to (12.125)

$$J(\theta - \rho \nabla J(\theta)) - J(\theta) = -\rho \|\nabla J(\theta)\|^2 + o(\rho).$$

■

An increment h as above is in the direction opposite to that of the gradient. Lemma 459 shows that this direction is a *direction of descent*, i.e. in this direction the cost function J decreases, at least for a small increment.

The "gradient method" is an iterative minimization method of the function J. From a point $\theta^{(0)} \in \Omega$, we will construct a sequence $\left(\theta^{(k)}\right)_{k \geq 0}$ of points of Ω, called a *minimizing sequence*, of the following manner:

$$\theta^{(k+1)} = \theta^{(k)} - \rho_k \nabla J\left(\theta^{(k)}\right)$$

where $\rho_k \geq 0$, while $\nabla J\left(\theta^{(k)}\right) \neq 0$. Recall that if there exists $\theta^{(k)}$ such that $\nabla J\left(\theta^{(k)}\right) = 0$ and if both Ω and J are convex, then $J\left(\theta^{(k)}\right)$ is a global minimum of J (see section 12.6.4, Theorem 455), thus $\theta^{(k)}$ is the value of the vector of parameters θ we are looking for.

In certain cases, the "step" ρ_k is chosen to be constant (with respect to k) to reduce the calculations: this is the *fixed-step gradient method*.

We can also, at the cost of more important and extensive calculations (but for better efficiency), optimize the step ρ_k by minimizing at each iteration k the function of a single variable

$$\tilde{J}_k(\rho) = J\left(\theta^{(k)} - \rho \nabla J\left(\theta^{(k)}\right)\right)$$

(with $\rho \geq 0$). This is what we call a *unidirectional minimization* (made in the direction $d_k = -\nabla J\left(\theta^{(k)}\right) \neq 0$). Denote as ρ_k^* the *optimal step*, if it exists.

THEOREM 460.– *Suppose that $\Omega = \mathbb{R}^n$ and that J is of class C^2 and is elliptic (section 12.6.4). Then the unidirectional minimization problem admits a unique solution and we can write*

$$\rho_k^* = \arg\min_{\rho \geq 0} \tilde{J}_k(\rho).$$

PROOF. Since the function J is coercive, so is obviously $\tilde{J}_k$ too. As a result, according to Theorem 456 (section 12.6.4), $\tilde{J}_k$ admits a global minimum in the unbounded closed convex set $[0,+\infty)$. In addition,

$$\begin{aligned}
\frac{d\tilde{J}_k}{d\rho}(\rho) &= -dJ\left(\theta^{(k)} - \rho\nabla J\left(\theta^{(k)}\right)\right)\nabla J\left(\theta^{(k)}\right), \qquad (12.126)\\
\frac{d^2\tilde{J}_k}{d\rho^2}(\rho) &= d^2J\left(\theta^{(k)} - \rho\nabla J\left(\theta^{(k)}\right)\right)\left(\nabla J\left(\theta^{(k)}\right), \nabla J\left(\theta^{(k)}\right)\right)\\
&= \left(\nabla J\left(\theta^{(k)}\right)\right)^T HJ\left(\theta^{(k)} - \rho\nabla J\left(\theta^{(k)}\right)\right)\nabla J\left(\theta^{(k)}\right)\\
&\geq \delta\left\|\nabla J\left(\theta^{(k)}\right)\right\|^2 > 0,
\end{aligned}$$

therefore $\tilde{J}_k$ is strictly convex (according to Proposition 454 of section 12.6.4). The minimum of $\tilde{J}_k$ in $[0,+\infty)$ is thus strict and unique (according to Theorem 455). ■

REMARK 461.– *The optimal step ρ_k^* belongs necessarily to $(0,+\infty)$. Indeed, if we had $\rho_k^* = 0$, we would have $J\left(\theta^{(k)} - \rho\nabla J\left(\theta^{(k)}\right)\right) > J\left(\theta^{(k)}\right)$ for any $\rho > 0$, which is impossible according to Lemma 459. As a result, ρ_k^* is characterized by the Euler equation $\frac{d\tilde{J}_k}{d\rho}(\rho_k^*) = 0$, which yields*

$$\nabla J\left(\theta^{(k+1)}\right)^T \nabla J\left(\theta^{(k)}\right) = 0$$

according to (12.126). *In other words, the* directions of two successive descents are orthogonal.

One can show the following result ([29], Theorem 8.4-3):

THEOREM 462.– *Under the hypotheses of Theorem 460, the optimal-step gradient method converges (i.e. $\left(J\left(\theta^{(k)}\right)\right) \to \min J$).*

The point is now to explicitly determine the optimal step ρ_k^*. This is not possible in the general case and it is then necessary to make use of a search by dichotomies. One exception to this situation is that where the cost function J is quadratic (and elliptic), i.e. is of the form

$$J(\theta) = \frac{1}{2}\theta^T A\theta + b^T\theta$$

where A is a symmetric real positive definite matrix. Let

$$w_k = A\theta^{(k)} + b = \nabla J\left(\theta^{(k)}\right).$$

It is easy to verify that

$$\rho_k^* = \frac{\|w_k\|^2}{w_k^T A w_k}.$$

Newton-Raphson method

The Newton-Raphson method – like all methods below – is a method of descent. Let $J : \mathbb{R}^n \to \mathbb{R}$ be a function of class C^2 and assume that J is elliptic. We have by applying the Taylor-Young formula in the neighborhood of $\theta^{(k)}$ (section 12.6.3)

$$J(\theta) = J_k(\theta) + o\left(\left\|\theta - \theta^{(k)}\right\|^2\right)$$

where $J_k(\theta)$ is the "second order approximation" of J in the neighborhood of $\theta^{(k)}$, i.e. setting $H_k = HJ\left(\theta^{(k)}\right)$,

$$J_k(\theta) = J\left(\theta^{(k)}\right) + \nabla J\left(\theta^{(k)}\right)\left(\theta - \theta^{(k)}\right) + \frac{1}{2}\left(\theta - \theta^{(k)}\right)^T H_k\left(\theta - \theta^{(k)}\right).$$

The function J_k is obviously elliptic and thus admits a strict global minimum at the point $\theta^{(k+1)}$ characterized by the Euler equation $\nabla J_k\left(\theta^{(k+1)}\right) = 0$ (section 12.6.4, Corollary 458). We have

$$\nabla J_k(\theta) = \nabla J\left(\theta^{(k)}\right) + H_k\left(\theta - \theta^{(k)}\right),$$

and therefore

$$\theta^{(k+1)} = \theta^{(k)} - H_k^{-1}\nabla J\left(\theta^{(k)}\right). \qquad (12.127)$$

REMARK 463.– *(i) Contrary to the gradient method, the Newton-Raphson method does not require unidirectional minimization (nevertheless, see (iii) below). (ii) In the case where J is quadratic (and also elliptic), the Newton-Raphson converges in one iteration, whereas in general the gradient method converges in an infinite number of iterations [29]. For a non-quadratic function J, when θ is close to the optimum, $J_k(\theta)$ is a good approximation of $J(\theta)$, thus the Newton-Raphson method converges rapidly. It is however to be avoided when θ is still far away from the optimum and we would then prefer the gradient algorithm. (iii) In order to avoid the "blocked" situations far away from the optimum, we can improve the Newton-Raphson method by making a unidirectional minimization at each iteration. In this case,* (12.127) *is replaced by*

$$\begin{cases} \theta^{(k+1)} = \theta^{(k)} - \rho_k^* H_k^{-1}\nabla J\left(\theta^{(k)}\right) \\ \rho_k^* = \arg\min_{\rho \geq 0} J\left(\theta^{(k)} - \rho H_k^{-1}\nabla J\left(\theta^{(k)}\right)\right) \end{cases}$$

(with of course $\rho_k^* = 1$ in the case of a quadratic function). We arrive at a method of descent with a direction given by

$$d_k = -H_k^{-1}\nabla J\left(\theta^{(k)}\right).$$

Newton-Gauss method

The big disadvantage of the Newton-Raphson method is the large amount of calculations necessary for the determination of the Hessian matrix H_k at each iteration k. This amount can be reduced in the case of a quadratic criterion, i.e.

$$J(\theta) = \frac{1}{2} \sum_{i \in I} j_i(\theta)^2$$

where I is a finite set of indices and where the functions $j_i : \mathbb{R}^n \to \mathbb{R}$ are of class C^2. We get

$$dJ(\theta) = \sum_{i \in I} j_i(\theta)\, dj_i(\theta),$$

$$HJ(\theta) = \sum_{i \in I} \left[\nabla j_i(\theta) \nabla j_i(\theta)^T + Hj_i(\theta)\, j_i(\theta) \right].$$

In the above expression, the calculation of the $Hj_i(\theta)$'s is disadvantageous. The *Newton-Gauss method* neglects these terms, which anyway are zero if the functions j_i are linear. We thus obtain the approximation

$$\boxed{H_k \simeq \sum_{i \in I} \nabla j_i\left(\theta^{(k)}\right)\left(\nabla j_i\left(\theta^{(k)}\right)\right)^T}. \tag{12.128}$$

Besides simplifying the calculations, another advantage is that the matrix on the right-hand side of (12.128) is always symmetric real non-negative definite, which is an essential point for d_k to actually be a direction of descent (see below).

Quasi-Newton methods

Let $J : \mathbb{R}^n \to \mathbb{R}$ be a function of class C^1. Consider iterations of the form

$$\boxed{\theta^{(k+1)} = \theta^{(k)} + \rho_k d_k} \tag{12.129}$$

$\rho_k > 0$, where each direction d_k is of the form

$$\boxed{d_k = -G_k \nabla J\left(\theta^{(k)}\right)} \tag{12.130}$$

G_k being a symmetric real matrix. We have the following:

THEOREM 464.– *A sufficient condition for d_k to be a direction of descent is that G_k be positive definite.*

PROOF. We have, according to the Taylor-Young formula

$$J\left(\theta^{(k)}+\rho d_k\right)=J\left(\theta^{(k)}\right)-\rho\nabla J\left(\theta^{(k)}\right)^T G_k\nabla J\left(\theta^{(k)}\right)+o\left(\rho\right),$$

thus $J\left(\theta^{(k)}+\rho d_k\right) < J\left(\theta^{(k)}\right)$ if $\nabla J\left(\theta^{(k)}\right) \neq 0$, $G_k > 0$ and if $\rho > 0$ is sufficiently small. ■

We call an iterative method of the form (12.129), (12.130) a *quasi-Newton method* whenever $G_k > 0$.

Suppose J is of class C^2 and is elliptic. Then we can optimize the step ρ_k by a unidirectional minimization, which leads to a "quasi-Newton method with optimal step". Following the same rationale as in the proof of Theorem 460 and in Remark 461, we indeed obtain the following:

THEOREM 465.– *Let* $\tilde{J}_k\left(\rho\right) = J\left(\theta^{(k)}+\rho d_k\right)$. *The unidirectional minimization problem* $\tilde{J}_k\left(\rho\right)\rightarrow\min$ *admits a unique solution*

$$\boxed{\rho_k^* = \arg\min_{\rho\geq 0}\tilde{J}_k\left(\rho\right) > 0}$$

characterized by the Euler condition which is equivalent to

$$\boxed{\nabla J\left(\theta^{(k+1)}\right)^T d_k = 0}.$$

Two quasi-Newton methods have been discussed so far:

– the gradient method, which corresponds to $G_k = I_n$;

– the Newton-Raphson method, which corresponds to $G_k = H_k^{-1}$ (a matrix which is positive definite if J is elliptic).

Intermediate strategies can be considered, for example

$$\boxed{G_k = \begin{cases} I_n, 1\leq k < k_0 \\ H_k^{-1}, k\geq k_0 \end{cases}}$$

where k_0 is such that $\theta^{(k_0)}$ is "sufficiently close" to the optimum.

This method makes it possible to have a faster convergence in the neighborhood of the optimum than the gradient method. On the other hand, for points further away from the optimum, we avoid possible difficulties such as, for example, getting a singular Hessian matrix in the case where J is not elliptic. In order to avoid a singularity in the calculations at the neighborhood of the optimum, we can take the precaution of replacing here above H_k by $H_k + \varepsilon_k I_n$, where $\varepsilon_k > 0$ is a sufficiently small real number.

By replacing again H_k by its approximation (12.128) from the Newton-Gauss method, we obtain the *Levenberg-Marquardt method* which is very efficient.

12.7. Probabilistic notions

This section consists of summaries and complements. For more details on the theory of measure and on that of probabilities, see, e.g. [103] and [83], respectively.

12.7.1. *Probability space*

σ-algebras and measurability

Let Ω be a non-empty set. A σ-algebra $\mathcal{F}$ over Ω is a set of parts of Ω containing the empty set and which is stable by passing into complement as well as by countable union; the pair $(\Omega, \mathcal{F})$ is called a *probabilizable space* (this notion is synonymous to that of *measurable space*).

An intersection of σ-algebras is again a σ-algebra. Let $\mathcal{E}$ be a non-empty set of parts of Ω; the smallest σ-algebra containing $\mathcal{E}$ is called the σ-algebra *generated* by $\mathcal{E}$. If Ω is a topological space (section 12.1.1), the σ-algebra generated by the open subsets of Ω is the *Borel* σ-algebra of Ω (and the elements of this σ-algebra are called the *Borel sets* of Ω).

Let $(\Omega, \mathcal{F})$ and $(\mathcal{Y}, \mathcal{T})$ be two probabilizable spaces; a function $Y : \Omega \to \mathcal{Y}$ is said to be $(\mathcal{F}, \mathcal{T})$-measurable (or $\mathcal{F}$-measurable if $\mathcal{Y}$ is a topological space and $\mathcal{T}$ is a Borel σ-algebra of $\mathcal{Y}$) if for any $B \in \mathcal{T}$, $Y^{-1}(B) \in \mathcal{F}$. It is immediate that $Y^{-1}(\mathcal{T})$ is a σ-algebra, it is also the smallest among the σ-algebras $\mathcal{G}$ for which Y is $(\mathcal{G}, \mathcal{T})$-measurable.

More generally, let $(\mathcal{Y}_i, \mathcal{T}_i)_{i \in I}$ be a family of probabilizable spaces and $(Y_i)_{i \in I}$ be a family of functions $Y_i : \Omega \to \mathcal{Y}_i$. It is also clear that $\bigcap_{i \in I} Y_i^{-1}(\mathcal{T}_i)$ is a σ-algebra, and it is the smallest among all the σ-algebras $\mathcal{G}$ for which each of the Y_i's $(i \in I)$ is $(\mathcal{G}, \mathcal{T})$-measurable.

DEFINITION 466.– *We call $Y^{-1}(\mathcal{T})$ the σ-algebra* generated *by Y and $\bigcap_{i \in I} Y_i^{-1}(\mathcal{T}_i)$ the σ-algebra* generated *by the family $(Y_i)_{i \in I}$.*

Probability

Let $(\Omega, \mathcal{F})$ be a probabilizable space. A *measure of probability* (or, more succinctly, a *probability*) P on such a space is a positive measure such that $P\{\Omega\} = 1$, which we also write as

$$\boxed{P\{\Omega\} = \int_\Omega dP(\omega) = 1}.$$

Then the triple $(\Omega, \mathcal{F}, P)$ is called a *probability space.*

REMARK 467.– *A simple and important case is that when the probability measure P is defined by a density p; in this case, $\Omega = \mathbb{R}^n$, $\mathcal{F}$ is the Borel σ-algebra $\mathcal{B}_n$ of $\mathbb{R}^n$, $dP(x) = p(x)\,dx$, where $p(x) \geq 0$, dx is the Lebesgue measure on $\mathbb{R}^n$, and*

$$\int_{\mathbb{R}^n} p(x)\,dx = 1.$$

12.7.2. *Random variable*

General notions

Let $(\Omega, \mathcal{F}, P)$ be a probability space; a *random variable* with values in a probabilizable space $(\mathcal{Y}, \mathcal{T})$ is an $(\mathcal{F}, \mathcal{T})$-measurable mapping $Y : \Omega \to \mathcal{Y}$. (A random variable is thus a function $\Omega \ni \omega \mapsto Y(\omega) \in \mathcal{Y}$; the dependence of Y with respect to ω expresses the "random draws" that we can perform from Y and, for fixed ω, a "realization" of Y is a value $Y(\omega)$.) A random variable is said to be real (resp., complex) if the probabilizable space $(\mathcal{Y}, \mathcal{T})$ is the set of real (resp., complex numbers) numbers equipped with its Borel σ-algebra. An event is "almost sure" if its probability is equal to 1; in particular, two random variables X and Y, with values in the same probabilizable space $(\mathcal{Y}, \mathcal{T})$, are *almost surely equal* if $P\{X = Y\} = 1$.

DEFINITION 468.– *(i) Let $\mathcal{F}_1$ and $\mathcal{F}_2$ be two sub-σ-algebras of $\mathcal{F}$; these sub-σ-algebras are said to be* independent *if for any $A_1 \in \mathcal{F}_1$ and any $A_2 \in \mathcal{F}_2$, $P\{A_1 \cap A_2\} = P\{A_1\} P\{A_2\}$. (ii) Let Y be a random variable and $\mathcal{G}$ be a sub-σ-algebra of $\mathcal{F}$. We say that Y is* independent *of $\mathcal{G}$ if the latter and the σ-algebra generated by Y are independent. (iii) Let Y_1 and Y_2 be two random variables; these are said to be* independent *if the σ-algebras they generate are independent.*

The *probability law* of a random variable Y with values in $(\mathcal{Y}, \mathcal{T})$ is the image probability of P by Y, i.e. the probability ν_Y defined on $(\mathcal{Y}, \mathcal{T})$ by

$$\boxed{\nu_Y(B) = P\{Y^{-1}(B)\}, \quad B \in \mathcal{T}}.$$

Suppose Y_1 and Y_2 are two random variables with values in the probabilizable spaces $(\mathcal{Y}_1, \mathcal{T}_1)$ and $(\mathcal{Y}_2, \mathcal{T}_2)$ respectively. For any $B_1 \in \mathcal{T}_1$ and any $B_2 \in \mathcal{T}_2$,

$$\nu_{(Y_1,Y_2)}(B_1 \times B_2) = P\{Y_1 \in B_1, Y_2 \in B_2\} = P\left\{Y_1^{-1}(B_1) \cap Y_2^{-1}(B_2)\right\}.$$

These random variables are independent if and only if for all sets B_1 and B_2 as above,

$$P\left\{Y_1^{-1}(B_1) \cap Y_2^{-1}(B_2)\right\} = P\left\{Y_1^{-1}(B_1)\right\} P\left\{Y_2^{-1}(B_2)\right\}.$$

We deduce the following:

PROPOSITION 469.– *The above random variables Y_1 and Y_2 are independent if and only if*

$$\boxed{\nu_{(Y_1,Y_2)}(B_1 \times B_2) = \nu_{Y_1}(B_1)\, \nu_{Y_2}(B_2), \quad \forall B_1 \in \mathcal{T}_1, \forall B_2 \in \mathcal{T}_2}.$$

Random variables of the first order

Let X be a random variable with values in the measurable space $(\mathbf{K}^n, \mathcal{B}_n)$ where $\mathcal{B}_n$ is the Borel σ-algebra of $\mathbf{K}^n$ ($\mathbf{K} = \mathbb{R}$ or $\mathbb{C}$); abusing the language, we will say that X has its values in $\mathbf{K}^n$.[18] This random variable is said to be of *the first order* if

$$E[\|X\|] = \int_\Omega \|X(\omega)\|\, dP(\omega) < +\infty$$

where $\|.\|$ is the standard Euclidean or Hermitian norm (section 12.1.2). By identifying two almost surely equal random variables (which we will do in the sequel[19]), the set of random variables of the first order with values in $\mathbf{K}^n$ is denoted as $L^1(\Omega, \mathcal{F}, P; \mathbf{K}^n)$, or $L^1(\Omega, \mathcal{F}, P)$ if $n = 1$ and $\mathbf{K} = \mathbb{R}$; this is a Banach space equipped with the norm $\|X\|_1 \triangleq E[\|X\|]$. The quantity

$$\boxed{E[X] = \int_\Omega X(\omega)\, dP(\omega)}$$

is called the *expectation* of X. If $\mathbf{K} = \mathbb{R}$, $E[X]$ is expressed as a function of the probability law ν_X of X according to

$$E[X] = \int_{\mathbb{R}^n} x d\nu_X. \qquad (12.131)$$

18. We can always come back to the case where $\mathbf{K} = \mathbb{R}$ by identifying $\mathbb{C}$ with $\mathbb{R}^2$, thus $\mathbb{C}^n$ with $\mathbb{R}^{2n}$, equipped with its Borel σ-algebra.

19. This is similar to what was done in section 12.2.2. This time, however, the measure considered is the probability P, instead of the Lebesgue measure.

Random variables of the second order

The random variable X is said to be *of the second order* if $\|X\|^2$ is a random variable of the first order. The set of random variables of the second order with values in $\mathbf{K}^n$ is denoted as $L^2(\Omega, \mathcal{F}, P; \mathbf{K}^n)$, or $L^2(\Omega, \mathcal{F}, P)$ if $n = 1$ and $\mathbf{K} = \mathbb{R}$; this is a Hilbert space equipped with the scalar product $\langle X, Y \rangle = E[X^* Y]$ (where $(.)^*$ designates the conjugate-transpose: see section 13.5.3); the associated Hilbert norm is denoted as $\|X\|_2$. We have

$$L^2(\Omega, \mathcal{F}, P; \mathbf{K}^n) \subset L^1(\Omega, \mathcal{F}, P; \mathbf{K}^n).$$

If $n = 1$, $|E[X]|^2 \leq E\left[|X|^2\right]$; the quantity $\sigma_X^2 = E\left[|X|^2\right] - |E[X]|^2$ is the *variance* of X and σ_X is its *standard deviation*; if $n \geq 1$, the matrix

$$Q_X = E\left[(X - E[X])(X - E[X])^*\right]$$

is called the *covariance matrix* of X.

Let X and Y be two real random variables of the second order. Then XY is a random variable of the first order and according to (12.131)

$$E[XY] = \int_{\mathbb{R}^2} xy d\nu_{(X,Y)}(x, y).$$

PROPOSITION 470.– *(i) If the random variables X and Y are independent, then $E[XY] = E[X]E[Y]$. (ii) If in addition one of them is centered, then $E\left[|X+Y|^2\right] = E\left[|X|^2\right] + E\left[|Y|^2\right]$.*

PROOF. (i) According to Proposition 469, $d\nu_{(X,Y)}(x, y) = d\nu_X(x)\, d\nu_Y(y)$, thus

$$E[XY] = \int_{\mathbb{R}} x d\nu_X(x) \int_{\mathbb{R}} d\nu_Y(y) = E[X]\, E[Y].$$

We have

$$E\left[|X+Y|^2\right] = E\left[|X|^2\right] + E\left[|Y|^2\right] + 2E[XY];$$

as a result, (ii) is an immediate consequence of (i). ∎

Distribution function and probability density

Let $X = (X_1, ..., X_n)$ be a random variable with values in the probabilizable space $(\mathbb{R}^n, \mathcal{B}_n)$ where $\mathcal{B}_n$ is the Borel σ algebra of $\mathbb{R}^n$. The *distribution function* of X, denoted as F_X, is defined on $\mathbb{R}^n$ by

$$F_X(x_1, ..., x_n) = P\{X_1 < x_1, ..., X_n < x_n\}.$$

The distribution function of X determines its probability law.

Suppose that the probability law of X admits a density δ_X, called the *probability density* of X, that is to say that the measure ν_X is *absolutely continuous* with respect to the Lebesgue measure dx, i.e.

$$d\nu_X = \delta_X \, dx \tag{12.132}$$

(Remark 467, section 12.7.1). We then have

$$F_X(x_1, ..., x_n) = \int_{\mathcal{X}(x)} \delta_X(\xi) \, d\xi \tag{12.133}$$

where $x = (x_1, .., x_n)$, $\mathcal{X}(x) = \prod_{1 \leq i \leq n} (-\infty, x_i)$ and $\xi = (\xi_1, ..., \xi_n)$; on the other hand, if X is of the first order, we then have, according to (12.131), (12.132),

$$\boxed{E[X] = \int_{\mathbb{R}^n} x \delta_X(x) \, dx.} \tag{12.134}$$

The following result is a direct consequence of Proposition 469 and of (12.133) :

PROPOSITION 471.– *Let X and Y be two independent variables with values in $\mathbb{R}^n$ and $\mathbb{R}^m$ respectively. (i) They are independent if and only if the distribution function $F_{(X,Y)}$ of the pair (X, Y) satisfies, for any $(x, y) \in \mathbb{R}^n \times \mathbb{R}^m$, the equality*

$$F_{(X,Y)}(x, y) = F_X(x) F_Y(y).$$

(ii) If (X, Y) admits a probability density $\delta_{(X,Y)}$, then X and Y admit probability densities δ_X and δ_Y respectively, the former given by

$$\delta_X(x) = \int_{\mathbb{R}^m} \delta_{(X,Y)}(x, y) \, dy$$

and the latter by a similar expression. In addition, X and Y are independent if and only if for any $(x, y) \in \mathbb{R}^n \times \mathbb{R}^m$,

$$\delta_{(X,Y)}(x, y) = \delta_X(x) \delta_Y(y).$$

Gaussian random variable

The above random variable is said to be *Gaussian* if its probability density is Gaussian. That is,

$$\boxed{\delta_X(x) = \frac{1}{\sqrt{(2\pi)^n \det Q_X}} \exp\left\{-(1/2)(x - \bar{x})^T Q_X^{-1} (x - \bar{x})\right\}}$$

where $\bar{x} = E[X]$ and Q_X is the covariance matrix (assumed invertible) of X. If $\bar{x} = 0$ and $Q_X = I_n$, then the corresponding probability law is called the *reduced Gaussian (or normal) law*.

Gaussian random variables are particularly important due to the *central limit theorem*:

THEOREM 472.– *Let* (X_n) *be a sequence of independent real random variables of the second order, with expectation* $\bar{x}$*, standard deviation* σ*, and the same probability law. Then, as* $n \to +\infty$*, the random variable*

$$Y_n = \frac{1}{\sigma\sqrt{n}} \sum_{k=1}^{n} (X_k - n\bar{x})$$

converges in law *towards the reduced normal law (i.e. the probability law of* Y_n *weakly* converges to the reduced normal law – see [83] for more details). If in addition each* X_n *has a probability density, then the probability density of* Y_n *converges* uniformly *to that of the reduced normal law.*

12.7.3. *Conditional expectation*

Case of random variables of the second order

The conditional expectation can be defined for random variables of the first order by way of a slightly difficult approach and is not necessary for this book. We thus will only consider the case of random variables of the second order.

Let $(\Omega, \mathcal{F}, P)$ be a probability space, let $X \in L^2(\Omega, \mathcal{F}, P; \mathbf{K}^n)$ and let $\mathcal{G}$ be a sub-σ-algebra of $\mathcal{F}$.

LEMMA 473.– *(i) There exists a unique random variable* $\hat{X} \in L^2(\Omega, \mathcal{G}, P; \mathbf{K}^n)$ *such that*

$$\left\| X - \hat{X} \right\|_2 = \min_{Y \in L^2(\Omega, \mathcal{G}, P; \mathbf{K}^n)} \| X - Y \|_2 .$$

(ii) This random variable is characterized by the relation $E[Y^* X] = E\left[Y^* \hat{X}\right]$, $\forall Y \in L^2(\Omega, \mathcal{G}, P; \mathbf{K}^n)$. *(iii) The mapping* $X \mapsto \hat{X}$ *is* $\mathbf{K}$*-linear from* $L^2(\Omega, \mathcal{F}, P; \mathbf{K}^n)$ *onto* $L^2(\Omega, \mathcal{G}, P; \mathbf{K}^n)$ *and has a norm* 1.

PROOF. As already seen, $L^2(\Omega, \mathcal{F}, P; \mathbf{K}^n)$ and $L^2(\Omega, \mathcal{G}, P; \mathbf{K}^n)$ are Hilbert spaces, and $L^2(\Omega, \mathcal{G}, P; \mathbf{K}^n) \subset L^2(\Omega, \mathcal{F}, P; \mathbf{K}^n)$. It suffices therefore to apply the orthogonal projection theorem (Theorem 429, section 12.1.2). ■

DEFINITION 474.– *(i) The above random variable* $\hat{X}$ *is called the* conditional expectation *of* X *with respect to the* σ*-algebra* $\mathcal{G}$ *and is denoted as* $E[X \mid \mathcal{G}]$. *(ii) Let* Y *be a random variable with values in the probabilizable space* $(\mathcal{Y}, \mathcal{T})$ *and consider the* σ*-algebra* $\mathcal{G} = Y^{-1}(\mathcal{T})$ *generated by* Y; $E[X \mid \mathcal{G}]$ *is also called the* conditional expectation *of* X *with respect to the random variable* Y *and is denoted as* $E[X \mid Y]$. *(iii) By giving to* Y *a value* $y \in \mathcal{Y}$, $E[X \mid Y]$ *becomes the element of* $\mathbf{K}^n$ *denoted as* $E[X \mid Y = y]$.

Let us now establish several properties of the conditional expectation.

THEOREM 475.– *(i) The mapping* $X \mapsto E[X \mid \mathcal{G}]$ *is* $\mathbf{K}$*-linear. (ii) If* X *is* $\mathcal{G}$*-measurable, then* $E[X \mid \mathcal{G}] = X$. *(iii)* $E[E[X \mid \mathcal{G}]] = E[X]$. *(iv) Given two* σ*-algebras* $\mathcal{H}$ *and* $\mathcal{G}$ *such that* $\mathcal{H} \subset \mathcal{G} \subset \mathcal{F}$, *we have*

$$E[E[X \mid \mathcal{G}] \mid \mathcal{H}] = E[X \mid \mathcal{H}] = E[E[X \mid \mathcal{H}] \mid \mathcal{G}].$$

PROOF. (i) is a reformulation of Lemma 473(iii). (ii) If X is $\mathcal{G}$-measurable, then $X \in L^2(\Omega, \mathcal{G}, P; \mathbf{K}^n)$, thus $E[X \mid \mathcal{G}] = X$. (iii) Let $Y = e_j$, the jth vector of the canonical basis of $\mathbf{K}^n$. It is clear that $Y \in L^2(\Omega, \mathcal{G}, P; \mathbf{K}^n)$, thus according to Lemma 473(ii), $E[X_j] = E\left[\hat{X}_j\right]$. Proceeding this way for any $j \in \{1, ..., n\}$, we obtain $E[X] = E\left[\hat{X}\right]$. (iv): We can obtain the orthogonal projection $E[X \mid \mathcal{H}]$ of X onto $L^2(\Omega, \mathcal{H}, P; \mathbf{K}^n)$ by first taking its orthogonal projection $E[X \mid \mathcal{G}]$ onto $L^2(\Omega, \mathcal{G}, P; \mathbf{K}^n)$, and then the orthogonal projection of $E[X \mid \mathcal{G}]$ onto $L^2(\Omega, \mathcal{H}, P; \mathbf{K}^n) \subset L^2(\Omega, \mathcal{G}, P; \mathbf{K}^n)$, which proves the first equality. The second immediately follows from (ii) since $E[X \mid \mathcal{H}]$ is $\mathcal{G}$-measurable. ■

Suppose now that X and Y are two random variables of the second order, assumed to be real for the sake of simplicity, and let $\mathcal{G}$ be a sub-σ-algebra of $\mathcal{F}$.

PROPOSITION 476.– *(i) If* Y *is* $\mathcal{G}$*-measurable and essentially bounded (with respect to the probability measure* P*), then* $E[XY \mid \mathcal{G}] = Y E[X \mid \mathcal{G}]$. *(ii) If* X *is independent of* $\mathcal{G}$, *then* $E[X \mid \mathcal{G}] = E[X]$.

PROOF. (i) We have $Y E[X \mid \mathcal{G}] \in L^2(\Omega, \mathcal{G}, P)$ and for any $Z \in L^2(\Omega, \mathcal{G}, P)$, $E[Y E[X \mid \mathcal{G}] Z] = E[XYZ]$ according to Lemma 473(ii), which proves the result. (ii) X is independent of any $Y \in L^2(\Omega, \mathcal{G}, P)$, thus $E[XY] = E[X] E[Y] = E[E[X] Y]$, and the result follows from Lemma 473(ii). ■

Case where a probability density exists

Suppose now that $\mathbf{K} = \mathbb{R}$, and that (X, Y) is a random variable with values in $\mathbb{R}^n \times \mathbb{R}^m$, which admits a probability density $(x, y) \mapsto \delta_{(X,Y)}(x, y)$ (section 12.7.2).

We call the function defined by

$$\delta_X^{Y=y}(x) = \frac{\delta_{(X,Y)}(x,y)}{\delta_Y(y)}$$

the *conditional density* of X with respect to $Y = y$; according to Proposition 471(ii),

$$\delta_Y(y) = \int_{\mathbb{R}^n} \delta_{(X,Y)}(x,y)\,dx.$$

According to the same proposition, the random variables X and Y are thus independent if and only if $\delta_X^{Y=y}(x) = \delta_X(x)$ for any $(x,y) \in \mathbb{R}^n \times \mathbb{R}^m$. In the general case, this is of course not true; instead,

$$\boxed{E[X \mid Y = y] = \int_{\mathbb{R}^n} x\delta_X^{Y=y}(x)\,dx}$$

which is similar to (12.134).

Chapter 13

Appendix 2: Algebra

13.1. Commutative rings and fields

13.1.1. *Generalities*

The notion of ring

A *ring* $\mathbf{R}$ is a set equipped with an addition $+$, such that $(\mathbf{R}, +)$ is an additive group, and with a multiplication $\times$ which is distributive relative to the addition and satisfies the usual properties; we suppose that there exists a *unit element*, denoted by 1, such that $1a = a1 = a$ for any $a \in \mathbf{R}$. Recall that any element a has an opposite $-a$ (such that $a + (-a) = 0$) since $\mathbf{R}$ is an additive group. The invertible elements of $\mathbf{R}$ (i.e. the elements u for which there exists an inverse denoted by u^{-1}, such that $u\,u^{-1} = u^{-1}\,u = 1$) are called the *units* of that ring. Two elements a and b are said to be *associates* (or *associated*) if there exist units v and v' such that $a = vbv'$. For example, in the ring $\mathbb{Z}$ of rational integers, where the units are -1 and 1, n and $-n$ are associates. Let us review a few examples of rings and the notions that go with them.

Rings of matrices

We denote the set of square matrices of order n with real entries as $\mathbb{R}^{n \times n}$; this is a ring. Its units are the matrices whose determinant is non-zero (its unit element is the identity matrix). Note that for two matrices A and B, we have in general $A\,B \neq B\,A$; the multiplication of matrices is thus non-commutative. On the other hand, the product AB can be zero while neither A nor B is zero; A and B are then called *zero divisors* (*A on the left, B on the right*).

DEFINITION 477.– *A ring is said to be* commutative *when its multiplication is commutative; a ring that has no zero divisors is said to be* integral, *or to be a* domain. *Therefore, a commutative integral ring is called a* commutative domain.

From the above, it follows that $\mathbb{R}^{n\times n}$ is a non-commutative and non-integral ring.

In what follows, all rings are commutative domains, except when otherwise stated.

Polynomial rings

A typical example of ring is the set $\mathbf{K}[s]$ of all polynomials in the indeterminate s with coefficients belonging to the field $\mathbf{K}$ ($\mathbf{K}=\mathbb{R}$ or $\mathbb{C}$). The units are the polynomials of degree zero, i.e. the constant non-zero polynomials (by convention, the degree of the zero polynomial is $-\infty$). If $g(s)$ and $h(s)$ are two polynomials, we have the equality $g(s)\,h(s) = h(s)\,g(s)$: the ring $\mathbf{K}[s]$ is thus commutative. In addition, if $g(s)\,h(s) = 0$, we must have $g(s) = 0$ or $h(s) = 0$, so that $\mathbf{K}[s]$ is integral. A polynomial is said to be *monic* if the coefficient of its highest degree term is 1. It is clear that two monic polynomials which are associated are necessarily equal.

A polynomial $p(s) \in \mathbf{K}[s]$ of degree $n \geq 0$ has n roots in $\mathbb{C}$. For example, the polynomial

$$p(s) = \left(s^2+1\right)(s-r)^k \qquad (k \geq 1)$$

has $k+2$ roots in $\mathbb{C}$, forming the set $\{i, -i, r, ..., r\}$, where r is repeated k times. The set of the *distinct roots* of this polynomial is $\{i, -i, r\}$, but the root r has an *order of multiplicity* (or, for short, a *multiplicity*) equal to k.

Rings of formal power series

Take a third example: the set $\mathbf{K}[[s]]$ of *formal power series* in s with coefficients belonging to $\mathbf{K}$, which is composed of all elements of the form

$$a(s) = \sum_{n\geq 0} a_n\, s^n . \tag{13.1}$$

This power series is *formal* in the sense that we do not concern ourselves with the question of convergence: this "formal power series" is nothing but the sequence of coefficients (a_n), except that we define on formal series an addition and a multiplication in the following manner:

$$\sum_{n\geq 0} a_n\, s^n + \sum_{n\geq 0} b_n\, s^n = \sum_{n\geq 0} (a_n + b_n)\, s^n ,$$

$$\sum_{n\geq 0} a_n\, s^n . \sum_{n\geq 0} b_n\, s^n = \sum_{n\geq 0} c_n\, s^n$$

where the sequence (c_n) is the *convolution product* of (a_n) and (b_n) (see section 12.2.1); since the two last sequences are only defined in the set of natural numbers $\mathbb{N}$ (but can be extended to the set of rational integers $\mathbb{Z}$ by positively supported sequences), the sequence (c_n) is well-defined.

The ring $\mathbf{K}[[s]]$ is commutative and one can show that it is integral [25]. A unit of $\mathbf{K}[[s]]$ is a formal power series of the form (13.1) such that $a_0 \neq 0$. Its inverse is thus of the form $\sum_{n \geq 0} b_n s^n$ with $b_0 = \frac{1}{a_0}$.

Rings of holomorphic functions

Let Ω be a non-empty open subset of $\mathbb{C}$. The set $\mathcal{O}(\Omega)$ of all *holomorphic functions* in Ω (section 12.4.1) is a commutative domain.

Fields, field of fractions

Let $\mathbf{R}$ be a ring. If non-zero elements of $\mathbf{R}$ are invertible, $\mathbf{R}$ is called a division ring. A *field* is a commutative division ring.

Consider a commutative domain $\mathbf{R}$. The set of all elements of the form $\frac{b}{a}$, where $b \in \mathbf{R}$ and a belongs to the complement of $\{0\}$ in $\mathbf{R}$ (denoted by $\mathbf{R} \setminus \{0\}$ or by $\mathbf{R}^{\times}$) is a field denoted as $\mathrm{Q}(\mathbf{R})$, and called the *field of fractions* of $\mathbf{R}$.

The field of fractions of $\mathbb{Z}$ is $\mathbb{Q}$, i.e. the field of rational numbers; $\frac{3}{2}, -\frac{5}{3}$, etc., are elements of $\mathbb{Q}$. The field of fractions of $\mathbf{K}[s]$ is the field $\mathbf{K}(s)$ of rational functions in s with coefficients in the field $\mathbf{K}$ (for more details about rational functions, see section 13.6.1). The field of fractions of $\mathbf{K}[[s]]$ is the field $\mathbf{K}((s))$ of *Laurent series* with coefficients in the field $\mathbf{K}$, and the elements of which are of the form $\sum_{n \geq \nu} a_n s^n$, where $\nu \in \mathbb{Z}$ and $a_n \in \mathbf{K}$.

The field of fractions of $\mathcal{O}(\Omega)$ (see above) is the field $\mathcal{M}(\Omega)$ of all *meromorphic functions* in Ω (section 12.4.1). A polynomial can be considered as a special meromorphic function in $\mathbb{C}$. Then, a root of this polynomial is nothing but a zero of this meromorphic function. Likewise, a rational function can be considered as a special meromorphic functions in $\mathbb{C}$.

** Differential field*

A *differential field* is a field $\mathbf{K}$ equipped with a derivation $\delta : a \rightarrow a^{\delta}$ satisfying Leibniz's rule: $(a\,b)^{\delta} = ab^{\delta} + a^{\delta}b$ [31]. For example, the field $\mathbb{C}(t)$ of rational functions in the variable t and with complex coefficients, equipped with the usual derivative with respect to t, is a differential field. A *constant* of $\mathbf{K}$ is an element $a \in \mathbf{K}$ whose derivative a^{δ} is zero. The set of constants of $\mathbf{K}$ is a field $\mathbf{k} \subset \mathbf{K}$, called the *field of constants* of $\mathbf{K}$. The derivative a^{δ} is often denoted as $\dot{a}$.

Vector space

Let $\mathbf{K}$ be a field[1], called the field of scalars (for example, $\mathbb{R}$ or $\mathbb{C}$). A *vector space* over $\mathbf{K}$ (also called a $\mathbf{K}$-vector space) is a set E equipped with an addition from $E \times E$

1. Or a division ring.

into E satisfying the usual properties (in such a way that E is an additive group) as well as with a multiplication from $\mathbf{K} \times E$ into E. We can add or subtract two vectors, that is two elements of E, and we can multiply a vector by a scalar.

The most classic example of $\mathbf{K}$-vector space is $\mathbf{K}^n$, the set of all elements of the form $(x_1, ..., x_n)$, $x_i \in \mathbf{K}$, $i \in \{1, ..., n\}$. This vector space is detailed in section 13.3.1.

The sets $\mathbf{K}[s]$, $\mathbf{K}[[s]]$, $\mathbf{K}(s)$ and $\mathbf{K}((s))$ are also $\mathbf{K}$-vector spaces (of infinite dimension, contrary to $\mathbf{K}^n$).

Algebra

Let $\mathbf{K}$ be a field and let E be a set satisfying the following conditions: (i) E is a *ring* (resp., a *commutative* ring); (ii) E is $\mathbf{K}$-vector space; (iii) the multiplication of the ring E is compatible with the multiplication from $\mathbf{K} \times E$ into E, in the sense where

$$\lambda (ab) = (\lambda a) b = a (\lambda b)$$

for any $a \in E, b \in E, \lambda \in \mathbf{K}$.

Such a set E is called an *unitary* $\mathbf{K}$-algebra (resp., a *commutative unitary* $\mathbf{K}$-algebra). The adjective "unitary" means that there exists a unit element in this algebra[2].

For example, the sets $\mathbf{K}[s]$, $\mathbf{K}[[s]]$, $\mathbf{K}(s)$ and $\mathbf{K}((s))$ are unitary commutative $\mathbf{K}$-algebras. The set $\mathbf{K}^{n \times n}$ is a *non-commutative* unitary $\mathbf{K}$-algebra.

* This definition of a $\mathbf{K}$-algebra extends to the case where $\mathbf{K}$ is a *commutative* ring, replacing condition (ii) by (ii'): E is a $\mathbf{K}$-module. *

13.1.2. *Divisibility*

Ideal

An ideal in a (commutative) ring $\mathbf{R}$ is a subgroup $\mathfrak{I}$ of the additive group $\mathbf{R}$ such that for any $r \in \mathbf{R}$ and any $a \in \mathfrak{I}$, $ra \in \mathfrak{I}$. Such an ideal $\mathfrak{I}$ is said to be *generated* by a set $\mathcal{J} \subset \mathbf{R}$ if any element of $\mathfrak{I}$ is of the form $\sum_{\lambda} r_\lambda a_\lambda$, $a_\lambda \in \mathcal{J}$, where (r_λ) is a finitely supported sequence of elements of $\mathbf{R}$. We say that $\mathcal{J}$ is a *generating set* of $\mathfrak{I}$, and this ideal is said to be *finitely generated* if $\mathcal{J}$ is finite. A *principal ideal* in $\mathbf{R}$ is an ideal generated by a unique element a and is denoted as (a) or $\mathbf{R}a$. It is clear that $(1) = \mathbf{R}$.

---

2. One can define non-unitary algebras, which do not admit a unit element (and thus are not rings). One example of a non-unitary algebra is given in section 12.2.2 (algebra $\mathcal{K}_+$).

Multiple and divisor

A *multiple* of an element a of $\mathbf{R}$ is an element of the form ba, $b \in \mathbf{R}$.

Let $a \neq 0$ and b be two elements of $\mathbf{R}$; we say that a is a *divisor* of b, which we denote as $a \mid b$, if b is a multiple of a, which is equivalent to $(b) \subset (a)$. Then, there exists a unique element c such that $b = ca$, and that we write $c = \frac{b}{a}$. For example, in the ring $\mathbb{Z}$, $3 = \frac{6}{2}$.

lcm

Let a and b be two non-zero elements of a ring $\mathbf{R}$. The set of multiples of a is the ideal (a). The set of all common multiples of a and b is thus $(a) \cap (b)$. Suppose this ideal is principal, i.e. of the form (m). This element m is a *least common multiple* (lcm) of a and b, since every common multiple of a and b is a multiple of m. Any other lcm of a and b is associated with m. We denote $\operatorname{lcm}(a, b)$ the set of all lcm's of a and b.

This extends to any number of non-zero elements $a_i, i \in \{1, ..., n\}$: an lcm of these elements (if it exists) is an element m such that $(m) = \cap_{i=1}^{n} (a_i)$.

If $\mathbf{R} = \mathbf{K}[s]$, $\operatorname{lcm}(a, b)$ designates the unique lcm of a and b which is a monic polynomial.

gcd and GCD domain

Let a and b be two non-zero elements of a ring $\mathbf{R}$. An element d is called a *greatest common divisor* (gcd) of a and b if d is a divisor of a and b and if any divisor of a and b divides d. Every other gcd of a and b is associated with d; the set consisting of all these elements is denoted as $\gcd(a, b)$.

If $\mathbf{R} = \mathbf{K}[s]$, $\gcd(a, b)$ designates the unique gcd of a and b which is a monic polynomial.

One can show the following ([31], section 3.1, Exercises 7 & 8):

LEMMA 478.– *(i) If non-zero elements a and b of a ring $\mathbf{R}$ admit an* lcm *m, they admit a* gcd *d such that $ab = dm$ (the existence of a* gcd *does not in general imply that of an* lcm*). (ii) In a ring $\mathbf{R}$, all pairs of non-zero elements admit a* lcm *if and only if all pairs of non-zero elements admit a* gcd.

The above lemma leads to the following:

DEFINITION 479.– *A* GCD domain *is a (commutative) domain $\mathbf{R}$ such that any two non-zero elements of this ring admit a* gcd.

Coprimeness

Two non-zero elements a and b of a ring $\mathbf{R}$ are said to be *coprime* (or *relatively prime*) if they have no common divisors except units, in other words if one of their gcd's is 1.

Atoms and primes

An *atom* in a ring $\mathbf{R}$ is a non-zero element which is only divisible by the units of $\mathbf{R}$ and by itself. A *prime element* in $\mathbf{R}$ (or a *prime*, for short) is a non-zero element p which is a nonunit and cannot divide the product of two elements of $\mathbf{R}$ without dividing one of them (* in other words, it is an element p such that the principal ideal (p) is prime *). Every prime in $\mathbf{R}$ is an atom and the converse holds true if $\mathbf{R}$ is a GCD domain ([10], section VI.13, Remark after Proposition 14).

In the ring $\mathbb{Z}$, the primes are the prime numbers $2, 3, 5, 7, 11$, etc. and their associates $-2, -3, -5, -7, -11$, etc. In $\mathbb{C}[s]$, the primes are the polynomials of the first degree. In $\mathbb{R}[s]$, the primes are the polynomials of the first degree and those polynomials of the second degree which do not have real roots.

A *representative system of primes* in a domain $\mathbf{R}$ is a family (p_i) of primes in $\mathbf{R}$ such that every prime in $\mathbf{R}$ (if any) is associated with one and only one p_i. If $\mathbf{R} = \mathbf{K}[s]$ ($\mathbf{K} = \mathbb{R}$ or $\mathbb{C}$), the chosen representative system always consists of monic polynomials.

Unique factorization into primes, UFD's

Let $\mathbf{R}$ be a ring, let (p_i) be a representative system of primes in $\mathbf{R}$ and let $a \in \mathbf{R}^{\times}$. If a can be uniquely written in the form

$$a = v \prod_i p_i^{\alpha_i} \tag{13.2}$$

where v is a unit of $\mathbf{R}$ and the non-negative integers α_i's are zero except for a finite number of them, we say that a admits *unique factorization into primes* given by (13.2); the quantity $l(a) = \sum_i \alpha_i$ is then called the *length* of a.

DEFINITION 480.– *A commutative domain* $\mathbf{R}$ *is called a* unique factorization domain *(UFD) if every non-zero element of* $\mathbf{R}$ *admits unique factorization into primes.*

One can prove the following ([91], section X.1): if the ring $\mathbf{R}$ is a UFD, then $\mathbf{R}[X]$ (where X designates an indeterminate) is again a UFD. We deduce by induction that $\mathbf{K}[X_1, ..., X_n]$ is a UFD.

PROPOSITION 481.– *A UFD is a GCD domain.*

PROOF. In a UFD, let a be given by (13.2) and let $b = v' \prod_i p_i^{\beta_i}$; then

$$d = \prod_i p_i^{\min(\alpha_i, \beta_i)}, \quad m = \prod_i p_i^{\max(\alpha_i, \beta_i)}$$

are a gcd and an lcm of a and b respectively. ■

Bézout domains

DEFINITION 482.– *A (commutative)* Bézout domain *is a commutative domain in which any finitely generated ideal is principal.*

Suppose $\mathbf{R}$ is a Bézout domain and let a, b be two non-zero elements of $\mathbf{R}$; $(a)+(b)$ is the smallest ideal containing (a) and (b). This ideal is generated by a and b, thus it is principal, and there exists d such that

$$(a) + (b) = (d).$$

It is clear that d is a gcd of a and b. We have thus proven the following:

PROPOSITION 483.– *A Bézout domain is a GCD domain.*

Let $a_1, ..., a_n$ be non-zero elements of a Bézout domain $\mathbf{R}$; an element d is a gcd of these elements if and only if $(d) = \sum_{1 \leq i \leq n} (a_i)$.

The following is proved in ([103] section 15.15):

PROPOSITION 484.– *Let Ω be an open connected subset of $\mathbb{C}$. The ring $\mathcal{O}(\Omega)$ is a Bézout domain. In particular, the ring $\mathcal{O}(\mathbb{C})$ of all entire functions (section 12.4.1) is a Bézout domain.*

13.1.3. *Principal ideal domains*

DEFINITION 485.– *(i) A commutative domain is called a (commutative)* principal ideal domain *if every ideal in this ring is principal.*
(ii) A Noetherian ring *is a ring in which every ideal is finitely generated.*

THEOREM 486.– *The following conditions are equivalent: (i)* $\mathbf{R}$ *is a principal ideal domain; (ii)* $\mathbf{R}$ *is a Noetherian ring which is a Bézout domain; (iii)* $\mathbf{R}$ *is both a UFD and a Bézout domain.*

PROOF. The equivalence between (i) and (ii) is obvious. For the equivalence between (i) and (iii), see ([11], section VII.2, Exercise 17 and section VII.3, Exercise 10). ■

A field $\mathbf{F}$ is a trivial example of a principal ideal domain, for its only ideals are $\{0\} = (0)$ and $\mathbf{F} = (1)$.

Euclidean domain

DEFINITION 487.– *(A) An Euclidean (commutative) domain is a commutative domain* $\mathbf{R}$ *in which we have defined an* Euclidean function. *That is to say, a mapping* θ *from* $\mathbf{R}$ *into* $\mathbb{N} \cup \{-\infty\}$ *(where* $\mathbb{N}$ *is the set of non-negative integers) such that: (i)* $\theta(0) = -\infty$*; (ii) for any elements* a *and* b *of* $\mathbf{R}^{\times}$, $\theta(ab) \geq \theta(a)$; *(iii) for any* a *in* $\mathbf{R}$ *and* b *in* $\mathbf{R}^{\times}$*, there exist elements* q *and* r *of* $\mathbf{R}$ *such that* $a = bq + r$ *and* $\theta(r) < \theta(b)$ *(*Euclidean division *of* a *by* b : q *is the quotient and* r *the remainder). (B) A commutative domain is said to be* strongly Euclidean *if condition (ii) is replaced by the following stronger condition (ii'): for any elements* a *and* b *of* $\mathbf{R}$, $\theta(a-b) \leq \max\{\theta(a), \theta(b)\}$, *and for any element* a *and* b *of* $\mathbf{R}^{\times}$, $\theta(ab) = \theta(a) + \theta(b)$*; in this case, the Euclidean function* θ *is called a* degree.

REMARK 488.– *(i) The quotient and remainder of the Euclidean division in the commutative domain* $\mathbf{R}$ *are unique if and only if the ring is* strongly Euclidean. *(ii) An element* v *of a Euclidean ring is a unit if and only if* $\theta(v) = \theta(1)$; *if* $a \neq 0$ *is not a unit,* $\theta(a) > \theta(1)$. *(iii) The ring* $\mathbb{Z}$ *of rational integers is Euclidean (with* $\theta(n) = |n|$ *if* $n \neq 0$ *) but not strongly Euclidean. The ring* $\mathbf{K}[s]$ *is the most typical example of a strongly Euclidean domain (*θ *is then the* degree *of a polynomial in the usual sense:* $\theta = d^{\circ}$*).*

THEOREM 489.– *An Euclidean domain is a principal ideal domain.*

PROOF. Let $\mathfrak{I}$ be a non-zero ideal in a Euclidean domain $\mathbf{R}$. Among the non-zero elements of $\mathfrak{I}$, consider one of the elements b of minimum degree. Let x be any element of $\mathfrak{I}$ and perform the Euclidean division of x by b; we have $x = bq + r$ with $\theta(r) < \theta(b)$, thus $\theta(r) = -\infty$, in other words $r = 0$ and $x \in (b)$. We thus have $\mathfrak{I} = (b)$. ∎

Non-Euclidean principal ideal domains

There exist principal ideal domains that are not Euclidean domains. A typical example is $\mathbf{K}[[s]]$. We cannot define the degree of a formal power series. On the other hand, we can define its *order*. The order $\omega(a)$ of the formal power series (13.1), assumed to be non-zero, is the smallest integer k such that $a_k \neq 0$. We put $\omega(0) = +\infty$. A unit of $\mathbf{K}[[s]]$ is a formal power series of zero order. A formal power series $a(s)$ of order k is of the form $s^k v(s)$, where $v(s)$ is a unit in $\mathbf{K}[[s]]$. It is clear that $a(s) \mid b(s)$ if and only if $\omega(a) \leq \omega(b)$. All ideals in $\mathbf{K}[[s]]$ are of the form (s^k) and it follows that $\mathbf{K}[[s]]$ is a principal ideal domain. Furthermore, $\mathbf{K}[[s]]$ admits (among the *proper ideals*, i.e. those that are different from $\mathbf{K}[[s]]$) a unique *maximal ideal* which is (s); a ring that admits a unique maximal ideal is said to be *local*. The ring $\mathbf{K}[[s]]$ is thus *a principal ideal domain which is local.*

13.1.4. *Matrices over commutative rings*

Matrix calculations

The set of matrices having p rows and m columns (said to be of size $p \times m$) with entries in a commutative ring $\mathbf{R}$, is denoted as $\mathbf{R}^{p \times m}$. A square matrix of size $n \times n$ is said to be of order n.

The basic operations on matrices (addition and multiplication of two matrices, multiplication of a matrix by a scalar) are assumed to be known. They are defined in the same manner over a commutative ring $\mathbf{R}$ as over $\mathbb{R}$ or $\mathbb{C}$. The identity matrix of order n is denoted as I_n. The element of the ith row and jth column of a matrix A is denoted as a_{ij}. Let $A \in \mathbf{R}^{p \times m}$; the matrix $A^T \in \mathbf{R}^{m \times p}$, the element of the ith row and the jth column of which is a_{ji}, is called the *transpose* of A.

Diagonal sum of two matrices

Let $A_1 \in \mathbf{R}^{p_1 \times m_1}$ and $A_2 \in \mathbf{R}^{p_2 \times m_2}$. The matrix

$$A = \begin{bmatrix} A_1 & 0 \\ 0 & A_2 \end{bmatrix},$$

where the zeros designate blocks of zeros of appropriate sizes, is denoted by $A = A_1 \oplus A_2$ and is called the *diagonal sum* of A_1 and A_2. We can define $A_1 \oplus A_2 \oplus A_3 = (A_1 \oplus A_2) \oplus A_3$, and so on.

Determinant

Let us go over a few properties of determinants again:

On $\mathbf{R}^{n \times n}$, the determinant is the unique n-linear alternating form with respect to the columns of the matrices of this set and such that $\det I_n = 1$; this remains true if we replace columns by rows.

As a consequence, if we denote by $A_{\bullet j}$, $1 \leq j \leq n$, the columns of the matrix A, $\det A = 0$ if and only if the columns of A are $\mathbf{R}$-linearly dependent, in other words if there exist elements $c_j \in \mathbf{R}$, $1 \leq j \leq n$, not all zeros, such that $\sum_{j=1}^{n} c_j A_{\bullet j} = 0$; this is also true for the rows $A_{i \bullet}$.

Another consequence (obtained by exchanging rows and columns) is that $\det A = \det A^T$.

If A and B both belong to $\mathbf{R}^{n \times n}$, $\det(A\,B) = \det(B\,A) = \det A \det B$.

Consider the matrix of order n

$$A = \begin{bmatrix} A_1 & 0 \\ A_3 & A_2 \end{bmatrix}, \tag{13.3}$$

where $A_1 \in \mathbf{R}^{n_1 \times n_1}$, $A_2 \in \mathbf{R}^{n_2 \times n_2}$, $A_3 \in \mathbf{R}^{n_2 \times n_1}$ with $n_1 + n_2 = n$. Then $\det A = \det A_1 \det A_2$. In particular, $\det (A_1 \oplus A_2) = \det A_1 \det A_2$.

Finally, if $\mathbf{R}$ is a field and

$$A = \begin{bmatrix} X & Y \\ Z & T \end{bmatrix}, \tag{13.4}$$

where X and T are square matrices and X is non-singular,

$$\boxed{\det A = \det X \det \left(T - Z X^{-1} Y\right)}. \tag{13.5}$$

Minors and cofactors

Let $A \in \mathbf{R}^{p \times m}$. A *minor* of order n of A (where $n \leq \min\{p, m\}$) is the determinant of a square submatrix of A of order n (obtained by suppressing rows and columns of A). A *principal minor* is a minor formed by suppressing rows and columns with same indices.

Now let $A \in \mathbf{R}^{n \times n}$; denote by m_{ij} the minor of order $n - 1$ obtained by suppressing the ith row and jth column of A. The *cofactor* of a_{ij} is $\alpha_{ij} = (-1)^{i+j} m_{ij}$. The matrix $\boldsymbol{\alpha}$ with entries α_{ij}, $1 \leq i \leq n$, $1 \leq j \leq n$, is called the *cofactor matrix* of A.

We have

$$\boxed{\det A = \sum_{j=1}^{n} a_{ij} \alpha_{ij} = \sum_{i=1}^{n} a_{ij} \alpha_{ij}}$$

(developments of the determinant along the ith row, and then the jth column, which are particular cases of the *Laplace expansion* of A : see [10], section III.8.6). The above equalities make it possible to calculate the determinant of a square matrix of order n from the determinants of square matrices of order $n - 1$ and thus, by repeating this procedure, to calculate the determinant of any square matrix, because the determinant of a square matrix of order 1 (i.e. of an element of $\mathbf{R}$) is equal to the matrix itself.

Rank

The *rank* r of a matrix $A \in \mathbf{R}^{p \times m}$ is the maximal order of the non-zero minors of A.

This rank is equal to the number of rows and the number of columns of A that are linearly independent. Of course, $r \leq \min\{p, m\}$.

If $r = m$, A is said to be *right-regular* or *full column rank*; if $r = p$, A is said to be *left-regular* or *full row rank*; if A is either right- or left-regular, it is said to be *semiregular*. Finally, a square matrix A that is both left- and right-regular is said to be *regular* (or *non-singular*).

An element of a commutative domain $\mathbf{R}$ can be considered as an element of the field of fractions $\mathbf{F} = \mathrm{Q}(\mathbf{R})$ of this ring (we then say that we have "embedded $\mathbf{R}$ in $\mathbf{F}$"). This makes $A \in \mathbf{R}^{p \times m}$ an element of $\mathbf{F}^{p \times m}$. We thus can *a priori* distinguish the rank of A over $\mathbf{R}$ (that is when A is considered as an element of $\mathbf{R}^{p \times m}$) and the rank of A over $\mathbf{F}$ (when A is considered as an element of $\mathbf{F}^{p \times m}$); in fact, one can show that *these two notions coincide*. For the calculation of the rank, it is more efficient to work over the field $\mathbf{F}$, since the notion of rank over a field is more familiar and convenient than that of rank over a ring. We will see later efficient methods of calculation of the rank of a matrix over a field (Corollary 501 of sections 13.2.3 and 13.5.7). A matrix is *regular* over $\mathbf{R}$ if and only if it is *invertible* over $\mathbf{F}$.

EXAMPLE 490.– *Consider the case where* $\mathbf{R} = \mathbf{K}[s]$. *A matrix* $A(s) \in \mathbf{K}[s]^{p \times m}$ *is called a* polynomial matrix. *According to the above, the rank of* A *over* $\mathbf{K}[s]$ *and its rank over* $\mathbf{K}(s)$ *coincide. A very different notion is the rank of* $A(s)$ *over* $\mathbf{K}$ *when the variable* s *takes on a particular value. Consider the following matrix:*

$$A(s) = \begin{bmatrix} s & 1 \\ 0 & 1 \end{bmatrix}. \tag{13.6}$$

We have $\det A(s) = s$*, thus the rank of* $A(s)$ *over* $\mathbf{R}$ *(denoted by* $\mathrm{rk}_{\mathbf{R}}\, A(s)$*) is equal to* 2*, because* $\det A(s)$ *is not the* zero polynomial. *On the other hand, over* $\mathbf{K} = \mathbb{R}$ *or* $\mathbb{C}$*, the rank of* $A(s)$ *is equal to* 2 *if* $s \neq 0$ *and to* 1 *if* $s = 0$*. The value* $s = 0$ *is a root of* $\det A(s)$.

Sylvester inequality and Sylvester domain

Let $A \in \mathbf{R}^{p \times n}$ and $B \in \mathbf{R}^{n \times m}$; if $\mathbf{R}$ is a field, one can show the following, called the Sylvester inequality (see for example [26])

$$\boxed{\mathrm{rk}(A) + \mathrm{rk}(B) - n \leq \mathrm{rk}(AB) \leq \min\{\mathrm{rk}(A), \mathrm{rk}(B)\};} \tag{13.7}$$

in addition, the above two inequalities become equalities if $p = n$ and A is invertible. If $\mathbf{R}$ is a ring, certain precautions have to be taken.

DEFINITION 491.– *A* Sylvester domain *is a commutative domain* $\mathbf{R}$ *over which whenever two matrices* $A \in \mathbf{R}^{p \times n}$ *and* $B \in \mathbf{R}^{n \times m}$ *are such that* $AB = 0$*, then* $\mathrm{rk}(A) + \mathrm{rk}(B) \leq n$.

We now can state the following ([31], Chapter 5):

THEOREM 492.– *(i) The inequality on the right side of* (13.7) *is valid over any commutative domain* $\mathbf{R}$ *and the one on the left side is valid (for any matrices* $A \in \mathbf{R}^{p \times n}$ *and* $B \in \mathbf{R}^{n \times m}$*) if and only if* $\mathbf{R}$ *is a Sylvester domain. (ii) A Bézout domain and a field are Sylvester domains.*

Inverse of a square matrix

Consider the cofactor matrix $\boldsymbol{\alpha}$ of a matrix $A \in \mathbf{R}^{n \times n}$. Its transpose $\boldsymbol{\alpha}^T$ is called the *classical adjoint matrix* or the *adjugate matrix* of A. One can show that:

$$\boldsymbol{\alpha}^T A = A \boldsymbol{\alpha}^T = \det A \,.\, I_n$$

A necessary and sufficient condition for A to be invertible over $\mathbf{F}$ (that is as an element of $\mathbf{F}^{n \times n}$, and thus to have an inverse in this set) is $\det A \neq 0$, since then

$$\boxed{A^{-1} = \frac{1}{\det A} \boldsymbol{\alpha}^T}. \tag{13.8}$$

In other words, a necessary and sufficient condition for A to be invertible over $\mathbf{F}$ is that A be non-singular.

If the submatrices A_1 and A_2 of the matrix A defined by (13.3) are invertible over $\mathbf{F}$, then A is invertible over $\mathbf{F}$ and

$$A^{-1} = \begin{bmatrix} A_1^{-1} & 0 \\ -A_2^{-1} A_3 A_1^{-1} & A_2^{-1} \end{bmatrix}. \tag{13.9}$$

Finally, we state the "Inversion lemma":

LEMMA 493.– *Let* A*,* B*,* C*,* D *be matrices with entries in a field* $\mathbf{F}$*, of size* $n \times n$*,* $n \times m$*,* $m \times m$ *and* $m \times n$ *respectively, and such that* A *and* C *are invertible over* $\mathbf{F}$*. Then*

$$(A + B C D)^{-1} = A^{-1} - A^{-1} B \left(D A^{-1} B + C^{-1}\right)^{-1} D A^{-1}.$$

Invertible matrices over a ring

Let $\mathbf{R}$ be a commutative domain. According to (13.8), a matrix $A \in \mathbf{R}^{n \times n}$ is invertible over $\mathbf{R}$ (or, more succinctly, *invertible*) if and only if its determinant is a *unit* in $\mathbf{R}$. The set of invertible matrices in $\mathbf{R}^{n \times n}$ is denoted as $\mathrm{GL}_n(\mathbf{R})$; it is a multiplicative group, called the *general linear group* of square matrices of order n over $\mathbf{R}$. The subgroup of $\mathrm{GL}_n(\mathbf{R})$ formed of matrices of determinant equal to 1 (called *unimodular matrices*) is the *special linear group* of the square matrices of order n over $\mathbf{R}$ and is denoted as $\mathrm{SL}_n(\mathbf{R})$ ([10], n°III.8.3).

It is important to make a difference between an *invertible matrix over* $\mathbf{F}$ and an *invertible matrix over* $\mathbf{R}$. For example, the matrix (13.6) is invertible over $\mathbf{F} = \mathbf{K}(s)$ and its inverse is

$$A^{-1}(s) = \begin{bmatrix} \frac{1}{s} & -\frac{1}{s} \\ 0 & 1 \end{bmatrix}.$$

It is clear that $A^{-1}(s) \in \mathbf{K}(s)^{2\times 2}$ but that $A^{-1}(s) \notin \mathbf{K}[s]^{2\times 2}$. The matrix $A(s)$ is thus not invertible over $\mathbf{R}$.

Equivalence of matrices

Two matrices $A \in \mathbf{R}^{p\times m}$ and $B \in \mathbf{R}^{p\times m}$ are said to be *left-equivalent* (resp., *right-equivalent*), which we denote as $A \overset{l}{\sim} B$ (resp., $A \overset{r}{\sim} B$) if there exists a matrix $U \in \mathrm{GL}_p(\mathbf{R})$ (resp., $V \in \mathrm{GL}_m(\mathbf{R})$) such that $B = U^{-1}A$ (resp., $B = AV$). The matrices A and B are then of the same rank. Two matrices $A \in \mathbf{R}^{p\times m}$ and $B \in \mathbf{R}^{p\times m}$ are said to be *equivalent*, which we denote as $A \sim B$, if there exist two matrices $U \in GL_p(\mathbf{R})$ and $V \in GL_m(\mathbf{R})$ such that $B = U^{-1}AV$.

Elementary and secondary operations

Elementary operations

The 3 matrices U_1, U_2 and U_3 below are invertible matrices of $\mathbf{R}^{2\times 2}$:

$$U_1 = \begin{bmatrix} 0 & 1 \\ 1 & 0 \end{bmatrix}, U_2 = \begin{bmatrix} 1 & \alpha \\ 0 & 1 \end{bmatrix}, U_3 = \begin{bmatrix} 1 & 0 \\ 0 & \upsilon \end{bmatrix}$$

where $\alpha \in \mathbf{R}$ and υ is a unit of $\mathbf{R}$.

Let $A \in R^{2\times m}$; it is immediate that: U_1A is obtained from A by permuting rows 1 and 2 of this matrix; U_2A is obtained by adding to the first row of A the second row multiplied by α; U_3A is obtained by multiplying the second row of A by υ.

In a more general manner, we define 3 types of elementary operations on the rows of a matrix $A \in \mathbf{R}^{p\times m}$: (i) permuting two rows of A; (ii) adding to the ith row of A the jth row ($j \neq i$) multiplied by an element of $\mathbf{R}$; (iii) multiplying a row of A by a unit in $\mathbf{R}$.

Secondary operations

A *secondary operation on the rows* is defined in the following manner:

(iv) Left-multiply two rows of A by a matrix $U_4 \in \mathrm{GL}_2(\mathbf{R})$.

Each of the *elementary or secondary operations on the rows* corresponds to the left-multiplication by an invertible matrix over $\mathbf{R}$.

We define in the same manner 3 types of *elementary operations on the columns* and one type of *secondary operations*; each of them corresponds to a *right-multiplication* by a particular invertible matrix.

The importance of elementary and secondary operations will become obvious in section 13.2.

13.1.5. *Bézout equation*

General case

Let a, b and c be 3 elements of a Bézout domain $\mathbf{R}$, where a and b are not both zero. We call *Bézout equation* an equation of the form

$$\boxed{a\,x + b\,y = c} \tag{13.10}$$

with unknowns x and y. This equation plays a very important role in control theory.

THEOREM 494.– *(i) The Bézout equation (13.10) admits a solution if and only if* $c \in (d)$ *where* $d \in \gcd(a, b)$. *Suppose* (x_0, y_0) *is a solution of this equation. All the other solutions are of the form* $x = x_0 + q\frac{b}{d}$, $y = y_0 - q\frac{a}{d}$, *where* q *spans* $\mathbf{R}$ *(parameterization of the solutions by* q*). (ii) If* $\mathbf{R}$ *is a strongly Euclidean domain, there exists a unique solution* (x, y) *such that* $d^\circ(y) < d^\circ\left(\frac{a}{d}\right)$.

PROOF. (i) It is clear that the Bézout equation (13.10) has no solution if $d \nmid c$. Suppose now that $d \mid c$; let α, β and γ be such that $a = \alpha d$ (that is $\alpha = \frac{a}{d}$), $b = \beta d$ and $c = \gamma d$. Then (13.10) is equivalent to

$$\alpha x + \beta y = \gamma \tag{13.11}$$

where α and β are coprime. Thus there exist elements z and w of $\mathbf{R}$ such that $\alpha z + \beta w = 1$, and therefore $\alpha(\gamma z) + \beta(\gamma w) = \gamma$. The Bézout equation thus admits a solution $(\gamma z, \gamma w)$. If now (x_0, y_0) is any solution of (13.10) or, in an equivalent manner, of (13.11), every solution (x, y) satisfies

$$\alpha(x - x_0) + \beta(y - y_0) = 0.$$

Taking into account the coprimeness of α and β, $\alpha \mid (y - y_0)$ and $\beta \mid (x - x_0)$. Thus there exists an element q of $\mathbf{R}$ such that $x - x_0 = q\beta$ and $y - y_0 = -q\alpha$, which in turn provides the parameterization of the solutions. (ii) We have from the above $y_0 = q\alpha + y$. If the domain $\mathbf{R}$ is strongly Euclidean, we obtain this expression by performing the Euclidean division of y_0 by α, q designating the quotient and y the remainder; thus there exists a unique solution (x, y) such that $d^\circ(y) < d^\circ(\alpha)$. ■

Bézout equation over $\mathbf{K}[s]$

Sylvester theorem

Let be two polynomials $a(s) = a_0 s^n + a_1 s^{n-1} + ... + a_n$ and $b(s) = b_0 s^m + b_1 s^{m-1} + ... + b_m$, where $a_0 \neq 0$ and $b_0 \neq 0$. The *Sylvester matrix* $\Sigma(a,b)$ of these two polynomials is the square matrix of order $n+m$ defined by

$$\Sigma(a,b) = \begin{bmatrix} a_0 & 0 & & b_0 & 0 & & \vdots \\ a_1 & \ddots & & b_1 & b_0 & 0 & \vdots \\ a_2 & \ddots & a_0 & \vdots & b_1 & \ddots & 0 \\ \vdots & \ddots & a_1 & b_m & \vdots & \ddots & b_0 \\ a_n & \ddots & a_2 & 0 & b_m & \vdots & b_1 \\ 0 & \ddots & \vdots & \vdots & 0 & \vdots & \vdots \\ \cdots & 0 & a_n & 0 & \cdots & 0 & b_m \end{bmatrix}$$

or the transpose of this matrix.

There are m columns comprising coefficients a_i and n columns comprising coefficients b_i. The determinant of this matrix is called the *Sylvester resultant*. The *Sylvester theorem* states that it is non-zero (in other words, the Sylvester matrix is invertible) if and only if the polynomials $a(s)$ and $b(s)$ are coprime ([10], n°IV.6.6).

Sylvester system associated with a Bézout equation

Consider the Bézout equation (13.10). According to Theorem 494, either this equation does not admit a solution, or it comes down to the Bézout equation (13.11) where the polynomials $\alpha(s)$ and $\beta(s)$ are coprime.

In the second case, which will henceforth be the only one of interest to us, we are thus led back to resolving an equation of the form (13.10) where the polynomials $a(s)$ and $b(s)$ are coprime. Write

$$\begin{cases} a(s) = a_0 s^{\boldsymbol{\alpha}} + a_1 s^{\boldsymbol{\alpha}-1} + ... + a_{\boldsymbol{\alpha}} \\ b(s) = b_0 s^{\boldsymbol{\beta}} + b_1 s^{\boldsymbol{\beta}-1} + ... + b_{\boldsymbol{\beta}} \\ c(s) = c_0 s^{\gamma} + c_1 s^{\gamma-1} + ... + c_{\gamma} \\ x(s) = x_0 s^{\boldsymbol{\xi}} + x_1 s^{\boldsymbol{\xi}-1} + ... + x_{\boldsymbol{\xi}} \\ y(s) = y_0 s^{\boldsymbol{\upsilon}} + y_1 s^{\boldsymbol{\upsilon}-1} + ... + y_{\boldsymbol{\upsilon}} \end{cases}$$

where a_0 and c_0 are non-zero.

The unknowns are $x_0, ..., x_{\boldsymbol{\xi}}, y_0, ..., y_{\boldsymbol{\upsilon}}$ and are thus of the number $\boldsymbol{\xi} + \boldsymbol{\upsilon} + 2$. Suppose, without loss of generality, that $\boldsymbol{\alpha} + \boldsymbol{\xi} \geq \boldsymbol{\beta} + \boldsymbol{\upsilon}$. The equations are obtained by equating the coefficients of terms of the same degree in (13.10). We necessarily have $\gamma+1=\boldsymbol{\alpha}+\boldsymbol{\xi}+1$, and thus $\gamma=\boldsymbol{\alpha}+\boldsymbol{\xi}$. As a result, there are as many equations as there are unknowns if and only if $\boldsymbol{\xi}+\boldsymbol{\upsilon}+2=\alpha+\boldsymbol{\xi}+1$, that is

$$\boldsymbol{\upsilon}=\boldsymbol{\alpha}-1. \tag{13.12}$$

According to Theorem 494, with $d=1$, we know there exists a unique solution $(x(s), y(s))$ such that $\boldsymbol{\upsilon}\geq d^{\circ}(y)$ satisfies condition (13.12). It is this solution that we will determine below.

By equating the coefficients of the same degree in (13.10), we obtain the linear system (13.13) below, which we will call the "Sylvester system" associated with the Bézout equation under consideration.

$$\begin{bmatrix} a_0 & 0 & \cdots & \cdots & \cdots & \cdots & \cdots & 0 \\ a_1 & a_0 & 0 & & b_0 & & & \vdots \\ \vdots & a_1 & \ddots & \ddots & b_1 & \ddots & & \vdots \\ \vdots & \vdots & \ddots & a_0 & \vdots & \ddots & b_0 & 0 \\ a_{\boldsymbol{\alpha}} & \vdots & \ddots & a_1 & b_{\boldsymbol{\beta}} & \vdots & b_1 & b_0 \\ 0 & a_{\boldsymbol{\alpha}} & \ddots & \vdots & 0 & \ddots & \vdots & b_1 \\ \vdots & \ddots & \ddots & \vdots & \vdots & \ddots & b_{\boldsymbol{\beta}} & \vdots \\ 0 & \cdots & 0 & a_{\boldsymbol{\alpha}} & 0 & \cdots & 0 & b_{\boldsymbol{\beta}} \end{bmatrix} \begin{bmatrix} x_0 \\ x_1 \\ \vdots \\ x_{\boldsymbol{\xi}} \\ y_0 \\ y_1 \\ \vdots \\ y_{\boldsymbol{\upsilon}} \end{bmatrix} = \begin{bmatrix} c_0 \\ c_1 \\ \vdots \\ c_{\boldsymbol{\alpha}} \\ c_{\boldsymbol{\alpha}+1} \\ \vdots \\ \vdots \\ c_{\boldsymbol{\alpha}+\boldsymbol{\xi}} \end{bmatrix} \tag{13.13}$$

The square matrix M of this system is of order $\boldsymbol{\xi}+2+\boldsymbol{\upsilon}$. The number of columns that contain the coefficients a_i (resp., b_i) is $\boldsymbol{\xi}+1$ (resp., $\boldsymbol{\upsilon}+1$). The matrix M is of the form

$$M=\begin{bmatrix} \mathcal{A} & 0 \\ \star & \Sigma(a,b) \end{bmatrix}$$

where $\mathcal{A}$ is a square lower triangular matrix, of order $(\boldsymbol{\alpha}-\boldsymbol{\beta})$, and whose all diagonal entries are equal to a_0. This submatrix $\mathcal{A}$ is thus invertible or empty ($\boldsymbol{\alpha}-\boldsymbol{\beta}$ is the number of coefficients above the first coefficient b_0 that are equal to 0 when one looks at the matrix M from left to right). As a result, the Sylvester system (13.13) admits a unique solution since $\Sigma(a,b)$ is invertible (according to the Sylvester theorem), the polynomials $a(s)$ and $b(s)$ being coprime by assumption.

The system (13.13) can be simplified when the matrix $\mathcal{A}$ is non-empty, that is when $\boldsymbol{\alpha}>\boldsymbol{\beta}$ (which is often the case). Then, the polynomials $a(s), c(s)$ and $x(s)$ can be assumed to be monic; the first equation of the Sylvester system becomes trivial and we can remove the term x_0 (equal to 1) from the unknowns by suppressing the first row and the first column of the matrix M, as well as the first element of the vector of unknowns and of the vector in the right-hand member where the c_i's are replaced by $c_i - a_i$ for $1 \leq i \leq \alpha$. We thus obtain the form as indicated in section 6.4.5.

13.2. Matrices over principal ideal domains

In this section, unless otherwise stated, $\mathbf{R}$ is a *commutative principal ideal domain*. However, we will use finer hypotheses on a case-by-case basis.

13.2.1. *Invertible matrices over principal ideal domains*

One can show that any left-multiplication by an invertible matrix over $\mathbf{R}$ is a composition of a finite number of elementary and secondary operations on the rows (each of these operations can of course be used more than once): see ([91], section X.7, Prop. 16). In the more particular case where $\mathbf{R}$ is an *Euclidean domain*, one can show that the use of secondary operations is not necessary ([91], section X.7, Theorem 17); it can nevertheless provide simplifications.

13.2.2. *Hermite form*

Column Hermite form

THEOREM 495.– *Let* $\mathbf{R}$ *be a commutative Bézout domain.*
(i) A matrix $A \in \mathbf{R}^{p \times m}$ *is left-equivalent to an upper triangular matrix, called a* column Hermite form *of* A :

$$\begin{bmatrix} T & * \\ 0 & 0 \end{bmatrix} \tag{13.14}$$

where T *is a square upper triangular matrix.*
(ii) There exists a permutation matrix $P \in \mathrm{GL}_m(\mathbf{R})$ *such that the column Hermite form of* $A\,P$ *is* (13.14) *where* T *is regular of order equal to* $r = \mathrm{rk}_{\mathbf{R}}\, A$.

PROOF. This proof is important because it is constructive, that is in *concrete cases*, we determine the column Hermite form of a matrix by repeating the rationale below. We will limit ourselves to the case where $\mathbf{R} = \mathbf{K}[s]$ (for the case where $\mathbf{R}$ is a commutative Bézout domain, see [67] or [22]).

1) Suppose the first column is non-zero. By permuting rows, we are led to the case where the coefficient a_{11} is such that $-\infty < d^\circ(a_{11}) \leq d^\circ(a_{j1})$, for any index $j \in \{1, ..., p\}$. Multiply the first row by the inverse of the leading coefficient of a_{11},

and then perform the Euclidean division of the a_{j1}'s $(j \neq 1)$ by a_{11}, i.e. write $a_{j1} = a_{11}\, q_{j1} + \bar{a}_{j1}$, where $d^\circ(\bar{a}_{j1}) < d^\circ(a_{11})$. From the jth row $(j \neq 1)$, subtract the first one multiplied by q_{j1}, then write $\bar{a}_{11} = a_{11}$. The first column of the matrix then obtained is composed of the $\bar{a}_{j1}$ and $d^\circ(\bar{a}_{j1}) < d^\circ(\bar{a}_{11})$ if $j \neq 1$. We can continue the process until the only non-zero term of the first column is the first one (which is with index $(1,1)$ in the matrix).

2) Having done that, if the second column is non-zero, proceed in the same manner for the submatrix obtained by suppressing the first row and the first column. The first column obtained for this submatrix eventually has the same property as the first column of the whole matrix. Its first coefficient is with index $(2,2)$ in the latter.

3) We now perform the Euclidean division of the coefficient with index $(1,2)$ (in the whole matrix) by that with index $(2,2)$, then we subtract from the first row the second row multiplied by the quotient of this division, and we obtain $d^\circ(\bar{a}_{12}) < d^\circ(\bar{a}_{22})$.

4) Continuing this way, we obtain the form desired for part (i).

5) (ii) is now obvious. ■

Row Hermite form

By exchanging the roles of the rows and the columns, we obtain the following:

THEOREM 496.– *Let* $\mathbf{R}$ *be a commutative Bézout domain.*
(i) A matrix $A \in \mathbf{R}^{p\times m}$ *is right-equivalent to a lower triangular matrix, called a* row Hermite form *of* A :

$$\begin{bmatrix} T & 0 \\ * & 0 \end{bmatrix} \tag{13.15}$$

where T *is a square lower triangular matrix.*
(ii) There exists a permutation matrix $P \in \mathrm{GL}_p(\mathbf{R})$ *such that the row Hermite form of* PA *is* (13.15) *where* T *is regular of order equal to* $r = \mathrm{rk}_{\mathbf{R}}\, A$.

13.2.3. *Smith form*

THEOREM 497.– *A matrix* $A \in \mathbf{R}^{p\times m}$*, of rank* r*, is equivalent to a matrix of the form*

$$\boxed{\Sigma = \mathrm{diag}(\alpha_1, ..., \alpha_r, 0, ..., 0)}$$

where the $\alpha_i, 1 \leq i \leq r$, *are all non-zero and are such that* $\alpha_i \mid \alpha_{i+1}, 1 \leq i \leq r-1$.[3] *These diagonal elements* α_i *are uniquely determined up to associates.*

DEFINITION 498.– *The matrix* Σ *is the* Smith form *of* A *and the elements* $\alpha_i, 1 \leq i \leq r$, *are its invariant factors. Two matrices* $A \in \mathbf{R}^{p \times m}$ *and* $B \in \mathbf{R}^{p \times m}$ *are equivalent if and only if they have the same Smith form (the Smith form of a matrix is thus also called its* Smith normal form, *or, for short, its* normal form*).*

Let us define an *elementary divisor ring* (in the case where such a ring is a commutative domain).

DEFINITION 499.– *An* elementary divisor ring *is a Bézout domain* $\mathbf{R}$ *in which whenever* $(a) + (b) + (c) = \mathbf{R}$, *there exist* p *and* q *such that* $(p\,a) + (p\,b + q\,c) = \mathbf{R}$.

REMARK 500.– *The statement of Theorem 497 remains valid when* $\mathbf{R}$ *is an elementary divisor ring. All principal ideal domains domains are elementary divisor rings but here are two examples elementary divisor rings which are not principal ideal domains: (i) the ring* $\mathcal{O}(\mathbb{C})$ *of entire functions ([67], section 5); (ii) the ring* $\mathcal{E} = \mathbb{R}(s)[e^{-s}] \cap \mathcal{O}(\mathbb{C})$ *(the elements of* $\mathcal{E}$ *are called* pseudo-polynomials; $\mathcal{E}$ *was introduced in [54] and used in [86], [55] for the study of systems with commensurate delays).*

Proof of the theorem

We will only give here the constructive part of the proof. For a proof of uniqueness of the Smith form (guaranteed under the condition that each invariant factor can be replaced by an associate), see for example ([10], Chapter VII). We thus will show how one can construct the Smith form of a matrix A. We limit ourselves to the case where $\mathbf{R}$ is an Euclidean domain. If $A = 0$, there is nothing to prove.

Obtaining a diagonal form

If $A \neq 0$, by a permutation of columns we are led to the case where the first column of A is non-zero. Proceeding as in Theorem 495, we arrive, using elementary or secondary operations on rows, to the case where this first column has a unique non-zero element a_{11}.

Now do the same for the first row: the Euclidean division of a_{1j} $(j > 1)$ by a_{11} is written as $a_{1j} = a_{11}\, q_{1j} + \bar{a}_{1j}$, where $d^\circ(\bar{a}_{1j}) < d^\circ(a_{11})$. Using elementary or secondary operations on columns, we thus obtain $(a_{11}, \bar{a}_{12}, ..., \bar{a}_{1m})$ as the first row.

3. Abusing the language, $\mathrm{diag}(\alpha_1, ..., \alpha_r, 0, ..., 0)$ represents a matrix (square or not, depending on the context), the elements of the principal diagonal of which are $\alpha_1, ..., \alpha_r, 0, ..., 0$ and all the other entries are zero.

By a permutation of columns, we bring to $(1,1)$ position, among these elements, one of those that are non-zero and of minimum degree.

We can now re-iterate the above whole process (by first operating on the rows, and then on the columns); these iterations will necessarily stop because each time the degree of the term with index $(1,1)$ decreases. We thus finish by obtaining a matrix $\bar{A}$, equivalent to A, and of the form

$$\bar{A} = \beta_1 \oplus \bar{A}_1.$$

Do the same with the submatrix $\bar{A}_1$, and so on. We finally obtain the form $\operatorname{diag}(\beta_1, ..., \beta_r, 0, ..., 0)$ where the β_i's, $1 \le i \le r$, are all non-zero. Nonetheless, they in general do not satisfy the required divisibility property.

Obtaining the divisibility property

Consider the submatrix $B = \operatorname{diag}(\beta_1, \beta_2)$. Let $\gamma_1 \in \gcd(\beta_1, \beta_2)$, in such a way that (according to the Bézout theorem) there exist x and y in $\mathbf{R}$ such that $x\beta_1 + y\beta_2 = \gamma_1$. There exists $\bar{\beta}_1$ in $\mathbf{R}$ such that $\beta_1 = \gamma_1 \bar{\beta}_1$. We obtain

$$\begin{bmatrix} 1 & y \\ 0 & 1 \end{bmatrix} \begin{bmatrix} \beta_1 & 0 \\ 0 & \beta_2 \end{bmatrix} \begin{bmatrix} x & 1 \\ 1 & 0 \end{bmatrix} = \begin{bmatrix} \gamma_1 & \beta_1 \\ \beta_2 & 0 \end{bmatrix}$$
$$\sim \begin{bmatrix} \gamma_1 & 0 \\ 0 & \bar{\beta}_1 \beta_2 \end{bmatrix}$$

thus we can lower the degree of β_1 except if $\beta_1 \mid \beta_2$ for in this case $\gamma_1 = \beta_1$. By continuing this procedure, we finally obtain $\operatorname{diag}(\beta_1, ..., \beta_r, 0, ..., 0) \sim \operatorname{diag}(\alpha_1, ..., \alpha_r, 0, ..., 0)$ where $\alpha_i \mid \alpha_{i+1}, 1 \le i \le r-1$. The constructive part of the theorem is thus proven.

Equivalence over a field

COROLLARY 501.– *A matrix $A \in \mathbf{F}^{p \times m}$ (where $\mathbf{F}$ is a field), of rank r, is equivalent to the matrix*

$$\Sigma = \begin{bmatrix} I_r & 0 \\ 0 & 0 \end{bmatrix}.$$

Indeed, we can apply the previous result, but as divisibility is trivial in a field, the invariant factors can be taken as 1. We obtain this form by using only the first part of the above procedure, which itself also becomes considerably simpler. This method is an efficient way of calculating the rank of a matrix over a field. We will see an even more efficient method – from a numerical point of view – later (section 13.5.7) when $\mathbf{F} = \mathbb{C}$ or $\mathbb{R}$.

Direct calculation of invariant factors

The invariant factors of a matrix $A \in \mathbf{R}^{p\times m}$, of rank r, satisfy the following property, which can be employed for their direct calculation (which in certain cases can be advantageous, especially when A is triangular):

PROPOSITION 502.– *Let $\alpha_1, ..., \alpha_r$ be the invariant factors of A. Then α_1 is a* gcd *of the elements of A and for any integer $q \in \{1, ..., r\}$, $\prod_{i=1}^{q} \alpha_i$ is a* gcd *of the non-zero minors of order q of A. (In particular, if $A \in \mathbf{R}^{n\times n}$ is non-singular, $\prod_{i=1}^{n} \alpha_i =$* det A.*)*

Example

Let $\mathbf{R} = \mathbf{K}[s]$ and $A = \text{diag}(\beta_1, \beta_2, \beta_3)$ where

$$\beta_1 = s+2,\ \beta_2 = (s+1)(s+2),\ \beta_3 = (s+1)(s+3).$$

We get from Proposition 502

$$\begin{aligned}
\alpha_1 &= \text{gcd}(\beta_1, \beta_2, \beta_3) = 1, \\
\alpha_2 &= (s+1)(s+2), \\
\alpha_3 &= (s+1)(s+2)(s+3).
\end{aligned}$$

Marking out elementary operations

The advantage of using row and column elementary and secondary operations is that it allows one to easily determine the invertible matrices U and V such that $U^{-1} A V = \Sigma$, where Σ is the Smith form of A. Indeed, let

$$M = \begin{bmatrix} A & I_p \\ I_m & 0 \end{bmatrix}.$$

By operating on the first p rows and the first m columns of M, we obtain

$$\begin{bmatrix} U^{-1} & 0 \\ 0 & I_m \end{bmatrix} M \begin{bmatrix} V & 0 \\ 0 & I_p \end{bmatrix} = \begin{bmatrix} \Sigma & U^{-1} \\ V & 0 \end{bmatrix}.$$

This method can also be applied, of course, to determine an invertible matrix that will allow one to transform (by left- or right-equivalence) a matrix A into one of its Hermite forms.

Example

Over $\mathbf{R} = \mathbf{K}[s]$, let $A = \begin{bmatrix} s & s & s^2 \\ s & s^2 & s^4 \end{bmatrix}$; then

$$M = \begin{bmatrix} s & s & s^2 & 1 & 0 \\ s & s^2 & s^4 & 0 & 1 \\ 1 & 0 & 0 & 0 & 0 \\ 0 & 1 & 0 & 0 & 0 \\ 0 & 0 & 1 & 0 & 0 \end{bmatrix}$$

$$\sim \begin{bmatrix} s & 0 & 0 & 1 & 0 \\ 0 & s(s-1) & 0 & -1 & 1 \\ 1 & -1 & s^2 & 0 & 0 \\ 0 & 1 & -s(s+1) & 0 & 0 \\ 0 & 0 & 1 & 0 & 0 \end{bmatrix}; \text{ from which}$$

$$\Sigma = \begin{bmatrix} s & 0 & 0 \\ 0 & s(s-1) & 0 \end{bmatrix}, U^{-1} = \begin{bmatrix} 1 & 0 \\ -1 & 1 \end{bmatrix}, V = \begin{bmatrix} 1 & -1 & s^2 \\ 0 & 1 & -s(s+1) \\ 0 & 0 & 1 \end{bmatrix}.$$

13.2.4. *Elementary divisors*

Let $\alpha_{j+1}, ..., \alpha_r$ be the *non-invertible* invariant factors of a matrix A defined over a principal ideal domain $\mathbf{R}$. If we factorize each of these invariant factors into primes, they take the form of products of pairwise coprime terms of the form p_k^n, where the p_k's are primes in $\mathbf{R}$.

DEFINITION 503.– *These p_k^n's are the* elementary divisors *of A. The number of times we encounter one of these elementary divisors in all the above decompositions into primes is its* multiplicity *(or its* order of multiplicity*).*

Example

Over $\mathbf{R} = \mathbf{K}[s]$, let $A \sim \text{diag}(\alpha_1, \alpha_2, \alpha_3, 0, ..., 0)$, with

$$\begin{aligned} \alpha_1(s) &= (s-1)^2(s-2), \quad \alpha_2(s) = (s-1)^2(s-2)^2, \\ \alpha_3(s) &= (s-1)^2(s-2)^2(s-3) \end{aligned}$$

The elementary divisors are:

$\delta_1(s) = (s-1)^2$ (of order 3)

$\delta_2(s) = s-2$ (of order 1)

$\delta_3(s) = (s-2)^2$ (of order 2)

$\delta_4(s) = s-3$ (of order 1).

Calculation of invariant factors from elementary divisors

As seen in the above example (and as a result of the definition), the calculation of the elementary divisors from the invariant factors is immediate.

Conversely, knowing the elementary divisors of a matrix, we can calculate its invariant factors. It suffices to construct a table in which each row is composed of the powers of the same prime, repeated a number of times equal its order of multiplicity, these powers being in decreasing order. The product of the columns thus provides, in reverse order, the invariant factors.

Using the previous example:

$(s-1)^2$	$(s-1)^2$	$(s-1)^2$
$(s-2)^2$	$(s-2)^2$	$s-2$
$s-3$	1	1
$\alpha_3(s)$	$\alpha_2(s)$	$\alpha_1(s)$

13.2.5. *Smith zeros*

Consider the case where $\mathbf{R} = \mathbb{C}[s]$. Let $A \in \mathbb{C}[s]^{n\times m}$, and let $\alpha_1(s), ..., \alpha_r(s)$ be its invariant factors. Let $\alpha(s) = \prod_{i=1}^{r} \alpha_i(s)$. The roots of $\alpha(s)$ are called the *Smith zeros* of A.

In the above example, the Smith zeros of A form the set

$$\{1,1,1,1,1,1,2,2,2,2,2,3\}.$$

Indeed, the number of times a Smith zero z is repeated in this set is equal to its order of multiplicity when z is considered as a root of the polynomial $\alpha(s)$.

Let z be a Smith zero of A and let

$$(s-z)^{\nu_1}, ..., (s-z)^{\nu_k}$$

$(1 \leq \nu_1 \leq ... \leq \nu_k)$ be the elementary divisors of A that are multiples of $s-z$, each of these elementary divisors repeated a number of times equal to its order of multiplicity. We adopt the following definition [19]:

DEFINITION 504.– *The integers $\nu_1, ..., \nu_k$ are the* structural indices *of the Smith zero z (or of the matrix A at z), $\rho = \nu_k$ is the* order *of this zero and $\sum_{i=1}^{k} \nu_i$ is its degree.*

Notice that if $A \in \mathbb{C}[s]$ (that is if this polynomial matrix is only a polynomial), the Smith zeros of this matrix are the roots of this polynomial. Let z be one of these Smith zeros; it has only one structural index, equal to its order, to its degree, and to the order of multiplicity of the root z.

Let us continue with the example of section 13.2.4: the Smith zero 1 has structural indices $\{2,2,2\}$, its order is 2, its degree is 6. The Smith zero 2 has structural indices $\{1,2,2\}$, its order is 2, its degree is 5. The Smith zero 3 has a unique structural index 1; its order and its degree are both equal to 1.

13.2.6. *Divisibility of matrices*

Left-divisor

Let $\mathbf{R}$ be a ring, let $E \in \mathbf{R}^{n \times k}$, $E \neq 0$, and let $L \in \mathbf{R}^{n \times n}$. We say that L is a *left-divisor* of E if there exists a matrix $X \in \mathbf{R}^{n \times m}$ such that $E = L\,X$ (which implies $\operatorname{rk} L \geq \operatorname{rk} E$ according to the Sylvester inequality: see Theorem 492, section 13.1.4).

Now let $A \in \mathbf{R}^{n \times n'}$ and $B \in \mathbf{R}^{n \times m}$. We say that $L \in \mathbf{R}^{n \times n}$ is a *common left-divisor* of A and B if L is a left-divisor of the matrix $E \triangleq \begin{bmatrix} A & B \end{bmatrix}$.

Greatest common left-divisor

Let A, B and E be as above; $L \in \mathbf{R}^{n \times n}$ is a greatest common left-divisor (gcld) of A and B if L is a left-divisor of E and if any left-divisor of E is a left-divisor of L.

THEOREM 505.– *Suppose that $\mathbf{R}$ is a Bézout domain and let $r \triangleq \operatorname{rk}\begin{bmatrix} A & B \end{bmatrix}$. (i) There exists a matrix $L \in \mathbf{R}^{n \times n}$, of rank r, such that $\begin{bmatrix} A & B \end{bmatrix} \overset{r}{\sim} \begin{bmatrix} L & 0 \end{bmatrix}$, and L is a* gcld *of A and B. (ii) The set of all* gcld*'s of A and B is the set of all matrices which are right-equivalent to L.*

PROOF. According to Theorem 495 (section 13.2.2), we know that $\begin{bmatrix} A & B \end{bmatrix} \overset{r}{\sim} \begin{bmatrix} L & 0 \end{bmatrix}$, where $L \in \mathbf{R}^{n \times n}$ is obviously of rank r. We easily deduce that L is a gcld of A and B. For a detailed proof of (ii), see ([114], section 4.1, Lemma 2). ■

Left-primeness and coprimeness

DEFINITION 506.– *(i) Let $E \in \mathbf{R}^{n \times k}$ be a matrix of rank n; E is said to be* left-prime *if E is right-invertible over $\mathbf{R}$. (ii) Let $A \in \mathbf{R}^{n \times n'}$ and $B \in \mathbf{R}^{n \times m}$ be such that $E \triangleq \begin{bmatrix} A & B \end{bmatrix}$ is of rank n; the matrices A and B are said to be* left-coprime *if E is left-prime.*

REMARK 507.– *If $\mathbf{R}$ is not a Bézout domain, there exist other weaker forms of left-primeness of a matrix (see for example [121]).*

Right-divisibility

What has just been said about left-divisibility transposes in an obvious way to right-divisibility. In particular, E is *right-prime* if and only if E^T is left-prime; the matrices A and C are right-coprime if and only if A^T and C^T are left-coprime. Likewise, Theorem 505 transposes in an obvious way to the case of right-coprimeness (the details are left to the reader).

13.2.7. *Coprime factorizations*

Let $\mathbf{R}$ be a commutative domain, let $\mathbf{F} = \mathrm{Q}(\mathbf{R})$ be its field of fractions (see section 13.1.1) and let $G \in \mathbf{F}^{p \times m}$.

DEFINITION 508.– *A pair of matrices* $(D_l, N_l) \in \mathbf{R}^{p \times p} \times \mathbf{R}^{p \times m}$ *is a* left-coprime factorization *of* G *over* $\mathbf{R}$ *if:* D_l *is invertible over* $\mathbf{F}$, $G = D_l^{-1} N_l$, *and the matrices* $\{D_l, N_l\}$ *are left-coprime.*

In the same manner, we define a right-coprime factorization (N_r, D_r) of $G = N_r D_r^{-1}$. Left-coprime factorizations have the property below (the corresponding statement for right-coprime factorizations is left to the reader).

LEMMA 509.– *Suppose that a left-coprime factorization* (D_l, N_l) *of* G *over* $\mathbf{R}$ *exists; then* (D_l', N_l') *is a left-coprime factorization of* G *over* $\mathbf{R}$ *if and only if there exists an invertible matrix* $U \in \mathbf{R}^{p \times p}$ *such that* $\begin{bmatrix} D_l & N_l \end{bmatrix} = U \begin{bmatrix} D_l' & N_l' \end{bmatrix}$.

PROOF. If (D_l', N_l') is a left-coprime factorization of G over $\mathbf{R}$, $\begin{bmatrix} D_l' & N_l' \end{bmatrix}$ is right-invertible and $G = D_l'^{-1} N_l'$. Thus there exist matrices X' and Y' over $\mathbf{R}$ such that

$$D_l' X' + N_l' Y' = I,$$

and so, left-multiplying this relation by $D_l D_l'^{-1}$,

$$D_l X' + Y' = D_l D_l'^{-1},$$

therefore $D_l D_l'^{-1} \in \mathbf{R}^{p \times p}$. By symmetry, $D_l' D_l^{-1} \in \mathbf{R}^{p \times p}$. As a result, $D_l D_l'^{-1} = U$ is invertible and $\begin{bmatrix} D_l & N_l \end{bmatrix} = U \begin{bmatrix} D_l' & N_l' \end{bmatrix}$. The converse is obvious. ■

DEFINITION 510.– *A matrix* $G \in \mathbf{F}^{p \times m}$ *admits a* doubly-coprime factorization *over* $\mathbf{R}$ *if, with the above notation,*

$$\begin{bmatrix} * & * \\ D_l & N_l \end{bmatrix} \begin{bmatrix} N_r & * \\ D_r & * \end{bmatrix} = I$$

where the asterisks denote submatrices with entries in $\mathbf{R}$.

The following is related to the doubly-coprime factorization :

DEFINITION 511.– *Let* $E \in \mathbf{R}^{n \times k}$ *be a matrix of rank* n*;* E *is said to be* completable *if there exists a matrix* $*$ *such that* $\begin{bmatrix} E \\ * \end{bmatrix}$ *is invertible over* $\mathbf{R}$.

Let $\mathbf{R}$ be a commutative domain, let $\mathbf{F} = \mathrm{Q}(\mathbf{R})$ be its field of fractions, and let $\mathbf{M}(\mathbf{F})$ be the set of all matrices of finite size with entries in $\mathbf{F}$. The result below can be easily proven ([114], section 8.1) and justifies Definition 510(ii):

LEMMA 512.– *Suppose that* $G \in \mathbf{M}(\mathbf{F})$ *admits a left-coprime factorization* (D_l, N_l) *over* $\mathbf{R}$*; then* G *admits a right-coprime factorization over* $\mathbf{R}$ *if and only if* $E \triangleq \begin{bmatrix} D_l & N_l \end{bmatrix}$ *is completable (in other words if* G *admits a doubly-coprime factorization over* $\mathbf{R}$*).*

DEFINITION 513.– *A commutative domain* $\mathbf{R}$ *having the following property is called an* Hermite ring*: any row* $A = \begin{bmatrix} a_1 & \cdots & a_n \end{bmatrix}$ *such that* $\sum_{1 \leq i \leq n} (a_i) = \mathbf{R}$ *is completable.*

All Sylvester domains (Definition 491, section 13.1.4) are Hermite rings but the converse does not hold ([31], section 5.5). One can prove the following ([114], section 8.1):

THEOREM 514.– *(A) The following conditions are equivalent: (i)* $\mathbf{R}$ *is a Bézout domain; (ii) every matrix* $G \in \mathbf{M}(\mathbf{F})$ *admits both a left-coprime factorization and a right-coprime factorization over* $\mathbf{R}$*; (iii) every matrix* $G \in \mathbf{M}(\mathbf{F})$ *admits a doubly coprime factorization over* $\mathbf{R}$*. (B) The following conditions are equivalent: (i)* $\mathbf{R}$ *is a Hermite ring; (ii) if* $G \in \mathbf{M}(\mathbf{F})$ *admits either a left- or a right-coprime factorization over* $\mathbf{R}$*, then it admits a doubly-coprime factorization over* $\mathbf{R}$*.*

13.2.8. *Bézout matrix equations*

Left Bézout equation

Let $\mathbf{R}$ be a ring, let $A \in \mathbf{R}^{n \times n'}$ and $B \in \mathbf{R}^{n \times m}$ be two matrices such that $\operatorname{rk}\begin{bmatrix} A & B \end{bmatrix} = n$, and let $C \in \mathbf{R}^{n \times n}$. The following equation is called a *left Bézout matrix equation*:

$$\boxed{AX + BY = C} \tag{13.16}$$

where the matrices $X \in \mathbf{R}^{n' \times n}$ and $Y \in \mathbf{R}^{m \times n}$ are the unknowns.

THEOREM 515.– *Suppose* $\mathbf{R}$ *is a Bézout domain, and let* L *be a* gcld *of* A *and* B. *Then equation (13.16) admits a solution if and only if* C *is a right-multiple of* L.

PROOF. There exist matrices $A' \in \mathbf{R}^{n\times n'}$ and $B' \in \mathbf{R}^{n\times m}$ such that $A = L\,A'$ and $B = L\,B'$, thus for any matrices $X \in \mathbf{R}^{n'\times n}$ and $Y \in \mathbf{R}^{m\times n}$, $A\,X + B\,Y = L\,(A'\,X + B'\,Y)$; the condition is therefore necessary. Conversely, suppose that $C \in \mathbf{R}^{n\times n}$ is a right-multiple of L, and thus of the form $C = L\,M$. We know that there exists an invertible matrix V such that $\begin{bmatrix} A & B \end{bmatrix} V = \begin{bmatrix} L & 0 \end{bmatrix}$. By decomposing V as

$$V = \begin{bmatrix} V_{11} & V_{12} \\ V_{21} & V_{22} \end{bmatrix}$$

with the proper sizes, we obtain

$$\begin{cases} A\,V_{11} + B\,V_{21} = L, \\ A\,V_{12} + B\,V_{22} = 0. \end{cases}$$

As a result, $A\,V_{11}\,M + B\,V_{21}\,M = L\,M = C$, and $(V_{11}M, V_{21}M)$ is a solution of (13.16), which completes the proof. ■

Right Bézout equation

We leave it to the reader to transpose Theorem 515 to the case of a "right Bézout matrix equation" $X\,A + Y\,B = C$.

13.3. Homomorphisms of vector spaces

13.3.1. *Vector spaces*

The notion of $\mathbf{K}$-vector space is classic when $\mathbf{K}$ is a field[4], which the reader can assume to be $\mathbb{R}$ or $\mathbb{C}$ (for further details, see section 13.4.1).

Basis

Here, we are only concerned with *finite-dimensional* $\mathbf{K}$-vector spaces.

A $\mathbf{K}$-vector space E is said to be of dimension n when any basis of E has n elements. To make this notion clear: a finite sequence $(e_i)_{1\leq i\leq n}$ of elements of E is a basis of E if any vector x of E can be expressed in a unique manner as a $\mathbf{K}$-linear combination of vectors of this basis, that is

$$x = \sum_{i=1}^{n} x_i\,e_i. \qquad (13.17)$$

4. A vector space can be defined over a division ring.

The $x_i \in \mathbf{K}$ in (13.17) are the *components* of x in the basis under consideration. The column-matrix $X = [x_1, ..., x_n]^T$ is the *representative matrix* (or the *representative*, for short) of x in this basis.

What we have seen so far shows that the choice of a basis makes it possible to represent any vector of E by an element of $\mathbf{K}^n$. Using a term specified later, E is "isomorphic" to $\mathbf{K}^n$. Therefore, $\mathbf{K}^n$ is the "model" of all $\mathbf{K}$-vector spaces of dimension n. The "canonical basis" $(\gamma_i)_{1 \leq i \leq n}$ of $\mathbf{K}^n$ is defined in the following manner:

$$\gamma_i = (0, ..., 0, 1, 0, ..., 0), \quad i \in \{1, ..., n\}$$

where the 1 is located at the ith position. In this basis, the representative of $(x_1, ..., x_n)$ is the column matrix $[x_1, ..., x_n]^T$, which we will also call a "column vector".

Quotient space

Let E_1 be a subspace of E. We can consider the following equivalence relation $\mathcal{R}$:

$$x \mathcal{R} y \Leftrightarrow x - y \in E_1.$$

The set consisting of the equivalence classes is again a $\mathbf{K}$-vector space, called the *quotient space* E/E_1.

Direct sum

Let E_1, E_2 be two subspaces of the same vector space E. We denote as $E_1 + E_2$ the subspace of E consisting of all elements $x = x_1 + x_2$ $(x_1 \in E_1, x_2 \in E_2)$; $E_1 + E_2$ is called the *sum* of E_1 and E_2.

If $E_1 \cap E_2 = 0$ (where 0 denotes the subspace of E consisting of zero alone), the above expression of x is unique and the sum of E_1 and E_2 is said to be "direct"; it is then denoted as $E_1 \oplus E_2$.

Let E_1, E_2 be two vector spaces. The product $E_1 \times E_2$ is the set of all pairs (x_1, x_2) where $x_1 \in E_1$ and $x_2 \in E_2$; obviously, $E_1 \times E_2$ has a structure of vector space. By identifying x_1 and x_2 with $(x_1, 0)$ and $(0, x_2)$ respectively, the product $E = E_1 \times E_2$ (sometimes called the *external direct sum* of E_1 and E_2) can be identified with $E_1 \oplus E_2$ where E_1 and E_2 are considered as subspaces of E.

Let be the direct sum

$$E = E_1 \oplus E_2; \tag{13.18}$$

E_2 is called a *supplement* of E_1. (Such a supplement is not unique.) Then let $x = x_1 + x_2$, where $x_1 \in E_1$ and $x_2 \in E_2$; x_1 and x_2 are uniquely determined in function of x. The vector x_1 is called the *component* of x in E_1 and the map $\mathbf{p}_1 : x \mapsto x_1$ is called the *projection* of E onto E_1 parallel to E_2.

Let $\mathcal{B}_1 = (\varepsilon_i)_{1 \leq i \leq n_1}$ be a basis of E_1 and $\mathcal{B}_2 = (\eta_i)_{1 \leq i \leq n_2}$ a basis of E_2; then $(\varepsilon_1, ..., \varepsilon_{n_1}, \eta_1, ..., \eta_{n_2})$ is a basis $\mathcal{B}$ of $E = E_1 \oplus E_2$; we denote this basis by $\mathcal{B}_1 \uplus \mathcal{B}_2$ and we call it the *concatenation* of bases $\mathcal{B}_1$ and $\mathcal{B}_2$ (this is not a simple union since the order of the basis vectors is important, as specified above). As a result:

$$\dim (E_1 \oplus E_2) = \dim E_1 + \dim E_2. \tag{13.19}$$

The following is classic and important :

THEOREM 516.– *Let E be a $\mathbf{K}$-vector space; every subspace of E admits a supplement.*

COROLLARY 517.– (Theorem of the incomplete basis). *Let E be a $\mathbf{K}$-vector space of dimension n and let $(x_i)_{1 \leq i \leq \rho}$ be a sequence of ρ linearly independent vectors ($\rho < n$). There exists a sequence $(x_i)_{\rho+1 \leq i \leq n}$ of vectors of E such that $(x_i)_{1 \leq i \leq n}$ is a basis of E.*

PROOF. Let E_0 be the subspace of E generated by the vectors x_i $(1 \leq i \leq \rho)$ and let $\tilde{E}_0$ be a supplement of E_0. Let $x_{\rho+1}$ be a non-zero vector belonging to $\tilde{E}_0$. The vectors x_i $(1 \leq i \leq \rho + 1)$ are $\mathbf{K}$-linearly independent. Continue this construction and suppose we have determined k $\mathbf{K}$-linearly independent vectors $x_{\rho+1}, ..., x_{\rho+k}$ (with $\rho + k < n$). Let E_k be the subspace of E generated by the vectors x_i $(1 \leq i \leq \rho + k)$ and let $\tilde{E}_k$ be a supplement of E_k. Let $x_{\rho+k+1}$ be a non-zero vector belonging to $\tilde{E}_k$. The vectors x_i $(1 \leq i \leq \rho + k + 1)$ are $\mathbf{K}$-linearly independent. This construction is complete when $\rho + k + 1 = n$. ■

13.3.2. *Homomorphisms and matrices*

Homomorphisms

Let E and F be two $\mathbf{K}$-vector spaces, such that $\dim E = n$ and $\dim F = m$.

A $\mathbf{K}$-linear mapping $\mathbf{u}$ from E into F is also called a homomorphism (of vector spaces) from E into F. For any $x \in E$, the vector $\mathbf{u}(x) \in F$ is often denoted as $\mathbf{u}.x$ or $\mathbf{u}\, x$.

The kernel of $\mathbf{u}$, denoted as $\ker \mathbf{u}$, is the subspace of E defined by

$$\boxed{\ker \mathbf{u} = \{x \in E : \mathbf{u}.x = 0\}}.$$

The image of $\mathbf{u}$, denoted by $\operatorname{im} \mathbf{u}$, is the vector subspace of F defined by

$$\boxed{\operatorname{im} \mathbf{u} = \{\mathbf{u}.x\,;x \in E\}}.$$

The homomorphism $\mathbf{u}$, if it is injective (that is if $\ker \mathbf{u} = 0$), is called a *monomorphism.*

If it is surjective (i.e. if $\operatorname{im} \mathbf{u} = F$), it is called an *epimorphism* (from E onto F).

If it is both injective and surjective, i.e. bijective, it is called an *isomorphism* (from E onto F). In this case, the vector spaces E and F are said to be *isomorphic* (which we denote as $E \cong F$). Two such vector spaces have exactly the same properties: making calculations in either one amounts to doing in the other one by "transporting" the calculations with the isomorphism $\mathbf{u}$ (or the inverse isomorphism $\mathbf{u}^{-1}$); this is why these spaces are often *identified*.

Canonical epimorphism

Let E_1 be a subspace of E. The mapping ϕ that maps a vector x of E to its class in E/E_1 is an epimorphism: it is called the *canonical epimorphism* from E onto E/E_1. Obviously, $\phi.x = 0$ if and only if $x \in E_1$; in other words, $\ker \phi = E_1$.

Inclusion

Consider the mapping $\mathbf{i}$ from E_1 into E which, to a vector x in E_1, associates this same element x, now considered to be in E; $\mathbf{i}$ is a monomorphism and is called the *inclusion* (or the *canonical injection*) from E_1 into E.

Canonical decomposition of a homomorphism

The diagram in Figure 13.1 below is commutative; it represents the *canonical decomposition* of a homomorphism $\mathbf{u}$ from E into F. In this diagram, $\tilde{\mathbf{u}}$ is the epimorphism from E onto $\operatorname{im} \mathbf{u}$, such that $\tilde{\mathbf{u}}.x = \mathbf{u}.x, \forall x \in E$; ϕ is the canonical epimorphism from E onto $E/\ker \mathbf{u}$; $\mathbf{i}$ is the inclusion from $\operatorname{im} \mathbf{u}$ into F; $\cong$ designates an isomorphism.

Application to a direct sum

Consider the direct sum (13.18) and the projection $\mathbf{p}_1$ from E into E_1 parallel to E_2. The canonical decomposition of this epimorphism is represented by the diagram in Figure 13.2 below. We have $E/E_2 \cong E_1$ and thus, according to (13.19) (see section 13.3.1):

$$\dim (E/E_2) + \dim E_2 = \dim E.$$

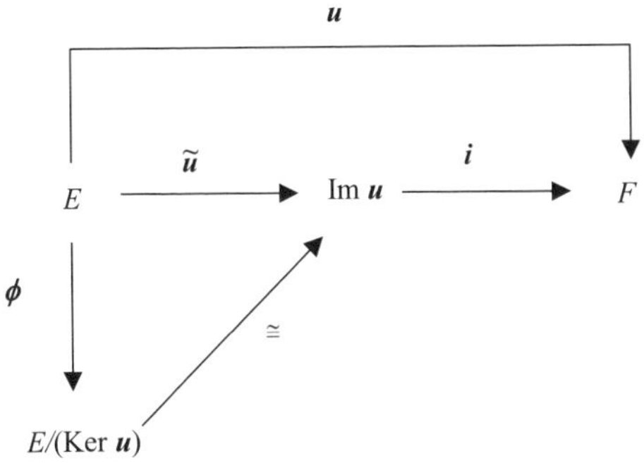

Figure 13.1. *Canonical decomposition of a homomorphism*

REMARK 518.– *Let* $\mathbf{u}$ *be a homomorphism from* E *into* F*, and let* E_1 *and* F_1 *be subspaces of* E *and* F *respectively, such that* $\mathbf{u}(E_1) \subset F_1$*. If* x_1 *and* x_2 *have the same canonical image in* E/E_1*, then* $\mathbf{u}\,x_1$ *and* $\mathbf{u}\,x_2$ *have the same canonical image in* F/F_1*. Therefore, there exists a homomorphism* $\bar{\mathbf{u}} : E/E_1 \to F/F_1$ *called the* homomorphism induced *by* $\mathbf{u}$*, making the diagram below commutative, where* ϕ *and* ψ *denote the canonical epimorphism from* E *onto* E/E_1 *and from* F *onto* F/F_1*, respectively:*

$$\begin{array}{ccc} E & \xrightarrow{\mathbf{u}} & F \\ \downarrow \phi & & \downarrow \psi \\ E/E_1 & \xrightarrow{\bar{\mathbf{u}}} & F/F_1 \end{array}$$

Note that if $F_1 = \{0\}$*, then* $F/F_1 = F$*, and the condition* $\mathbf{u}(E_1) \subset F_1$ *(which is necessary and sufficient for* $\bar{\mathbf{u}}$ *to exist) is equivalent to* $E_1 \subset \ker \mathbf{u}$.

Rank of a homomorphism

The *rank* of a homomorphism $\mathbf{u} : E \to F$ is defined by: $\operatorname{rk} \mathbf{u} = \dim(\operatorname{im} \mathbf{u})$.

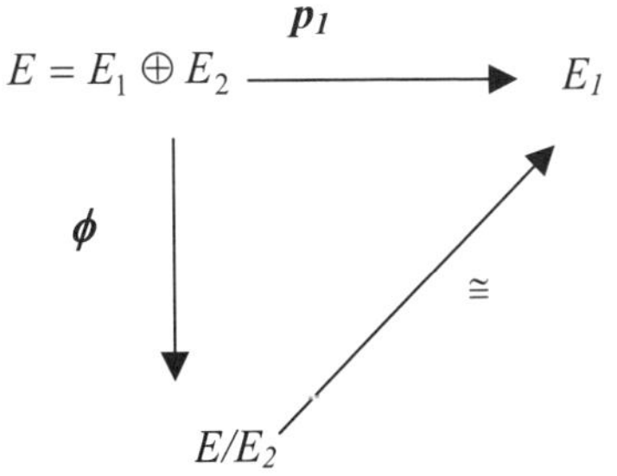

Figure 13.2. *Decomposition of a projection*

Since $\mathrm{im}\,\mathbf{u} \cong E/\ker \mathbf{u}$, we have

$$\boxed{\mathrm{rk}\,\mathbf{u} + \dim(\ker \mathbf{u}) = \dim E}. \tag{13.20}$$

Matrix representation

Matrix of a homomorphism

Let $\mathbf{u} : E \to F$ be a homomorphism, $(e_i)_{1 \leq i \leq n}$ be a basis of E and $(\varepsilon_i)_{1 \leq i \leq m}$ be a basis of F. In these bases, we associate a matrix $A \in \mathbf{K}^{m \times n}$ with $\mathbf{u}$ in the following manner: the element a_{ij} of the ith row and jth column of A is the ith component of $\mathbf{u}.e_j$ in the basis $(\varepsilon_i)_{1 \leq i \leq m}$.[5]

Let $x \in E$ and $y = \mathbf{u}.x \in F$; let $X = [x_1, ..., x_n]^T$ be the representative of x in the basis $(e_i)_{1 \leq i \leq n}$, and let $Y = [y_1, ..., y_m]^T$ be the representative of y in the basis $(\varepsilon_i)_{1 \leq i \leq m}$. We have:

$$y = \mathbf{u}.x = \mathbf{u}.\sum_{j=1}^{m} x_j\, e_j = \sum_{j=1}^{m} x_j\, \mathbf{u}.e_j;$$

as a result,

$$y_i = \sum_{j=1}^{m} a_{ij}\, x_j,$$

so that

$$\boxed{Y = A\,X}. \tag{13.21}$$

Change of basis

Consider a new basis $(e_i')_{1 \leq i \leq n}$ of E. The change of basis matrix P from $(e_i)_{1 \leq i \leq n}$ to $(e_i')_{1 \leq i \leq n}$ is the matrix of the identity of E (denoted by I_E), from E equipped with the basis $(e_i')_{1 \leq i \leq n}$ onto E equipped with the basis $(e_i)_{1 \leq i \leq n}$. The entry p_{ij} of this matrix $P \in \mathbf{K}^{n \times n}$ is thus the ith component, in the basis $(e_i)_{1 \leq i \leq n}$, of e_j' (in other words, the jth column of P consists of the components of e_j' in the basis $(e_i)_{1 \leq i \leq n}$).

If X and X' are the representatives of a same vector $x \in E$ in the bases $(e_i)_{1 \leq i \leq n}$ and $(e_i')_{1 \leq i \leq n}$ respectively, we have according to (13.21)

$$\boxed{X = P\,X'}. \tag{13.22}$$

Every change of basis matrix is invertible and, conversely, every invertible matrix can be considered as a change of basis matrix.

5. A different convention is used in section 13.4.

Change of basis and equivalence

Let us do the same in F : let $(\varepsilon'_i)_{1\leq i\leq m}$ be a new basis of F and let $Q \in \mathbf{K}^{m\times m}$ be the change of basis matrix. Last, let A' be the matrix representing the homomorphism $\mathbf{u}$ in the new bases and let Y' be the representative of $y = \mathbf{u}.x$ in the basis $(\varepsilon'_i)_{1\leq i\leq m}$. We have $Y' = A'\,X'$, and therefore, according to (13.21) and (13.22)

$$\boxed{A' = Q^{-1}\,A\,P}. \tag{13.23}$$

Thus, two matrices represent the same homomorphism if and only if they are equivalent (section 13.1.4). In particular, the simplest possible form of a matrix representing a homomorphism (of vector spaces) is furnished by Corollary 501 (section 13.2.3). The reader can verify that the notion of rank, as defined in the present paragraph, is identical to that defined in section 13.1.4 (in the case where the ring considered is a field).

Transposition

Let $\mathbf{u} : E \to F$ (where the $\mathbf{K}$-vector spaces E and F are finite-dimensional), and let E^* and F^* be the dual spaces of E and F, respectively (section 12.1.2). There exists a unique homomorphism ${}^t\mathbf{u} : F^* \to E^*$ such that for any $x \in E$ and any $y^* \in F^*$, $\langle {}^t\mathbf{u}\,y^*, x\rangle = \langle y^*, \mathbf{u}\,x\rangle$, and ${}^t\mathbf{u}$ is called the transpose of $\mathbf{u}$.

Let $(e_i)_{1\leq i\leq n}$ be a basis of E. The linear forms e_i^* defined by $\langle e_i^*, e_j\rangle = \delta_{ij}$ $(1 \leq i \leq n, 1 \leq j \leq n)$ where δ_{ij} is the *Kronecker index* (defined by $\delta_{ij} = 0$ if $i \neq j$, $\delta_{ii} = 1$), form a basis of E^*, and $(e_i^*)_{1\leq i\leq n}$ is called the *dual basis* of $(e_i)_{1\leq i\leq n}$.

Let $(e_i)_{1\leq i\leq n}$ and $(\varepsilon_i)_{1\leq i\leq m}$ be bases of E and F, respectively. In these bases, $\mathbf{u}$ is represented by a matrix A. We can easily verify that in the dual bases $(\varepsilon_i^*)_{1\leq i\leq m}$ and $(e_i^*)_{1\leq i\leq n}$, ${}^t\mathbf{u}$ is represented by the matrix A^T, i.e. the transpose of matrix A (section 13.1.4).

13.3.3. *Endomorphisms of vector spaces*

Endomorphisms and automorphisms

Let E be a $\mathbf{K}$-vector space of dimension n. An *endomorphism* $\mathbf{u}$ of E is a homomorphism from E into itself.

This endomorphism is injective if and only if $\dim(\ker \mathbf{u}) = 0$, which, according to (13.20), is the same as saying that $\operatorname{rk}\mathbf{u} = n$, or that $\mathbf{u}$ is surjective. One such endomorphism is called an *automorphism* of E.

Matrix representation

Once a basis of E is chosen, an endomorphism $\mathbf{u}$ is represented by a matrix $A \in \mathbf{K}^{n\times n}$. This matrix A is invertible if, and only if $\mathbf{u}$ is an automorphism.

Let E_1 be a subspace of E, of dimension n_1. We say that E_1 is *invariant* under $\mathbf{u}$ if $\mathbf{u}.E_1 \subset E_1$ (where $\mathbf{u}.E_1 = \{\mathbf{u}.x_1 : x_1 \in E_1\}$).

Then let E_2 be a supplement of E_1 and let be a basis of E that is the concatenation of a basis of E_1 and a basis of E_2. In this new basis, the matrix A representing $\mathbf{u}$ takes the form

$$A = \begin{bmatrix} A_1 & * \\ 0 & * \end{bmatrix} \tag{13.24}$$

where $A_1 \in \mathbf{K}^{n_1\times n_1}$. This matrix A_1 represents, in the basis of E_1 considered, the restriction $\mathbf{u}_1$ of $\mathbf{u}$ to E_1 (thinking of this restriction as an endomorphism of E_1).

Diagonal sum of endomorphisms

Let $E_1, ..., E_q$ be subspaces of a $\mathbf{K}$-vector space E, such that $E = E_1 \oplus ... \oplus E_q$. In addition, for any $i \in \{1, ..., q\}$, let $\mathbf{u}_i$ be an endomorphism of E_i. On the other hand, let x be a vector of E and let x_i be its component in E_i ($i \in \{1, ..., q\}$). We can define the endomorphism $\mathbf{u}$ of E which associates $\mathbf{u}_i.x_i$ with each x_i for any $i \in \{1, ..., q\}$. This endomorphism is called the *diagonal sum* of the endomorphisms $\mathbf{u}_i$ and is denoted as $\mathbf{u}_1 \oplus ... \oplus \mathbf{u}_q$ (by analogy with a diagonal sum of matrices).

Conversely, let $\mathbf{u}$ be an endomorphism of $E = E_1 \oplus ... \oplus E_q$ such that for every $i \in \{1, ..., q\}$, E_i is invariant under $\mathbf{u}$. Then, denoting the restriction of $\mathbf{u}$ to E_i as $\mathbf{u}_i$ (thinking of this restriction as an endomorphism of E_i), we have $\mathbf{u} = \mathbf{u}_1 \oplus ... \oplus \mathbf{u}_q$.

Consider a basis $\mathcal{B}$ of E, of the form $\biguplus_{1\leq i\leq q} \mathcal{B}_i$, where for each $i \in \{1, ..., q\}$, $\mathcal{B}_i$ is a basis of E_i. Each endomorphism $\mathbf{u}_i$ is represented, in the basis $\mathcal{B}_i$, by a square matrix A_i; and according to (13.24), the endomorphism $\mathbf{u} = \mathbf{u}_1 \oplus ... \oplus \mathbf{u}_q$ is represented, in the basis $\mathcal{B}$, by the matrix $A_1 \oplus ... \oplus A_q$.

Change of basis and similarity

Let $\mathbf{u}$ be an endomorphism of a $\mathbf{K}$-vector space E of dimension n, $\mathcal{B}$ be a basis of E and $A \in \mathbf{K}^{n\times n}$ be the matrix representing $\mathbf{u}$ in this basis. Let $\mathcal{B}'$ be another basis of E, P be the change of basis matrix from $\mathcal{B}$ to $\mathcal{B}'$, and $A' \in \mathbf{K}^{n\times n}$ be the matrix representing $\mathbf{u}$ in the basis $\mathcal{B}'$. According to (13.23), we have

$$\boxed{A' = P^{-1} A P}. \tag{13.25}$$

Two matrices A and A' of $\mathbf{K}^{n\times n}$ satisfying such a relation are said to be *similar* [10] or *conjugate* [31] (which we denote as $A \approx A'$). According to (13.25),

$$\det A' = \det P^{-1} \det A \det P = \det A, \tag{13.26}$$

and thus the determinant of a square matrix depends only on the endomorphism $\mathbf{u}$ that this matrix represents. We can call it the *determinant of this endomorphism* and denote it as $\det \mathbf{u}$.

Eigenvalues and eigenvectors

An element $\lambda \in \mathbf{K}$ is called an *eigenvalue* of the endomorphism $\mathbf{u}$ if there exists a non-zero vector $x \in E$ such that

$$\mathbf{u}.x = \lambda x,$$

in other words

$$\boxed{(\lambda I_E - \mathbf{u}).x = 0}. \tag{13.27}$$

In this case, x is called an *eigenvector* associated with the eigenvalue λ.

It is clear that there exist vectors $x \neq 0$ satisfying (13.27) if and only if $\ker(\lambda I_E - \mathbf{u}) \neq 0$, thus if

$$\det(\lambda I_E - \mathbf{u}) = 0. \tag{13.28}$$

We call the polynomial

$$\boxed{p_{\mathbf{u}}(s) = \det(s I_E - \mathbf{u})}.$$

the *characteristic polynomial* of $\mathbf{u}$.

As a result, $\lambda \in \mathbf{K}$ is an eigenvalue of the endomorphism $\mathbf{u}$ if and only if λ is a root of its characteristic polynomial.

One can prove the following ([10], n°III.8.1):

LEMMA 519.– *The characteristic polynomial $p_{\mathbf{u}}(s)$ can be written as*

$$p_{\mathbf{u}}(s) = s^n + \sum_{k=1}^{n} (-1)^k \Delta_k s^{n-k}$$

where $\Delta_1 = \operatorname{Tr} \mathbf{u}$ (the trace *of $\mathbf{u}$, i.e. the sum of all eigenvalues of $\mathbf{u}$) and $\Delta_n = \det \mathbf{u}$. Let A be the matrix representing $\mathbf{u}$ in any basis of E; then Δ_k is the sum of the principal minors of order k of A.*

We assume in what follows that the characteristic polynomial of the endomorphism considered has its roots in $\mathbf{K}$ (which is obviously ensured if $\mathbf{K} = \mathbb{C}$ or, more generally, if $\mathbf{K}$ is an algebraically closed field).

The set of all eigenvalues of $\mathbf{u}$ is called its *spectrum* and is denoted as $\operatorname{Sp}(\mathbf{u})$.

A useful lemma

Let $A \in \mathbf{K}^{n \times m}$ and $B \in \mathbf{K}^{m \times n}$ be two matrices.

LEMMA 520.– *(i) The non-zero eigenvalues of* $A\,B$ *and those of* $B\,A$ *coincide. (ii)* $I_n + A\,B$ *is invertible if and only if* $I_m + B\,A$ *is invertible and in that case* $B\left(I_n + A\,B\right)^{-1} = \left(I_m + B\,A\right)^{-1} B$.

PROOF. (i) $\lambda \in \mathbf{K}$ is an eigenvalue of $A\,B$ if and only if there exists a non-zero vector $x \in \mathbf{K}^n$ such that $\left(\lambda I_n - A\,B\right) x = 0$. Suppose $\lambda \neq 0$ and write $y = B\,x$. Then $y \neq 0$ and $\left(\lambda I_m - B\,A\right)\, y = 0$, which shows that λ is an eigenvalue of $B\,A$. By symmetry, the converse holds too. (ii) The matrix $I_n + A\,B$ is invertible if and only if -1 is not an eigenvalue of $A\,B$, and this holds if and only if $I_m + B\,A$ is invertible according to (i). Let then $v \in \mathbf{K}^n$ be any vector and $y = B\,\left(I_n + A\,B\right)^{-1} v = B\,u$ with $u = \left(I_n + A\,B\right)^{-1} v$. We thus have $u + A\,y = v$, from which $B\,A\,y = B\,v - B\,u = B\,v - y$, and finally $y = \left(I_m + B\,A\right)^{-1} B\,v$. Since $B\left(I_n + A\,B\right)^{-1} v = \left(I_m + B\,A\right)^{-1} B\,v$ for any $v \in \mathbf{K}^n$, we get the desired identity. ■

Multiplicities of an eigenvalue

Algebraic multiplicity

Let $\lambda \in \mathbf{K}$ be a root of $p_{\mathbf{u}}\left(s\right)$ and let σ be the order of multiplicity of this root, so that

$$p_{\mathbf{u}}\left(s\right) = \left(s - \lambda\right)^{\sigma} \pi_{\lambda}\left(s\right)$$

where $\pi_{\lambda}\left(s\right)$ is a polynomial which is not divisible by $s - \lambda$. We call σ the *algebraic multiplicity* of the eigenvalue λ.

Eigenspace and geometric multiplicity

The eigenspace associated with the eigenvalue λ is

$$E_{\lambda} = \ker\left(\lambda I_E - \mathbf{u}\right).$$

The *geometric multiplicity* of the eigenvalue λ is $\rho = \dim E_{\lambda}$.

LEMMA 521.– *The inequality* $\rho \leq \sigma$ *is always satisfied.*

PROOF. Let $\left(\varepsilon_i\right)_{1 \leq i \leq n}$ be a basis of E, the first ρ elements of which form a basis of E_{λ}; in this basis, $\mathbf{u}$ is represented, according to (13.24), by a matrix of the form

$$A = \begin{bmatrix} \Lambda & * \\ 0 & B \end{bmatrix}$$

where $\Lambda = \operatorname{diag}\left(\lambda, ..., \lambda\right)$, λ repeated ρ times. As a result, $\det\left(sI_n - A\right) = \left(s - \lambda\right)^{\rho} \det\left(s\,I_{n-\rho} - B\right)$, thus $\left(s - \lambda\right)^{\rho}$ divides $p_{\mathbf{u}}\left(s\right)$, which proves that $\rho \leq \sigma$. ■

Diagonalizable endomorphisms

An endomorphism $\mathbf{u}$ of E is said to be *diagonalizable* if there exists a basis of E in which $\mathbf{u}$ is represented by a diagonal matrix A.

It is immediate that, in this case, the eigenvalues of $\mathbf{u}$ are found on the diagonal of A and that the image of any vector ε of this basis by $\mathbf{u}$ is of the form $\mathbf{u}.\varepsilon = \lambda\,\varepsilon$, thus the basis in question is constituted of eigenvectors. As a result, a necessary condition for $\mathbf{u}$ to be diagonalizable is that $\mathbf{u}$ has n $\mathbf{K}$-linearly independent eigenvectors. It is clear that this is also a sufficient condition.

Let us look at this in more detail.

Let $\lambda_1, ..., \lambda_k$ be the distinct eigenvalues of $\mathbf{u}$, $\sigma_1, ..., \sigma_k$ be their algebraic multiplicities and $\rho_1, ..., \rho_k$ be their geometric multiplicities.

First, we have $\sum_{i=1}^{k} \sigma_i = d^\circ\left(p_{\mathbf{u}}\right) = n$.

On the other hand,

$$E_{\lambda_i} \cap E_{\lambda_j} = 0 \quad (i \neq j)\,. \tag{13.29}$$

Indeed, if $x \in E_{\lambda_i} \cap E_{\lambda_j}, i \neq j$, we have $\mathbf{u}\,x = \lambda_i\,x = \lambda_j\,x$, thus $(\lambda_i - \lambda_j)\,x = 0$, from which $x = 0$ because $\lambda_i \neq \lambda_j$.

For any $i \in \{1, ...k\}$, E_{λ_i} is invariant under $\mathbf{u}$ and the restriction $\mathbf{u}_{\lambda_i}$ of $\mathbf{u}$ to E_{λ_i} is represented (whatever the basis of E_{λ_i} considered) by the diagonal matrix $\lambda_i I_{\rho_i}$.

As a result:

If $\rho_i = \sigma_i, \forall i \in \{1, ..., k\}$, then $E = E_{\lambda_1} \oplus ... \oplus E_{\lambda_k}$, $\mathbf{u} = \mathbf{u}_{\lambda_1} \oplus ... \oplus \mathbf{u}_{\lambda_k}$ and this endomorphism is represented by the diagonal matrix

$$A = \left(\lambda_1 I_{\rho_1}\right) \oplus ... \oplus \left(\lambda_k I_{\rho_k}\right)$$

in a basis constituted of eigenvectors.

If there exists an index $j \in \{1, ..., k\}$ such that $\rho_j < \sigma_j$, there will *not* be n $\mathbf{K}$-linearly independent eigenvectors and $\mathbf{u}$ is not diagonalizable.

We thus have obtained the following:

THEOREM 522.– *An endomorphism* $\mathbf{u}$ *of* E *is diagonalizable if and only if the geometric multiplicity of each of its eigenvalues is equal to its algebraic multiplicity. This holds if and only if there exists a basis of* E *that consists of eigenvectors of* $\mathbf{u}$*, and then* $\mathbf{u}$ *is represented in this basis by a diagonal matrix.*

Polynomials of endomorphisms

Let $p(s) = s^m + p_1 s^{m-1} + \cdots + p_m$ be a polynomial of $\mathbf{K}[s]$. On the other hand, let $\mathbf{u}$ be an endomorphism of the $\mathbf{K}$-vector space E. Write

$$p(\mathbf{u}) = \mathbf{u}^m + p_1 \mathbf{u}^{m-1} + \cdots + p_m I_E$$

where $\mathbf{u}^i$ is the ith iteration of $\mathbf{u}$, defined by induction according to $\mathbf{u}^0 = I_E$ and $\mathbf{u}.\mathbf{u}^i = \mathbf{u}^{i+1}, i \geq 0$; $p(\mathbf{u})$ is an endomorphism of E.

Minimal polynomial

THEOREM 523.– *There exists a non-empty subset $\mathfrak{A}$ of $\mathbf{K}[s]$ consisting of those polynomials $p(s)$ that annihilate $\mathbf{u}$, i.e. which are such that $p(\mathbf{u}) = 0$. This set $\mathfrak{A}$ is an ideal in $\mathbf{K}[s]$. There exists a unique polynomial $q_{\mathbf{u}}(s)$ in $\mathfrak{A}$ which is monic and of minimal degree.*

PROOF. Let $\{e_1, ..., e_n\}$ be a basis of E and consider the ith element e_i of this basis. The $n+1$ elements $e_i, \mathbf{u}.e_i, ..., \mathbf{u}^n.e_i$ are $\mathbf{K}$-linearly dependent since $\dim E = n$. Therefore there exist $n+1$ coefficients γ_{ij} of $\mathbf{K}$, not all zero, such that $\sum_{j=1}^{n+1} \gamma_{ij}\, \mathbf{u}^{j-1}.e_i = 0$; thus, putting

$$\gamma_i(s) = \sum_{j=1}^{n+1} \gamma_{ij} s^{j-1}$$

we obtain $\gamma_i(\mathbf{u}).e_i = 0$. Therefore $\gamma(s) = \prod_{i=1}^{n} \gamma_i(s)$ is a polynomial such that for any $i \in \{1, ..., n\}$, $\gamma(\mathbf{u}).e_i = 0$ and thus $\gamma(\mathbf{u}).x = 0, \forall x \in E$. As a result, $\gamma(\mathbf{u}) = 0$. Thus, the set $\mathfrak{A}$ is non-empty since $\gamma(s) \in \mathfrak{A}$. It is clear that $\mathfrak{A}$ is an additive group.

If $\omega(s) \in \mathfrak{A}$ and if $\varphi(s) \in \mathbf{K}[s]$ is a multiple of $\omega(s)$, i.e. there exists $\psi(s) \in \mathbf{K}[s]$ such that $\varphi(s) = \omega(s)\psi(s)$, then $\varphi(\mathbf{u}) = \omega(\mathbf{u})\psi(\mathbf{u}) = 0$, and so $\varphi(s) \in \mathfrak{A}$. Therefore $\mathfrak{A}$ is an ideal in $\mathbf{K}[s]$; and since this ring is a principal ideal domain, the ideal $\mathfrak{A}$ is principal, and there exists a unique monic polynomial $q_{\mathbf{u}}(s)$ such that $\mathfrak{A} = (q_{\mathbf{u}})$. ■

DEFINITION 524.– *The polynomial $q_{\mathbf{u}}(s)$ is called the* minimal polynomial *of* $\mathbf{u}$.

LEMMA 525.– *The roots of the minimal polynomial are the eigenvalues of* $\mathbf{u}$.

PROOF. Let us temporarily add the hypothesis that $\mathbf{K}$ contains the eigenvalues of $\mathbf{u}$ *and* the roots of $q_{\mathbf{u}}$. 1) If $\mu \in \mathbf{K}$ is a root of $q_{\mathbf{u}}$ and is not an eigenvalue of $\mathbf{u}$, we can write $q_{\mathbf{u}}(s) = (s - \mu)\, q'_{\mathbf{u}}(s)$, where $q'_{\mathbf{u}}(s) \in \mathbf{K}[s]$. It is clear that $\mathbf{u} - \mu I_E$ is an automorphism, thus $q_{\mathbf{u}}(\mathbf{u}) = 0$ implies $q'_{\mathbf{u}}(\mathbf{u}) = 0$, which is impossible since

$d^\circ(q'_{\mathbf{u}}) < d^\circ(q_{\mathbf{u}})$. 2) Conversely, for any eigenvalue λ of $\mathbf{u}$, the associated eigenspace E_λ is only annihilated by the multiples of $s-\lambda$, thus $s-\lambda$ must be a divisor of $q_{\mathbf{u}}(s)$. ■

The following proposition is an obvious consequence of Lemma 525:

PROPOSITION 526.– *The factorization of $q_{\mathbf{u}}(s)$ into primes is of the form*

$$q_{\mathbf{u}}(s) = \prod_{i=1}^{k} (s-\lambda_i)^{\beta_i}$$

where $\beta_i \geq 1, i \in \{1,...,k\}$.

13.3.4. * *Jordan form*

This section provides a constructive proof of the reduction to Jordan form[6].

Generalized eigenspaces

Consider the expression of the minimal polynomial $q_{\mathbf{u}}(s)$ of an endomorphism $\mathbf{u}$, given by Proposition 526.

THEOREM 527.– *Let $\tilde{E}_{\lambda_i} = \ker(\lambda_i I_E - \mathbf{u})^{\beta_i}$. These spaces $\tilde{E}_{\lambda_i}$ have the following properties: (i) $\tilde{E}_{\lambda_i}$ is invariant under $\mathbf{u}$; (ii) $\tilde{E}_{\lambda_i} \cap \tilde{E}_{\lambda_j} = 0$ if $i \neq j$; (iii) $E = \tilde{E}_{\lambda_1} \oplus ... \oplus \tilde{E}_{\lambda_k}$; (iv) $\tilde{E}_{\lambda_i} \supset E_{\lambda_i}$, and $\tilde{E}_{\lambda_i} = E_{\lambda_i}$ if and only if $\beta_i = 1$.*

PROOF. We have $s(\lambda_i - s)^{\beta_i} = (\lambda_i - s)^{\beta_i} s$, thus $\mathbf{u}(\lambda_i I_E - \mathbf{u})^{\beta_i} = (\lambda_i I_E - \mathbf{u})^{\beta_i} \mathbf{u}$. Therefore, if x is a vector of E such that $(\lambda_i I_E - \mathbf{u})^{\beta_i} x = 0$, then $(\lambda_i I_E - \mathbf{u})^{\beta_i} \mathbf{u} x = 0$, which means that $\tilde{E}_{\lambda_i}$ is invariant under $\mathbf{u}$. (ii) Suppose that $x \in \tilde{E}_{\lambda_i} \cap \tilde{E}_{\lambda_j}, i \neq j$. Since the polynomials $(\lambda_i - s)^{\beta_i}$ and $(\lambda_j - s)^{\beta_j}$ are coprime, there exist, according to the Bézout theorem, polynomials $v(s)$ and $w(s)$ in $\mathbf{K}[s]$ such that

$$v(s)(\lambda_i - s)^{\beta_i} + w(s)(\lambda_j - s)^{\beta_j} = 1.$$

As a result,

$$v(\mathbf{u})(\lambda_i I_E - \mathbf{u})^{\beta_i}.x + w(\mathbf{u})(\lambda_j I_E - \mathbf{u})^{\beta_j}.x = x,$$

6. The approach taken here is the most traditional. Another more abstract approach is presented in section 13.4.2.

thus $x = 0$. (iii) Let $\tilde{E} = \tilde{E}_{\lambda_1} \oplus ... \oplus \tilde{E}_{\lambda_k}$; if $\tilde{E} \neq E$, this subspace $\tilde{E}$ will admit a supplement in E (section 13.3.1, Theorem 516). Now suppose that $x \neq 0$ is a vector belonging to this supplement: it is clear that we would have $q_{\mathbf{u}}(\mathbf{u}).x \neq 0$, which is impossible. (iv) is obvious. ∎

DEFINITION 528.– *If* $\beta_i > 1$, $\tilde{E}_{\lambda_i}$ *is called the* generalized eigenspace *associated with the eigenvalue* λ_i.

According to Theorem 527, we have

$$\boxed{\mathbf{u} = \tilde{\mathbf{u}}_{\lambda_1} \oplus ... \oplus \tilde{\mathbf{u}}_{\lambda_k}}$$

where $\tilde{\mathbf{u}}_{\lambda_i}$ is the restriction of $\mathbf{u}$ to $\tilde{E}_{\lambda_i}$. To further study the endomorphism $\mathbf{u}$, we are now led to study any one of its restrictions $\tilde{\mathbf{u}}_{\lambda_i}$.

The problem comes down to the case where $\lambda_i = 0$ by writing $\tilde{\mathbf{v}}_{\lambda_i} = \tilde{\mathbf{u}}_{\lambda_i} - \lambda_i I_{\tilde{E}_{\lambda_i}}$. The minimal polynomial of $\tilde{\mathbf{v}}_{\lambda_i}$ is obviously s^{β_i}. This study is carried out below, with a simplified notation.

Nilpotent endomorphisms

So, let $\mathbf{v}$ be an endomorphism of a $\mathbf{K}$-vector space E of dimension n, the minimal polynomial of which is $q_{\mathbf{v}}(s) = s^m, m \leq n$.

We thus have $\mathbf{v}^m = 0$: this endomorphism is said to be *nilpotent*. On the other hand, $\mathbf{v}^r \neq 0, \forall r \in \{0, ..., m-1\}$.

Cyclic subspace

Since $\mathbf{v}^{m-1} \neq 0$, there exists a vector e_1 of E such that $\mathbf{v}^{m-1} e_1 \neq 0$.

The vectors $e_1, \mathbf{v}.e_1, ..., \mathbf{v}^{m-1}.e_1$ are $\mathbf{K}$-linearly independent. Indeed, suppose there exist scalars $\xi_1, ..., \xi_m$, not all zero, such that

$$\sum_{i=1}^{m} \xi_i \, \mathbf{v}^{m-i} \, e_1 = 0, \tag{13.30}$$

and let ξ_j be the first non-zero scalar of the list $(\xi_i)_{1 \leq i \leq m}$. By applying the operator $\mathbf{v}^{j-1}$ to (13.30), we obtain $\xi_j \, \mathbf{v}^{m-1} \, e_1 = 0$, which is impossible.

Let E_1 be the subspace generated by the vectors e_1, $\mathbf{v}.e_1, ..., \mathbf{v}^{m-1}.e_1$; this subspace admits a basis $\{e_1, \mathbf{v}.e_1, ..., \mathbf{v}^{m-1}.e_1\}$, and is obviously invariant under $\mathbf{v}$. A subspace having this double property (i.e. a subspace that is invariant under $\mathbf{v}$ and admits a basis having the above particular form) is said to be $\mathbf{v}$-cyclic, and we call $\{e_1, \mathbf{v}.e_1, ..., \mathbf{v}^{m-1}.e_1\}$ a $\mathbf{v}$-*cyclic basis* of E_1. We call e_1 a $\mathbf{v}$-*cyclic generator* of this basis, as well as of E_1.

The restriction $\mathbf{v}_1$ of $\mathbf{v}$ to E_1 is represented, in the cyclic basis considered, by the square matrix of order m

$$J_{0,m} = \begin{bmatrix} 0 & 0 & \cdots & 0 \\ 1 & \ddots & \ddots & \vdots \\ \vdots & \ddots & \ddots & \vdots \\ 0 & \cdots & 1 & 0 \end{bmatrix}$$

all the elements of which are zero, except those on the subdiagonal, which are 1.

The matrix $J_{\lambda,m} = J_{0,m} + \lambda I_m$ is called the *Jordan block* of order m relative to λ.

Decomposition into cyclic subspaces

If $n = m$, the entire vector space E is $\mathbf{v}$-cyclic and there is nothing more to say about it.

Suppose $m < n$. It is clear that $\operatorname{im}\mathbf{v}^r \subset \operatorname{im}\mathbf{v}^{r'}$ if $r \geq r'$ and $\operatorname{im}\mathbf{v}^m = 0$, thus there exists an integer $m_2 \in \{1, ..., m\}$ such that $\operatorname{im}\mathbf{v}^{m_2} \subset E_1$ and $\operatorname{im}\mathbf{v}^{m_2-1} \cap E_1 \neq 0$. Thus, let ε_2 be a non-zero vector of E such that $\mathbf{v}^{m_2-1}.\varepsilon_2 \notin E_1$ and let $\mu_0, ..., \mu_{m-1}$ be the components of $\mathbf{v}^{m_2}.\varepsilon_2$ in the basis $\{e_1, \mathbf{v}.e_1, ..., \mathbf{v}^{m-1}.e_1\}$ of E_1. Writing that $\mathbf{v}^{m-m_2}\,\mathbf{v}^{m_2}\,\varepsilon_2 = 0$, we immediately show (using the fact that the vectors $\mathbf{v}^{m-m_2}\,e_1, ..., \mathbf{v}^{m-1}.e_1$ are $\mathbf{K}$-linearly independent) that $\mu_i = 0$ $(1 \leq i \leq m_2 - 1)$, thus

$$\mathbf{v}^{m_2}.\varepsilon_2 = \mu_{m_2}\mathbf{v}^{m_2}e_1 + ... + \mu_{m-1}\mathbf{v}^{m-1}e_1.$$

Now let

$$e_2 = \varepsilon_2 - \left(\mu_{m_2}e_1 + ... + \mu_{m-1}\mathbf{v}^{m-m_2-1}e_1\right). \tag{13.31}$$

Then $\mathbf{v}^{m_2-1}.e_2 \notin E_1$ and $\mathbf{v}^{m_2}.e_2 = 0$. Thus, $\{e_2, \mathbf{v}\,e_2 ..., \mathbf{v}^{m_2-1}.e_2\}$ is a $\mathbf{v}$-cyclic basis of a $\mathbf{v}$-cyclic subspace E_2, and it is easily shown that $E_1 \cap E_2 = 0$.

Continuing this way, we construct a finite number ν of $\mathbf{v}$-cyclic subspaces E_i, such that $\dim E_{i+1} \leq \dim E_i$ and

$$E = E_1 \oplus ... \oplus E_\nu.$$

The restriction $\mathbf{v}_i$ of $\mathbf{v}$ to E_i is represented, in a $\mathbf{v}$-cyclic basis of E_i, by the lower triangular Jordan block J_{0,m_i} and we have

$$\mathbf{v} = \mathbf{v}_1 \oplus ... \oplus \mathbf{v}_\nu. \tag{13.32}$$

This endomorphism is thus represented, in a basis of E that consists of the concatenation of the $\mathbf{v}$-cyclic bases of the $\mathbf{v}$-cyclic subspaces $E_i, 1 \leq i \leq \nu$, by the lower triangular matrix

$$J_{0,m_1} \oplus ... \oplus J_{0,m_\nu}. \tag{13.33}$$

REMARK 529.– *(i) In the basis* $\{f_{i,1} = \mathbf{v}^{mi-1}.e_i, ..., f_{i,m_i} = e_i\}$ *of* E_i, $\mathbf{v}_i$ *is represented by the upper triangular Jordan block* J_{0,m_i}^T. *We have indeed the recurrence relation*

$$f_{i,j} = \mathbf{v}.f_{i,j+1}, \quad 1 \leq j \leq m_i - 1. \tag{13.34}$$

Therefore, $J_{0,m_i} \approx J_{0,m_i}^T$. *(ii) It is clear that* $f_{i,1} \in \ker \mathbf{v}$; *but we cannot generate, starting from* any *element of* $\ker \mathbf{v}$, *a* $\mathbf{v}$*-cyclic subspace* E_i *by the recursive procedure* (13.34) : *this procedure needs to be followed in the other direction, from a non-zero element of* $\ker \mathbf{v}^{m_i-1}$.

Similarity invariants and elementary divisors

Let $\mathbf{u}$ be an endomorphism of a $\mathbf{K}$-vector space E of dimension n and let $A \in \mathbf{K}^{n \times n}$ be a matrix representing $\mathbf{u}$ in a basis of E. The invariant factors of $sI_n - A$ are unchanged under change of basis, thus they only depend on the endomorphism $\mathbf{u}$. Among them, those which are not invertible (that is, not equal to 1) are called the *similarity invariants* of $\mathbf{u}$ (or of A).

Likewise, the elementary divisors of $s\,I_n - A$ depend only on the endomorphism $\mathbf{u}$ and are also called the *elementary divisors of this endomorphism* (or of A). If $\mathbf{K}$ contains all roots of $p_{\mathbf{u}}(s)$, these elementary divisors are of the form $(s - \lambda)^l$, where λ is an eigenvalue of $\mathbf{u}$.

It is clear that the Jordan block J_{0,m_i} has a unique similarity invariant which is s^{m_i}. Therefore, the *similarity invariants* of (13.32) are $s^{m_\nu}, ..., s^{m_1}$ and they coincide with the *elementary divisors* of this nilpotent endomorphism.

Jordan theorem

Let $\mathbf{u}$ be an endomorphism of a finite-dimensional $\mathbf{K}$-vector space E, such that all roots of its characteristic polynomial $p_{\mathbf{u}}(s)$ belong to $\mathbf{K}$. The *Jordan theorem* below is a direct consequence of the above:

THEOREM 530.– *The vector space E can be expressed as a direct sum of* $\mathbf{u}$*-cyclic subspaces* $E_{\lambda,l}$, *where* $E_{\lambda,l}$ *is associated with the elementary divisor* $(s - \lambda)^l$ *of* $\mathbf{u}$ $(\lambda \in \mathrm{Sp}(\mathbf{u}))$. *Choosing a* $\mathbf{u}$*-cyclic basis in each of these subspaces,* $\mathbf{u}$ *is represented by a diagonal sum of Jordan blocks* $J_{\lambda,l}$, *each of these blocks appearing in this diagonal sum a number of times equal to the multiplicity order of the elementary divisor* $(s - \lambda)^l$. *The number of blocks* $J_{\lambda,l}$ *appearing in this diagonal sum, for the* *same eigenvalue λ but different values of l, is equal to the geometric multiplicity of the eigenvalue λ. The matrix obtained in this way is lower triangular, and is called the* Jordan form *of* $\mathbf{u}$. *Two matrices having the same Jordan form are similar.*

DEFINITION 531.– *Let us take a basis of E in which the endomorphism $\mathbf{u}$ is represented by its Jordan form. The vectors of this basis are called the* generalized eigenvectors *of* $\mathbf{u}$, *and a generalized eigenvector x for which there exists $\lambda \in \mathrm{Sp}(\mathbf{u})$ and $k \geq 1$ such that $x \in \ker(\lambda I_E - \mathbf{u})^k$ is said to be* associated *with the eigenvalue λ.*

COROLLARY 532.– *An endomorphism is diagonalizable if and only if all its Jordan blocks are of order* 1, *that is if the roots of its minimal polynomial are all simple.*

COROLLARY 533.– *Let $\mathbf{u}$ be an endomorphism;* $\det \mathbf{u}$ *is the product of the eigenvalues of* $\mathbf{u}$ *(repeating each eigenvalue in this product a number of times equal to its algebraic multiplicity).*

From Theorem 530 and Remark 529(i), we deduce the following result:

PROPOSITION 534.– *Every matrix $A \in \mathbf{K}^{n \times n}$ is similar to its transpose.*

The proof of the following corollary is now easy:

COROLLARY 535.– *Let E be a finite-dimensional* $\mathbf{K}$*-vector space,* $\mathbf{u}$ *be an endomorphism of E and* ${}^t\mathbf{u}$ *be that endomorphism of E^* which is the transpose of* $\mathbf{u}$ *(section 13.3.2). The endomorphisms* $\mathbf{u}$ *and* ${}^t\mathbf{u}$ *have the same similarity invariants and in particular each eigenvalue of* $\mathbf{u}$ *is an eigenvalue of* ${}^t\mathbf{u}$. *Let λ and μ be eigenvalues of* $\mathbf{u}$, *let x be a generalized eigenvector of* $\mathbf{u}$ *associated with λ (Definition 531) and let y^* be a generalized eigenvector of* ${}^t\mathbf{u}$ *associated with μ; if $\lambda \neq \mu$, then we have* $\langle y^*, x\rangle = 0$.

Cayley-Hamilton theorem

Let $\alpha_{j+1}(s), \ldots, \alpha_n(s)$ be the similarity invariants of $\mathbf{u}$ (where j is the number of invariant factors of $s\,I_n - A$ that are equal to 1). According to Theorem 530, the fact that $\alpha_i \mid \alpha_{i+1}$ $(j+1 \leq i \leq n-1)$ and equality (13.26) (section 13.3.3), we have the following result:

PROPOSITION 536.– *The minimal polynomial $q_{\mathbf{u}}(s)$ and the characteristic polynomial $p_{\mathbf{u}}(s)$ of an endomorphism $\mathbf{u} : E \rightarrow E$ (where $\dim E = n$) can be expressed as a function of its similarity invariants $\alpha_i(s)$ $(j+1 \leq i \leq n)$ in the* *following manner:*

$$\boxed{q_{\mathbf{u}}(s) = \alpha_n(s)} \tag{13.35}$$

$$\boxed{p_{\mathbf{u}}(s) = \prod_{i=j+1}^{n} \alpha_i(s)} \tag{13.36}$$

Therefore, the minimal polynomial $q_{\mathbf{u}}(s)$ divides the characteristic polynomial $p_{\mathbf{u}}(s)$ (i.e. $p_{\mathbf{u}}(s) \in \mathcal{A}$: see Theorem 523 in section 13.3.3). We thus obtain the following result, called the *Cayley-Hamilton theorem*:

THEOREM 537.– *For any endomorphism* $\mathbf{u}$ *of a finite-dimensional* $\mathbf{K}$*-vector space* E*, the equality* $p_{\mathbf{u}}(\mathbf{u}) = 0$ *is satisfied.*

Example of reduction to Jordan form

Let

$$A = \begin{bmatrix} 0 & 0 & 0 & 0 & 1 \\ 1 & 0 & 0 & 1 & 1 \\ -1 & 1 & 0 & -1 & 0 \\ 0 & 0 & 0 & 0 & -1 \\ 0 & 0 & 0 & 0 & 0 \end{bmatrix}.$$

We have $p_A(s) = \det(s\, I_5 - A) = s^5$; thus the matrix A has a unique eigenvalue $\lambda = 0$, which is of algebraic multiplicity 5. Nevertheless, we can easily verify that $\dim \ker A = 2$, thus the eigenvalue $\lambda = 0$ is of geometric multiplicity 2. Thus A is not diagonalizable according to Theorem 522 (section 13.3.3), and is similar to the diagonal sum of two Jordan blocks according to Theorem 530. *A priori*, these may be either of order 1 and 4, or of order 2 and 3.

The reader can as an exercise verify that the Smith form of $sI_5 - A$ is $\operatorname{diag}\left(1, 1, 1, s^2, s^3\right)$, thus the similarity invariants of A are s^2 and s^3; it follows that the elementary divisors of A are s^2 and s^3, therefore

$$A \approx J_{0,3} \oplus J_{0,2}.$$

The minimal polynomial is $q_A(s) = s^3$. In addition,

$$E = E_1 \oplus E_2$$

where E_1 and E_2 are cyclic subspaces of dimension 2 and 3, respectively.

In order to determine E_1, let us choose a generator e_1 such that $A^2 e_1 \neq 0$.[7] We have

$$A^2 = \begin{bmatrix} 0 & 0 & 0 & 0 & 0 \\ 0 & 0 & 0 & 0 & 0 \\ 1 & 0 & 0 & 1 & 1 \\ 0 & 0 & 0 & 0 & 0 \\ 0 & 0 & 0 & 0 & 0 \end{bmatrix}.$$

The first column of A^2 is non-zero, thus we can take $e_1 = [1, 0, ..., 0]^T$; $A\, e_1$ and $A^2\, e_1$ are then the first column of A and A^2, respectively, i.e. $[0, 1, -1, 0, 0]^T$ and $[0, 0, 1, 0, 0]^T$. Therefore, $\{e_1, A\, e_1, A^2\, e_1\}$ is a cyclic basis of E_1.

Let us now choose a vector ε_2 such that $A\, \varepsilon_2$ does not belong to E_1. The vector $\varepsilon_2 = [0, 0, 0, 0, 1]^T$ satisfies this condition, since $A\, \varepsilon_2 = [1, 1, 0, -1, 0]^T$ (fifth column of A). On the other hand, we have $A^2\, \varepsilon_2 = A^2\, e_1$; let us take, according to (13.31), $e_2 = \varepsilon_2 - e_1 = [-1, 0, 0, 0, 1]^T$.

The change of basis matrix then obtained is

$$P = \left[e_1, A\, e_1, A^2\, e_1, e_2, A\, e_2\right] = \begin{bmatrix} 1 & 0 & 0 & -1 & 1 \\ 0 & 1 & 0 & 0 & 0 \\ 0 & -1 & 1 & 0 & 1 \\ 0 & 0 & 0 & 0 & -1 \\ 0 & 0 & 0 & 1 & 0 \end{bmatrix}$$

The reader can verify that $P^{-1}\, A\, P = J_{0,3} \oplus J_{0,2}$.

Note that the determination of P is much easier by the previous determination of the structure of the Jordan form of A, i.e. by the calculation of its elementary divisors.

13.4. * The language of modules

13.4.1. *General notions*

Definition

The notion of *module* is a very practical way of efficiently introducing the key concepts of systems theory [42]. A module is analogous to a vector space, but instead of the scalars being elements of a field, these are elements of a ring.

7. Here, by the same abuse of language as usual, we identify a vector of $\mathbb{R}^5$ with its representative in the canonical basis of this space.

Let E be a vector space defined over a field $\mathbf{K}$ ($\mathbf{K}$ can be, for example, the field of real or complex numbers), let v be an element of E and let $\lambda \neq 0$ be an element of $\mathbf{K}$. If $\lambda\, v = 0$, then $v = 0$, since $(1/\lambda)\ \lambda\, v = 0$. This rationale is no longer valid if λ has no inverse, which could be the case if $\mathbf{K}$ were a ring, not a field.

Now let $\mathbf{A}$ be any ring. An $\mathbf{A}$-module M is a set equipped with an addition $+ : M \times M \to M$ (which is commutative and makes M an abelian group) and with a multiplication $\mathbf{A} \times M \to M : (\lambda, m) \mapsto \lambda m$ (for a *left module*, i.e. when the scalars multiply with the elements of this module from the left) such that

- $\lambda\,(m_1 + m_2) = \lambda m_1 + \lambda m_2$
- $\lambda_1\,(\lambda_2 m) = (\lambda_1 \lambda_2)\, m$
- $(\lambda_1 + \lambda_2)\, m = \lambda_1 m + \lambda_2 m$
- $1m = m$

for any $\lambda, \lambda_1, \lambda_2 \in \mathbf{A}$ and any $m, m_1, m_2 \in M$. If $\mathbf{A}$ is a field or a division ring $\mathbf{K}$, then M is a $\mathbf{K}$-vector space.

Elementary notions

Let M be an $\mathbf{A}$-module. A *submodule* N of M is an $\mathbf{A}$-module included in M.

We can define a *quotient module* M/N (where N is a submodule of M), a *sum* $M_1 + M_2$ and a *direct sum* (or *external direct sum*) $M_1 \oplus M_2$ of two $\mathbf{A}$-modules M_1 and M_2, exactly as we did in section 13.3.1 for vector spaces.

A homomorphism of $\mathbf{A}$-module (also called an $\mathbf{A}$-homomorphism) is an $\mathbf{A}$-linear mapping $f : M \to N$ (where M and N are $\mathbf{A}$-modules). We likewise define a *monomorphism*, an *epimorphism* and an *isomorphism* of $\mathbf{A}$-modules (the latter is again denoted as $\cong$). One can show the following (see [91], section III.10 and [102], Chapter 2):

THEOREM 538.– *(i) The canonical decomposition of a homomorphism (see the commutative diagram in Figure 13.1), the notion of induced homomorphism (see Remark 518) and the notion of projection onto a submodule parallel to a supplementary submodule (see the commutative diagram in Figure 13.2) are still valid even in the case of a homomorphism of* $\mathbf{A}$*-modules. (ii) Let* M' *and* M'' *be submodules of an* $\mathbf{A}$*-module* M*. Then*

$$\frac{M'}{M' \cap M''} \cong \frac{M' + M''}{M''}.$$

(iii) Let M*,* M' *and* M'' *be* $\mathbf{A}$*-modules such that* $M'' \subset M' \subset M$*. Then*

$$M/M' \cong \frac{M/M''}{M'/M''}.$$

(iv) Let M be an $\mathbf{A}$-module and M' be a submodule of M. There exists a one-to-one correspondence between submodules S of M/M' and the intermediate submodules Σ between M' and M (that is such that $M' \subset \Sigma \subset M$) given by $S \mapsto \Sigma \triangleq \varphi^{-1}(S)$ where $\varphi : M \to M/M'$ is the canonical epimorphism.
(v) Let M, M', M'', N' and N'' be $\mathbf{A}$-modules such that $M = M' \oplus M''$, $N' \subset M'$ and $N'' \subset M''$. Then we have the canonical isomorphism

$$\frac{M' \oplus M''}{N' \oplus N''} \cong \frac{M'}{N'} \oplus \frac{M''}{N''}.$$

Finitely generated modules

An $\mathbf{A}$-module M is said to be of *finitely generated* if it is generated by a finite number of elements $m_1, ..., m_k$, i.e. for any $x \in M$, there exist $\lambda_1, ..., \lambda_k \in \mathbf{A}$ such that $x = \sum_{1 \leq i \leq k} \lambda_i m_i$ (this expression is in general non-unique). We then write $M = \left[(m_i)_{1 \leq i \leq k}\right]_{\mathbf{A}}$ or simply $[\mathbf{m}]_{\mathbf{A}}$, where $\mathbf{m}$ is the finite sequence $(m_i)_{1 \leq i \leq k}$ (which can be identified with the column matrix $[m_1 \; ... \; m_k]^T$).

Free modules

A basis of an $\mathbf{A}$-module M is also defined as in section 13.3.1. But whereas every vector space admits a basis, those modules that admit a basis are quite particular and are called *free modules.*

A finitely generated free module L is isomorphic to a direct sum $\bigoplus_{i=1}^{k} \mathbf{A}_i = \mathbf{A}^k$, where each $\mathbf{A}_i$ is a module isomorphic to $\mathbf{A}$ (when the latter is considered as a module over itself). The integer k is called the *rank* of the free module L.

Presentation of a module

LEMMA 539.– *Let M be an $\mathbf{A}$-module generated by k elements m_i, $1 \leq i \leq k$. Then M is isomorphic to the quotient of a free $\mathbf{A}$-module of rank k.*

PROOF. Let $(c_i)_{1 \leq i \leq k}$ be the canonical basis of $\mathbf{A}^k$ and let $\varphi : \mathbf{A}^k \to M$ the $\mathbf{A}$-homomorphism defined by $\varphi(c_i) = m_i$, $1 \leq i \leq k$. It is clear that φ is an epimorphism, thus $M \cong \mathbf{A}^k / \ker \varphi$ according to Theorem 538(i). ■

THEOREM 540.– *Let $M = [\mathbf{m}]_{\mathbf{A}}$ be a finitely generated $\mathbf{A}$-module, where $\mathbf{m} = [m_1 \; ... \; m_k]^T$. The following conditions are equivalent: (i) $M \cong \mathbf{A}^k / \ker \varphi$ and $\ker \varphi$ is finitely generated. (ii) There exists an $\mathbf{A}$-homomorphism $f : \mathbf{A}^q \to \mathbf{A}^k$ such that*

$M \cong \mathbf{A}^k / \operatorname{im} f \triangleq \operatorname{coker} f$.[8] *(iii) There exists a matrix* $R \in \mathbf{A}^{q \times k}$ *such that the* generators m_i, $1 \leq i \leq k$, *are only related by the equality*

$$\boxed{R\mathbf{m} = 0}. \tag{13.37}$$

PROOF. 1) $(i) \Rightarrow (ii)$: If M is finitely generated, then $M \cong \mathbf{A}^k / \ker \varphi$ according to Lemma 539. Suppose $\ker \varphi$ is generated by q elements. Applying Lemma 539 to the module $\ker \varphi \subset \mathbf{A}^k$, there exists an epimorphism $g : \mathbf{A}^q \to \ker \varphi$. Let $f : \mathbf{A}^q \to \mathbf{A}^k$ be the $\mathbf{A}$-homomorphism defined by $f(x) = g(x)$, $x \in \mathbf{A}^q$. Then $\operatorname{im} f = \ker \varphi$. 2) $(ii) \Rightarrow (iii)$: Let $(a_i)_{1 \leq i \leq q}$ and $(c_i)_{1 \leq i \leq k}$ be the canonical bases of $\mathbf{A}^q$ and $\mathbf{A}^k$, respectively, and write

$$f(a_i) = \sum_{j=1}^{k} r_{ij} c_j, \quad 1 \leq i \leq q; \tag{13.38}$$

let R be the matrix (r_{ij}) $(1 \leq i \leq q, 1 \leq j \leq k)$. Let $m_j = \varphi(c_j), 1 \leq j \leq k$, where $\varphi : \mathbf{A}^k \to \mathbf{A}^k / \operatorname{im} f$ is the canonical epimorphism. Then the generators m_i of $\mathbf{A}^k / \operatorname{im} f$ are related by the only equality (13.37). 3) $(iii) \Rightarrow (i)$: See Remark 542. ∎

DEFINITION 541.– *An* $\mathbf{A}$*-module* M *satisfying one of the three equivalent conditions in Theorem 540 is said to be* finitely presented. *The relation* (13.37) *is called a* presentation *of* M *and the matrix* R *appearing in this relation is called a* matrix of definition *(or* matrix of presentation*) of* M.

REMARK 542.– *If we represent the elements of* $\mathbf{A}^q$ *and* $\mathbf{A}^k$ *by* rows *in the canonical bases, then* R *is the matrix of* f *in these bases and* f *is identified with the right-multiplication by* R *(written* $f = \bullet R$*)*[9]*. According to* (13.38), $f(\mathbf{A}\, a_i)$ *is the submodule of* $\mathbf{A}^k$ *generated by the ith row of* R*. The module* $M = [\mathbf{m}]_{\mathbf{A}}$ *is related to the definition matrix* R *by the isomorphism* $M \cong \operatorname{coker}(\bullet R) = \mathbf{A}^k / \ker \varphi$*, where* $\varphi : \mathbf{A}^k \to \mathbf{A}^k / \ker \varphi$ *is the canonical epimorphism.*

A definition matrix of a finitely presented module is non-unique. We have indeed the following result ([31], section 0.6):

THEOREM 543.– *Let* M *be a finitely presented* $\mathbf{A}$*-module with definition matrix* $R \in \mathbf{A}^{q \times k}$*; then* $R' \in \mathbf{A}^{q' \times k'}$ *is again a definition matrix of* M *if and only if* $\operatorname{coker}(\bullet R') \cong \operatorname{coker}(\bullet R)$*. The matrices* R *and* R' *are then said to be* left-similar *and this implies* $k - q = k' - q'$ *(this integer is called the* characteristic *of* M*). If* $R' \sim R$*, then* R *and* R' *are left-similar, but the converse does not hold.*

8. $\mathbf{A}^k / \operatorname{im} f \triangleq \operatorname{coker} f$ is called the *cokernel* of f.

9. This convention is adopted throughout section 13.4 but only in that section. There are good reasons for this (see [15], [22]).

Suppose from here on that the ring $\mathbf{A}$ is a *commutative domain*. One can prove the following result ([11], Chapter I):

THEOREM 544.– *If* $\mathbf{A}$ *is Noetherian, then any finitely generated* $\mathbf{A}$*-module is finitely presented.*

The Noetherian condition is too restrictive for certain applications (especially, a Bézout domain is not Noetherian unless it is a principal ideal domain) and this is why the following is useful:

DEFINITION 545.– *A ring* $\mathbf{A}$ *is* coherent *if every finitely generated ideal in* $\mathbf{A}$ *is finitely presented.*

Bézout domains, Noetherian rings, are all coherent. We have the following result ([31], Theorem A.9, p. 555):

THEOREM 546.– *If the ring* $\mathbf{A}$ *is coherent, then the category of finitely presented* $\mathbf{A}$*-modules of is closed under taking finitely generated submodules, finite direct sums, kernels and cokernels, * and thus it is an abelian subcategory of the category of all* $\mathbf{A}$*-modules. **

One can easily prove the following:

THEOREM 547.– *Let* M *be an* $\mathbf{A}$*-module with definition matrix* $R \in \mathbf{A}^{q\times k}$*. The module is* M *free if and only if* R *is* completable.

Torsion

Let $\lambda \neq 0$ be an element of $\mathbf{A}$ and let m be an element of M. As already said, the equality $\lambda m = 0$ does not imply $m = 0$, unless λ is an invertible element of $\mathbf{A}$. An element of M satisfying such an equality is called a *torsion element* and an element that is not of torsion is called *free*. The only torsion element of a vector space is 0. The set of torsion elements of an $\mathbf{A}$-module M is a submodule of M, and is referred to as the *torsion submodule* of M, denoted by $\mathcal{T}(M)$. A module M such that $\mathcal{T}(M) = 0$ (that is $\mathcal{T}(M)$ is reduced to the element 0, or in other words M only contains free elements) is said to be *torsion-free*. For any $\mathbf{A}$-module M, the quotient $M/\mathcal{T}(M)$ is torsion-free. All free $\mathbf{A}$-modules are torsion-free, while the converse does not hold in general.

We have the following ([10], n°II.7.10 & Chapter VII), ([31], section 0.3):

THEOREM 548.– *(A) Let* M *be a finitely presented* $\mathbf{A}$*-module and* $R \in \mathbf{A}^{q\times k}$ *be a presentation matrix of* M*. The* $\mathbf{A}$*-module* M *is of torsion if and only if* R *is* *left-invertible over* $\mathbf{F} = \mathrm{Q}(\mathbf{A})$. *(B) If A is a* principal ideal domain *(resp., a* Bézout domain*), then any submodule (resp., any finitely generated submodule) of a free* $\mathbf{A}$-*module is free.*

Cyclic modules

A cyclic $\mathbf{A}$-module is a module Γ generated by a unique element m, i.e. $\Gamma = [m]_{\mathbf{A}} = \mathbf{A}\,m$. We call the *annihilator* of m (and we denote as $\mathrm{Ann}\,(m)$) the subset $\mathfrak{a}$ of $\mathbf{A}$ such that $\mathfrak{a}m = 0$, that is $\lambda m = 0$, $\forall \lambda \in \mathfrak{a}$. It is clear that $\mathfrak{a}$ is an ideal. Let $\varphi : \mathbf{A} \to \Gamma$ be the epimorphism defined by $\varphi(\lambda) = \lambda m$; we have $\ker \varphi = \mathfrak{a}$, thus $\Gamma \cong \mathbf{A}/\mathfrak{a}$. Conversely, let $\mathfrak{a}$ be an ideal in $\mathbf{A}$ and let $\psi : \mathbf{A} \to \mathbf{A}/\mathfrak{a}$ be the canonical epimorphism. The module $\mathbf{A}/\mathfrak{a}$ is cyclic, generated by $\psi(1)$.

Indecomposable modules

DEFINITION 549.– *An* $\mathbf{A}$*-module M is said to be* decomposable *if it is the direct sum of two submodules different from 0 and M, and is said to be* indecomposable *otherwise.*

We have the following ([10], Chapter I):

LEMMA 550.– *Let $\mathfrak{a}$ be an ideal in $\mathbf{A}$. The cyclic module $\mathbf{A}/\mathfrak{a}$ is* decomposable *if and only if there exist two ideals $\mathfrak{b}$ and $\mathfrak{c}$, different from 0 and $\mathbf{A}$, such that $\mathbf{A} = \mathfrak{b} + \mathfrak{c}$ and $\mathfrak{a} = \mathfrak{b} \cap \mathfrak{c}$; then, $\mathbf{A}/\mathfrak{a} \cong \mathbf{A}/\mathfrak{b} + \mathbf{A}/\mathfrak{c}$.*

13.4.2. *Modules over principal ideal domains*

Except when otherwise stated, $\mathbf{R}$ is a principal ideal domain in the sequel.

Primary decomposition of a cyclic module

Since any ideal in $\mathbf{R}$ is principal, it is generated by a unique element a and is thus of the form $\mathbf{R}\,a = (a)$. Let $\Gamma \cong \mathbf{R}/(a)$ be a cyclic $\mathbf{R}$-module.

(i) If $a = 0$, then $\Gamma \cong \frac{\mathbf{R}}{(0)} = \mathbf{R}$ is a free module of rank 1.

(ii) If $a \neq 0$, then the cyclic module Γ is torsion, and $\Gamma = 0$ if and only if a is a unit of $\mathbf{R}$.

In what follows, we will use the following result (valid when $\mathbf{R}$ is a Bézout domain):

PROPOSITION 551.– *(i) Let $b, c \in \mathbf{R}^{\times}$ be two coprime elements. Then, $\frac{\mathbf{R}}{(b\,c)} \cong \frac{\mathbf{R}}{(b)} \oplus \frac{\mathbf{R}}{(c)}$. (ii) Let p be a prime in $\mathbf{R}$ and n be a positive integer; then the module $\frac{\mathbf{R}}{(p^n)}$ is indecomposable.*

PROOF. (i) Let $a = bc$. Since b and c are coprime, we have $(b) + (c) = \mathbf{R}$ and $(b) \cap (c) = (a)$. Thus, Assertion (i) is a consequence of Lemma 550. (ii) Suppose $\frac{\mathbf{R}}{(p^n)}$ is decomposable. According to Lemma 550, there exist elements $b, c \in \mathbf{R}$ that are coprime and such that $p^n \in \operatorname{lcm}(b, c)$. This is obviously impossible. ■

Given a nonunit a of $\mathbf{R}$, let us consider the unique factorization of this element into primes (section 13.1.2), now written in the form

$$a = v \prod_i p_i^{k_i}, \quad k_i \geq 0.$$

According to Proposition 551, it is clear that

$$\frac{\mathbf{R}}{(a)} \cong \bigoplus_i \frac{\mathbf{R}}{\left(p_i^{k_i}\right)} \tag{13.39}$$

and that the modules $\frac{\mathbf{R}}{\left(p_i^{k_i}\right)}$ $(k_i \neq 0)$ are indecomposable. The terms $p_i^{k_i}$ which are not equal to 1 are called the *elementary divisors* of the cyclic module $M \cong \frac{\mathbf{R}}{(a)}$. A module that is a direct sum $\bigoplus_{j \in J} \frac{\mathbf{R}}{(p_i^j)}$ (J finite) is said to be *primary* (we will encounter such modules in the explicit decomposition in Theorem 556 below).

Canonical decomposition of a module

Let M be a finitely generated module over a principal ideal domain $\mathbf{R}$. According to Proposition 544 (section 13.4.1), M is finitely presented, thus $M = [\mathbf{m}]_{\mathbf{A}}$ is defined by a relation such that (13.37) where $R \in \mathbf{R}^{q \times k}$.

There exist matrices $U \in \mathrm{GL}_q(\mathbf{R})$ and $V \in \mathrm{GL}_k(\mathbf{R})$ such that

$$U^{-1} R V = \Sigma$$

where $\Sigma = \operatorname{diag}(\alpha_1, ..., \alpha_r, 0, ..., 0)$ is the Smith form of R (Theorem 497, section 13.2.3). The polynomials α_i, $1 \leq i \leq r$, are the invariant factors of R. They are *non-zero* and such that $\alpha_i \mid \alpha_{i+1}, 1 \leq i \leq r - 1$.

Write $\mathbf{m} = V\, \mathbf{v}$, where $\mathbf{v} = [v_1 \; ... \; v_k]^T$. Then $M = [\mathbf{v}]_{\mathbf{R}}$ and equation (13.37) of section 13.4.1 is equivalent to

$$\operatorname{diag}(\alpha_1, ..., \alpha_r, 0, ..., 0) \begin{bmatrix} v_1 \\ \vdots \\ v_r \\ v_{r+1} \\ \vdots \\ v_k \end{bmatrix} = 0. \tag{13.40}$$

If $q > r$, (13.40) (and thus (13.37)) includes $q - r$ trivial equations $0 = 0$, which can be suppressed. We thus can assume $q = r$; then equation (13.40) reduces to:

$$\alpha_i \, v_i = 0, \quad 1 \leq i \leq r. \tag{13.41}$$

For $1 \leq i \leq r$, v_i is a torsion element which generates the cyclic torsion module $[v_i]_{\mathbf{R}} \cong \frac{\mathbf{R}}{(\alpha_i)}$, and this module is zero if and only if α_i is invertible. Suppose that the number of invertible elements in the list $(\alpha_i)_{1 \leq i \leq r}$ is equal to j; then we only need to consider the α_i's, $j + 1 \leq i \leq r$. For $r + 1 \leq i \leq k$, v_i generates the free cyclic module $[v_i]_{\mathbf{R}} \cong \frac{\mathbf{R}}{(0)} = \mathbf{R}$. We have

$$M = [\mathbf{v}]_{\mathbf{R}} = \bigoplus_{j+1 \leq i \leq k} [v_i]_{\mathbf{R}} \, .$$

Here above, $\mathcal{T}(M) = \bigoplus_{j+1 \leq i \leq r} [v_i]_{\mathbf{R}}$, and $\Phi = \bigoplus_{r+1 \leq i \leq k} [v_i]_{\mathbf{R}}$ is a free module of rank $k - r$ since $(v_i)_{r+1 \leq i \leq k}$ is a basis of Φ (setting $\Phi = 0$ if $r = k$). We have thus obtained the decomposition

$$\boxed{M = \mathcal{T}(M) \oplus \Phi} \tag{13.42}$$

where Φ is a free submodule of M, such that $\Phi \cong \mathbf{R}^{k-r}$.

In addition,

$$\boxed{\mathcal{T}(M) \cong \bigoplus_{j+1 \leq i \leq r} \frac{\mathbf{R}}{(\alpha_i)}}. \tag{13.43}$$

DEFINITION 552.– *The elements α_i, $j + 1 \leq i \leq r$ (uniquely determined up to associates) are called the* invariant factors *of the module M. The integer $k - r$ is called the* rank *of the module M.*

REMARK 553.– *The terminology of ([10], Chapter VII) is slightly different and is justified by the following rationale: (i) Rather than the elements $\alpha_i, j+1 \leq i \leq r$, we can, in a more "intrinsic" manner, consider the ideals generated by these elements, i.e. the $\mathfrak{a}_{k-i+1} = (\alpha_i)$, $j + 1 \leq i \leq r$. (iii) In addition, $\Phi \cong \bigoplus_{r+1 \leq i \leq k} \frac{\mathbf{R}}{(0)}$; the $k - r$ zero ideals $\mathfrak{a}_i$ $(1 \leq i \leq k - r)$ in this sum can be taken into account in the same way as the above non-zero ideals $\mathfrak{a}_i$ $(k - r + 1 \leq i \leq k - j)$. The invariant factors of M, in the sense of ([10], Chapter VII), are in the end the principal ideals $\mathfrak{a}_1, \ldots, \mathfrak{a}_{k-j}$ (some of which may be zero); these are such that $\mathfrak{a}_1 \subset \mathfrak{a}_2 \ldots \subset \mathfrak{a}_{k-j} \neq \mathbf{R}$ and*

$$\boxed{M \cong \bigoplus_{1 \leq i \leq k-j} \mathbf{R}/\mathfrak{a}_i} \tag{13.44}$$

(the principal ideals $\mathfrak{a}_i$ are uniquely determined by these conditions).

We thus have the following:

THEOREM 554.– *Let* $\mathbf{R}$ *be a principal ideal domain and* M *be a finitely generated* $\mathbf{R}$*module (or, more generally, let* $\mathbf{R}$ *be an elementary divisor ring and* M *be a finitely presented* $\mathbf{R}$*-module). Then* M *is the direct sum of its torsion submodule* $\mathcal{T}(M)$ *and of a free submodule (of the same rank as* M*). In addition,* $\mathcal{T}(M)$ *decomposes according to* (13.43) *as a function of its non-zero invariant factors* α_i *or* (α_i) $(j+1 \leq i \leq r)$ *and* M *decomposes according to* (13.44) *as a function of all its invariant factors* $\mathfrak{a}_i$ $(1 \leq i \leq k-j)$ *("canonical decomposition of the module* M*").*

PROPOSITION 555.– *Let* $\mathbf{R}$ *be a Bézout domain and* M *a be a finitely generated* $\mathbf{R}$*-module. (i)* M *decomposes according to* (13.42) *(where* Φ *is a free module). (ii) The module* M *is free if and only if it is torsion-free. (iii) Suppose now that* M *is finitely presented;* M *has a left-regular definition matrix* R*,* M *is free if and only if* R *is equivalent to* $\begin{bmatrix} I_r & 0 \end{bmatrix}$ *and* M *is torsion if and only if* R *is regular.*

PROOF. (i): See ([31], section 5.1, Theorem 1.3). (ii) is an obvious consequence of (i). (iii): According to Theorem 495 (section 13.2.2), the problem comes down to the case where the definition matrix $R \in \mathbf{R}^{q \times k}$ of M is in column Hermite form, and then to the case $q = r$ $(r = \operatorname{rk} B)$ by suppressing the $q - r$ trivial equations $0 = 0$. The module M is free if and only if R is completable, thus equivalent to $\begin{bmatrix} I_r & 0 \end{bmatrix}$. Besides, according to Theorem 548(i), M is torsion if and only if R, now of rank r, is left-invertible over $\mathrm{Q}(\mathbf{R})$, thus invertible over that field, and finally regular over $\mathbf{R}$. ■

Theory of elementary divisors

The complete theory of *elementary divisors* of a finitely generated module M over a principal ideal domain $\mathbf{R}$ can immediately be deduced from Theorem 554 and from the primary decomposition of a cyclic module. The result may be stated as follows:

THEOREM 556.– *Let* (p_i) *be a representative system of primes in* $\mathbf{R}$ *(section 13.1.2). There exist integers* $\mu(0) \geq 0$*,* $n_i \geq 0$ *and* $\mu(\pi_i) \geq 0$*, where* $\pi_i = p_i^{n_i}$*, with the following properties: these integers are uniquely determined and are zero except for a finite number of them, and in the decomposition* (13.42)*,*

$$\boxed{\mathcal{T}(M) \cong \bigoplus_i \left(\frac{\mathbf{R}}{(\pi_i)}\right)^{\mu(\pi_i)}, \quad \Phi \cong \mathbf{R}^{\mu(0)}}.$$

DEFINITION 557.– *The* $\pi_i = p_i^{n_i}$*,* $i \in I$ *(or the ideals* (π_i) *they generate) are called the* elementary divisors *of module* M*, and the integers* $\mu(\pi_i)$ *are their* multiplicities. *If* $\mu(0) > 0$*, this integer (which is the rank of* M*) is called the multiplicity of the elementary divisor* 0.

REMARK 558.– *Just as Theorem 554 is a reformulation of Theorem 497 (section 13.2.3), Theorem 556 is a reformulation (even more precise) of what has been stated in beginning of section 13.2.4. This is why the theory of modules presented in this section can be considered as a* language *– which proves to be extremely useful.*

Smith zeros of a module

Let $R \in \mathbf{R}^{q \times k}$, where $\mathbf{R} = \mathbb{C}[\partial]$. The Smith zeros of R and their multiplicities have been defined in section 13.2.5. These quantities only depend on the elementary divisors of R, which themselves only depend on $\mathcal{T}(M)$, where $M \cong \operatorname{coker}(\bullet R)$. These zeros can thus be called, in a more intrinsic manner, the *Smith zeros of the module* $\mathcal{T}(M)$ [19]. Definition 504 (section 13.2.5) remains valid, *mutatis mutandis*. For any finitely generated torsion $\mathbf{R}$-module T, $\mathcal{Z}(T)$ denotes the set of its Smith zeros (each of them repeated a number of times equal to its degree) and $\varepsilon(T)$ denotes the set of its elementary divisors (each of them repeated a number of times equal to its multiplicity). The following is proven in [19]:

LEMMA 559.– *Let T_1 and T_2 be two finitely generated torsion $\mathbf{R}$-modules. (i) If $T_2 \subset T_1$, then $\mathcal{Z}(T_1) = \mathcal{Z}(T_2) \dot{\cup} \mathcal{Z}(T_1/T_2)$ where $\dot{\cup}$ designates the "disjoint union"*[10]. *(ii) If $T_1 \cap T_2 = 0$, then $\mathcal{Z}(T_1 \oplus T_2) = \mathcal{Z}(T_1) \dot{\cup} \mathcal{Z}(T_2)$ and $\varepsilon(T_1 \oplus T_2) = \varepsilon(T_1) \dot{\cup} \varepsilon(T_2)$.*

Complement on quotient modules

Let $\mathbf{R}$ be a principal ideal domain, M be a finitely generated $\mathbf{R}$-module and F be a free submodule of M.

LEMMA 560.– *(i) There exists a maximal free submodule Φ_F of M which contains F. (ii) We have $M = \mathcal{T}(M) \oplus \Phi_F$.*

PROOF. * (i) : Let $\mathcal{L}_F$ be the set of all free submodules of M that contains F, ordered by inclusion. This set is inductive; indeed, let $\mathcal{C}$ be a chain of $\mathcal{L}_F$, i.e. a totally ordered subset, and $F_1 = \bigcup_{L \in \mathcal{L}_F} L$; then the submodule F_1 of M is torsion-free, and it is free according to Proposition 555(ii); it contains F, thus it belongs to $\mathcal{C}$ and is its largest element. Statement (i) therefore derives from the Zorn Lemma. (ii): Suppose $\mathcal{T}(M) \oplus \Phi_F \neq M$ and let $m \in M$, $m \notin \mathcal{T}(M) \oplus \Phi_F$; then $\Phi_F \neq [m]_{\mathbf{R}} + \Phi_F$ and m is a free element, thus the module $[m]_{\mathbf{R}} + \Phi_F$ is torsion-free, and so is free; this module thus belongs to $\mathcal{L}_F$, which contradicts the fact that Φ_F is a maximal element of this set. * ■

Let N be a submodule of M; according to Proposition 555(i), there exists a free submodule F of N such that $N = \mathcal{T}(N) \oplus F$. We have $\mathcal{T}(N) \subset \mathcal{T}(M)$ and

10. For example, $\{x, y\} \dot{\cup} \{x, z\} = \{x, x, y, z\}$ ([91], section I.8).

according to Lemma 560, there exists a free submodule Φ_F of M such that $M = \mathcal{T}(M) \oplus \Phi_F$ and $F \subset \Phi_F$. Writing $\Phi_N = \Phi_F$, we obtain the following:

PROPOSITION 561.– *There exists a free submodule Φ_N of M such that $M = \mathcal{T}(M) \oplus \Phi_N$ and $N = \mathcal{T}(N) \oplus (\Phi_N \cap N)$. If Φ'_N is another free submodule of M such that $M = \mathcal{T}(M) \oplus \Phi'_N$ and $N = \mathcal{T}(N) \oplus (\Phi'_N \cap N)$, then $\Phi'_N \cong \Phi_N$ and $\Phi'_N / (\Phi'_N \cap N) \cong \Phi_N / (\Phi_N \cap N)$.*

PROOF. We have $F = \Phi_N \cap F = \Phi_N \cap N$. Assume Φ'_N is free, $M = \mathcal{T}(M) \oplus \Phi'_N$ and $N = \mathcal{T}(N) \oplus (\Phi'_N \cap N)$. Then $\Phi'_N \cong M/\mathcal{T}(M) \cong \Phi_N$; in addition, $\mathcal{T}(N) \subseteq \mathcal{T}(M)$ and $\Phi_N \cap N \subseteq N$, $\Phi'_N \cap N \subseteq N$, thus by Theorem 538, both modules below are isomorphic to M/N :

$$\mathcal{T}(M)/\mathcal{T}(N) \oplus \Phi_N / (\Phi_N \cap N) \cong \mathcal{T}(M)/\mathcal{T}(N) \oplus \Phi'_N / (\Phi'_N \cap N),$$

hence $\Phi_N / (\Phi_N \cap N) \cong \Phi'_N / (\Phi'_N \cap N)$ by ([31], section 8.2, Corollary 2.5). ■

13.4.3. *Structure of endomorphisms*

In this paragraph, $\mathbf{R}$ is the principal ideal domain $\mathbf{K}[\partial]$, where $\mathbf{K} = \mathbb{R}$ or $\mathbb{C}$; ∂ is the indeterminate and thus in principle plays a purely formal role, but it is convenient to think of ∂ as the usual derivative d/dt.

Companion matrices associated with a cyclic module

Since $\mathbf{R}$ is a principal ideal domain, every ideal $\mathfrak{a}$ in $\mathbf{R}$ is principal, and thus is of the form $\mathbf{R}\,a = (a)$, $a \in \mathbf{R}$. As a result, if $\Gamma = [w]_{\mathbf{R}}$ is a cyclic $\mathbf{R}$-module, there exists a polynomial $a = a(\partial) \in \mathbf{R}$ (chosen to be monic) such that

$$\Gamma \cong \frac{\mathbf{R}}{(a)}$$

and $a(\partial)$ is such that

$$a(\partial)\, w = 0. \tag{13.45}$$

We say that equation (13.45) is a *non-trivial* homogenous differential equation if $0 \neq a(\partial)$ is a *non-invertible* polynomial (i.e. of degree ≥ 1).

Let Γ be the cyclic module with generator w defined by (13.45), where $a(\partial) \neq 0$ is a monic polynomial of degree $n \geq 1$, that is

$$a(\partial) = \partial^n + a_1 \partial^{n-1} + ... + a_n, \quad a_i \in \mathbf{K}.$$

Write $x_i = \partial^{n-i} w \ (1 \leq i \leq n)$ and $x = \begin{bmatrix} x_1 & \cdots & x_n \end{bmatrix}^T$. We get

$$\partial x = \begin{bmatrix} -a_1 & -a_2 & \cdots & \cdots & -a_n \\ 1 & 0 & & & 0 \\ 0 & \ddots & \ddots & & \vdots \\ \vdots & \ddots & \ddots & \ddots & 0 \\ 0 & \cdots & 0 & 1 & 0 \end{bmatrix} x, \tag{13.46}$$

$$\partial x^T = x^T \begin{bmatrix} -a_1 & 1 & 0 & \cdots & 0 \\ -a_2 & 0 & \ddots & \ddots & \vdots \\ \vdots & \ddots & \ddots & \ddots & 0 \\ \vdots & \ddots & \ddots & \ddots & 1 \\ -a_n & \cdots & 0 & 0 & 0 \end{bmatrix}. \tag{13.47}$$

The matrices $C_1(a)$ and $C_2(a)$ in the right-hand member of (13.46) and (13.47) are called the *companion matrices* of the polynomial $a(\partial)$. There exist two other companion matrices $C_3(a)$ and $C_4(a)$, obtained by writing the equations $\partial \eta = C_3(a)\,\eta$ and $\partial \eta^T = \eta^T C_4(a)$, with $\eta = \begin{bmatrix} x_n & \cdots & x_1 \end{bmatrix}^T$. The explicit expressions of $C_3(a)$ and $C_4(a)$ are left to the reader.

PROPOSITION 562.– *(i) The characteristic polynomial of the companion matrices $C_i(a), i \in \{1, ..., 4\}$, is $a = a(\partial)$. (ii) The cyclic module $\Gamma \cong \frac{\mathbf{R}}{(a)}$, considered as a $\mathbf{K}$-vector space (denoted as E_a) is of dimension $n = d^\circ(a)$.*

PROOF. (i) We obtain $\det(\partial I_n - C_1(a)) = a(\partial)$ by developing this determinant with respect to the first row. The other companion matrices of $a(\partial)$ are similar to $C_1(a)$ (since, according to Proposition 534 of section 13.3.4, a matrix is similar to its transpose), and thus have the same characteristic polynomial as this matrix. (ii) According to (13.46), the elements $x_1, ..., x_n$ are $\mathbf{K}$-linearly independent and $(x_i)_{1 \leq i \leq n}$ is a basis of E_a. ■

REMARK 563.– *(i) The mapping $\partial : M \to M$ is $\mathbf{K}$-linear, thus is an endomorphism A_a of E_a. Since $\{x_n, \partial x_n, ..., \partial^{n-1} x_n\}$ is a basis of E_a, this vector space is A_a-cyclic and the above basis is A_a-cyclic (section 13.3.4). The vector x_n is an A_a-cyclic generator of E_a (and of the above basis). Any matrix representing the endomorphism A_a of E_a is similar to a companion matrix of $a(\partial)$ and is said to be* cyclic. *(ii) Let $y = \sum_{1 \leq i \leq n} Y_i x_i \in E_a$ and let $Y = [Y_1 \ ... \ Y_n]$ be the row representing y in the basis $x = (x_i)_{1 \leq i \leq n} = [x_1 \ ... \ x_n]^T$ (see Remark 542, section 13.4.1). Since $y = Y x$, the endomorphism $\partial = A_a$ is the map $Y \mapsto Y C_1(a)$ in the basis x.*

REMARK 564.– *In what follows, $C(a)$ designates one of the companion forms $C_m(a)$, where $m \in \{1, ..., 4\}$ is chosen once and for all.*

Rational canonical form of an endomorphism

Consider the differential equation

$$\partial x = A x \tag{13.48}$$

where $A \in \mathbf{K}^{n \times n}$ represents, in the basis $x = (x_i)_{1 \leq i \leq n}$ of $E \cong \mathbf{K}^n$, an endomorphism $\mathbf{u}$ (adopting the convention in Remark 542 of section 13.4.1).

The differential equation (13.48) is also written as $R(\partial) x = 0$, where $R(\partial) = \partial I_n - A$. According to Corollary 555(ii) (section 13.4.2), it defines a finitely generated torsion $\mathbf{R}$-module M. The map $\partial : M \rightarrow M$ is identified with $\mathbf{u}$.[11] These observations can be completed by the following:

THEOREM 565.– *Let M be a finitely generated $\mathbf{R}$-module. The module M is torsion if and only if it is a finite-dimensional $\mathbf{K}$-vector space.*

PROOF. (i) If M is torsion, then $M = \mathcal{T}(M)$ satisfies the relation (13.43) of section 13.4.2 and each cyclic torsion module $\frac{\mathbf{R}}{(\alpha_i)}$ is a finite-dimensional $\mathbf{K}$-vector space according to Proposition 562. (ii) Conversely, let M be an $\mathbf{R}$-module which is a finite-dimensional $\mathbf{K}$-vector space and let $(x_i)_{1 \leq i \leq n}$ be a basis of M. For any $i \in \{1, ..., n\}$, ∂x_i is a $\mathbf{K}$-linear combination of the x_k, $1 \leq k \leq n$, thus there exists a matrix $A \in \mathbf{K}^{n \times n}$ such that the relation (13.48) is satisfied. The module M is thus torsion. ■

The torsion module $M = \mathcal{T}(M)$ defined by (13.48) admits decomposition (13.43). Let E_{α_i} be the cyclic module $\frac{\mathbf{R}}{(\alpha_i)}$ $(j + 1 \leq i \leq r)$, considered as a $\mathbf{K}$-vector space. According to Remark 563, $\partial : \frac{\mathbf{R}}{(\alpha_i)} \rightarrow \frac{\mathbf{R}}{(\alpha_i)}$ is an endomorphism A_{α_i} of E_{α_i} and the latter space is A_{α_i}-cyclic. In an A_{α_i}-cyclic basis, A_{α_i} is thus represented by a companion matrix $C(\alpha_i)$. We have therefore the following:

THEOREM 566.– *Let E be a finite-dimensional $\mathbf{K}$-vector space and let $\mathbf{u}$ be an endomorphism of E. (i) E can be expressed as a direct sum of $\mathbf{u}$-cyclic subspaces*

$$E = \bigoplus_{j+1 \leq i \leq r} E_{\alpha_i} \tag{13.49}$$

such that the restriction A_{α_i} of $\mathbf{u}$ to E_{α_i} is represented, in a $\mathbf{u}$-cyclic basis, by a companion matrix $C(\alpha_i)$. (ii) The polynomials $\alpha_i(\partial)$ $(j + 1 \leq i \leq r)$ are the

11. Here we take advantage of the fact that the indeterminate ∂ is identified with d/dt. An equivalent but more abstract approach consists in using the formalism of ([10], section VII.5).

similarity invariants *of* $\mathbf{u}$ *(section 13.3.4); if these are ordered in such a way that* $\alpha_i \mid \alpha_{i+1}$ $(j+1 \leq i \leq r-1)$*, the decomposition* (13.49) *is unique. (iii) In the concatenation of the above* $\mathbf{u}$*-cyclic bases, the endomorphism* $\mathbf{u}$ *is represented by the diagonal sum* $\bigoplus_{j+1 \leq i \leq r} C(\alpha_i)$ *(*"rational canonical form *of the endomorphism* $\mathbf{u}$*").*

DEFINITION 567.– *The number of terms of the sum* (13.39) *(i.e.* $r-j$*) is called the* cyclic index *of* $\mathbf{u}$*. The endomorphism* $\mathbf{u}$ *is said to be* cyclic *if its cyclic index is equal to* 1 *(that is if* $\mathbf{u}$ *is representable by a cyclic matrix in any basis and by a companion matrix in a* $\mathbf{u}$*-cyclic basis: see Remark 563).*

Recall that the minimal polynomial of $\mathbf{u}$ is $\alpha_r(\partial)$ and that its characteristic polynomial is $\prod_{j+1 \leq i \leq r} \alpha_i(\partial)$ (Proposition 536, section 13.3.4). We deduce from Theorem 566 the following:

COROLLARY 568.– *An endomorphism is cyclic if and only if its minimal polynomial is equal to its characteristic polynomial.*

Jordan form

Let $\mathbf{K} = \mathbb{C}$; then a prime polynomial $p(\partial)$ is of the form $\partial - \mathbf{r}$, $\mathbf{r} \in \mathbb{C}$, and an elementary divisor is of the form $\pi(\partial) = (\partial - \mathbf{r})^k$.

An indecomposable module $\frac{\mathbf{R}}{(\pi)}$ is a $\mathbb{C}$-vector space E_π. Let v be a generator of $\frac{\mathbf{R}}{(\pi)}$, such that $(\partial - \mathbf{r})^k v = 0$. Put $X_{k-i} = (\partial - \mathbf{r})^i v$, $1 \leq i \leq k$, and $X = \begin{bmatrix} X_1 & \cdots & X_k \end{bmatrix}^T$. We obtain $\partial X = J_{\mathbf{r},k} X$ where $J_{\mathbf{r},k}$ is the Jordan block of order k relative to $\mathbf{r}$ as defined in section 13.3.4; $J_{\mathbf{r},k}$ represents the endomorphism $\partial : \frac{\mathbf{R}}{(\pi)} \to \frac{\mathbf{R}}{(\pi)}$ in the basis X of E_π.

Theorem 530 (section 13.3.4) is now a direct consequence of Theorem 556 (section 13.4.2), by concatenating the bases constructed here above for each $\mathbb{C}$-vector space E_π, as π spans the set of elementary divisors of the endomorphism $\mathbf{u}$ (and taking into account their multiplicities).

13.5. Orthogonality and symmetry

In what follows, E is a Hilbert space of finite dimension n over $\mathbf{K}$, where $\mathbf{K} = \mathbb{R}$ or $\mathbb{C}$. This space is equipped with a scalar product $\langle ., . \rangle$ (see section 12.1.2). We will be careful not to confuse the latter with a duality bracket.

13.5.1. *Orthonormal basis*

Let $\mathcal{B} = (e_i)_{1 \leq i \leq n}$ be a basis of E. This basis is said to be *orthonormal* if $\langle e_i, e_j \rangle = \delta_{ij}$, where δ_{ij} is the Kronecker index (see section 13.3.2).

One shows, using the "Gram-Schmidt orthogonalization principle" (see, e.g. [91], as well as the proof of Theorem 579 at section 13.5.7) that we can always choose an orthonormal basis in E.

Let x and y be two vectors of E with components $x_1, ..., x_n$ and $y_1, ..., y_n$, respectively, in this basis. We have

$$\begin{aligned}\langle x, y \rangle &= \left\langle \sum_{i=1}^{n} x_i\, e_i, \sum_{j=1}^{n} y_j\, e_j \right\rangle = \sum_{i=1}^{n} \sum_{j=1}^{n} \bar{x}_i\, y_j\, \langle e_i, e_j \rangle \\ &= \sum_{i=1}^{n} \bar{x}_i\, y_i = X^* Y \end{aligned} \qquad (13.50)$$

where $X = [x_1 \; ... \; x_n]^T, Y = [y_1 \; ... \; y_n]^T$ and where $X^* = [\bar{x}_1 \; ... \; \bar{x}_n]^T$ (conjugate transpose of X).

The quantity in the right-hand side of (13.50) is the "standard scalar product" on $\mathbf{K}^n$ (the vector $(x_1, ..., x_n)$ of this space is identified with the column matrix $X = [x_1, ..., x_n]^T$ which represents it in the canonical basis) and (13.50) makes it possible to identify the space E with $\mathbf{K}^n$ equipped with the standard scalar product.

13.5.2. *Orthogonality*

Two vectors x and y are said to be orthogonal ($x \perp y$) if $\langle x, y \rangle = 0$.

If F is a subspace of E, the set of vectors of E which are orthogonal to all vectors of F is a subspace of E, called the orthogonal of F (and denoted as $F^{\perp}$). It is immediate that $F \cap F^{\perp} = 0$. One can also show that $F + F^{\perp} = E$. Thus, F and $F^{\perp}$ are supplementary. We call $F^{\perp}$ the *orthogonal supplement* of F, and E the "orthogonal direct sum" of F and $F^{\perp}$, which we write as

$$E = F \overset{\perp}{\oplus} F^{\perp}.$$

One can also show that

$$F^{\perp\perp} = F.$$

Note that F does not have a unique supplement (aside from the trivial case where $F = 0$ or $F = E$), while F has by definition a unique *orthogonal supplement*.

13.5.3. *Adjoint endomorphism*

THEOREM 569.– *Let* $\mathbf{u}$ *be an endomorphism of* E. *There exists a unique endomorphism* $\mathbf{u}^*$ *of* E*, called the* adjoint *of* $\mathbf{u}$*, such that for any* $x, y \in E$

$$\boxed{\langle x, \mathbf{u}\, y\rangle = \langle \mathbf{u}^*\, x, y\rangle}. \tag{13.51}$$

More specifically, let $A = (a_{ij})$ *be the matrix representing* $\mathbf{u}$ *in an orthonormal basis* $\mathcal{B}$; $\mathbf{u}^*$ *is the endomorphism represented, in this same basis, by* $A^* \triangleq \bar{A}^T = (\bar{a}_{ji})$ *(conjugate transpose of* A*).*

PROOF. Let $z = \mathbf{u}\, y$. According to (13.21), the representative of z in the basis $\mathcal{B}$ is $Z = AY$. Thus, by (13.50),

$$\langle x, \mathbf{u}\, y\rangle = X^*\, A\, Y = (A^*\, X)^*\, Y$$

and the theorem is proven. ∎

Properties of the adjoint

THEOREM 570.– *(i) We have*

$$E = \ker \mathbf{u} \overset{\perp}{\oplus} \operatorname{im} \mathbf{u}^*. \tag{13.52}$$

(ii) The mapping $\mathbf{u} \to \mathbf{u}^*$ *is antilinear (section 12.1.2) and* $(\mathbf{u}^*)^* = \mathbf{u}$.
(iii) If λ *is an eigenvalue of* $\mathbf{u}$*, then* $\bar{\lambda}$ *is an eigenvalue of* $\mathbf{u}^*$ *and* $\dim\ker(\lambda\, I_E - \mathbf{u}) = \dim\ker\left(\bar{\lambda}\, I_E - \mathbf{u}^*\right)$.
(iv)

$$\det \mathbf{u}^* = \overline{\det \mathbf{u}}.$$

PROOF. (i) derives from the following observation: an element of $\ker \mathbf{u}$ is a vector y of E such that $\mathbf{u}\, y$ is orthogonal to all vectors x of E and by (13.51) this in turn means that y is orthogonal to $\operatorname{im} \mathbf{u}^*$. The proof of (ii) is easy. (iii) can be proved in the following manner: according to (13.52) we have for any $\lambda \in \mathbb{C}$, $E = \ker(\lambda\, I_E - \mathbf{u}) \overset{\perp}{\oplus} \operatorname{im}\left(\bar{\lambda}\, I_E - \mathbf{u}^*\right)$, thus $\operatorname{rk}\left(\bar{\lambda}\, I_E - \mathbf{u}^*\right) = n - \dim\ker(\lambda\, I_E - \mathbf{u})$. Now, $\dim\ker\left(\bar{\lambda}\, I_E - \mathbf{u}^*\right) = n - \operatorname{rk}\left(\bar{\lambda}\, I_E - \mathbf{u}^*\right)$ according to (13.20) (section 13.3.2) and thus $\dim\ker\left(\bar{\lambda}\, I_E - \mathbf{u}^*\right) = \dim\ker(\lambda\, I_E - \mathbf{u})$. (iv) is proved as follows: if the endomorphism $\mathbf{u}$ is represented in an orthonormal basis by the matrix A, we have $\det \mathbf{u} = \det A = \overline{\det \bar{A}} = \overline{\det \bar{A}^T}$ (see section 13.1.4 or Proposition 534), from which we get $\det \mathbf{u} = \overline{\det \mathbf{u}^*}$. ∎

13.5.4. *Unitary endomorphism*

An endomorphism $\mathbf{u}$ is said to be *unitary* if it preserves the scalar product, that is if for any $x, y \in E$

$$\langle \mathbf{u}\, x, \mathbf{u}\, y \rangle = \langle x, y \rangle .$$

This is equivalent to $\langle \mathbf{u}^* \, \mathbf{u}\, x, y \rangle = \langle x, y \rangle$, thus also to

$$\boxed{\mathbf{u}^* = \mathbf{u}^{-1}}. \tag{13.53}$$

An endomorphism is unitary if and only if it transforms an orthonormal basis into another orthonormal basis.

Let A be the matrix representing an endomorphism $\mathbf{u}$ in an orthonormal basis; it is clear that $\mathbf{u}$ is unitary if and only if

$$A^* = A^{-1}. \tag{13.54}$$

A matrix satisfying (13.54) is said to be *unitary*, and *orthogonal* if $\mathbf{K} = \mathbb{R}$ (in this case, $A^* = A^T$). According to the property *(iv)* in Theorem 570 and (13.53), if the endomorphism $\mathbf{u}$ is a unitary, we have

$$\det \mathbf{u} = \pm 1.$$

13.5.5. *Normal endomorphism*

DEFINITION 571.– *An endomorphism* $\mathbf{u}$ *of* E *is said to be* normal *if it commutes with its adjoint:* $\mathbf{u}\,\mathbf{u}^* = \mathbf{u}^*\,\mathbf{u}$.

THEOREM 572.– *Let* $\mathbf{u}$ *be a normal endomorphism. (i) For any eigenvalue* λ *of* $\mathbf{u}$, $\ker(\lambda\, I_E - \mathbf{u}) = \ker\left(\bar{\lambda}\, I_E - \mathbf{u}^*\right)$. *(ii) The endomorphism* $\mathbf{u}$ *is diagonalizable in an orthonormal basis. (iii) Conversely, a diagonalizable endomorphism* $\mathbf{u}$ *in an orthonormal basis is normal.*

PROOF. (i) Let λ be an eigenvalue of $\mathbf{u}$ and write $\mathbf{v} = \mathbf{u} - \lambda\, I_E$. It is clear that a vector x is an eigenvector of $\mathbf{u}$ associated with the eigenvalue λ if, and only if x is an eigenvector of $\mathbf{v}$ associated with the eigenvalue 0. One such vector can be taken to be unitary. In addition, $\mathbf{v}$ is a normal endomorphism. Then let $y = \mathbf{v}^*\, x$. We have

$$\|y\|^2 = \langle \mathbf{v}^*\, x, \mathbf{v}^*\, x \rangle = \langle \mathbf{v}\, \mathbf{v}^*\, x,\, x \rangle = \langle \mathbf{v}^*\, \mathbf{v}\, x, x \rangle = 0.$$

As a result, $y = 0$ and x is an eigenvector of $\mathbf{v}^*$ associated with the eigenvalue 0. Conversely, changing $\mathbf{v}$ to $\mathbf{v}^*$, it is clear that any eigenvector of $\mathbf{v}^*$ associated with the eigenvalue 0 is an eigenvector of $\mathbf{v}$ associated with this same eigenvalue. It follows that $\ker \mathbf{v} = \ker \mathbf{v}^*$. (ii) We thus have $(\ker \mathbf{v})^{\perp} = (\ker \mathbf{v}^*)^{\perp} = \operatorname{im} \mathbf{v}$ according to (13.52) (section 13.5.3). Therefore, the restriction of $\mathbf{v}$ to $\operatorname{im} \mathbf{v}$ is injective. The only elementary divisors of $\mathbf{v}$ that are multiples of s are therefore equal to s, because if there exists an integer $m > 1$ and a vector x such that $\mathbf{v}^m\, x = 0$, we have $\mathbf{v}\,\left(\mathbf{v}^{m-1}\, x\right) = 0$, thus $\mathbf{v}^{m-1}\, x = 0$; by induction, we obtain $\mathbf{v}\, x = 0$. The roots of the minimal polynomial $q_{\mathbf{u}}(s)$ are thus simple, which proves that $\mathbf{u}$ is diagonalizable (section 13.3.4, Corollary 532). Suppose that $\mathbf{v}$, which has an eigenvalue 0, also has a non-zero eigenvalue $\tilde{\mu}$. Let us show that $\ker \mathbf{v} \perp \ker (\mathbf{v} - \tilde{\mu}\, I_E)$. Let $x \in \ker \mathbf{v} = \ker \mathbf{v}^*$ and $y \in \ker (\mathbf{v} - \tilde{\mu}\, I_E)$. From the fact that $\mathbf{v}^* x = 0$, we get

$$\tilde{\mu}\, \langle x, y \rangle = \langle x, \tilde{\mu}\, y \rangle = \langle x, \mathbf{v}\, y \rangle = \langle \mathbf{v}^* x, y \rangle = 0$$

thus $\langle x, y \rangle = 0$. Consider again the endomorphism $\mathbf{u}$: we have just shown that if λ and μ are two distinct eigenvalues of $\mathbf{u}$, then $\ker (\lambda\, I_E - \mathbf{u}) \perp \ker (\mu\, I_E - \mathbf{u})$. According to the Gram-Schmidt orthogonalization principle, we can choose an orthonormal basis in each eigenspace of $\mathbf{u}$. The concatenation of these bases is an orthonormal basis of E in which $\mathbf{u}$ diagonalizes. (iii) Let $\mathbf{u}$ be an endomorphism, represented in an orthonormal basis by a diagonal matrix $\Lambda = \operatorname{diag}(\lambda_1, ..., \lambda_n)$. We have $\Lambda^* = \operatorname{diag}\left(\bar{\lambda}_1, ..., \bar{\lambda}_n\right)$, thus $\Lambda\,\Lambda^* = \Lambda^*\,\Lambda = \operatorname{diag}\left(|\lambda_1|^2, ..., |\lambda_n|^2\right)$. ■

13.5.6. *Self-adjoint endomorphism*

General case

DEFINITION 573.– *An endomorphism* $\mathbf{u}$ *of* E *is said to be* self-adjoint *if* $\mathbf{u}^* = \mathbf{u}$.

Let $\mathbf{u}$ be an endomorphism represented, in an orthonormal basis $\mathcal{B}$, by a matrix A; it is clear that $\mathbf{u}$ is self-adjoint if and only if $A^* = A$. One such matrix A is said to be *Hermitian* (and *symmetric* if $\mathbf{K} = \mathbb{R}$; in this case, $A^* = A^T$).

THEOREM 574.– *A self-adjoint endomorphism is diagonalizable in an orthonormal basis and all its eigenvalues are real.*

PROOF. A self-adjoint endomorphism $\mathbf{u}$ is normal, thus is diagonalizable in an orthonormal basis. In addition, by Theorem 572(i) (section 13.5.5), for any eigenvalue λ of $\mathbf{u}$, $\ker (\lambda\, I_E - \mathbf{u}) = \ker \left(\bar{\lambda}\, I_E - \mathbf{u}\right)$, thus $\lambda = \bar{\lambda}$. ■

Non-negative or positive self-adjoint endomorphisms

DEFINITION 575.– *A self-adjoint endomorphism* $\mathbf{u}$ *is said to be* non-negative *(resp.,* positive*),*[12] *which we can write as* $\mathbf{u} \geq 0$ *(resp.,* $\mathbf{u} > 0$*) if all its eigenvalues are*

12. Such an endomorphism is also called non-negative definite (resp., positive definite).

non-negative (resp., positive); a matrix A representing $\mathbf{u}$ in an orthonormal basis is said to be Hermitian (symmetric if $\mathbf{K} = \mathbb{R}$) non-negative (resp., positive) definite, which we write as $A \geq 0$ (resp., $A > 0$).

It is clear that a positive self-adjoint endomorphism is invertible and is thus an automorphism (see section 13.3.3).

Let $\mathbf{u}$ be a self-adjoint endomorphism and

$$x = \sum_{k=1}^{n} x_k \, e_k$$

be a vector of E, where $\{e_1, ..., e_k\}$ is an orthonormal basis of E formed by the eigenvectors of $\mathbf{u}$, e_k being associated with the eigenvalue λ_k. We have

$$\langle x, \mathbf{u}\, x \rangle = \left\langle \sum_{k=1}^{n} x_k \, e_k, \sum_{k=1}^{n} x_k \, \lambda_k \, e_k \right\rangle = \sum_{k=1}^{n} \lambda_k \, |x_k|^2 . \tag{13.55}$$

As a result,

– $\mathbf{u}$ is non-negative if and only if $\langle x, \mathbf{u}\, x \rangle \geq 0$ for any $x \in E$.

– $\mathbf{u}$ is positive if and only if $\langle x, \mathbf{u}\, x \rangle > 0$ for any vector $x \neq 0$ of E.

By (13.55), if $\mathbf{u}$ is a non-negative self-adjoint endomorphism, we have the inequality

$$\langle x, \mathbf{u}\, x \rangle \leq \lambda_{\max}(\mathbf{u}) \, \|x\|^2$$

where $\lambda_{\max}(\mathbf{u})$ is the largest eigenvalue of $\mathbf{u}$. In addition, the two sides of this inequality become equal when x is an eigenvector associated with the eigenvalue $\lambda_{\max}(\mathbf{u})$. Therefore,

$$\max_{\|x\|=1} \sqrt{\langle x, \mathbf{u}\, x \rangle} = \sqrt{\lambda_{\max}(\mathbf{u})}. \tag{13.56}$$

The proof of the following is straightforward:

PROPOSITION 576.– *Let be the Hermitian matrix*

$$A = \begin{bmatrix} A_{11} & A_{12} \\ A_{12}^* & A_{22} \end{bmatrix}$$

where $A_{11} > 0$. Then

$$A = \begin{bmatrix} I & 0 \\ A_{12}^* \, A_{11}^{-1} & I \end{bmatrix} \begin{bmatrix} A_{11} & 0 \\ 0 & C \end{bmatrix} \begin{bmatrix} I & 0 \\ A_{12}^* \, A_{11}^{-1} & I \end{bmatrix}^*,$$

$$C = A_{22} - A_{12}^* \, A_{11}^{-1} \, A_{12}$$

and $A \geq 0$ (resp., $A > 0$) if and only if $C \geq 0$ (resp., $C > 0$).

Square roots of a Hermitian matrix

We will define the *Hermitian square root* of an endomorphism $\mathbf{u} \geq 0$ (or, in an equivalent manner, of the Hermitian matrix $Q \geq 0$ which represents it in an orthonormal basis). One such matrix Q diagonalizes in an orthonormal basis, thus there exists an unitary matrix U such that $Q = U^* \Lambda U$, where $\Lambda = \operatorname{diag}\{\lambda_1, ..., \lambda_n\}$ and the λ_i's are the eigenvalues of Q. Since $Q \geq 0$, the λ_i are real numbers ≥ 0. Write $\sqrt{\Lambda} = \operatorname{diag}\left\{\sqrt{\lambda_1}, ..., \sqrt{\lambda_n}\right\}$. The Hermitian square root of Q is defined by

$$\sqrt{Q} = U^* \sqrt{\Lambda}\, U.$$

This terminology is justified by the following two properties: (i) $\sqrt{Q} \geq 0$; (ii) $\sqrt{Q}\sqrt{Q} = Q$.

More generally, a square matrix Q of order n is Hermitian (resp. symmetric real) non-negative definite if and only if there exists a matrix E with complex (resp., real) entries such that $Q = E^* E$ (resp., $Q = E^T E$) and this matrix E can be chosen to be *left-regular*, that is having r rows and n columns, where $r = \operatorname{rk} Q$.

The above latter assertion can be proved as follows: let Q be a Hermitian non-negative definite matrix of order n and of rank r, let $\lambda_1, ..., \lambda_r$ be the non-zero eigenvalues of Q and let $\Lambda^0 = \operatorname{diag}\{\lambda_1, ..., \lambda_r\}$. There exists a unitary matrix U such that $Q = U^* \left(\Lambda^0 \oplus 0\right) U = E^* E$ where

$$E = \begin{bmatrix}\Lambda^0 & 0\end{bmatrix}.$$

Any matrix E such that $Q = E^* E$ is called a *square root* of Q. For example, $\begin{bmatrix}\sqrt{2} & 0\end{bmatrix}$ and $\begin{bmatrix}-\sqrt{2} & 0\end{bmatrix}$ are two left-regular square roots of the matrix $\begin{bmatrix}2 & 0\\ 0 & 0\end{bmatrix}$ and its unique symmetric square root is $\begin{bmatrix}\sqrt{2} & 0\\ 0 & 0\end{bmatrix}$. As a complement, see Corollary 586 (section 13.5.7).

13.5.7. *Singular values*

Norm of operators in Hilbert spaces

Let E and F be two finite-dimensional Hilbert spaces and let $\mathbf{u}$ be a homomorphism from E into F. According to section 12.1.3, the norm of $\mathbf{u}$ (induced by the norms of E and F) is defined by[13]

$$\boxed{\|\mathbf{u}\| \triangleq \max_{\|x\|=1} \|\mathbf{u}\, x\|}. \tag{13.57}$$

13. The function $x \rightarrow \|\mathbf{u}\, x\|$ is continuous in the compact set $\|x\| = 1$, and thus admits a maximum in this set. We thus can replace the sup by max.

THEOREM 577.– *For every homomorphism* $\mathbf{u} : E \rightarrow F$, *we have the equality*

$$\boxed{\|\mathbf{u}\| = \|\mathbf{u}^*\| = \sqrt{\lambda_{\max}\left(\mathbf{u}^*\,\mathbf{u}\right)} \triangleq \bar{\sigma}\left(\mathbf{u}\right)}.$$

PROOF. We have $\|\mathbf{u}\,x\|^2 = \langle \mathbf{u}\,x, \mathbf{u}\,x\rangle = \langle \mathbf{u}^*\,\mathbf{u}\,x, x\rangle$ and $\mathbf{u}^*\,\mathbf{u}$ is clearly a non-negative self-adjoint endomorphism of E. We have thus for any $x \neq 0$

$$\|\mathbf{u}\,x\| = \sqrt{\langle \mathbf{u}^*\,\mathbf{u}\,x, x\rangle}.$$

In addition, according to (13.56) (section 13.5.6), the maximum (13.57) is attained by putting x as a unitary eigenvector of $\mathbf{u}^*\,\mathbf{u}$ associated with the largest eigenvalue of this endomorphism. For the equality $\|\mathbf{u}\| = \|\mathbf{u}^*\|$, see ([35], (11.5.2)). ■

Suppose that the above spaces E and F are of dimension n and m respectively and that orthonormal bases are chosen in these two spaces. In these bases, the homomorphism $\mathbf{u}$ is represented by a matrix $A \in \mathbf{K}^{m \times n}$. The "operator norm" of this matrix is defined by

$$\boxed{\|A\| = \|\mathbf{u}\|}$$

and this quantity $\bar{\sigma}\left(\mathbf{u}\right)$ is equally denoted as $\bar{\sigma}\left(A\right)$. This norm is clearly multiplicative (see section 12.1.3).

Note that a unitary endomorphism has an operator norm equal to 1; this also holds for a unitary (or orthogonal) matrix.

One can easily show the following result (see for example [35], (11.1.3)):

PROPOSITION 578.– *Let* $A \in \mathbf{K}^{n \times n}$; *every eigenvalue* λ *of* A *satisfies* $|\lambda| \leq \bar{\sigma}\left(A\right)$.

Singular value decomposition

THEOREM 579.– *Let* $A \in \mathbf{K}^{m \times n}$. *There exist* unitary *(or* orthogonal *if* $\mathbf{K} = \mathbb{R}$*) matrices* $U \in \mathbf{K}^{m \times m}$ *and* $V \in \mathbf{K}^{n \times n}$ *such that*

$$U^*\,A\,V = \operatorname{diag}\left(\sigma_1, ..., \sigma_p\right) \in \mathbf{K}^{m \times n}, \quad p = \min\left(m, n\right), \tag{13.58}$$

where $\sigma_1 \geq ... \geq \sigma_p \geq 0$. *These quantities* σ_i *are uniquely determined. We have* $\sigma_1 = \bar{\sigma}\left(A\right)$ *and the largest integer* r *such that* $\sigma_r \neq 0$ *is the rank of* A.

PROOF. According to the above, there exists a unitary column vector $u_1 \in \mathbf{K}^n$ such that $\|A\,u_1\| = \sigma_1$, where $\sigma_1 = \bar{\sigma}\left(A\right)$. There thus exists a unitary column vector v_1 in $\mathbf{K}^m$ such that $A\,u_1 = \sigma_1\,v_1$. The *Gram-Schmidt orthogonalization principle*, applied to $\mathbf{K}^n$, can be expressed as follows: there exists a matrix $V_1 \in \mathbf{K}^{n \times (n-1)}$ such that

$V = \begin{bmatrix} v_1 & V_1 \end{bmatrix}$ is a unitary matrix. Similarly, there exists a matrix $U_1 \in \mathbf{K}^{m \times (m-1)}$ such that $U = \begin{bmatrix} u_1 & U_1 \end{bmatrix}$ is a unitary matrix. One easily shows that $U^* A V$ has the following structure :

$$U^* A V = \begin{bmatrix} \sigma_1 & w^* \\ 0 & B \end{bmatrix} \triangleq A_1.$$

We thus have $\|A_1\| \leq \|U^*\| \|A\| \|V\| = \|A\|$. On the other hand, $A = U A_1 V^*$, thus $\|A\| \leq \|U\| \|A_1\| \|V^*\| = \|A_1\|$. Consequently, $\|A\| = \|A_1\|$. The column vector $z = \begin{bmatrix} \sigma_1 & w^T \end{bmatrix}^T$ is such that

$$\|A_1 z\|^2 \geq \left(\sigma_1^2 + w^* w\right)^2 = \left(\sigma_1^2 + w^* w\right) \|z\|^2$$

which implies that $\|A_1\|^2 \geq \sigma_1^2 + w^* w$. But since $\|A_1\| = \|A\| = \sigma_1$, this implies $w = 0$, from which $A_1 = \sigma_1 \oplus B$ with $\|B\| \leq \sigma_1$. Continuing this way, we obtain the result by induction. ■

DEFINITION 580.– *The quantities σ_i $(1 \leq i \leq p)$ are called the* singular values *of A. The expression (13.58) is called the* singular value decomposition *of A.*[14]

PROPOSITION 581.– *(i) Let $A \in \mathbf{K}^{m \times n}$ be a matrix of rank r and $\sigma_i(A)$, $1 \leq i \leq r$ be its non-zero singular values. We have $\sigma_i(A) = \sqrt{\lambda_i(A A^*)} = \sqrt{\lambda_i(A^* A)}$, $1 \leq i \leq r$, where $\lambda_i(.)$ is the ith eigenvalue of the matrix in parentheses (these eigenvalues arranged in decreasing order). (ii) A and A^* have the same singular values. (iii) Let $A \in \mathbf{K}^{n \times n}$ be an invertible matrix. Then $\sigma_i\left(A^{-1}\right) = \frac{1}{\sigma_{n-i+1}(A)}$. In particular, $\bar{\sigma}\left(A^{-1}\right) = \frac{1}{\underline{\sigma}(A)}$, where $\bar{\sigma}$ and $\underline{\sigma}$ denote the largest and smallest singular value respectively.*

PROOF. (i) Let $\Sigma = \operatorname{diag}(\sigma_1, ..., \sigma_r)$. According to (13.58), we have $U^* A V = \begin{bmatrix} \Sigma & 0 \\ 0 & 0 \end{bmatrix}$, thus $U^* A A^* U = U^* A V (U^* A V)^* = \begin{bmatrix} \Sigma^2 & 0 \\ 0 & 0 \end{bmatrix}$; similarly, $V^* A^* A V = (U^* A V)^* U^* A V$, thus (i) is proven. (ii) is an immediate consequence of (i). (iii) Let σ_i, $1 \leq i \leq n$, be the singular values of A, arranged in decreasing order (they are all non-zero). According to (13.58), there exist unitary matrices U and V such that $U^* A V = \operatorname{diag}\{\sigma_i\}$, thus $V^* A^{-1} U = \operatorname{diag}\left\{\sigma_i^{-1}\right\}$ (since $U^{-1} = U^*$ and $V^{-1} = V^*$). The singular values of A^{-1}, arranged in decreasing order, are thus $\sigma_n^{-1}, ..., \sigma_1^{-1}$, which proves (iii). ■

14. One can define the singular value decomposition of a homomorphism, but the presentation is facilitated by choosing bases and by reasoning on the matrix representing this homomorphism in the chosen bases, as what we have done here.

Pseudo-inverse

General case

Let $A \in \mathbf{K}^{m \times n}$ be a matrix of rank $r \leq \min(n, m)$. Let $\Sigma = \operatorname{diag}(\sigma_1, ..., \sigma_r)$, where $\sigma_1 \geq ... \geq \sigma_r$ are the non-zero singular values of A. The singular value decomposition theorem (Theorem 579) shows that there exist two unitary matrices $U \in \mathbf{K}^{m \times m}$ and $V \in \mathbf{K}^{n \times n}$ such that

$$A = U(\Sigma \oplus 0) V^*.$$

DEFINITION 582.– *We call a* pseudo-inverse *of A a matrix $A^\dagger \in \mathbf{K}^{n \times m}$ such that*

$$\boxed{A^\dagger = V\left(\Sigma^{-1} \oplus 0\right) U^*}.$$

Left-inverse

PROPOSITION 583.– *Suppose $r = n$ (which implies $m \geq n$). (i) The matrix A is left-invertible and $A^\dagger$ is a left-inverse of A, that is $A^\dagger A = I_n$; a left-inverse is non-unique if $m > n$. (ii) We have $A^\dagger = (A^* A)^{-1} A^*$.*

PROOF. (i) We can write A in the form

$$A = U \begin{bmatrix} \Sigma \\ 0 \end{bmatrix} V^*,$$

therefore

$$A^\dagger = V \begin{bmatrix} \Sigma^{-1} & 0 \end{bmatrix} U^*.$$

We deduce that

$$A^\dagger A = V \begin{bmatrix} \Sigma^{-1} & 0 \end{bmatrix} U^* U \begin{bmatrix} \Sigma \\ 0 \end{bmatrix} V^* = I_n.$$

In addition, there is non-uniqueness in the case $m > n$ because in that case there exist several solutions L to the equation $LA = I_n$, since there are $n\,m$ unknowns in L for n^2 equations. This proves (i). (ii) On the other hand,

$$A^* A = V \begin{bmatrix} \Sigma & 0 \end{bmatrix} U^* U \begin{bmatrix} \Sigma \\ 0 \end{bmatrix} V^* = V\, \Sigma^2\, V^*$$

thus

$$(A^* A)^{-1} A^* = V\, \Sigma^{-2}\, V^* V \begin{bmatrix} \Sigma & 0 \end{bmatrix} U^* = V \begin{bmatrix} \Sigma^{-1} & 0 \end{bmatrix} U^* = A^\dagger$$

which proves (ii). ■

Right-inverse

Interchanging the roles of A and A^*, we obtain the following:

PROPOSITION 584.– *Suppose that* $r = m$ *(which implies* $n \geq m$*). (i) The matrix* A *is right-invertible and* $A^\dagger$ *is a right-inverse of* A*, that is* $A\, A^\dagger = I_m$*; a right-inverse is non-unique if* $n > m$*. (ii) We have* $A^\dagger = A^* \, (A\, A^*)^{-1}$.

Condition number of a matrix

DEFINITION 585.– *Let* $A \in \mathbf{K}^{n \times n}$ *be an invertible matrix. The* condition number *of the matrix* A *is the number* $\kappa(A)$ *defined by:*

$$\boxed{\kappa(A) = \frac{\bar{\sigma}(A)}{\underline{\sigma}(A)} = \|A\| \, \left\|A^{-1}\right\|}.$$

When $\kappa(A)$ is very large, the matrix A is said to be *ill-conditioned*. Such a matrix poses difficulties in digital computation: see [56].

Relations between square roots of a matrix

Let $A \in \mathbf{K}^{r \times n}$ and $B \in \mathbf{K}^{r \times n}$ be two square roots of the same matrix $Q \in \mathbf{K}^{n \times n}$, $Q \geq 0$, and, for the sake of simplicity, suppose that A and B are left-regular, that is $r = \operatorname{rk} Q$. The following result is a consequence of Theorem 579:

COROLLARY 586.– *(i) There exists a unitary matrix* $U \in \mathbf{K}^{r \times r}$ *such that* $B = U\, A$*. (ii) Conversely, if* $A \in \mathbf{K}^{r \times n}$ *is a square root of* $Q \geq 0$ *and if* $U \in \mathbf{K}^{r \times r}$ *is unitary, then* $B = U\, A$ *is again a square root of* Q*.*

PROOF. (i) According to Theorem 579, there exist two unitary matrices $\tilde{U}$ and V such that $A = \tilde{U}^* \begin{bmatrix} \Sigma & 0 \end{bmatrix} V$, where $\Sigma = \operatorname{diag}(\sigma_1, ..., \sigma_r)$ is the diagonal matrix formed by the non-zero singular values of A. We thus have

$$A^* A = V^* \begin{bmatrix} \Sigma^* \\ 0 \end{bmatrix} \tilde{U} \tilde{U}^* \begin{bmatrix} \Sigma & 0 \end{bmatrix} V = V^* \begin{bmatrix} \Sigma^2 & 0 \\ 0 & 0 \end{bmatrix} V.$$

Since $A^* A = B^* B$, there exists a unitary matrix W such that $B = W^* \begin{bmatrix} \Sigma & 0 \end{bmatrix} V = W^* \tilde{U} A = U\, A$ where $U \triangleq W^* \tilde{U} \in \mathbf{K}^{r \times r}$ is unitary. (ii) is obvious. ■

13.6. Fractions and special rings

13.6.1. *Rational functions*

Set $\mathbf{K}(s)^{p\times m}$

A *rational function* $f(s)$, with coefficients in the field $\mathbf{K} = \mathbb{R}$ or $\mathbb{C}$, is the quotient of two polynomials with coefficients in $\mathbf{K}$:

$$f(s) = \frac{N(s)}{D(s)} \tag{13.59}$$

where $D(s)$ is not the zero polynomial. The set of these rational functions is denoted by $\mathbf{K}(s)$. This set is both a field and a $\mathbf{K}$-vector space, thus a $\mathbf{K}$-algebra.

The set of matrices of rational fractions with p rows and m columns is denoted by $\mathbf{K}(s)^{p\times m}$. This set is both a $\mathbf{K}$-vector space and a $\mathbf{K}(s)$-vector space. In addition, we can form the product of any two elements of $\mathbf{K}(s)^{n\times n}$ and this product belongs to this set which is thus a ring (this ring is non-commutative if $n > 1$), and therefore a $\mathbf{K}(s)$-algebra; this algebra is unitary, with unit element I_n.

Irreducible rational functions

The rational function (13.59) is *irreducible* if $N(s)$ and $D(s)$ have no common factor. We can come back to that case, by canceling the common factors (if any). Therefore, all rational functions are assumed to be irreducible in the sequel.

Poles, zeros and residues

A rational function is a meromorphic function of a special kind, thus a zero, a pole and their multiplicities, as well as the residue at a pole, are defined for a rational function as for a meromorphic function (section 12.4.1). More specifically, the *zeros* of $f(s)$ are the roots $z_1, ..., z_m$ of $N(s)$ in $\mathbb{C}$ (with $m = d^\circ(N)$), while the *poles* of this rational function are the roots $p_1, ..., p_n$ of $D(s)$ in $\mathbb{C}$ (with $n = d^\circ(D)$). Let Z (resp., P) be the set of *distinct* zeros (resp., poles) of $f(s)$.

The *relative degree* of $f(s)$ is the rational integer $\delta(f) = n - m$, while $-\delta(f)$ is sometimes called the *degree* of f (see [10], Chapter IV), generalizing the notion of degree of a polynomial.

If f and g are two rational functions, then $\delta(f\,g) = \delta(f) + \delta(g)$.

Behavior at infinity

A rational fraction $f(s)$ is said to be *proper* (resp., *strictly proper*, *biproper*) if $\delta(f) \geq 0$ (resp., $\delta(f) > 0$, $\delta(f) = 0$). A rational function that is not proper is said to be *improper*. It is clear that

$f(s)$ is proper if and only if $\lim_{|s|\to+\infty} |f(s)| < +\infty$;

$f(s)$ is strictly proper if and only if $\lim_{|s|\to+\infty} |f(s)| = 0$;

$f(s)$ is biproper if and only if $\lim_{|s|\to+\infty} |f(s)| \in \mathbb{R} \setminus \{0\}$.

More generally, a matrix $G(s)$ belonging to $\mathbb{R}(s)^{p\times m}$ is said to be *proper* (resp., *strictly proper*) if all its elements are proper (resp., strictly proper). It is said to be *biproper* if $p = m$, $G(s)$ is invertible and proper, and $G^{-1}(s)$ is proper.

Decomposition into simple elements in $\mathbb{C}(s)$

Performing the Euclidean division of $N(s)$ by $D(s)$ (see section 13.1.3), we obtain $N(s) = D(s)\,Q(s) + R(s)$, where the quotient $Q(s)$ is a polynomial and where the remainder $R(s)$ is a polynomial such that $d^\circ(R) < d^\circ(D)$. As a result,

$$f(s) = Q(s) + \frac{R(s)}{D(s)} = Q(s) + g(s) \tag{13.60}$$

where $g(s) = \frac{R(s)}{D(s)}$ is a strictly proper rational function. This decomposition is unique.

On the other hand,

$$g(s) = \sum_{p\in P} \sum_{1\le j\le n_P} \frac{\alpha_{pj}}{(s-p)^j} \tag{13.61}$$

where $\alpha_{pj} \in \mathbb{C}$. Recall that the complex number α_{p1} is the *residue* $\operatorname{Res}(f;p)$ of f at the pole p (section 12.4.1).

13.6.2. *Algebra* $\Re\mathcal{H}_\infty$

Definition

We denote by $\Re\mathcal{H}_\infty$ the set of rational functions with real coefficients, which

i) are proper,

ii) have no poles in the closed right half-plane $\bar{\mathbb{C}}^+ = \{s : \operatorname{Re}(s) \ge 0\}$.

The ring $\Re\mathcal{H}_\infty$

The sum of two elements of $\Re\mathcal{H}_\infty$ belongs to $\Re\mathcal{H}_\infty$. The element $f(s) \in \Re\mathcal{H}_\infty$ has an opposite $-f(s)$. The product of two elements of $\Re\mathcal{H}_\infty$ belongs to $\Re\mathcal{H}_\infty$. As a result, $\Re\mathcal{H}_\infty$ is a ring. This ring is a commutative domain.

Let f be a non-zero element of $\Re\mathcal{H}_\infty$, let $m_+(f)$ be the number of its zeros in $\bar{\mathbb{C}}^+$ and let $\delta(f)$ be its relative degree. Write $\theta(f) = m_+(f) + \delta(f)$. One can prove the following ([114], section 2.1):

THEOREM 587.– *The function θ is a* degree *on the ring $\Re\mathcal{H}_\infty$. As a result, $\Re\mathcal{H}_\infty$ is a strongly Euclidean domain, (and thus a principal ideal domain).*

Examples

1) Let

$$f(s) = \frac{(s-1)\, s}{(s+1)^3}.$$

We have $m_+(f) = 2$, $\delta(f) = 1$, thus $\theta(f) = 3$.

2) Let

$$g(s) = \frac{(s+2)(s+3)}{(s+1)^2}$$

We have $\theta(g) = 0$, thus g is a unit of $\Re\mathcal{H}_\infty$ (see section 13.1.3).

The algebra $\Re\mathcal{H}_\infty$

On the other hand, $\Re\mathcal{H}_\infty$ is an $\mathbb{R}$-vector space, and thus is a commutative $\mathbb{R}$-algebra.

The space of operators $\Re\mathcal{H}_\infty$

We will now leave pure algebra and add some analytical considerations.

THEOREM 588.– *The mapping $\|.\|_\infty$, from $\Re\mathcal{H}_\infty$ into $\mathbb{R}^+$, defined by*

$$\boxed{\|f\|_\infty = \sup_{\mathrm{Re}(s)\geq 0} |f(s)|} \tag{13.62}$$

is a norm on the vector space $\Re\mathcal{H}_\infty$. This norm is multiplicative (see section 12.1.3), which makes $\Re\mathcal{H}_\infty$ a "normed algebra". Last, we have the equality

$$\|f\|_\infty = \sup_{\omega\geq 0} |f(i\omega)|. \tag{13.63}$$

PROOF. Let $f \in \Re\mathcal{H}_\infty$. Since f has no poles in the closed right half-plane and is a proper rational function, the quantity (13.62) is finite, thus $\|f\|_\infty \in \mathbb{R}^+$. If $\alpha \in \mathbb{R}$, it is clear that $\|\alpha f\|_\infty = |\alpha|\, \|f\|_\infty$. If $g \in \Re\mathcal{H}_\infty$, the triangle inequality $\|f+g\|_\infty \leq \|f\|_\infty + \|g\|_\infty$ obviously holds. Suppose that $\|f\|_\infty = 0$. Let $f_\infty = \lim_{|s|\to+\infty} f(s)$, in such a way that $h(s) = f(s) - f_\infty$ is a strictly proper rational function. We have $f(s) = f_\infty + h(s) = 0, \forall s$. In addition, $\lim_{|s|\to+\infty} h(s) = 0$, therefore $f_\infty = 0$ and $f = h$. According to section 12.4.4, f is the Laplace transform of a function $u \in L_1$. The Fourier transform of u is the function $\omega \mapsto f(i\omega) = 0$, thus $u = 0$ and finally

$f = 0$. In addition, it is clear that $\|f\,g\|_\infty \leq \|f\|_\infty \|g\|_\infty$. The expression (13.63) is an easy consequence of the maximum modulus principle (Theorem 444). ∎

Let $\tilde{\Sigma}$ be the convolution operator $u \mapsto g * u$, where $g \in \mathcal{S}'_+$ is a distribution whose Laplace transform $\hat{g}(s)$ is a proper rational function. One can prove the following [34], [115]:

THEOREM 589.– *The following five conditions are equivalent: (i) $\hat{g}(s) \in \Re\mathcal{H}_\infty$; (ii) the kernel g is of the form $g = g_0\delta + h$, where $g_0 \in \mathbb{R}$ and $h \in L_1$; (iii) for any $p \in [1, +\infty]$, $\tilde{\Sigma}$ is a continuous linear operator from L_p into L_p; (iv) $\tilde{\Sigma}$ is a continuous linear operator from L_2 into L_2, of which the norm (induced by that of L_2 and denoted as $\gamma_2\left(\tilde{\Sigma}\right)$) is equal to $\|\hat{g}\|_\infty$; (v) $\tilde{\Sigma}$ is a continuous linear operator from L_∞ into L_∞ whose norm (induced by that of L_∞ and denoted as $\gamma_\infty\left(\tilde{\Sigma}\right)$) is equal to $|g_0| + \|h\|_1$.*

The set $\Re\mathcal{H}_\infty^{p\times m}$ is that of all matrices of size $p \times m$, the entries of which belong to $\Re\mathcal{H}_\infty$. It is an $\mathbb{R}$-vector space, and it is a non-commutative $\mathbb{R}$-algebra if $p = m > 1$.

One can easily verify that the mapping $\|.\|_\infty$, from $\Re\mathcal{H}_\infty^{p\times m}$ into $\mathbb{R}^+$, defined by

$$\boxed{\|G\|_\infty = \sup_{\mathrm{Re}(s)\geq 0} \bar{\sigma}\left(G(s)\right) = \sup_{\omega\geq 0} \bar{\sigma}\left(G(i\omega)\right)} \qquad (13.64)$$

is a norm on $\Re\mathcal{H}_\infty^{p\times m}$.

An element $G(s) \in \Re\mathcal{H}_\infty^{p\times m}$ is the transfer matrix of a convolution operator $\tilde{\Sigma}$ from L_2^m into L_2^p (with the convention of section 2.5.1). One can show [122] that the norm of this operator (induced by the norms of L_2^m and L_2^p and denoted as $\gamma_2\left(\tilde{\Sigma}\right)$) is

$$\gamma_2\left(\tilde{\Sigma}\right) = \|G\|_\infty. \qquad (13.65)$$

13.6.3. **Algebra* $\mathcal{H}_\infty$

The normed algebra $\Re\mathcal{H}_\infty$ is a subalgebra of a *Banach algebra* (i.e. a complete normed algebra) denoted as $\mathcal{H}_\infty$, and $\Re\mathcal{H}_\infty = \mathcal{H}_\infty \cap \mathbb{R}(s)$. The space $\mathcal{H}_\infty$ is called a *Hardy space*. A function of the complex variable $f(s)$ belongs to $\mathcal{H}_\infty$ if, and only if (i) f is analytic in the right half-plane and (ii) $\|f\|_\infty \triangleq \sup_{\mathrm{Re}(s)>0} |f(s)| < +\infty$. We then have $\|f\|_\infty = \operatorname{ess.sup}_{\omega\geq 0} |f(i\omega)|$ ([53], Theorem 2.3.1).

On the other hand, $\mathcal{H}_\infty$ is a GCD domain [109] which is neither a UFD, nor a Bézout domain, but has the following property, as shown in [98]:

THEOREM 590.– *$\mathcal{H}_\infty$ is a coherent Sylvester domain.*

We denote by $\mathcal{H}_\infty^{p\times m}$ the $\mathbb{R}$-vector space (which is an $\mathbb{R}$-algebra if $p = m$) consisting of matrices of size $p \times m$ and whose all entries belong to $\mathcal{H}_\infty$. The norm defined on this space is:

$$\boxed{\|G\|_\infty = \sup_{\mathrm{Re}(s)>0} \bar{\sigma}\left(G\left(s\right)\right)}.$$

It is immediately clear that $\mathcal{H}_\infty^{p\times m}$ is a Banach space, and a Banach algebra if $p = m$.

13.6.4. **Classification of rings*

We have the following classification of the rings encountered so far:

$$\begin{array}{ccccccccccccc} & & & & UFD & \rightarrow & GCD & & & & & & \\ & & & \nearrow & & & \uparrow & \nwarrow & & & & & \\ ED & \rightarrow & PID & \rightarrow & EDR & \rightarrow & BD & \rightarrow & CSD & \rightarrow & SD & \rightarrow & HR \\ & & & \searrow & & & \downarrow & \swarrow & & & & & \\ & & & & NR & \rightarrow & CR & & & & & & \end{array}$$

where $\rightarrow$ signifies "is more restrictive than", and where *ED*, *PID*, *EDR*, *BD*, *CSD*, *SD*, *HR*, *NR*, *CR*, *UFD* and *GCD* respectively means Euclidean domain, principal ideal domain, elementary divisor ring, Bézout domain, coherent Sylvester domain, Sylvester domain, Hermite ring, Noetherian ring, coherent ring, unique factorization domain and GCD domain.

13.6.5. ** Change of rings*

Ring of fractions

Let $\mathbf{R}$ be a ring and S be a multiplicative part of $\mathbf{R}$ (that is to say if $s_1, s_2 \in S$, then $s_1 s_2 \in S$) not containing zero. We denote by $S^{-1}\,\mathbf{R}$ the set of elements of the form r/s, $r \in \mathbf{R}$, $s \in S$; one easily shows that this set is a ring, called *the ring of fractions of* $\mathbf{R}$ *with denominator in* S. The following theorem summarizes the results proven in ([11], Chapter VII, section 1-3), ([102], Chapter 4), [60], ([31], Theorem 5.11), ([79], Theorem 3.8).

THEOREM 591.– *If the ring* $\mathbf{R}$ *is a principal ideal domain (resp., a Bézout domain, a unique factorization domain, a Noetherian ring, a coherent ring, a Sylvester coherent domain, an elementary divisor ring), then so is* $S^{-1}\,\mathbf{R}$.

Extension of the ring of scalars

Let M be an $\mathbf{R}$-module and $\mathbf{A}$ be an $\mathbf{R}$-algebra (see section 13.1.1). Then the *tensor product* $\mathbf{A}\otimes_{\mathbf{R}} M$ is the set of all finite linear combinations $\sum a_i \otimes m_i$, $a_i \in \mathbf{A}$, $m_i \in M$; $\mathbf{A}\otimes_{\mathbf{R}} M$ is an $\mathbf{A}$-module. The *functor* $\mathbf{A}\otimes_{\mathbf{R}} -$, from the category of $\mathbf{R}$-modules into that of $\mathbf{A}$-modules, is exact covariant[15] and is called the functor "extension of the ring of scalars". The $\mathbf{R}$-homomorphism $M \ni m \rightarrow 1 \otimes m \in \mathbf{A}\otimes_{\mathbf{R}} M$ is said to be canonical (this is not an epimorphism).

Restriction of the ring of scalars

Let $\mathbf{A}$ be an $\mathbf{R}$-algebra and M be an $\mathbf{A}$-module. The set M has a canonical structure of $\mathbf{R}$-module; this module is denoted as $M_{[\mathbf{R}]}$. The functor $M \mapsto M_{[\mathbf{R}]}$ is exact covariant and is called the functor "restriction of the ring of scalars".

This relation between the extension and the restriction of the ring of scalars is detailed in ([10], n°II.5.2). The reader should be aware that these operations are not inverse of one another. For example, since 2 is a unit of $\mathbb{Q}$, $\mathbb{Q}\otimes_{\mathbb{Z}} (\mathbb{Z}/2\mathbb{Z}) = 0$, whose restriction to $\mathbb{Z}/2\mathbb{Z}$ remains zero.

Modules of fractions

The ring of fractions $\mathbf{A} = S^{-1}\mathbf{R}$ is an $\mathbf{R}$-algebra and when $\mathbf{A}$ is defined in that way, $\mathbf{A}\otimes_{\mathbf{R}} M$ is denoted as $S^{-1}M$. The tensor product $a \otimes m$ ($a \in \mathbf{A}$, $m \in M$) can be denoted as $a\,m$ ([11], section II.2), the canonical homomorphism $\theta : M \rightarrow S^{-1}M$ is written as $m \rightarrow \frac{1}{1}m$ (where $\frac{1}{1}$ is the unit element of $\mathbf{A}$) and according to ([102], Theorem 3.71)

$$\boxed{\ker\theta = \{m \in M : \exists s \in S,\ s\,m = 0\}}. \tag{13.66}$$

Let $\mathbf{A} = \mathbf{K}$, the field of fractions of $\mathbf{R}$ (see section 13.1.1); the tensor product $\mathbf{K}\otimes_{\mathbf{R}} M$ is a $\mathbf{K}$-vector space denoted by $M_{(\mathbf{K})}$. Since $S = \mathbf{R} \setminus \{0\}$, we have according to (13.66)

$$\boxed{\ker\theta = \mathcal{T}(M)}. \tag{13.67}$$

In particular, if M is torsion-free, it is canonically isomorphic to (and identified with) $\theta(M) \subset M_{(\mathbf{K})}$ according to Theorem 538(ii). Therefore, considering elements $m_1, ..., m_k$ of M, these are $\mathbf{R}$-linearly independent if and only if $\theta(m_1), ..., \theta(m_k)$ are $\mathbf{K}$-linearly independent. On the other hand, if $M' \subset M$, $M'_{(\mathbf{K})}$ is identified with a subspace of $M_{(\mathbf{K})}$ and $(M/M')_{(\mathbf{K})}$ is identified with $M_{(\mathbf{K})}/M'_{(\mathbf{K})}$ ([10], n°II.7.10).

15. The notions of category and functor are detailed in [102] (as far as categories of modules are concerned); see also [91]. A succinct account is given in [22].

Chapter 14

Solutions of Exercises

14.1. Exercises of Chapter 1

Solution of Exercise 1

Applying the mesh rule and the nodal rule, we find:

$$C\,\frac{dV}{dt} = \frac{R_1+R_2}{R_1\,R_2}\,i + (R_2\,C + \frac{L}{R_1})\,\frac{di}{dt} + L\,C\,\frac{d^2\,i}{dt^2}.$$

Solution of Exercise 2

The simplest approach is to apply the Lagrange equations, which yield:

$$(M+2m)\,\ddot{y} + \sum_{i=1}^{2}(m\,l_i\,\cos\theta_i\,\ddot{\theta}_i - m\,l_i\,\sin\theta_i\,\dot{\theta}_i^2) - f = 0$$

$$l_i\,\ddot{\theta}_i - g\,\sin\theta_i + \ddot{y}\,\cos\theta_i = 0,\ i = 1, 2.$$

Solution of Exercise 3

Volume is conserved: $S\,\frac{dh}{dt} = Q_1 + Q_2 - \sigma\sqrt{2\,g\,h}$.

Mass is conserved: $\frac{d}{dt}\,(c_s\,S\,h) = c_1\,Q_1 + c_2\,Q_2 - c_s\,\sigma\sqrt{2\,g\,h}$, which together with the previous equation yields:

$$S\,h\,\frac{dc_s}{dt} + c_s\,(Q_1 + Q_2) = c_1\,Q_1 + c_2\,Q_2.$$

Solution of Exercise 4

Volume is conserved: $S\,\frac{dh}{dt} = Q_1 + Q_2 - \sigma\sqrt{2\,g\,h}$.

Energy is conserved: $S\,h\,\frac{dT_s}{dt} + Q_1\,(T_s - T_1) + Q_2\,(T_s - T_2) = 0$.

We note that the equations are the same as those in the previous exercise, the concentrations being replaced by the temperatures.

14.2. Exercises of Chapter 2

Solution of Exercise 53

We have from (1.19), assuming zero initial conditions and writing $u = f$:

$$\left(m_1\,s^2 + k\right)\,\hat{z}_1 - k\,\hat{z}_2 = \hat{u}, \quad \left(m_2\,s^2 + k\right)\,\hat{z}_2 - k\,\hat{z}_1 = 0.$$

Therefore, putting $y = z_2$:

$$G\,(s) = \frac{\hat{y}}{\hat{u}} = \frac{\frac{k}{m_1 m_2}}{s^2\left(s^2 + k\,\frac{m_1+m_2}{m_1 m_2}\right)}.$$

The system thus has no transmission zeros, while the transmission poles are $\{0, 0, i\omega_0, -i\omega_0\}$ with $\omega_0 = \sqrt{k\,\frac{m_1+m_2}{m_1 m_2}}$.

Solution of Exercise 54

The output, this time, is ω, and with zero initial conditions, $\hat{\omega}\,(s) = s\hat{\theta}\,(s)$. The transfer function of the new control system considered is therefore $H\,(s) = sG\,(s)$, where $G\,(s)$ is given by (2.33), i.e.

$$H\,(s) = \frac{k}{s^2 + 2\varsigma\,\omega_0\,s + \omega_0^2}.$$

Thus, the system has no transmission zeros and has transmission poles $\{p_1, p_2\}$ where p_1 and p_2 are the roots of the polynomial $s^2 + 2\varsigma\,\omega_0\,s + \omega_0^2$.

Solution of Exercise 55

Case (a): we have

$$\begin{aligned}
G\,(s) &= \frac{1}{(s-1)^3\,(s+1)}\begin{bmatrix} (s+1)^2\,(s-1) & (s-1)^{\ 2} \\ 0 & (s+1)^3 \end{bmatrix} \\
&\sim \frac{1}{(s-1)^3\,(s+1)}\begin{bmatrix} 1 & 0 \\ 0 & (s-1)\,(s+1)^5 \end{bmatrix} \\
&= \begin{bmatrix} \frac{1}{(s-1)^3\,(s+1)} & 0 \\ 0 & \frac{(s+1)^4}{(s-1)^2} \end{bmatrix}.
\end{aligned}$$

The distinct poles of $G(s)$ are thus $s=-1$ (simple pole) and $s=1$; the latter has structural indices $\{2,3\}$, thus its order is 3 and its degree is 5. The transmission order of the system is $5+1=6$. On the other hand, $G(s)$ has a unique zero $s=-1$; this one here has a unique structural index of 4 (thus its order and its degree are both equal to 4).

Case (b) and (c): the Smith–MacMillan form of $G(s)$ is

$$(b)\quad \begin{bmatrix} \frac{1}{(s+1)^2\,(s+2)^2} & 0 \\ 0 & s+2 \end{bmatrix};\quad (c)\quad \begin{bmatrix} \frac{1}{(s+1)^2\,(s+2)} & 0 \\ 0 & s \end{bmatrix}$$

and we come to the same conclusion as above.

Solution of Exercise 56

Equilibrium positions of the inverted double pendulum:

The first equation yields $f^*=0$.

The second equation yields $\sin\theta_i^*=0,\ i=1,2$, from which we get $\theta_i^*=0$ or π (mod. 2π).

Linearization of the system at $y^*=0, \theta_1^*=\theta_2^*=0$:

$$\begin{cases} (M+2m)\,\frac{d^2y}{dt^2}+m\,l_1\,\frac{d^2\theta_1}{dt^2}+m\,l_2\,\frac{d^2\theta_2}{dt^2}-f=0 \\ l_i\,\frac{d^2\theta_i}{dt^2}-g\,\theta_i+\frac{d^2y}{dt^2}=0,\ i=1,2. \end{cases}$$

Solution of Exercise 57

Equilibrium of volume: $Q_1^*+Q_2^*-\sigma\sqrt{2\,g\,h^*}=0$.

Equilibrium of concentration $c_s : c_s^*\,(Q_1^*+Q_2^*)=c_1\,Q_1^*+c_2\,Q_2^*$. Using the first equation, we obtain $c_1\,Q_1^*+c_2\,Q_2^*=c_s^*\,\sigma\sqrt{2\,g\,h^*}$.

Note that, except if $c_1=c_2$, the values of the different variables at equilibrium are entirely determined by h^* and c_s^*.

Linearization of the equations:

$$\begin{cases} \frac{d}{dt}\Delta h+\frac{\sigma}{S}\sqrt{\frac{g}{2\,h^*}}\,\Delta h=\frac{1}{S}\,(\Delta Q_1+\Delta Q_2) \\ \frac{d}{dt}\Delta c_s+2\,\frac{\sigma}{S}\sqrt{\frac{g}{2\,h^*}}\,\Delta c_s=\frac{c_1-c_s^*}{S\,h^*}\,\Delta Q_1+\frac{c_2-c_s^*}{S\,h^*}\,\Delta Q_2. \end{cases}$$

Solution of Exercise 58

The linearized equations of the inverted pendulum can be put in the form $D(\partial)\, y = N(\partial)\, u$ with

$$D(\partial) = \begin{bmatrix} (M+m)\,\partial^2 & m\,l\,\partial^2 \\ \partial^2 & l\,\partial^2 - g \end{bmatrix}, \quad N(\partial) = \begin{bmatrix} 1 \\ 0 \end{bmatrix}$$

and $y = [z \;\; \theta]^T$, $u = f$. We have $\det D(\partial) \neq 0$, thus the rank of the system is equal to 1. There is thus one input, which is the control and is naturally the force f. In this representation, which is a left form, there is no latent variables.

We can introduce as intermediate variables velocities $v = \frac{dz}{dt}$ and $q = \frac{d\theta}{dt}$, which are then latent variables.

Solution of Exercise 59

There are clearly two inputs, which we will pick as ΔQ_1 and ΔQ_2. It is necessary to regulate Δh, in such a way that the tank does not overflow nor become empty. The aim of the mixer is to make it possible to regulate the concentration. The two controlled variables are thus Δh and Δc_s.

Solution of Exercise 60

(i) The equilibrium conditions are $f(x^*, u^*) = 0$, $y^* = g(x^*, u^*)$. (ii) Write $\Delta x = x - x^*$, $\Delta u = u - u^*$, $\Delta y = y - y^*$. We obtain $\partial \Delta x = A\,\Delta x + B\,\Delta u$, $\Delta y = C\,\Delta x + D\,\Delta u$ with $A = \frac{\partial f}{\partial x}(x^*, u^*)$, $B = \frac{\partial f}{\partial u}(x^*, u^*)$, $C = \frac{\partial g}{\partial x}(x^*, u^*)$, $D = \frac{\partial g}{\partial u}(x^*, u^*)$.

14.3. Exercises of Chapter 3

Solution of Exercise 70

The static gain of the system is $G(0) = 0$. Write $K = \frac{|G(i\omega)|^2}{k^2}$ and $\varpi = \left(\frac{\omega}{\omega_0}\right)^2$. We obtain

$$K = \frac{\varpi}{(\varpi - 1)^2 + 4\varsigma^2\,\varpi}.$$

We easily verify that $\frac{d\,K}{d\,\varpi}$ becomes zero when $\varpi = 1$, thus when $\omega_r = \omega_0$. The maximum of $|G(i\omega)|$ is $|G(i\omega_r)| = \frac{k}{2\varsigma}$.

Solution of Exercise 71

(a) The step response is that of a first order system, the transfer function of which is

$$G(s) = \frac{k}{1 + \tau s}.$$

The static gain is $k = 0.1$. The time constant τ is the abscissa at which the intersection of the tangent at the origin and the horizontal axis is located, and the ordinate of which is the static gain. We have thus $\tau \simeq 0.1$ s.

(b) The step response is that of a second order system that has no zero, and the transfer function of which is thus of the form

$$G(s) = \frac{k\,\omega_0^2}{s^2 + 2\varsigma\,\omega_0\,s + \omega_0^2}.$$

The static gain is $k = 10$. The overshoot is of the order of 37%, which corresponds to a damping coefficient $\varsigma \simeq 0.3$. On the other hand, the maximum of the curve is attained at $t = 1$ s, thus $\frac{\pi}{\omega_p} = 1$, from which we also have $\omega_0 = \frac{\pi}{\sqrt{1-\varsigma^2}} \simeq 3.3$ rad/s.

(c) This system has no resonance. The slope of the amplitude goes from 0 to -40 dB/dec in the neighborhood of 0.1 rad/s. The argument goes from $0°$ to $-180°$ and is $-90°$ for $\omega = 0.1$ rad/s approximately. It is therefore a stable second order system with undamped natural angular frequency $\omega_0 = 0.1$ rad/s and with a damping coefficient ς between 0.7 and 1. It is difficult to be more specific regarding ς. The static gain is equal to 10 (i.e. 20 dB).

Solution of Exercise 72

(a) We have

$$G(s) = \lambda \frac{\prod_{k=1}^{n-r} (z_k - s)}{\prod_{k=1}^{n} (p_k - s)}.$$

(b) The system is with "negative start" if and only if $\rho < 0$. (c) We have

$$\rho = \frac{y^{(r)}(0^+)}{G(0)} = \frac{\prod_{k=1}^{n-r} \left(-\frac{1}{z_k}\right)}{\prod_{k=1}^{n} \left(-\frac{1}{p_k}\right)}$$

since, according to the Initial value theorem (section 12.3.4),

$$y^{(r)}\left(0^+\right) = \lim_{s \in \mathbb{R}, s \to +\infty} s^r\, G(s).$$

(d) Consider the denominator $\prod_{k=1}^{n}\left(-\frac{1}{p_k}\right)$. If a pole p_k is real, we have $-\frac{1}{p_k} > 0$ since this pole belongs to the left half-plane (for the system is stable). If a pole p_k is complex, the conjugate term $-\frac{1}{\bar{p}_k}$ also appears in the product. We have $\left(-\frac{1}{p_k}\right)\left(-\frac{1}{\bar{p}_k}\right) = \frac{1}{|p_k|^2} > 0$. As a result, the denominator is positive. This reasoning also applies to the terms of the numerator, except those that correspond to non-negative real zeros. (e) According to (b) and (d), the system Σ is with "negative start" if and only if the number of non-negative real zeros is odd. (This result is due to Vidyasagar [113]; we should take care not to confuse a system with "negative start" with a non-minimal phase system: there exist non-minimum phase systems whose step response is monotonic increasing, for example the system with transfer function $G(s) = \frac{s^2+2\varsigma\omega_0 s+\omega_0^2}{\omega_0^2(s+1)^2}$ where $\omega_0 = 10$ and $\varsigma = -0.1$.) (f) This result is coherent with the step responses in Figures 3.14 and 3.15: the systems with transfer functions $e_n(s)$ and $\epsilon_n(s)$ are with "negative start" for $n = 1$ and $n = 3$ and with "positive start" for $n = 2$. The integer n is also the number of non-negative real zeros of these transfer functions (as easily verified).

14.4. Exercises of Chapter 4

Solution of Exercise 100

(i)–(iii) Bode, Nyquist and Black plots:

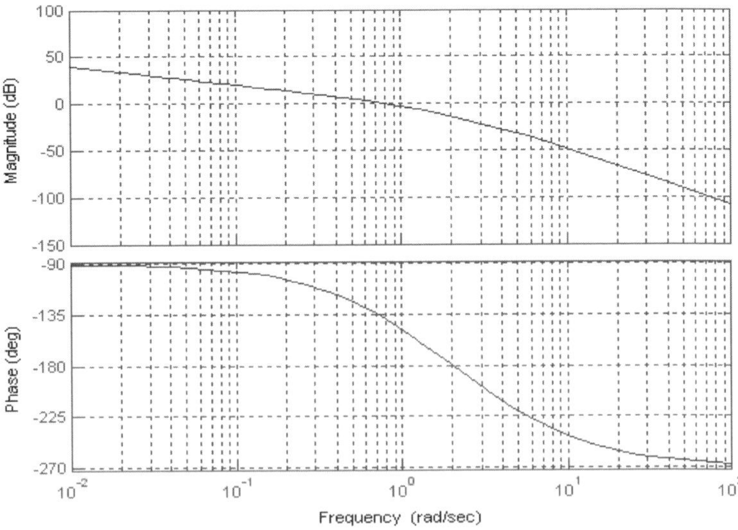

Figure 14.1. *Bode plot – Exercise 100*

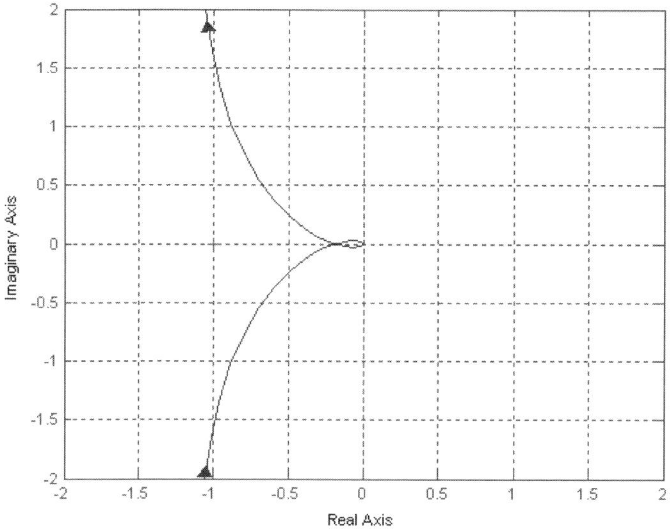

Figure 14.2. *Nyquist plot – Exercise 100*

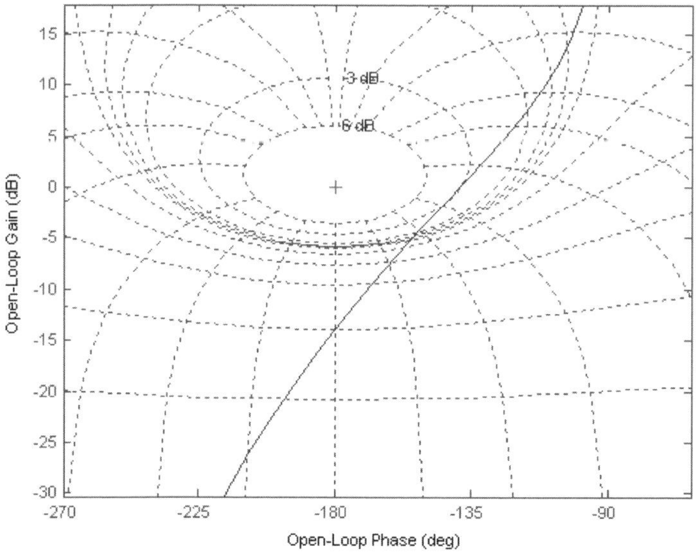

Figure 14.3. *Black plot – Exercise 100*

(iv) The necessary gain to obtain a phase margin of $60°$ is -6.3 dB, which is about 0.48. The delay margin is therefore of 2.4 t.u. (t.u. = time unit). The gain margin is $(-20.3 \text{ dB}, +\infty)$.

Solution of Exercise 101

Questions (i)–(iv) are easy: the feedback system is unstable. (v) This example shows that the "Nyquist criterion" obtained with a Bromwich contour that encircles the imaginary poles on the right is false.

Solution of Exercise 102

Question (i) is easy: the feedback system is unstable. (ii) This example shows that the proposed statement is incorrect.

Solution of Exercise 105

We have

$$\begin{aligned} L_{i,\gamma}(s) &= L_{o,\gamma}(s) = P_\gamma(s) = \begin{bmatrix} \frac{1}{s+1} & \frac{-\gamma}{s+1} \\ 0 & \frac{s+2}{s+1} \end{bmatrix} \\ g_\gamma(s) &= |I_2 + L_{o,\gamma}(s)| - 1 = \frac{2s+3}{(s+1)^2}. \end{aligned}$$

Since the Nyquist plot of $g_\gamma(s)$ (that is the MIMO Nyquist plot of $L_{o,\gamma}(s)$) is entirely located inside the right half-plane, the distance from this plot to -1 is equal to 1, whatever the value of γ is. (ii) Suppose A_0 is replaced by A_ε, the control variable being $u = -y$. The feedback system equation becomes $\dot{y} = \begin{bmatrix} -2 & \gamma \\ \varepsilon & -2 \end{bmatrix} y = F\, y$, where $|s\, I_2 - F| = s^2 + 4s + 4 - \gamma\varepsilon$. The feedback system is stable on condition that the two eigenvalues of F lie in the left half-plane, that is $\gamma\varepsilon < 4$. This feedback system is thus unstable whenever $\gamma \geq \frac{4}{\varepsilon}$. (iii) We have $Mm_{i,\gamma} = Mm_{o,\gamma} = \inf_{\omega \geq 0} \underline{\sigma}(I_2 + L_{o,\gamma}(i\omega))$ and the reader can verify, after a few calculations, that this quantity tends to 0 as $\gamma \to +\infty$. (iv) As a result, $Mm_{i,\gamma} = Mm_{o,\gamma}$ is a good indicator of the lack of robustness of the closed-loop system for large values of γ, whereas this lack of robustness is absolutely not reflected by the MIMO Nyquist plot.

14.5. Exercises of Chapter 5

Solution of Exercise 107

(i) Yes, since $\mathbf{P}$ is an integrator system (see section 5.1.2). (ii) We use a lead. We need a phase lead $\varphi_d = 29°$ at $\omega_0 = 1$ rad/s, thus $\alpha = 2.9$ and $\tau = 0.59$ s . We deduce from there that $K_{PD}(s) = 0.86\,\frac{1+1.70\,s}{1+0.59\,s}$. (iii) No (as for the controller determined at (ii)). (iv) The PID controller is such that $\varphi_d - \varphi_I = 29°$. Choosing $S = \varphi_d + \varphi_I = 90°$, we obtain $\varphi_d = 59.5°$ and $\varphi_I = 30.5°$. The characteristics of the PID controller are finally as follows: $k = 1$; $T_I = 5.1$ s; $T_d = 0.95$ s; $N = 3.5$.

14.6. Exercises of Chapter 6

Solution of Exercise 121

(i) If $\delta_0 > 0$, the Sylvester system is written as

$$\begin{bmatrix} 1 & 0 & \cdots & \cdots & 0 & 0 & \cdots & \cdots & 0 \\ a_1 & 1 & \ddots & & \vdots & \vdots & \ddots & & \vdots \\ \vdots & a_1 & \ddots & \ddots & \vdots & b_0 & & \ddots & \vdots \\ a_n & & \ddots & \ddots & 0 & b_1 & \ddots & & 0 \\ 0 & \ddots & & \ddots & 1 & \vdots & \ddots & 0 & \vdots \\ \vdots & & \ddots & & a_1 & b_n & & b_1 & b_0 \\ \vdots & & & \ddots & \vdots & 0 & \ddots & \vdots & b_1 \\ \vdots & & & & a_n & \vdots & \ddots & b_n & \vdots \\ 0 & \cdots & \cdots & \cdots & 0 & 0 & \cdots & 0 & b_n \end{bmatrix} \begin{bmatrix} \sigma_1 \\ \sigma_2 \\ \vdots \\ \vdots \\ \sigma_{n+\delta_0-1} \\ r_0 \\ r_1 \\ \vdots \\ r_n \end{bmatrix} = \begin{bmatrix} c_1 - a_1 \\ \vdots \\ c_n - a_n \\ c_{n+1} \\ \vdots \\ \vdots \\ \vdots \\ \vdots \\ c_{2n+\delta_0} \end{bmatrix}.$$

The order of the matrix on the left is still $2n+\delta_0$ and the number of zeros below the last a_n (resp., above the first b_0) is equal to 1 (resp., $\delta_0 - 1$). (ii) If $\delta_0 = 0$, it is necessary to go back to (13.13) in section 13.1.5. We have $A_{cl}(\partial) = c_0\, \partial^{2n} + c_1\, \partial^{2n-1} + ... + c_{2n}$ where $c_0 \neq 1$ if $b_0 \neq 0$.

Solution of Exercise 123

1) $R(\partial) = 21\,\partial^2 + 31\,\partial + 10$; $S(\partial) = \partial^3 + 14\,\partial^2 + 24\,\partial$; $T(\partial) = (\partial + 1)(\partial + 10)$. 2) See section 6.3.5. 3) $\delta\left(\frac{R}{S}\right) = 1$.

Solution of Exercise 124

(i) $R(\partial) = (\partial + 1)^2$; $S(\partial) = \partial(\partial + 2)$; $T(\partial) = \partial + 1$. (ii) $S_o(s) = \frac{s}{s+1}$; $Mm = 1$. (iii) The transfer function between r and y is $G = \frac{B\,T}{A_{cl}} = \frac{1}{(s+1)^2}$. The step response is the inverse Laplace transform of $\frac{1}{s}G(s) = \frac{1}{s\,(s+1)^2} = \frac{1}{s} - \frac{1}{s+1} - \frac{1}{(s+1)^2}$, which is $(1 - te^{-t} - e^{-t})\,\mathbf{1}(t)$ (see section 12.4.4). This response thus has no overshoot (and of course with no static error).

Solution of Exercise 125

(i) A reasonable value of α is of the order of 10 (but, since the system presents no particular difficulties – for it is open-loop stable and minimum phase – we can choose a smaller value of α). (ii) With $\alpha = 5$, we take $A_s(\partial) = (\partial + 1)(\partial + 5)$ and $A_{cl}(\partial) = A_s(\partial)(\partial + 1)^2$, from there we have $R(\partial) = 5\partial^2 + 10\partial + 5$, $S(\partial) = \partial\left(\partial^2 + 8\partial + 12\right)$, $T(\partial) = (\partial + 1)(\partial + 5)$. (iii) The advantage of the RST controller determined in Exercise 124 is its simplicity, but its disadvantage, compared

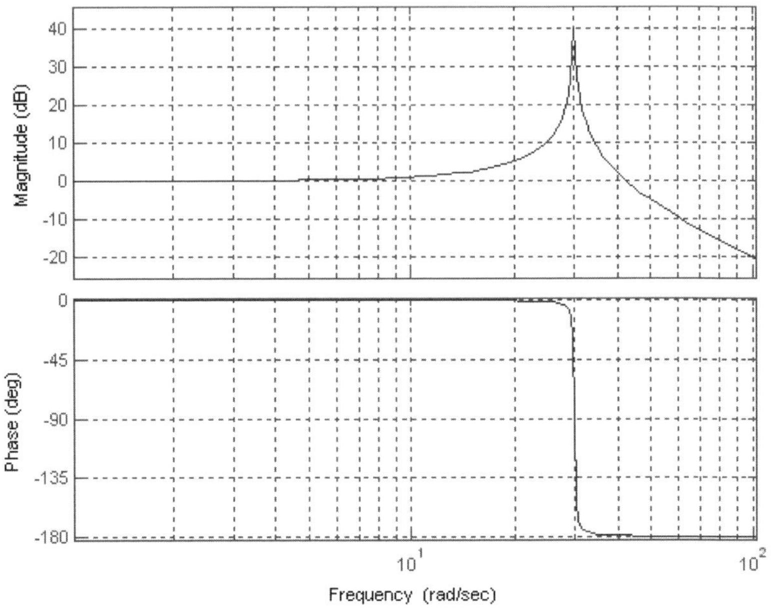

Figure 14.4. *Bode plot of E(s) – Exercise 126*

with the one determined in this exercise, is that is does not filter the measurement noise and that it generates a less rapid roll-off of the open-loop transfer function in high frequencies; this could mean less robustness against neglected dynamics at these frequencies.

Solution of Exercise 126

(i) The result is coherent, since the transfer function $L(s)$ resembles that of a pure integrator $\left(\frac{1}{s}\right)$ in the low frequencies, as this is the case for Exercise 124, with a slope which steepens around 5 rad/s (which corresponds to the absolute value of the pole added). (ii) The Bode plot of $E(s)$ is shown in Figure 14.4 for $\omega_0 = 30$ rad/s. The maximum is attained at the resonant angular frequency $\omega_r = \omega_0\sqrt{1 - 2\varsigma^2} \simeq \omega_0$ and with a value of $\frac{1}{2\varsigma\sqrt{1-\varsigma^2}} \simeq \frac{1}{2\varsigma} = 100$, i.e. 40 dB. (iii) With the RST controller determined in Exercise 125, the stability is preserved in presence of the modeling error, because for $\omega \geq 30$ rad/s, $|L(i\omega)|$ is less than -40 dB; thus $|P(i\omega)E(i\omega)|$ remains less than 0 dB at these frequencies (and the Nyquist criterion can be used to conclude stability). This is not the case with the RST controller determined in Exercise 124 (where $L(s) = \frac{1}{s}$).

Solution of Exercise 127

(i) $S(\partial) = \partial(\partial + 3)$, $R(\partial) = 4\partial^2 + 7\partial + 1$, $T(\partial) = (\partial + 1)^2$. (ii) $S(\partial) = \partial\left(\partial^2 + 4\partial + 7\right)$, $R(\partial) = 7\partial^2 + 12\partial + 1$, $T(\partial) = (\partial + 1)^3$. (iii) For Question (i), "too slow" a pole is chosen: it would be preferable to place it at -10 for example; for Question (ii), two "faster" poles of the closed-loop system must be chosen.

14.7. Exercises of Chapter 7

Solution of Exercise 208

(i) Let $\{A, B, C\}$ be a state representation of the DC motor, where x is the specified state vector. We obtain

$$\begin{aligned} A &= \begin{bmatrix} 0 & 1 & 0 \\ 0 & -\frac{\lambda}{J} & \frac{K}{J} \\ 0 & -\frac{K}{L} & -\frac{R}{L} \end{bmatrix}, \quad B = \begin{bmatrix} 0 \\ 0 \\ \frac{1}{L} \end{bmatrix} \\ C &= \begin{bmatrix} 1 & 0 & 0 \end{bmatrix}. \end{aligned}$$

(ii) Controllability and observability are verified using the Kalman criterion (Theorems 141 and 150). We have indeed

$$\Gamma(A, B) = \begin{bmatrix} 0 & 0 & \frac{K}{JL} \\ 0 & \frac{K}{JL} & * \\ \frac{1}{L} & * & * \end{bmatrix}$$

which is of rank 3 since $K \neq 0$ (see section 1.3). On the other hand,

$$\Omega(C, A) = \begin{bmatrix} 1 & 0 & 0 \\ 0 & 1 & 0 \\ 0 & 0 & \frac{K}{J} \end{bmatrix}$$

is of rank 3 for the same reason.

Solution of Exercise 209

(i) The state equation is $\dot{x} = A\,x + B\,u$ with

$$A = \begin{bmatrix} 0 & -\varepsilon\,\sigma^2 & 0 & 0 \\ 0 & 0 & 1 & 0 \\ 0 & (1+\varepsilon)\,\sigma^2 & 0 & 0 \\ 1 & 0 & 0 & 0 \end{bmatrix}, \quad B = \begin{bmatrix} 1 \\ 0 \\ -1 \\ 0 \end{bmatrix}.$$

(ii) The characteristic polynomial of A is $p_A(s) = s^2\left(s^2 + (\varepsilon+1)\,\sigma^2\right)$, thus the system poles are $\left\{0, 0, \pm\sqrt{1+\varepsilon}\,\sigma\right\}$. The system is therefore unstable. (iii) We easily verify, using the Kalman criterion, that the system is controllable. (iv) If $C = \begin{bmatrix} * & * & * & 0 \end{bmatrix}$, then the system is unobservable; it does have a structure such as (7.13) with $A_{\bar{o}} = 0$. The system is not detectable according to Proposition 171 (section 7.2.3) and Definition 185 (section 7.3). (v) If $C = \begin{bmatrix} 0 & 0 & 0 & 1 \end{bmatrix}$, the

Kalman criterion shows that the system is observable. (vi) The transmission zeros are equal to the invariant zeros according to Theorem 179(i) (section 7.2.6) and are thus the values $s \in \mathbb{C}$ for which the Rosenbrock matrix (7.20) "looses its rank". We obtain $\{i.z.\} = \{\pm\sigma\}$. When $\varepsilon \ll 1$, two system poles approach the two $i.z.$'s and the system tends towards unobservability.

Solution of Exercise 210

(i) The state equation is $\dot{x} = A\,x + B\,u$ with

$$A = \begin{bmatrix} 0 & 0 & -\lambda & \lambda \\ 0 & 0 & -\lambda\,\rho & \lambda\,\rho \\ 1 & 0 & 0 & 0 \\ 0 & 1 & 0 & 0 \end{bmatrix}, \quad B = \begin{bmatrix} 1 \\ 0 \\ 0 \\ 0 \end{bmatrix}.$$

(ii) The characteristic polynomial of A is $p_A(s) = s^2\left(s^2 + \lambda(\rho+1)\right)$. Since $\lambda(\rho+1) = k\frac{m_1+m_2}{m_1 m_2}$, the system poles are $\left\{0, 0, \pm i\sqrt{k\frac{m_1+m_2}{m_1 m_2}}\right\}$; these values are the Smith zeros of the polynomial matrix $D(s)$ according to (2.13), which is clearly coherent according to Definition 17 (section 2.3.7). (iii) It is easy to verify, using the Kalman criterion, that the system is controllable. (iv) If $C = \begin{bmatrix} 0 & 1 & 0 & 0 \end{bmatrix}$, the Kalman criterion shows that the system is unobservable. The interpretation is immediate: by measuring only the velocity $\dot{z}_2$, we cannot determine the position z_2. (v) Since the system has no $i.d.z.$'s, we have $\{h.m.\} = \{o.d.z.\}$ according to Corollary 176 (section 7.2.5). The $o.d.z.$'s are determined using Proposition 170 (section 7.2.3). The $o.d.z$'s are indeed the Smith zeros of the matrix

$$\begin{bmatrix} s\,I_n - A \\ C \end{bmatrix}.$$

In this case, we easily show that $\{o.d.z.\} = \{0\}$. Thus, $\{h.m.\} = \{0\}$. (vi) With $C = \begin{bmatrix} 0 & 0 & 0 & 1 \end{bmatrix}$, we show, using the Kalman criterion, that the system is observable. (vii) With this choice, the system is minimal and its $t.z.$'s are thus identical to its $i.z.$'s, i.e. to the the Smith zeros of the Rosenbrock matrix. Since the latter is square, its Smith zeros are roots of its determinant. We see that this is a constant, and so the system has no transmission zeros. Another faster way of getting to this conclusion is to use the result obtained in Exercise 53.

Solution of Exercise 211

(i) The controllable canonical form is $\{A, B, C\}$ with

$$A = \begin{bmatrix} -\frac{R}{L} & -\frac{1}{L\,C} \\ 1 & 0 \end{bmatrix}, \quad B = \begin{bmatrix} 1 \\ 0 \end{bmatrix}$$

$$C = \begin{bmatrix} \frac{1}{L} & 0 \end{bmatrix}.$$

(ii) The observability matrix is

$$\Omega = \begin{bmatrix} \frac{1}{L} & 0 \\ -\frac{R}{L^2} & -\frac{1}{L^2\,C} \end{bmatrix}.$$

If the capacity tends toward $+\infty$, the matrix Ω becomes singular and the system becomes unobservable (the charge q of the capacitor is no longer observable).

Solution of Exercise 212

(i) We can eliminate $\frac{d^2 y}{dt^2}$ of the equations of motion of the inverted double pendulum and we obtain the state representation $\dot{x} = A\,x + B\,u$ where

$$A = \begin{bmatrix} 0 & 0 & 1 & 0 \\ 0 & 0 & 0 & 1 \\ a_1 & a_2 & 0 & 0 \\ a_3 & a_4 & 0 & 0 \end{bmatrix}, \quad B = \begin{bmatrix} 0 \\ 0 \\ 1/l_1 \\ 1/l_2 \end{bmatrix}$$

with $a_1 = \frac{(M+m)g}{M\,l_1}$, $a_2 = \frac{m\,g}{M\,l_1}$, $a_3 = \frac{m\,g}{M\,l_2}$, $a_4 = \frac{(M+m)g}{M\,l_2}$. (ii) We study the controllability using the Kalman criterion. Since the matrix $\Gamma(A, B)$ is square, it suffices to examine on what conditions its determinant is non-zero. This condition is $l_1 \neq l_2$ (the reader is requested to wonder why). (iii) The Kalman criterion shows that the system is observable when the output is x_1.

Solution of Exercise 216

(i) $\{o.d.z.\} = \{1\}$ and $\{i.z.\} = \varnothing$. (ii) $\{i.d.z.\} = \{1\}$ and $\{i.z.\} = \varnothing$. (iii) These results are coherent with Theorem 179 and show that we cannot expect to improve the latter.

Solution of Exercise 217

(i) is easy. For (ii), we proceed as in section 7.4.3 and we obtain

$$A_c = \begin{bmatrix} -4 & -3 & 0 & 0 \\ 1 & 0 & 0 & 0 \\ 0 & 0 & -6 & -8 \\ 0 & 0 & 1 & 0 \end{bmatrix}, \quad B_c = \begin{bmatrix} 1 & 0 \\ 0 & 0 \\ 0 & 1 \\ 0 & 0 \end{bmatrix},$$

$$C_c = \begin{bmatrix} 1 & 3 & 1 & 4 \\ 1 & 1 & 1 & 2 \end{bmatrix}.$$

This form is not unique: if x designates the state of the above canonical form, x' is equally the state of the controllable canonical form with $x' = \begin{bmatrix} x_3 & x_4 & x_1 & x_2 \end{bmatrix}^T$. This is due to the fact that two of the controllability indices are equal; indeed, $\{\mu_1, \mu_2\} = \{2, 2\}$.

Solution of Exercise 218

The whole exercise consists in using the duality relations of Corollary 153 (section 7.1.4) and the observations that follow this corollary. As a result : (i) The observability indices of (C, A) are the controllability indices of $\left(A^T, C^T\right)$. (ii) The matrices B_{oo} and B_o have no particular form. The canonical forms to be determined are thus those of (C, A). We have $(C_{oo}, A_{oo}) = \left(A_{cc}^T, C_{cc}^T\right)$ and $(C_o, A_o) = \left(A_c^T, C_c^T\right)$, where $\left(A_{cc}^T, C_{cc}^T\right)$ (resp., $\left(A_c^T, C_c^T\right)$) is the canonical form of controllability (resp., the controllable canonical form) of $\left(A^T, C^T\right)$.

Solution of Exercise 219

(i) We easily verify that $G(s) = D^{-1}(s) N(s)$ and that the matrices $\{D, N\}$ are left-coprime. (ii) Let $H(s) = G^T(s)$. We have $H(s) = N^T(s) D^{-T}(s)$ where $\left\{N^T, D^T\right\}$ are right-coprime. We determine a realization of $H(s)$ in controllable canonical form by proceeding as in section 7.4.3 and we deduce from there a realization of $G(s)$ in observable canonical form by using the results of Exercise 218. This realization $\{A_o, B_o, C_o\}$ is obtained by choosing as the output $y' = (y_2, y_1)$, and is given by

$$A_o = \begin{bmatrix} -4 & 1 & 0 & 0 & 0 \\ -4 & 0 & 0 & 0 & 0 \\ 0 & 0 & -4 & 1 & 0 \\ 1 & 0 & -5 & 0 & 1 \\ 2 & 0 & -2 & 0 & 0 \end{bmatrix}, \quad B_o = \begin{bmatrix} 1 & -1 \\ 0 & 0 \\ 0 & -1 \\ 0 & 0 \\ 0 & 0 \end{bmatrix},$$

$$C_o = \begin{bmatrix} 1 & 0 & 0 & 0 & 0 \\ 0 & 0 & 1 & 0 & 0 \end{bmatrix}.$$

The observability indices are $\{\omega_1, \omega_2\} = \{2, 3\}$. Since these are all different, this canonical form of observability is unique.

Solution of Exercise 220

(i) The controllability matrix of $\mathbf{\Sigma}$ is

$$\Gamma = \begin{bmatrix} 0 & 0 & 0 \\ 1 & -3 & 9 \\ 0 & -4 & 8 \end{bmatrix}.$$

Since $\operatorname{rk}\Gamma = 2$, $\mathbf{\Sigma}$ is uncontrollable and has one $i.d.z.$ (for $1 = 3 - 2$). The observability matrix of $\mathbf{\Sigma}$ is

$$\Omega = \begin{bmatrix} 1 & -1 & -1 & 2 & 1 & -5 \\ 0 & 1 & 0 & -3 & 0 & 9 \\ 0 & 0 & 0 & 0 & 0 & 0 \end{bmatrix}^T.$$

As a result, $\operatorname{rk}\Omega = 2$, $\boldsymbol{\Sigma}$ is unobservable and has one $o.d.z.$ (for $1 = 3 - 2$). (ii) The system poles are the eigenvalues of A which is lower triangular; these are thus the diagonal elements of A, which are $\{-1, -3, 1\}$. (iii) The first equation of the system is $\partial x_1 = -x_1$. Thus, -1 is an $i.d.z.$ and it is the only one according to (i). On the other hand, $\{A, B, C\}$ is directly decomposed according to observability and has a unique $o.d.z.$ which is equal to 1. According to Proposition 173 (section 7.2.4), $\{i.o.d.z.\} = \varnothing$. As a result, according to Corollary 176 (section 7.2.5), $\{h.m.\} = \{-1, 1\}$. At last, according to Theorem 179(i) (section 7.2.6), $\{t.p.\} = \{s.p.\} \setminus \{h.m.\} = \{-3\}$. (iv) The system $\boldsymbol{\Sigma}$ is thus stabilisable (for all $i.d.z.$'s belongs to the left half-plane) but not detectable (the $o.d.z.$ belongs to the right half-plane). (v) The Rosenbrock matrix is

$$R(s) = \begin{bmatrix} s+1 & 0 & 0 & 0 \\ -1 & s+3 & 0 & -1 \\ -1 & 4 & s-1 & 0 \\ 1 & 0 & 0 & 0 \\ -1 & 1 & 0 & 0 \end{bmatrix} \sim \begin{bmatrix} 0 & 0 & 0 & 0 \\ 0 & 0 & 0 & 1 \\ 0 & 0 & s-1 & 0 \\ 1 & 0 & 0 & 0 \\ 0 & 1 & 0 & 0 \end{bmatrix}.$$

Its rank is 4 for $s \neq 1$ and is 3 for $s = 1$, thus $\{i.z.\} = \{1\}$.

Solution of Exercise 221

1) (i) The system is neither controllable nor observable. (ii) Its poles are $\{-1, 1, 0\}$. (iii) (a) The transfer function is

$$G(s) = C\,(s\,I_3 - A)\,B = \frac{1}{s}.$$

(b) To find the $i.d.z.$'s, we form the matrix

$$\begin{bmatrix} s\,I_3 - A & B \end{bmatrix} = \begin{bmatrix} s+1 & 0 & 0 & 0 \\ -1 & s-1 & 0 & 0 \\ 0 & 0 & s & 1 \end{bmatrix}$$

$$\sim \begin{bmatrix} 1 & 0 & 0 & 0 \\ 0 & 1 & 0 & 0 \\ 0 & 0 & (s-1)(s+1) & 1 \end{bmatrix}.$$

As a result, $\{i.d.z.\} = \{1, -1\}$ (see Proposition 166, section 7.2.2). (c) To find its $o.d.z.$'s, we form the matrix

$$\begin{bmatrix} s\,I_3 - A \\ C \end{bmatrix} = \begin{bmatrix} s+1 & 0 & 0 \\ -1 & s-1 & 0 \\ 0 & 0 & s \\ -1 & 0 & 1 \end{bmatrix} \sim \begin{bmatrix} 1 & 0 & 0 \\ 0 & 1 & 0 \\ 0 & 0 & s-1 \\ 0 & 0 & 0 \end{bmatrix}.$$

As a result, $\{o.d.z.\} = \{1\}$ (see Proposition 170, section 7.2.3). (d) According to (a), $\{t.p.\} = \{0\}$, thus $\{h.m.\} = \{s.p.\} \setminus \{t.p.\} = \{-1, 1\}$. Consequently, $\{i.o.d.z.\} = \{i.d.z.\} \dot{\cup} \{o.d.z\} \setminus \{h.m.\} = \{1\}$. (iv) The system is neither stabilizable nor detectable. (v) To find the invariant zeros, we form the Rosenbrock matrix

$$R(s) = \begin{bmatrix} s\,I_3 - A & -B \\ C & 0 \end{bmatrix} = \begin{bmatrix} s+1 & 0 & 0 & 0 \\ -1 & s-1 & 0 & 0 \\ 0 & 0 & s & -1 \\ -1 & 0 & 1 & 0 \end{bmatrix} \sim \begin{bmatrix} 1 & 0 & 0 & 0 \\ 0 & 1 & 0 & 0 \\ 0 & 0 & 1 & 0 \\ 0 & 0 & 0 & (s-1)(s+1) \end{bmatrix};$$

as a result, $\{i.z.\} = \{1, -1\}$. This result can equally be obtained, more simply, by applying Theorem 179(iv) (section 7.2.6), knowing that $\{t.z.\} = \emptyset$.

Solution of Exercise 222

(i) Apply the Popov–Belevitch–Hautus test (Corollary 139, section 7.1.3,). (ii) Using the controllability $\leftrightarrow$ observability duality (Corollary 153, section 7.1.4), we obtain the following criterion : $\mathbf{\Sigma}$ is observable if and only if there exists no (right) eigenvector of A that will annihilate C, that is, no column vector $v \neq 0$ such that $A\,v = \lambda\,v$ and $C\,v = 0$.

Solution of Exercise 223

According to the Taylor's formula,

$$P(t) = \sum_{j=0}^{2n+1} p_j\,(t - t_0)^j$$

with $p_j = P^{(j)}(t_0)/j!$. It is immediate that $p_0 = y_0$ and $p_j = 0$ for $1 \leq j \leq n$. Consequently, for $0 \leq \beta \leq n$,

$$P^{(\beta)}(t_1) = \sum_{j=n+1}^{2n+1-\beta} \frac{j!}{(j-\beta)!} \Delta^{j-\beta} p_j.$$

The $n+1$ coefficients p_j $(n+1 \leq j \leq 2n+1)$ are thus obtained by solving the system of $n+1$ linear equations

$$\begin{aligned} \sum_{j=n+1}^{2n+1} \Delta^j p_j &= y_1 - y_0, \\ \sum_{j=n+1}^{2n+1-\beta} \frac{j!}{(j-\beta)!} \Delta^{j-\beta} p_j &= 0, \quad 1 \leq \beta \leq n. \end{aligned}$$

Solution of Exercise 224

(i) We have $z_1 = z_2 + (m_2/k)\ \ddot{z}_2$ and $f = (m_1 + m_2)\ \ddot{z}_2 + (m_1 m_2/k)\ z_2^{(4)}$. As a result, z_2 is a flat output of the system. This flat output is the most natural possible. (ii) The equilibrium conditions are $z_2(t_0) = z_{20}$, $z_2(t_1) = z_{21}$, $z_2^{(\beta)}(t_i) = 0, 1 \leq \beta \leq 4$, $i \in \{0, 1\}$.

14.8. Exercises of Chapter 8

Solution of Exercise 263

Let be the system $\dot{x} = A\,x + B\,u$, $y = C\,x + D\,u$ and the state feedback control $u = v - K\,x$. The Rosenbrock matrix of the feedback system is

$$\begin{bmatrix} s\,I_n - A + B\,K & -B \\ C - D\,K & D \end{bmatrix} \sim \begin{bmatrix} s\,I_n - A & -B \\ C & D \end{bmatrix};$$

the equivalence is obtained by subtracting $\begin{bmatrix} -B \\ D \end{bmatrix} K$ from $\begin{bmatrix} s\,I_n - A + B\,K \\ C - D\,K \end{bmatrix}$, using columns elementary operations.

Solution of Exercise 264

(i) We can easily show that Σ is controllable and observable using the Kalman criterion. (ii) $G(s) = C\,(s\,I_2 - A)^{-1}\,B = \frac{-5}{s^2+s-2}$. Since Σ has no hidden modes, its invariant zeros coincide with its transmission zeros – there are none –, and its poles coincide with its transmission poles, which are $\{1, -2\}$ (Theorem 179(i), section 7.2.6). (iii) The answer is yes, since Conditions (i) and (ii) in Proposition 250 hold.

(iv) System (8.33) is thus controllable and we can place the poles of the closed-loop system using the control (8.34). Writing $K = \begin{bmatrix} k_1 & k_2 & k_3 \end{bmatrix}$ and identifying $\det(s\,I_3 - F + G\,K)$ term by term with $(s+2)\,(s+1)^2 = s^3 + 4s^2 + 5s + 2$, we obtain

$$\det(s\,I_3 - F + G\,K) = s^3 + (k_1 - 2k_2 + 1)\,s^2 + (-3k_1 + k_2 - 2)\,s - 5k_3$$

which yields

$$\begin{cases} k_1 - 2k_2 = 3 \\ 3k_1 - k_2 = -7 \\ 5k_3 = -2 \end{cases}$$

and finally we get $k_1 = -17/5 = -3.4$, $k_2 = -16/5 = -3.2$, $k_3 = -2/5 = -0.4$. The poles of $L_i\,(s)$ coincide with the eigenvalues of F, which are $\{-2, 0, 1\}$. The control is found to be

$$u = -K_p\,x - K_i \int e\,(t)\,dt + const.$$

with $K_p = \begin{bmatrix} k_1 & k_2 \end{bmatrix}$ and $K_i = k_3$. (v) Since the closed-loop system is stable, the Nyquist plot (taking into account the "half-circle at infinity") encloses in the direct sense point -1 a number of times which is equal to the number of poles of $L_i\,(s)$ belonging to $\bar{\mathbb{C}}_+$, i.e. 2 times. Since Rule 111 (section 6.3.5) is being applied, the modulus margin is equal to 1. We can deduce from these considerations that the Nyquist plot of $L_i\,(s)$ has the shape shown in Figure 14.5.

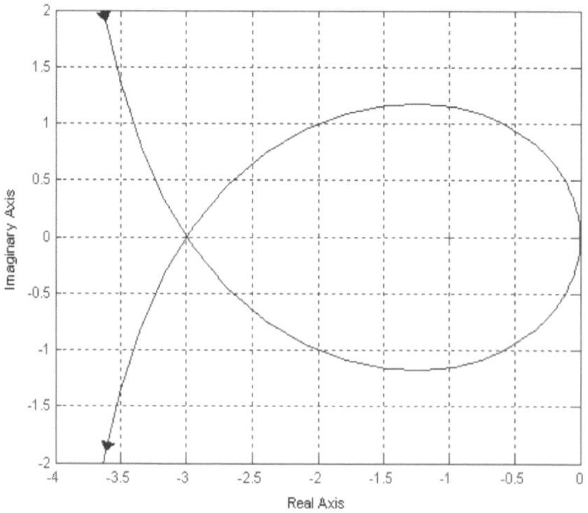

Figure 14.5. *Nyquist plot of Li(s) – Exercise 264*

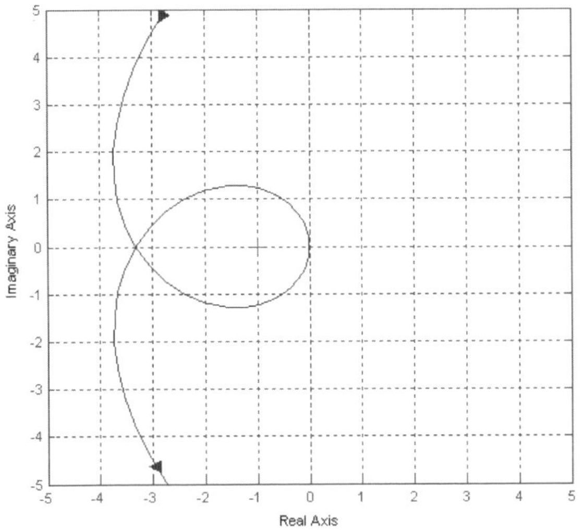

Figure 14.6. *Nyquist plot of Li(s) – Exercise 265*

Solution of Exercise 265

(i) We have shown in Exercise 209 that Σ is controllable and observable (with $C = \begin{bmatrix} 0 & 0 & 0 & 1 \end{bmatrix}$). (ii) The poles of Σ are $\{0, 0, \pm\sqrt{11}\}$, its invariant zeros are $\{\pm\sqrt{10}\}$. (iii) Same answer as in Exercise 264, for the same reasons. (iv) $K = \begin{bmatrix} -65.2 & -270.4 & -85.2 & -104.4 & -72 \end{bmatrix}$. (v) The Nyquist plot of $L_i(s)$ has the shape shown in Figure 14.6, since $Mm_i = 1$.

Solution of Exercise 266

(i) The state equations directly express the discharge equations (see section 1.4) and $C = \begin{bmatrix} 0 & \alpha_3 \end{bmatrix}$.

(ii) Using the Kalman criterion, we immediately show that the system is controllable and observable.

(iii) The system poles are $\{-\alpha_2, -\alpha_3\}$, they are located in the left half-plane, and thus the system is stable.

(iv) Forming the Rosenbrock matrix, we show that the system has no finite zeros. The static gain is $G(0)$ with $G(s) = C(s I_2 - A)^{-1} B$, and so $G(0) = -C A^{-1} B = 0.01$.

(v) The control law we are looking for is of the form (8.35) with $K_p = \begin{bmatrix} 12 & 30.5 \end{bmatrix}$ and $K_i = 0.625$. The choice of this pole placement is justified by Rule 111 (section 6.3.5).

Solution of Exercise 267

(i) Taking as a state vector $x = \begin{bmatrix} \theta & v & h \end{bmatrix}^T$, we obtain a system $\{A, B, C\}$ with

$$A = \begin{bmatrix} -\frac{1}{\tau_1} & 0 & 0 \\ \sigma & -\frac{1}{\tau_2} & 0 \\ 0 & 1 & 0 \end{bmatrix}, \quad B = \begin{bmatrix} 1 & 0 \\ 0 & \frac{1}{\tau_2} \\ 0 & 0 \end{bmatrix},$$
$$C = \begin{bmatrix} 0 & 0 & 1 \end{bmatrix}.$$

(ii) System poles: $\left\{-\frac{1}{\tau_1}, -\frac{1}{\tau_2}, 0\right\}$. The system is unstable (and, more precisely, "marginally stable": see Definition 183, section 7.3).

(iii) Since $u = 0$, the matrix B is replaced by its second column b_2. The system is observable but not controllable (and $\{i.d.z.\} = \left\{-\frac{1}{\tau_1}\right\}$). Since the first equation is $\tau_1 \dot{\theta} = -\theta$, θ converges inevitably to 0.

(iv)(a) This time, the matrix B is replaced by its first column b_1. This has no influence on the observability, but the system is now controllable (the determinant of the controllability matrix is σ^2).

(b) We obtain a control law of the form (8.35) with $K_p = \begin{bmatrix} 19/10 & 1 \end{bmatrix}$ and $K_i = 1$.

Solution of Exercise 269

The only modification needed is that matrix G of System (8.33) is now written as: $G = \begin{bmatrix} B \\ D \end{bmatrix}$.

14.9. Exercises of Chapter 9

Solution of Exercise 297

(i) Let $\tilde{K} = \begin{bmatrix} \tilde{k}_1 \\ \tilde{k}_2 \end{bmatrix}$ be the observer gain. We have

$$\det\left(s\, I_2 - A + \tilde{K}\, C\right) = \begin{vmatrix} s + 2\tilde{k}_1 & \tilde{k}_1 - 2 \\ 2\tilde{k}_2 - 1 & s + \tilde{k}_2 + 1 \end{vmatrix}$$
$$= s^2 + s\left(2\tilde{k}_1 + \tilde{k}_2 + 1\right) + 3\tilde{k}_1 + 4\tilde{k}_2 - 2.$$

Identifying this polynomial term by term with $(s+5)^2 = s^2 + 10s + 25$, we obtain the equations

$$\begin{cases} 2\tilde{k}_1 + \tilde{k}_2 = 9 \\ 3\tilde{k}_1 + 4\tilde{k}_2 = 27 \end{cases}$$

hence $\tilde{k}_1 = 9/5$ and $\tilde{k}_2 = 27/5$. (ii) The observer poles lie on the real axis, they are negative and fast compared to the system poles, thus they comply with the rule expressed by Theorem 112 (since all system zeros are at infinity).

Solution of Exercise 298

(i) The observer gain is $\tilde{K} = \begin{bmatrix} 32 \\ 17 \end{bmatrix}$. (ii) Same reason as is in Exercise 297.

Solution of Exercise 299

(i) $\mathbf{\Sigma}$ is in a canonical controllable form, thus its characteristic polynomial can be "read" directly on the matrix A. Its poles are $\{1, -2\}$ and $\mathbf{\Sigma}$ is unstable. (ii) Transfer function:

$$G(s) = \frac{s+\beta}{(s-1)(s+2)}.$$

If $\beta \notin \{-1, 2\}$, the transmission poles are $\{1, -2\}$ and the set of its transmission zeros is $\{-\beta\}$. If $\beta = -1$ (resp., $\beta = 2$), $\mathbf{\Sigma}$ has a unique transmission pole -2 (resp., 1) and has no finite transmission zero. (iii) $\mathbf{\Sigma}$ is clearly controllable (see (i)) and is observable for $\beta \notin \{1, -2\}$. (iv) Using the Rosenbrock matrix, we find that $\mathbf{\Sigma}$ has one invariant zero $-\beta$ (regardless of the value of β). (v) Writing that $\det(s\,I_2 - A + B\,K) = s^2 + 3s + 2$, we find that $K = \begin{bmatrix} 2 & 4 \end{bmatrix}$. The equilibrium state is $x^* = -(A - B\,K)^{-1}\, B\,k_0$ and the corresponding output is $y^* = C\,x^*$. We thus need to have $-C\,(A - B\,K)^{-1}\, B\,k_0 = 1$, hence $k_0 = 2/\beta$. (vi)(a) Neither $\mathbf{\Sigma}$ nor the controller is an integrator system, thus there is a non-zero static error in the presence of a non-zero output disturbance. (b) The answer is yes, since the number of outputs is equal to the number of inputs and $s = 0$ is not an invariant zero of $\mathbf{\Sigma}$. (c) We obtain $K_p = \begin{bmatrix} 3 & 5 \end{bmatrix}$ and $K_i = 2$. (d) Rule 111 is being complied with. (vii)(a): See (9.14). (b) Identifying $\det\left(s\,I_2 - A + \tilde{K}\,C\right)$ term by term with $s^2 + 11s + 10$, we obtain $\tilde{K} = \begin{bmatrix} 9 & 1 \end{bmatrix}^T$.

Solution of Exercise 300

(i) The Kalman criterion shows that $\mathbf{\Sigma}$ is not controllable but observable. The poles, which are the eigenvalues of A, are $\{-1, 1\}$. (ii) The stabilizability can be studied using the Popov-Belevitch-Hautus test (Proposition 186). The unique $i.d.z.$ is

-1, thus $\mathbf{\Sigma}$ is stabilizable. (iii) Let $u = -\begin{bmatrix} k_1 & k_2 \end{bmatrix} x$. We obtain

$$\begin{aligned} \det(s\,I_2 - A + B\,K) &= s^2 + s\,(k_1 - k_2) + k_1 - k_2 - 1 \\ &= (s+1)\,(s + k_1 - k_2 - 1)\,. \end{aligned}$$

We thus have $\det(s\,I_2 - A + B\,K) = (s+1)\,(s-\lambda)$ if and only if $k_2 + 1 - k_1 = \lambda$. As a result, a solution exists but is not unique. For $\lambda = -1$, a parameterization of the solutions is $k_2 = k$, $k_1 = k + 1 - \lambda$, where k is an arbitrary parameter. (iv) The full-order observer is given by (9.3) where $D = 0$ and where $\tilde{K} = \begin{bmatrix} \tilde{k}_1 & \tilde{k}_2 \end{bmatrix}^T$ is such that $\det\left(s\,I_2 - A + \tilde{K}\,C\right) = (s+1)\,(s+10)$. After a term by term comparison, we obtain $\tilde{k}_1 = 11$ and $\tilde{k}_2 = -11$. (v) In order to use (9.34), put $x' = \begin{bmatrix} x_2 & x_1 \end{bmatrix}^T$. We obtain $\dot{x}' = A'\,x' + B'\,u$, $y = C'\,x'$ with

$$A' = \begin{bmatrix} -1/2 & -3/2 \\ -1/2 & 1/2 \end{bmatrix}, \quad B' = \begin{bmatrix} -1 \\ 1 \end{bmatrix}, \quad C' = \begin{bmatrix} 0 & 1 \end{bmatrix}.$$

The only state to be reconstructed is $x'_1 = x_2$. We have $A'_r = -1/2$ and $C'_r = -1/2$. Take $\varpi = -10$ as an observer pole. The gain $\tilde{K}_r$ of this observer has to be chosen such that $A'_r - \tilde{K}_r\,C'_r = \varpi$, thus $\tilde{K}_r = (A'_r - \varpi)\,/C'_r = -19$. According to (9.38), the minimal observer is given by

$$\begin{cases} \dot{z} = -10\,z + 198\,y + 18\,u, \\ \hat{x}_2 = z + 19\,y. \end{cases}$$

(vi) In the parameterization of (iv) with $k = 0$ we obtain $u = -k_1 x_1 = -2y$ and so an observer is not required.

Solution of Exercise 301

(i) $\mathbf{\Sigma}$ is defined by the left form $\left(\partial^2 + \partial - 1\right) y = u$. In case (a), $S(\partial) = \partial^2 + 3\partial$, $R(\partial) = 4\partial^2 + 7\partial + 1$, $T(\partial) = \partial + 1$. In case (b), $S(\partial) = \partial^3 + 4\partial^2 + 7\partial$, $R(\partial) = 7\partial^2 + 12\partial + 1$, $T(\partial) = \partial^2 + 2\partial + 1$. (ii) We have $J = 0$ and

$$F = \begin{bmatrix} -1 & 1 \\ 1 & 0 \end{bmatrix}, \quad G = \begin{bmatrix} 0 \\ 1 \end{bmatrix}, \quad H = \begin{bmatrix} 1 & 0 \end{bmatrix}.$$

(a) $u = -K_p\,x - K_i \int e(t)\,dt + const.$ with $K_p = \begin{bmatrix} 2 & 2 \end{bmatrix}$, $K_i = 1$. (b) The minimal observer is $\dot{z} = -z + y + u$, $\hat{x}_1 = y$, $\hat{x}_2 = z + y$. The full-order observer is

$$\partial\hat{x} = F\,\hat{x} + G\,u + L\,(y - H\,\hat{x}), \quad L = \begin{bmatrix} 1 \\ 1 \end{bmatrix}.$$

(c) The two state feedback/observer syntheses with integral action are of the form (9.30) and are identical to the RST controller in Question (i)(a) in the case of a minimal observer and to that of Question (i)(b) in the case of a full-order observer. (iii) The drawback of these controllers is that they do not place the closed-loop poles according to the conditions in Theorem 112, which can lead to a lack of robustness. To correct this flaw, we can choose observers that have faster poles (at -10 for example, in the case of the minimal observer, and at $\{-10, -10\}$ in the case of the full-order observer), since $\mathbf{\Sigma}$ has all its zeros at infinity.

Solution of Exercise 302

Let be the system

$$\left\{\begin{array}{r} \partial\, x = A\, x + B\, u + d_1 \\ y = C\, x + D\, u + d_2 \\ e = E\, y - r \end{array}\right.$$

with sizes as specified in section 9.2.1. We get $z = E\,(C\,x + D\,u + d_2) - r = H\,x + J\,u + d_3 - r$ where $H = E\,C$ and $J = E\,D$. If the state x is available for the control, the solution developed in section 8.3.2 remains valid (changing the notation) under the condition that the following hypotheses are satisfied:

– (A, B) is controllable;

– the transfer matrix $E\,(s\,I_n - A)^{-1}\,B + J$ is semiregular over $\mathbb{R}\,(s)$;

– $m \geq p$ and the system $\{A, B, H, J\}$ has no invariant zero that is a root of $\varphi\,(s)$.

In the present case, the control (8.38) has to be replaced by

$$v = -\left[\begin{array}{cc} K_p & K_\varepsilon \end{array}\right]\left[\begin{array}{c} \hat{\eta} \\ \eta_\varepsilon \end{array}\right]$$

where $\hat{\eta}$ is obtained by reconstructing the variable $\eta = \varphi\,(\partial)\,x$ by means of the observer

$$\partial\,\hat{\eta} = A\,\hat{\eta} + B\,v + \tilde{K}\,(\varphi\,(\partial)\,y - C\,\hat{\eta} - D\,v)$$

where $v = \varphi\,(\partial)\,u$ and where $\tilde{K}$ is a gain matrix such that $A - \tilde{K}\,C$ has all its eigenvalues appropriately chosen in the left half-plane. One such gain can be chosen under the condition that the following hypothesis is satisfied :

– (C, A) is observable.

Now let $\hat{x}$ be a variable such that $\varphi\,(\partial)\,\hat{x} = \hat{\eta}$. The control law solution of the problem is finally given by the relation (8.39) of section 8.3.2 and by $u = -K_p\hat{x} - K_\varepsilon x_\varepsilon$, which replaces (8.40).

14.10. Exercises of Chapter 10

Solution of Exercise 355

We obtain $x_d = 0$. The strict inequality is thus essential.

Solution of Exercise 356

(i) The poles of Σ are $\{-2, -2\}$. (ii) We have $A = -2\,I_2 + J_{0,2}$ (with the notation of section 13.3.4). We thus have $A_d = \exp(A\,T) = \exp(-2\,T)\,\exp(J_{0,2}\,T)$ and

$$\exp(J_{0,2}\,t) = I_2 + J_{0,2}\,T = \begin{bmatrix} 1 & 0 \\ t & 1 \end{bmatrix}$$

because $(J_{0,2})^2 = 0$. As a result,

$$A_d = \begin{bmatrix} e^{-0.4} & 0 \\ 0.2\,e^{-0.4} & e^{-0.4} \end{bmatrix} = \begin{bmatrix} 0.6703 & 0 \\ 0.1341 & 0.6703 \end{bmatrix}.$$

On the other hand, $B_d = \int_0^T e^{A\,t}\,dt\,B = \int_0^T e^{A\,t}\,B\,dt$ with

$$e^{A\,t}\,B = \begin{bmatrix} e^{-2t} \\ t\,e^{-2t} \end{bmatrix}.$$

We have $\int_0^T e^{-2t}\,dt = \frac{1}{2}\left(1 - e^{-2T}\right) = 0.1648$. Furthermore, $\int t\,e^{-2t}\,dt$ is of the form $e^{-2t}\,(a\,t + b)$. Differentiating this last expression and identifying the result with $t\,e^{-2t}$, we get $a = -1/2$ and $b = -1/4$. Therefore,

$$\int_0^T t\,e^{-2t}\,dt = \frac{1}{4}\left[1 - (2T + 1)\,e^{-2T}\right] = 0.0154,$$

and we obtain

$$B_d = \begin{bmatrix} 0.1648 \\ 0.0154 \end{bmatrix}.$$

Of course, $C_d = C$. (iii) We have

$$G_d(z) = C\,(z\,I_2 - A_d)^{-1}\,B_d = \frac{0.0154\,z + 0.0118}{z^2 - 1.3406\,z + 0.4493}.$$

The poles of Σ_d are the diagonal elements of A_d (since this matrix is lower triangular), i.e. $\{0.6703,\ 0.6703\} = \left\{e^{pT},\ e^{pT}\right\}$ with $p = -2$.

Solution of Exercise 357

(i)

$$A = \begin{bmatrix} 0 & 1 \\ 0 & 0 \end{bmatrix}, \quad B = \begin{bmatrix} 0 \\ 1 \end{bmatrix}, \quad C = \begin{bmatrix} 1 & 0 \end{bmatrix}.$$

(ii)

$$A_d = \begin{bmatrix} 1 & T \\ 0 & 1 \end{bmatrix}, \quad B_d = \begin{bmatrix} T^2/2 \\ T \end{bmatrix}$$

thus $\{A_d, B_d, C\}$ is not in observable canonical form. (iii) The transfer function of $\mathbf{\Sigma}_d$ is $G_d(z) = \frac{T^2(z+1)}{2(z-1)^2}$ (see Table (10.14)), thus $\mathbf{\Sigma}_d$ is described by the left form $D(q)\, y_d = N(q)\, u_d$ with

$$D(q) = (q-1)^2, \quad N(q) = \frac{T^2}{2}(q+1).$$

(iv) Taking into account a delay $\tau = nT$, the left form of $\mathbf{\Sigma}_d$ becomes

$$(q-1)^2 q^n y_d = \frac{T^2}{2}(q+1)\, u_d. \tag{14.1}$$

(v) The corresponding observable canonical form is, for $n = 3$,

$$\begin{aligned} A_d &= \begin{bmatrix} 2 & 1 & 0 & 0 & 0 \\ -1 & 0 & 1 & 0 & 0 \\ 0 & 0 & 0 & 1 & 0 \\ 0 & 0 & 0 & 0 & 1 \\ 0 & 0 & 0 & 0 & 0 \end{bmatrix}, \quad B_d = \begin{bmatrix} 0 \\ 0 \\ 0 \\ T^2/2 \\ T^2/2 \end{bmatrix}, \\ C_d &= \begin{bmatrix} 1 & 0 & 0 & 0 & 0 \end{bmatrix}. \end{aligned}$$

(vi) The reader is asked to carefully draw the suggested block diagram, which will make apparent a "delay line". (vii) The generalization to any n is then evident. (viii) Equation (14.1) is a classic left form of a discrete-time system. As a result, an RST controller can be easily designed using the invoked methods, and it is then a very effective *predictive control*.

Solution of Exercise 358

The following formulas are correct if λ is not an eigenvalue of $\breve{A}$:

$$\begin{aligned} A_d &= \left(\lambda I - \breve{A}\right)^{-1}\left(\lambda I + \breve{A}\right), \quad B_d = \lambda T\left(\lambda I - \breve{A}\right)^{-1}\breve{B}, \\ C &= \frac{2}{T}\breve{C}\left(\lambda I - \breve{A}\right)^{-1}, \quad D = \breve{C}\left(\lambda I - \breve{A}\right)^{-1}\breve{B} + \breve{D}. \end{aligned}$$

In general, $D \neq 0$.

Solution of Exercise 359

(i) is clear according to (10.37). (ii) We have $\tilde{A}_d = (1/T)(A_d - I_n)$ and $\tilde{B}_d = (1/T)\, B_d$. (iii) According to (10.37), $\tilde{A}_d \to A$ and $\tilde{B}_d \to B$. For small sampling periods, the quantization risks destroying all the information contained in matrices A_d and B_d according to (i), which is not the case if we use the delta formalism according to the above. (iv) Let be for example a controller represented by a state-space system using the operator delta, i.e. $\delta\, \eta_d = F\eta_d + Gz_d$, $u_d = H\eta_d + Jz_d$, where η_d is the state, z_d is the controller input, consisting of discretized measurements and of the reference signal, and where u_d is the discrete-time control calculated. (iv) Using the delta formalism, the calculations are organized as follows: (1) At instant $k\,T$, (a) we acquired signal z_d, (b) we calculate $u_d = \mu_d + J z_d$ where $\mu_d = H \eta_d$, a quantity which can be calculated before hand, (c) we apply the control. (2) Between instants $k\,T$ and $(k+1)\,T$, (a') we calculate $\nu_d \triangleq F\,\eta_d + G\,z_d$, (b') we calculate $\eta_{d+1} = T \nu_d + \eta_d$, (c') we calculate $\mu_{d+1} = H \eta_{d+1}$. At instant $(k+1)\,T$, we are ready to re-iterate the calculations with k changing to $k+1$. The calculations are thus simple and fast, which is in no way the case with the operator Δ; this operator is very useful for the *synthesis* of a discrete-time controller but not at all for its *implementation*.

Solution of Exercise 360

(i) $G_d(z) = \frac{1-a}{z-a} = \frac{B_d(z)}{A_d(z)}$ where $a = \exp(-T)$ and

$$\breve{G}(w) = G_d\left(\frac{1+(T/2)\,w}{1-(T/2)\,w}\right) = \frac{-\left(\frac{1-a}{1+a}\right) w + \frac{2}{T}\frac{1-a}{1+a}}{w + \frac{T}{2}\frac{1-a}{1+a}}.$$

(ii) For $T \to 0^+$, $1+a \to 2$, $1-a \sim T$, thus

$$\breve{G}(w) \sim \frac{1}{1+w} = G(w)$$

where $G(s)$ is the transfer function of $\boldsymbol{\Sigma}$. This is consistent with Remark 343(ii). (iii) It now remains to apply the method discussed in section 6.3.4 and to take into account the results of Exercise 121 (section 6.5). Thus we take $A_{cl}(\Delta) = (\Delta + 10/3)^3$ and the polynomials $\breve{S}(\Delta)$, $\breve{R}(\Delta)$ of the form

$$\breve{S}(\Delta) = \Delta^2 + \breve{\sigma}_1\,\Delta, \quad \breve{R}(\Delta) = \breve{r}_1\,\Delta + \breve{r}_2.$$

With $T = 0.2$, we get

$$\breve{G}(w) = \frac{-0.05\,w + 0.9992}{w^2 + 0.9992\,w}.$$

Solving the Sylvester system, we obtain $\breve{\sigma}_1 = 11.56$, $\breve{r}_1 = 25.61$, $\breve{r}_2 = 37.16$. On the other hand, $\breve{T}(\Delta) = \kappa\,(\Delta + 10/3)^2$ where κ is such that $\breve{T}(0) = \breve{R}(0) = \breve{r}_2$, which yields

$$\breve{T}(\Delta) = 3.34\,\Delta^2 + 22.30\,\Delta + 37.16.$$

(iv) The corresponding discrete-time RST controller is obtained by writing

$$\frac{\left[\begin{array}{cc} R_d(z) & T_d(z) \end{array}\right]}{S_d(z)} = \frac{\left[\begin{array}{cc} \breve{R}(w) & \breve{T}(w) \end{array}\right]}{\breve{S}(z)}\Big|_{w=\frac{2}{T}\frac{z-1}{z+1}}$$

which yields

$$\begin{aligned}
R_d(q) &= 1.3603\,q^2 + 0.3448\,q - 1.0155,\\
S_d(q) &= q^2 - 0.9278\,q - 0.0722,\\
T_d(q) &= 2.7583\,q^2 - 2.7583\,q + 0.6896.
\end{aligned}$$

We have $S_d(1) = 0$ (integrator system), $R_d(1) = T_d(1)$ (no static error) and $R_d(-1)$ (zero gain at Nyquist frequency).

(v) We again use the method of Example 352(ii). The transfer function of the pseudo-continuous system which we begin the computation with is given by

$$\breve{G}(w) = \frac{1-(T/2)\,w}{1+(T/2)\,w}\,G_d\left(\frac{1+(T/2)\,w}{1-(T/2)\,w}\right).$$

This pseudo-continuous system is defined by the left form with polynomials

$$\begin{aligned}
\breve{A}(\Delta) &= \Delta^2 + 10.9967\,\Delta + 9.9668,\\
\breve{B}(w) &= 0.0997\,\Delta^2 - 1.9934\,\Delta + 9.9668.
\end{aligned}$$

The characteristic polynomial of the pseudo-continuous closed-loop is (see section 6.3.5)

$$\begin{aligned}
A_{cl}(\Delta) &= A_c(\Delta)\,(\Delta + 10/3)\,(\Delta + 2/T),\\
A_c(\Delta) &= (\Delta + 10/3)\,(\Delta + 2/T).
\end{aligned}$$

Solving the Sylvester system, we get

$$\begin{aligned}
\breve{S}(\Delta) &= \Delta^3 + 16.24\,\Delta^2 + 168.70\,\Delta,\\
\breve{R}(\Delta) &= 27.72\,\Delta^2 + 314.38\,\Delta + 371.60,\\
\breve{T}(\Delta) &= 3.34\,\Delta^3 + 55.74\,\Delta^2 + 260.12\,\Delta + 371.60,
\end{aligned}$$

hence (after simplification of the root $q = 0$, common to all the polynomials) we finally get

$$\begin{aligned}
S_d(q) &= q^3 - 0.6813\,q^2 - 0.0722\,q - 0.2466,\\
R_d(q) &= 1.4585\,q^2 + 0.3448\,q - 1.1137,\\
T_d(q) &= 2.7583\,q^2 - 2.7583\,q + 0.6896.
\end{aligned}$$

For information only, the behavior of the controlled system is shown in Figure 14.7 when the following events occur: (a) unit step command at $t = 0$; (b) step disturbance

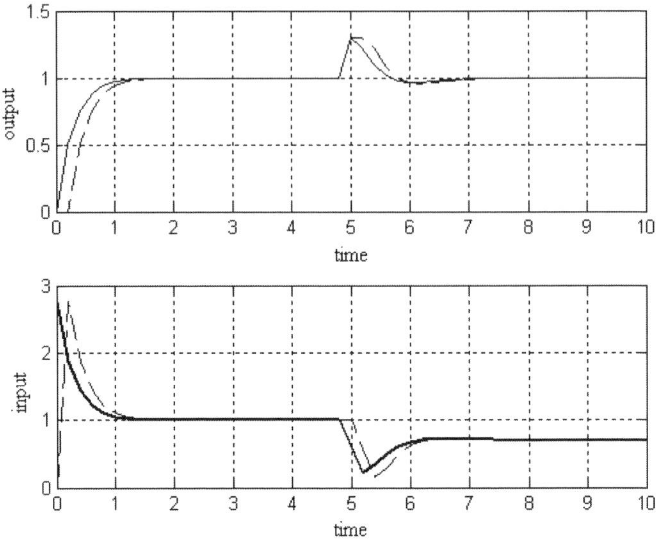

Figure 14.7. *Time-domain responses – Exercise 360*

adding at the output from $t = 5$. We can see the responses with the first controller (-) and with the second one (- -). The absolute value of the sensitivity function

$$\left| \frac{1}{1 + \frac{B_d(z)}{A_d(z)} \frac{R_d(z)}{S_d(z)}} \right|_{z=e^{i\omega T}},$$

expressed in dB, is also shown, as a function of ω, in Figure 14.8. This shows that with the two controllers, the modulus margin is correct (with, of course, an advantage for the first controller, because the introduction of a pure delay can only be disadvantageous).

Solution of Exercise 361

It suffices to replace the angular frequency ω_0 in the expression of $D_1(\Delta)$ by $\tilde{\omega}_0 = (2/T_e) \tan(T_e \omega_0 / 2)$.

Solution of Exercise 362

(i) Let $\mathbf{\Sigma}_d$ be the system obtained by discretization of System $\mathbf{\Sigma}$ with sampling period $T > 0$, where $\mathbf{\Sigma}$ is defined by $\dot{x} = A\,x + B\,u$. Then z is a pole of $\mathbf{\Sigma}_d$ if and only if $z = e^{sT}$ where s is a pole of $\mathbf{\Sigma}$, according to (10.15) (section 10.3.4) and Proposition 439 (section 12.4.2). As a result, z cannot belong to $(-\infty, 0)$. (ii) If $\mathbf{\Sigma}$ is a delay system, $\mathbf{\Sigma}_d$ has poles at 0 (see Exercise 357).

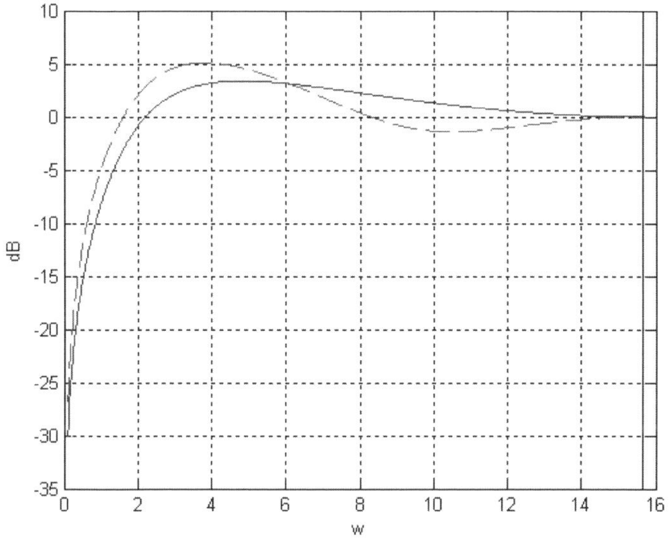

Figure 14.8. *Sensitivity function – Exercise 360*

Solution of Exercise 363

(i) The system $\mathbf{\Sigma}_d$, associated with the $\mathbf{R}$-module M, has an $i.d.z.$ which is -1, it is therefore not stabilizable. (ii) The equation of the $\mathbf{B}$-module $Q^{-1}M = \breve{M}$ is $q\,\breve{y}_d = \breve{u}_d$, and by restriction of the ring of scalars from $\mathbf{B}$ to $\mathbf{C}$ we obtain $(1 + \Delta/\lambda)\,\breve{y}_d = (1 - \Delta/\lambda)\,\breve{u}_d$, which defines the pseudo-continuous system $\breve{\mathbf{\Sigma}}$ (see Example 347). This system is controllable, which does not contradict Theorem 350 nor Proposition 346 since -1 is a pole of $\mathbf{\Sigma}_d$. (iii) The system $\mathbf{\Sigma}_d$, which has a pole -1, could not have been obtained by discretization of a continuous-time system according to Exercise 362.

14.11. Exercises of Chapter 11

Solution of Exercise 421

(i) We have

$$\begin{bmatrix} r_{yy}(0) & -r_{yu}(0) \\ -r_{yu}(0) & r_{uu}(0) \end{bmatrix} \begin{bmatrix} \hat{a} \\ \hat{b} \end{bmatrix} = \begin{bmatrix} -r_{yy}(1) \\ r_{yu}(-1) \end{bmatrix}.$$

(ii) We calculate $r'_{yy}(1)$, $r'_{yu}(0)$ and $r'_{yu}(-1)$ using Proposition 367. These quantities all have an absolute value of less than 1 and the results are consistent. (iii) We obtain $\hat{a} = -0.6091$ and $\hat{b} = 0.9636$. (iv) The pole of the identified system is $z = 0.6091$. It is inside the open unit circle, thus the identified system is stable. Its static gain is $\frac{\hat{b}\,z^{-1}}{1+\hat{a}\,z^{-1}}$ for $z = 1$, which is 2.4651.

Solution of Exercise 424

(i) Let θ_C be the vector whose components are the coefficients of the polynomial C of the ARMAX model. With this model, the transfer functions of the expression (11.32) (section 11.2.3) are of the form

$$G\left(q^{-1},\theta\right)=\frac{B\left(q^{-1},\theta_s\right)}{A\left(q^{-1},\theta_s\right)},\quad H\left(q^{-1},\theta\right)=\frac{C\left(q^{-1},\theta_C\right)}{A\left(q^{-1},\theta_s\right)},$$

and so the model cannot be put in the form (11.35) of section 11.2.5: the hypotheses of Theorem 408(i) are not in force. (ii) Suppose

$$\breve{G}\left(q^{-1}\right)=\frac{\breve{B}\left(q^{-1}\right)}{\breve{A}\left(q^{-1}\right)}=G\left(q^{-1},\breve{\theta}_s\right)$$

is the deterministic part of the "true" system, and

$$n\left(t\right)=\frac{\breve{C}\left(q^{-1}\right)}{\breve{D}\left(q^{-1}\right)}w\left(t\right)$$

is the "actual" colored measurement noise, so that

$$y\left(t\right)=\frac{\breve{B}\left(q^{-1}\right)}{\breve{A}\left(q^{-1}\right)}u\left(t\right)+\frac{\breve{C}\left(q^{-1}\right)}{\breve{D}\left(q^{-1}\right)}w\left(t\right).$$

In the quantity Q_2 of the proof of Theorem 408, the ratio $\tilde{H}=\breve{H}/H$ is of the form

$$\tilde{H}\left(q^{-1}\right)=\frac{\breve{C}\left(q^{-1}\right)A\left(q^{-1},\theta_s\right)}{\breve{D}\left(q^{-1}\right)C\left(q^{-1},\theta_C\right)}.$$

To minimize Q_2, we have to make this ratio equal to 1, and thus to obtain the equality

$$C\left(q^{-1},\theta_C\right)=\frac{\breve{C}\left(q^{-1}\right)}{\breve{D}\left(q^{-1}\right)}A\left(q^{-1},\theta_s\right).$$

In order to have at the same time $\theta_s=\breve{\theta}_s$, it is necessary that

$$C\left(q^{-1},\theta_C\right)=\frac{\breve{C}\left(q^{-1}\right)}{\breve{D}\left(q^{-1}\right)}\breve{A}\left(q^{-1}\right),$$

which is possible with a power series

$$C\left(q^{-1},\theta_C\right)=1+\sum_{\tau=1}^{+\infty}c\left(\tau\right)q^{-\tau}$$

but not with a polynomial. By truncating the above power series, we obtain a good approximation if a sufficient number of terms are kept. We thus can obtain a correct estimation of the deterministic part by using a polynomial $C\left(q^{-1},\theta_C\right)$ whose degree is large enough.

Solution of Exercise 425

Same principle as used in Exercise 424.

Solution of Exercise 426

We have

$$y(t) = G_i\left(q^{-1}, \theta\right) \tilde{u}(t) + H_c\left(q^{-1}\right) H_i\left(q^{-1}, \theta\right) w(t).$$

with $\tilde{u}(t) = G_c\left(q^{-1}\right) u(t)$, which is the input to be considered. We deal with the factor $H_c\left(q^{-1}\right)$ as in section 11.2.6.

Solution of Exercise 427

(i) Yes, according to Theorem 417. (ii) Using $\mathcal{H}\left(q^{-1}, \theta_\beta\right) = 1$ in (11.56), we also obtain a bicausal bistable second order transfer function $H\left(q^{-1}, \theta_s, \theta_\beta\right)$, thus $n_c = 2$ is the minimal number of coefficients which we will use for polynomial $C\left(q^{-1}\right)$ of the ARMAX model (i.e. $C\left(q^{-1}\right) = 1 + c_1 q^{-1} + c_2 q^{-2}$).

Bibliography

[1] B.D.O Anderson, J.B. Moore, *Optimal Filtering*, Prentice-Hall, Englewood Cliffs, NJ, 1979.

[2] B.D.O Anderson, J.B. Moore, *Optimal Control – Linear Quadratic Methods*, Prentice-Hall, Englewood Cliffs, NJ, 1989.

[3] K.J. Aström, B. Wittenmark, *Adaptive Control*, Addison-Wesley, 1989.

[4] K.J. Aström, B. Wittenmark, *Computer-Controlled Systems, Theory and Design*, 2nd ed., Prentice-Hall, Englewood Cliffs, NJ, 1990.

[5] J. Bass, *Fonctions de corrélation, fonctions pseudo-aléatoires et applications*, Masson, 1984.

[6] M. Benidir, B. Picinbono, "Extended table for eliminating the singularities in Routh's array", *IEEE Trans. Automat. Control*, 35(2), 218–222, 1990.

[7] S.P. Bhattacharyya, H. Chapellat, L. H. Keel, *Robust Control: The Parametric Approach*, Prentice-Hall, Englewood Cliffs, NJ, 1995.

[8] A. Blanc-Lapierre, B. Picinbono, *Fonctions aléatoires*, Masson, 1981.

[9] H.W. Bode, *Network Analysis and Feedback Amplifier Design*, Van Nostrand, 1945.

[10] N. Bourbaki, *Algebra I and II,* Springer, 1989–1990.

[11] N. Bourbaki, *Commutative Algebra, Chapters 1–7*, Springer, 1989.

[12] N. Bourbaki, *Functions of a Real Variable – Elementary Theory*, Springer, 2004.

[13] H. Bourlès, "Sur la robustesse des régulateurs linéaires quadratiques multivariables, optimaux pour une fonctionnelle de coût quadratique", *C.R Acad. Sci., Serie I*, 252, 971–974, 1981.

[14] H. Bourlès, "Semi-cancellable fractions in system theory", *IEEE Trans. Automat. Control*, 39(10), 2148–2153, 1994.

[15] H. Bourlès, "Structural properties of discrete and continuous linear time-varying systems: a unified approach", *Advanced Topics in Control Systems Theory – Lecture Notes from FAP 2004* (F. Lamnabhi-Lagarrigue, A. Loria and E. Panteley, eds), chap. 6, pp. 225–280, *Lecture Notes in Control and Information Sciences*, vol. 311, Springer,

2005. "Structural properties of linear systems – Part II: Structure at infinity", *Advanced Topics in Control Systems Theory – Lecture Notes from FAP 2005* (A. Loria, F. Lamnabhi-Lagarrigue and E. Panteley, eds), chap. 7, pp. 259–284, *Lecture Notes in Control and Information Sciences*, vol. 328, Springer, 2006.

[16] H. Bourlès, "Impulsive systems and behaviors in the theory of linear dynamical systems", *Forum Mathematicum*, 17(5), 781–808, 2005.

[17] H. Bourlès, F. Aïoun, "Approche H_∞ et μ-synthèse", *La Robustesse – analyse et synthèse de commandes robustes*, A. Oustaloup (co-ordinator), chap. 3, pp. 163–235, Hermès, Paris, 1994.

[18] H. Bourlès, F. Colledani, "W-Stability and Local Input-Output Stability Results", *IEEE Trans. on Autom. Control*, 40(6), pp. 1102–1108, 1995. H. Bourlès, "Addendum to 'W-Stability and Local Input-Output Stability Results' ", *IEEE Trans. on Automat. Control*, 45(6), 1220–1221, 2000.

[19] H. Bourlès, M. Fliess,"Finite poles and zeros of linear systems: an intrinsic approach", *Int. J. Control*, 68(4), 897–922, 1997.

[20] H. Bourlès, E. Irving, "La méthode LQG/LTR: une approche polynômiale temps continu/temps discret", *RAIRO APII*, 25, 545–592, 1991.

[21] H. Bourlès, B. Marinescu, "Poles and zeros at infinity of linear time-varying systems", *IEEE Trans. Automat. Control*, 44, 1981–1985, 1999.

[22] H. Bourlès, B. Marinescu, *Linear Time-Varying Systems: An Algebraic-Analytic Approach*, Springer-Verlag (forthcoming).

[23] H. Bourlès, U. Oberst, "Duality for differential-difference systems over Lie groups", *SIAM J. Control Optim.*, 48, 2051–2084, 2009.

[24] J.-P. Caron, J.-P. Hautier, *Modélisation et commande de la machine asynchrone*, Editions Technip, 1995.

[25] H. Cartan, *Elementary Theory of Analytic Functions of One or Several Variables*, Dover Publications Inc., 1995.

[26] C.-T. Chen, *Linear System Theory and Design*, Holt, Rinehart and Winston, 1984.

[27] J. Chen, "Multivariable gain-phase and sensitivity integral relations and design tradeoffs", *IEEE Trans. Automat. Control*, 43(3), 373–385, 1998.

[28] P. Chevrel, H. Bourlès, "Reduced order $\mathcal{H}_2$ and $\mathcal{H}_\infty$ observers", *32nd IEEE CDC*, 13–17 December San Antonio (Texas), pp. 2915–2916, 1993.

[29] P.G. Ciarlet, *Introduction à l'analyse numérique et à l'optimisation*, Masson, 1990. (English translation: *Introduction to Numerical Linear Algebra and Optimisation*, Cambridge University Press, 1989.)

[30] E.A. Coddington and N. Levinson, *Theory of Ordinary Differential Equations*, McGraw-Hill, 1955.

[31] P. Cohn, *Free Rings and their Relations*, Academic Press, 1985.

[32] P. Cohn, *Further Algebra and Applications*, Springer, 2003.

[33] R.F. Curtain, H.J. Zwart, *An Introduction to Infinite Dimensional Linear Systems Theory*, Springer, 1995.

[34] C.A. Desoer, M. Vidyasagar, *Feedback Systems: Input-Output Properties*, Academic Press, 1975.

[35] J. Dieudonné, *Eléments d'Analyse*, Vol. I to VI, Gauthier-Villars, 1969–1975. (English translation: *Treatise on Analysis*, Academic Press, 1969–1978.)

[36] J. Dieudonné, *Infinitesimal Calculus*, Kershaw Publishing, 1973.

[37] J. L. Doob, *Stochastic Processes*, John Wiley, 1953.

[38] J. C. Doyle, G. Stein, "Robustness with observers", *IEEE Trans. Automat. Control*, 24, 607–611, 1979.

[39] J.C. Doyle, B.A. Francis, A.R. Tannenbaum, *Feedback Control Theory*, MacMillan, 1992.

[40] J.A. Farrell, M.M. Polycarpou, *Adaptive Approximation Based Control-Unifying Neural, Fuzzy and Traditional Adaptive Approximation Approaches*, Wiley, 2006.

[41] A. Feintuch, R. Saeks, *System Theory – A Hilbert Space Approach*, Academic Press, 1982.

[42] M. Fliess, "Some structural properties of generalised linear systems", *Systems Control Lett.*, 15, 391–396, 1990.

[43] M. Fliess, "Remark on Willems' trajectory characterization of linear controllability", *Systems Control Lett.*, 19, 43–45, 1992.

[44] M. Fliess, "Some remarks on the Brunovsky canonical form", *Kybernetika*, 29(5), 417–422, 1993.

[45] M. Fliess, "Une interprétation algébrique de la transformation de Laplace et des matrices de transfert", *Linear Algebra Appl.*, 203–204, 429–442, 1994.

[46] M. Fliess, H. Bourlès, "Discussing some examples of linear system interconnections", *Systems Control Lett.*, 27, 1–7, 1996.

[47] M. Fliess, S.T. Glad, "An algebraic approach to linear and nonlinear control", *Essays on Control: Perspectives in the Theory and its Applications* (H. L. Trentelman and J. C. Willems, eds), Birkhäuser, chap. 8, 223–267, 1993.

[48] M. Fliess, J. Lévine, Ph. Martin, P. Rouchon, "Flatness and defect of non-linear systems: introducing theory and examples", *Int. J. Control*, 61(6), 1327–1361, 1995.

[49] A.L.D. Franco, H. Bourlès, E.R. de Pieri, H. Guillard, "Robust nonlinear control associating robust feedback linearization and H_∞ control", *IEEE Trans. Automat. Control*, 51(7), 1200–1207, 2006.

[50] J.S. Freudenberg, D.P. Looze, "Right half plane poles and zeros and design tradeoffs in feedback systems", *IEEE Trans. Automat. Control*, 30(6), 555–565, 1985.

[51] J. S. Freudenberg, D.P. Looze, *Frequency Domain Properties of Scalar and Multivariable Feedback Systems*, Springer, 1988.

[52] F.R. Gantmacher, *The Theory of Matrices*, vol. 1, Chelsea, 1959.

[53] J.B. Garnett, *Bounded Analytic Functions*, Academic Press, 1981.

[54] H. Glüsing-Lüerssen, "A behavioral approach to delay differential equations", *SIAM J. Control Opt.*, 35, 480–499, 1997.

[55] H. Glüsing-Lüerssen, *Linear Delay-Differential Systems with Commensurate Delays: An Algebraic Approach*, Springer, 2001.

[56] G.H. Golub, C.F. Van Loan, *Matrix Computations*, 2nd ed., The Johns Hopkins University Press, 1989.

[57] G.C. Goodwin, K.S. Sin, *Adaptive Filtering Prediction and Control*, Prentice-Hall, 1984.

[58] H. Gorecki, S. Fuksa, P. Grabowski, A. Korytowski, *Analysis and Synthesis of Time Delay Systems*, Wiley, 1989.

[59] J.K. Hale, S. M. Verduyn Lunel, *Introduction to Functional Differential Equations*, Springer, 1993.

[60] M. E. Harris, "Some results on coherent rings", *Proc. Amer. Math. Soc.*, 17, 474–479, 1966.

[61] P. Ioannou, G. Tao, "Frequency domain conditions for strictly positive real functions", *IEEE Trans. Automat. Control*, 32(1), 53–54, 1987.

[62] A. Isidori, *Nonlinear Control Systems* (3rd ed.), Springer, 1995.

[63] T. Kaczorek, *Linear Control Systems – Volume 1*, John Wiley, 1992.

[64] T. Kailath, *Linear Systems*, Prentice-Hall, 1980.

[65] R. E. Kalman, "Mathematical description of linear dynamical systems", *SIAM J. Control*, Ser. A, 1(2), 152–192, 1963.

[66] I. Kaminer, A.M. Pascoal, P.P. Khargonekar, E. Coleman, "A velocity algorithm for the implementation of gain-scheduled controllers", *Automatica*, 31(8), 1185–1191, 1995.

[67] I. Kaplansky, "Elementary divisors and modules", *Trans. Amer. Soc.*, 66, 464–491, 1949.

[68] V. Krishnan, *Nonlinear Filtering and Smoothing: An Introduction to Martingales, Stochastic Integrals and Estimation*, Wiley, 1984.

[69] V. Kucera, "A contribution to matrix quadratic equations", *IEEE Trans. Automat. Control*, 17(3), 344–347, 1972.

[70] P. Kundur, *Power System Stability and Control*, McGraw-Hill, 1994.

[71] H. Kwakernaak, "Optimal low-sensitivity linear feedback systems", *Automatica*, 5, 279, 1969.

[72] H. Kwakernaak, R. Sivan, *Linear Optimal Control Systems*, Wiley, 1972.

[73] T. Y. Lam, *Lectures on Modules and Rings*, Springer, 1999.

[74] I. D. Landau, *Commande des systèmes – Conception, identification et mise en œuvre*, Hermès, 2002.

[75] I. D. Landau, A. Karimi, "A Recursive Algorithm for ARMAX Model Identification in Closed Loop", *IEEE Trans. on Automat. Control*, 44(4), 840–843, 1999.

[76] L. D. Landau, E.M. Lifshitz, *Mechanics* (3rd ed.), Pergamon Press, 1976.

[77] P. de Larminat, Y. Thomas, *Automatique des systèmes linéaires* - vol. 1-3, Flammarion, 1975–1977.

[78] P. de Larminat, *Automatique – commande des systèmes linéaires* (2ème ed.), Hermès-science, 1996.

[79] M.D. Larsen, J. Lewis, T.S. Shores, "Elementary divisor rings and finitely presented modules", *Trans. Amer. Math. Soc.*, 187(1), 231–248, 1974.

[80] D.J. Leith, W.E. Leithead, "Survey of gain-scheduling analysis and design", *Int. J. Control*, 73(11), 1001–1025, 2000.

[81] L. Ljung, *System Identification –Theory for the User*, Prentice-Hall, 1987.

[82] L. Ljung, U. Forssell, "An alternative motivation for the indirect approach to closed-loop identification", *IEEE Trans. Automat. Control*, 44(11), 2206–2209, 1999.

[83] M. Loève, *Probability Theory* (3rd ed.), Van Nostrand Co., 1963.

[84] H. Logemann, "On the Nyquist criterion and robust stabilization of infinite-dimensional systems", *Robust Control of Linear Systems and Nonlinear Control* (M.A Kaashoek, J.H. van Schuppen and A.C.M. Ran, eds), vol. II, Birkhäuser, 627–634, 1990.

[85] H. Logemann, "Stabilization and regulation of infinite-dimensional systems using coprime factorizations", *LNCIS* 185, Springer, 102–139, 1993.

[86] J.J. Loiseau, "Invariant factor assignment for a class of time-delay systems", *Kybernetika*, 37(3), 265–275, 2001.

[87] D.G. Luenberger, "Observing the state of a linear system", *IEEE Trans. Mil. Electron.*, 8, 74–80, 1964.

[88] D.G. Luenberger, "An introduction to observers", *IEEE Trans. Automat. Control*, 16, 596–602, 1971.

[89] D.C. MacFarlane, K. Glover, *Robust Controller Design Using Coprime Factor Plant Descriptions*, Springer, 1990.

[90] J.M. Maciejowski, *Multivariable Feedback Design*, Addison-Wesley, 1989.

[91] S. MacLane, G. Birkhoff, *Algebra* (2nd ed.), MacMillan, 1979.

[92] B. Marinescu, H. Bourlès, "Robust predictive control with separation property: a reduced-state design for control systems with non-equal time delays", *Automatica*, 36, 555–562, 2000.

[93] C.L. Matson, P.S. Maybeck, "On an assumed convergence result in the LQG/LTR technique", *IEEE Trans. Automat. Control*, 36, 123–125, 1991.

[94] U. Oberst, "Multidimensional constant linear systems", *Acta Appl. Math.*, 20, 1–175, 1990

[95] R. Pallu de la Barrière, *Cours d'automatique théorique*, Dunod, 1966. (English translation: *Optimal Control Theory: A Course in Automatic Control Theory*, Dover Publications, 1980.)

[96] J.W. Polderman, J.C. Willems, *Introduction to Mathematical System Theory*, Springer, 1998.

[97] V.M. Popov, "Invariant description of linear, time-invariant controllable systems, *SIAM J. Control*, 10, 252–264, 1972.

[98] A. Quadrat, "The fractional representation approach to synthesis problems: an algebraic analysis viewpoint. Part I: (weakly) doubly coprime factorizations. Part II: internal stabilization", *SIAM J. Control Optim.*, 42, 266–299, 300–320, 2003.

[99] R. Rabah, B. Bergeon, "On state space representation for linear discrete-time systems in Hilbert spaces", *Kharkov University Vestnik*, 514(50), 53–62, 2001.

[100] H.H. Rosenbrock, *State-Space and Multivariable Theory*, Nelson, 1970.

[101] H.H. Rosenbrock, "Correction to 'the zeros of a system'", *Int. J. Control*, 20(3), 525–527, 1974.

[102] J.J. Rotman, *An Introduction to Homological Algebra*, Academic Press, 1979.

[103] W. Rudin, *Real and Complex Analysis* (3rd ed.), McGraw-Hill, 1987.

[104] W.J. Ruth, J.S. Shamma, "Research on gain scheduling", *Automatica*, 36, 1401–1425, 2000.

[105] C.E. Shannon, "Communications in the Presence of Noise", *Proc. IRE*, 37, pp. 10–21, 1949.

[106] L. Schwartz, *Théorie des distributions*, Hermann, 1966.

[107] S. Skogestad and I. Postlethwaite, *Multivariable Feedback Control –Analysis and Design*, Wiley, 2001.

[108] J.-J. Slotine, W. Li, *Applied Nonlinear Control*, Prentice-Hall, 1991.

[109] M.C. Smith, "On stabilization and the existence of coprime factorizations", *IEEE Trans. Automat. Control*, 34(9), 1005–1007, 1989.

[110] T. Söderström, P. Stoica, *System Identification*, Prentice-Hall, 1989.

[111] E.D. Sontag, *Mathematical Control Theory*, Springer, 1990.

[112] V. del Toro, *Basic Electric Machines*, Prentice-Hall, 1990.

[113] M. Vidyasagar, "On undershoot and nonminimum phase zeros", *IEEE Trans. Automat. Control*, 31, 440, 1986.

[114] M. Vidyasagar, *Control System Synthesis – A Factorization Approach*, MIT Press, 1987.

[115] M. Vidyasagar, *Nonlinear Systems Analysis* (2nd ed.), Prentice-Hall, 1993.

[116] J.C. Willems, *The Analysis of Feedback Systems*, MIT Press, 1971.

[117] J.C. Willems, "Paradigms and puzzles in the theory of dynamical systems", *IEEE Trans. Automat. Control*, 36, 259–294, 1991.

[118] W. M. Wonham, "Random differential equations in control theory", *Probabilistic Methods in Applied Mathematics*, vol. 2 (A.T. Bharucha-Reid, ed.), 131–220, Academic Press, 1970.

[119] W.M. Wonham, *Linear Multivariable Control – A Geometric Approach*, Springer, 1985.

[120] L.A. Zadeh, C.A. Desoer, *Linear System Theory*, McGraw-Hill, 1963.

[121] E. Zerz, "Primeness of multivariate polynomial matrices", *Systems Control Lett.*, 29, 139–145, 1996.

[122] K. Zhou, J.C. Doyle, K. Glover, *Robust and Optimal Control*, Prentice-Hall, 1996.

Index

L

M

P

Q, R